Fluid Dynamics for Global Environmental Studies

Fluid Dynamics for Global Environmental Studies

Springer

Department of Earth System Science
and Technology, Interdisciplinary Graduate
School of Engineering Sciences, Kyushu
University
Fukuoka
Japan

ISBN 978-4-431-56497-3 ISBN 978-4-431-56499-7 (eBook)
DOI 10.1007/978-4-431-56499-7

Library of Congress Control Number: 2017930950

Based on a translation from the Japanese language edition: *Chikyu Kankyo wo Manabu tameno Ryutai Rikigaku* by Department of Earth System Science and Technology, Kyushu University, Copyright © SEIZANDO-SHOTEN PUBLISHING CO., LTD. All Rights Reserved.

Printed on acid-free paper

This Springer imprint is published by Springer Nature
The registered company is Springer Japan KK
The registered company address is: Chiyoda First Bldg. East, 3-8-1 Nishi-Kanda, Chiyoda-ku, Tokyo 101-0065, Japan

Foreword

I am very pleased and proud to announce the publication of the second volume in the Green Asia Lecture Book Series, following the first one, entitled *Mathematical Analysis of Environmental System*, published in 2014.

Green Asia (Global Strategy for Green Asia) is one of the Programs for Leading Graduate Schools promoted by the Ministry of Education, Culture, Sports and Technology in Japan, in which we aim to establish a science and engineering leadership training program that promotes environmental and energy innovation to reach out from Asia to the rest of the world. One of the outstanding features of Green Asia as an advanced educational program is that all lectures are offered in English. Hence, the publication of cutting-edge books is one of the most important, visible, and tangible outputs of the Green Asia Program. Each of the volumes in the series deals with essential theories, fundamentals, practical applications, or upcoming topics, all of which are actually used in the program lectures.

This volume was originally published in Japanese and established an excellent reputation from many students for many years. It is because our excellent professors working for the Department of Earth System Science and Technology, Interdisciplinary Graduate School of Engineering Sciences, Kyushu University, were responsible for intelligibly describing key points of fluid dynamics and its application. The book has been carefully translated into English by the professors so as to maintain the flavor of the original. I am confident that this book can serve as a lighthouse for beginning students as well as for engineers and scientists.

It will be wonderful if our publication project can disseminate all of the brilliant content and approaches produced in the Green Asia Program to a worldwide audience.

Prof. Jun Tanimoto, Dr. Eng.
Director of Green Asia Education Center
and Head Coordinator of Advanced
Graduate Program in Global Strategy for Green Asia
Kyushu University
Professor
Interdisciplinary Graduate School of Engineering Sciences
Kyushu University

Kyushu University Program for Leading Graduate Schools
Advanced Graduate Program in Global Strategy for **Green Asia**

Preface to the English Edition

This book is an English edition of a Japanese book with the same title. A new chapter, "Space Plasma Environment," written by Prof. Tohru Hada, has been added in this edition to cover the plasma flow influenced by the Lorentz force that was not included in the Japanese edition.

This English edition is supported by the advanced graduate program in Global Strategy for Green Asia at Kyushu University. We would like to express our sincere gratitude to all those people who have supported this publication.

April 2015

Yoshinobu Wakata
Department of Earth System Science and Technology
Kyushu University Fukuoka Japan

Preface to the Original Japanese Edition

This book introduces the fundamental concepts of environmental fluid dynamics. It is intended for use by students, researchers, and specialists working in the research fields of geophysical fluids (such as atmosphere or ocean) and general fluid environment. Because the Earth is covered by atmosphere and oceans and is exposed to solar wind, the knowledge of fluid motion is essential for tackling its environmental issues. Although fundamental fluid mechanics is found in many existing textbooks, few of them clearly describe all the essential ideas, from the fundamentals of fluid dynamics to environmental science, with a careful explanation of the governing mathematics. This book has been developed to bridge that gap.

In recent years, with an improvement in the educational setting for environmental science, many students graduating from various university departments choose a career in environment-related fields and undertake specialized researches. Under these circumstances, many students have probably never acquired a fundamental knowledge of fluid mechanics before engaging in such research. The present book will be an invaluable resource for such students. Meanwhile, students who have previously studied fluid dynamics will come to realize that certain aspects of environmental fluid dynamics, such as stratification or rotation effects, are not truly covered by general fluid mechanics. Furthermore, general fluid dynamics treats flows around objects, whereas in environmental fluid dynamics, flows inside boundaries are also important. To accommodate these needs, fluid dynamics "repackaged for environmental sciences" should be learned again. Considering these issues, this book takes the reader from the derivation of the fundamental fluid dynamics equations through to environmental science in eight chapters. Intermediate calculations are carefully demonstrated, and fundamental concepts are explained as comprehensively as possible. Using this book, readers will acquire plentiful knowledge about the dynamics of fluid motion, which will assist them in their more advanced research of environmental science and technology. A home page related to this book will be established at http://www.esst.kyushu-u.ac.jp/textbook/, which allows the interaction of the authors and the readers of this book.

This book is based on a series of actual lectures, which are compulsory for obtaining a master's degree in the Department of Earth System Science and

Technology, Kyushu University. This fact is reflected in the selection of contents. Gravitational and Coriolis forces are given as early examples of external forces, although advanced electromagnetic flow under the Lorentz force is excluded in the present edition. Each chapter was written by the corresponding course instructor in FY2000. During the editing process, predicates and symbols, which may vary across disciplines, have been unified wherever possible. References are provided as footnotes within the text, while more general materials accessed while writing the text are compiled as a bibliography at the end of the book.

In addition to sincerely thanking the authors of the book, we offer sincere thanks to many staff who participated in the planning and operating of these lectures. We also express the heartfelt thanks to Seizando-Shoten Publishing Co., Ltd., for their valuable support in publishing this book.

February 2001

Jong-Hwan Yoon
Head, Department of Earth System Science and Technology
Interdisciplinary Graduate School of Engineering Sciences
Kyushu University
Fukuoka, Japan

The original version of the book was revised: Belated corrections in Copyright page of Frontmatter and Chapter 9 have been updated. The erratum to the book is available at 10.1007/978-4-431-56499-7_10

Supervisors for English Edition

(Alphabetical order)
Hada, Tohru (Chaps. 1, 9)
Hirose, Naoki (Chap. 7, Preface)
Hu, Changhong (Chaps. 2, 3, Appendix)
Ichikawa, Kaoru (Chaps. 1–8, Index)
Isobe, Atsuhiko (Chaps. 6, 8)
Kyozuka, Yusaku (Chaps. 2, 3)
Matsunaga, Nobuhiro (Chaps. 4–6)
Matsuno, Takeshi (Chap. 7, Index)
Nakamura, Masahiko (Chaps. 2, 3)
Okamoto, Hajime (Chap. 1)
Okamura, Makoto (Chaps. 4–6)
Senju, Tomoharu (Chap. 8)
Sugihara, Yuji (Chaps. 4–6)
Takemura, Toshihiko (Chap. 8)
Uno, Itsushi (Chap. 1)
Wakata, Yoshinobu (Chaps. 4–6)
Yamamoto, Masaru (Chaps. 4–6)
Yoshida, Shigeo (Chaps. 2, 3)

Authors' Curriculum of the Original Japanese Edition

Masayuki Oikawa
Emeritus Professor, Kyushu University
Masashi Kashiwagi
Professor, Graduate School of Engineering, Osaka University

Keita Iga
Associate Professor, Atmosphere and Ocean Research Institute and Graduate School of Science, The University of Tokyo
Kaoru Ichikawa
Associate Professor, Research Institute for Applied Mechanics and Interdisciplinary Graduate School of Engineering Sciences
Kyushu University
Tetsuo Yanagi
Emeritus Professor, Kyushu University

Contents

Chapter 1
Fundamental Equations in Fluid Dynamics

Fluid dynamics is a study of the movement of gases and liquids. It has an unbelievable range of applications. The Earth's atmosphere and oceanic movements are within the scope of fluid dynamics. This book mainly aims to describe dynamics of rotating and stratified fluids, as well as environmental fluid dynamics. This chapter is allocated for the explanation of fundamental equations in fluid dynamics. First, stress and stress tensors will be discussed, and then fluids will be characterized on their basis. Next, the mass conservation law, equation of motion (momentum conservation law), and energy equations as well as the specifications of fluid motion will be discussed. To complete the equation of motion for a viscous fluid, it is necessary to relate the deformation of fluid elements due to fluid motion with stress. Furthermore, the concept of vorticity and its governing equations will be discussed. In this book, incompressible flow will be considered in most cases; however, this chapter has been made as general as possible. In this book, appropriate vector notation and tensor notation will be used where necessary.

1.1 Fluid and Fluid Dynamics

Liquids and gases are generally called **fluids**. Compared with solids, which resist operations that try to deform them, fluids cannot resist such operations and become deformed without limit, although they do resist compression. This nature is a characteristic of fluids, and it also links them to the phenomenon of "flowing." In the next section we will adopt this as the mechanical definition of fluids. Whether something can be treated as a fluid depends on the time and space scales under consideration. For example, although the Earth's mantle is usually classified as a solid, when considering mantle convection in which deformation occurs over extremely long time and space scales, it can be thought of as a fluid.

Fluid dynamics is a study that assumes fluid as a continuum and discusses its macroscopic motion. Microscopically, substances have discontinuous structures composed of molecules. Substances generally comprise a vast number, for example

D.E.S.S.T. Kyushu Univ., *Fluid Dynamics for Global Environmental Studies*,
DOI 10.1007/978-4-431-56499-7_1

10^{23}, of molecules; hence, it is not realistic to explain the macroscopic motion of substances on the basis of the motion of individual molecules. In fluid dynamics, the microscopic structure of fluid substances is ignored and treated as a continuum, and its macroscopic movement is discussed. Phenomena are described using a small number of **macroscopic variables** such as velocity, density, pressure, and temperature. Considering fluid substances as a continuum in this way, there is also the benefit that air and water, whose microscopic structures are very different, can be treated in a unified manner. Fluid dynamics has a vast range of applications such as in applied engineering, biology, atmosphere, ocean, and space partly because fluids have a unified framework, which is not dependent on individual fluid molecules.

Let us consider the conditions for regarding an actual fluid as a continuum. Under the continuum hypothesis, we consider, for example, a fluid's density $\rho(\boldsymbol{r}, t)$ as a continuous function of position vector $\boldsymbol{r}$[1] and time t (usually including partial derivatives of an appropriate order). Let us look at gas density on a molecular basis. At time t, if the number of molecules within volume δV of length scale $\delta\ell$ with point $\boldsymbol{r}$ as the center is N and the molecular mass is m, the average density is $Nm/\delta V$. The limiting value of this average density when $\delta\ell \to 0$ is the density in the continuum approximation. However, it is clear that such a limiting value does not physically exist. This is because if $\delta\ell$ is decreased and becomes comparable with the size of a molecule, then the discontinuity of the substance structure is directly reflected, and the average density becomes discontinuous. In contrast, if $\delta\ell$ is increased and δV starts to contain plenty of gas molecules, the average density will begin to maintain a fixed value. This value is the actual local value of density. If $\delta\ell$ is increased further and becomes a size similar to the length of the scale of the macroscopic phenomenon under consideration L_{macro} (assuming that the density changes macroscopically), then the average density is influenced by the macroscopic change. Furthermore, if there is a sufficiently large range of $\delta\ell$ for which the average density takes a fixed value such as that mentioned above, then it can be said that an actual continuum approximation is valid.

Moreover, we usually assume that fluids are in a **locally thermal equilibrium** or a state close to it and for a volume δV, which is negligibly small relative to $(L_{\text{macro}})^3$ but still contains a large number of molecules, relations in equilibrium thermodynamics can be applied to thermodynamic quantities such as temperature.[2] For this purpose, molecules need to make sufficiently frequent collisions within a time period much shorter than the time scale T_{macro} of macroscopic phenomenon under consideration.

The following two conditions must be satisfied for these conditions to be true.

1. When the mean free path of the substance molecules, i.e., the average distance a molecule moves until a molecule collides with another molecule, is λ,

$$\lambda \ll L_{\text{macro}} .$$

[1] It will be stated simply as point $\boldsymbol{r}$ hereinafter.

[2] This will be explained with a little more detail in Sect. 1.9.

2. When the mean free time of the substance molecules, i.e., the average time until a molecule collides with another molecule, is τ,

$$\tau \ll T_{\mathrm{macro}} .$$

In air at 0 °C and 1 atmospheric pressure, $\lambda \sim 6 \times 10^{-6}$ cm and $\tau \sim 10^{-10}$ s; hence, everyday phenomena satisfy these conditions easily. Moreover, whether it can be deemed as a continuum depends on what type of phenomenon is considered. Thus, even for a collection of stars such as the Galaxy, if a phenomenon of a scale much greater than the average distance between stars is considered, then it can be treated as a continuum.

As mentioned above, physically, there is a lower limit of $\delta\ell$ for which the average density maintains as a fixed value; however, a continuum is a hypothetical medium supposing that $\delta\ell$ satisfying this quality can be considered to be as small as desired. Hereafter, fluids are assumed to be a continuum, and macroscopic variables that describe fluids, such as density, velocity, pressure, and temperature, are assumed to be continuous functions of a point in the flow region $\boldsymbol{r}$ and time t.

1.2 Stress and Stress Tensors

1.2.1 Body Force and Surface Force

Generally, two types of forces work on fluids. One is **body force** such as gravity, electromagnetic forces, and inertial forces. If the force per unit mass is $\boldsymbol{K}(\boldsymbol{r}, t)$, and the fluid density is $\rho(\boldsymbol{r}, t)$, then the force working on a body element δV at point $\boldsymbol{r}$ is

$$\boldsymbol{K}(\boldsymbol{r}, t)\rho\delta V . \tag{1.1}$$

The other is **surface force** that is acted by fluids on either side of a surface toward one another through the surface, and the force per unit surface area is called **stress**. In contrast to body force, which is generally a long-range force, surface force is a short-range force originating from a molecular basis. The mechanics of the molecular basis of surface force differ between air and water; however, as a continuum, such differences can be ignored and both the surface forces can be expressed in the same form.

1.2.2 Stress and Stress Tensors

Let us consider stress. Generally, stress is dependent on the direction of the surface through which it is acting on. As shown in Fig. 1.1, consider a small surface with

Fig. 1.1 Direction of surface and stress

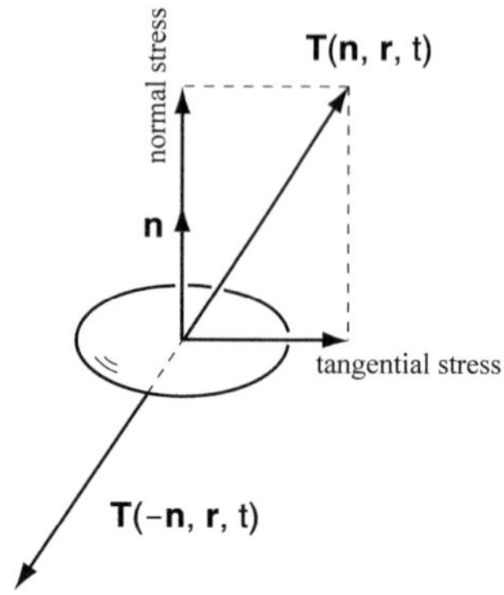

area δS and unit normal vector $\boldsymbol{n}$ at point $\boldsymbol{r}$ in the fluid at time t. The force acting on the fluid on the reverse by the fluid on the obverse side (define the obverse side of the surface as the side to which $\boldsymbol{n}$ points) through this surface is written as

$$\boldsymbol{T}(\boldsymbol{n}, \boldsymbol{r}, t)\delta S, \tag{1.2}$$

and the force per unit area $\boldsymbol{T}(\boldsymbol{n}, \boldsymbol{r}, t)$ is called stress acting on that surface at time t and point $\boldsymbol{r}$. The component of $\boldsymbol{T}(\boldsymbol{n}, \boldsymbol{r}, t)$ in the direction $\boldsymbol{n}$ is called normal stress, and the component parallel to the surface (more precisely, a tangential plane) is called tangential stress. $\boldsymbol{T}(\boldsymbol{n}, \boldsymbol{r}, t)$ is the stress exerted on the fluid on the reverse side by the fluid on the obverse side, and the stress exerted on the fluid on the obverse side by the fluid on the reverse side is expressed as $\boldsymbol{T}(-\boldsymbol{n}, \boldsymbol{r}, t)$ by replacing $\boldsymbol{n}$ with $-\boldsymbol{n}$. Therefore, by Newton's third law of motion, the following relation is established:

$$\boldsymbol{T}(-\boldsymbol{n}, \boldsymbol{r}, t) = -\boldsymbol{T}(\boldsymbol{n}, \boldsymbol{r}, t)\,. \tag{1.3}$$

Hence, if normal stress is positive, the fluids will be pulling against one another across the surface, and if it is negative, they will be pushing against one another. Hereafter, $\boldsymbol{r}$ and t in $\boldsymbol{T}(\boldsymbol{n}, \boldsymbol{r}, t)$ will be omitted.

Now let us consider the balance of the forces acting on a small tetrahedron as shown in Fig. 1.2. The forces acting on the tetrahedron are inertial forces, external forces, and surface forces exerted through the four surfaces. Inertia and external forces are body forces, and are of order $(\delta\ell)^3$, where the representative length of the tetrahedron is $\delta\ell$. In contrast, surface forces are of order of $(\delta\ell)^2$; hence, as $\delta\ell \to 0$, the contribution of body forces can be ignored, and the balance equation of the surface forces acting on the four surfaces is obtained:

$$\boldsymbol{T}(\boldsymbol{n})\delta S + \sum_{j=1}^{3} \boldsymbol{T}(-\boldsymbol{e}_j)\delta S_j = 0\,. \tag{1.4}$$

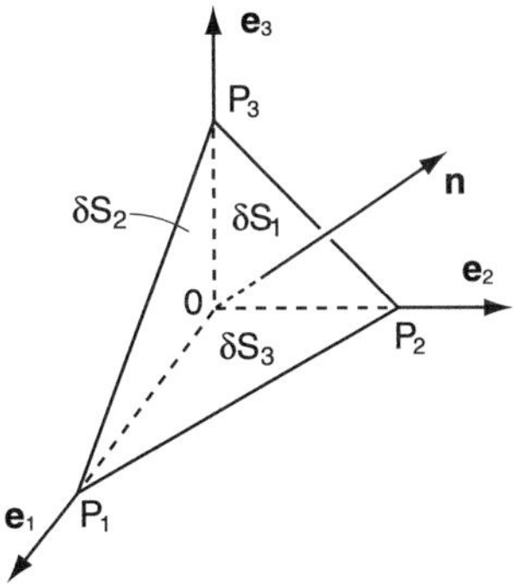

Fig. 1.2 Small tetrahedron with three surfaces perpendicular to the coordinate axes

The inertia force has been taken into account in the balance equation of the force; hence, note that this relationship also holds for moving fluids. Here δS is the area of $\triangle \mathrm{P_1P_2P_3}$, and δS_j is the area of the surface perpendicular to the x_j-axis. Moreover, $\boldsymbol{e}_j$ is a unit vector in the x_j-axis direction.[3] The x_i-component of vector $\boldsymbol{T}$ is expressed as T_i, and this can be written as[4]

$$\boldsymbol{T} = (T_1, T_2, T_3)\,.$$

This means that

$$\boldsymbol{T} = T_1\boldsymbol{e}_1 + T_2\boldsymbol{e}_2 + T_3\boldsymbol{e}_3 = \sum_{j=1}^{3} T_j\boldsymbol{e}_j\,. \tag{1.5}$$

Furthermore, by omitting the summation sign in (1.5), it is simplified as

$$\boldsymbol{T} = T_j\boldsymbol{e}_j\,. \tag{1.6}$$

In other words, unless otherwise specified, if the same Latin subscript appears two times, then agree to take a sum from 1 to 3 (Einstein's convention).

If this convention is used, (1.4) can be written as

$$\boldsymbol{T}(\boldsymbol{n})\delta S + \boldsymbol{T}(-\boldsymbol{e}_j)\delta S_j = 0\,. \tag{1.7}$$

If $\delta S_j = (\boldsymbol{n}\cdot\boldsymbol{e}_j)\delta S$ and (1.3) are used, from (1.7), the following is obtained:

$$\boldsymbol{T}(\boldsymbol{n}) = \boldsymbol{T}(\boldsymbol{e}_j)(\boldsymbol{n}\cdot\boldsymbol{e}_j)\,. \tag{1.8}$$

---

[3] x_1, x_2, x_3 indicates x, y, z, respectively, and $\boldsymbol{e}_1, \boldsymbol{e}_2, \boldsymbol{e}_3$ refers to the basic vector $\boldsymbol{i}, \boldsymbol{j}, \boldsymbol{k}$ of the xyz axes, respectively. Hereafter, both will be used as appropriate. Moreover, even if not stated specifically, it will be assumed that an appropriate xyz Cartesian coordinate system is in place.

[4] Under normal circumstances, it should probably be expressed as a column vector because of considerations relating to the amount of paper a row vector will be used. Instead of referring to it as vector with components T_i, it can be referred to simply as vector T_i.

Therefore, if the stresses $\boldsymbol{T}(\boldsymbol{e}_1)$, $\boldsymbol{T}(\boldsymbol{e}_2)$, $\boldsymbol{T}(\boldsymbol{e}_3)$ acting on the surfaces perpendicular to the three coordinate axes are known, then the stress $\boldsymbol{T}(\boldsymbol{n})$ acting on any surface will be known. If the following relation

$$\tau_{ij} = T_i(\boldsymbol{e}_j) \tag{1.9}$$

is defined, then because of $\boldsymbol{n} \cdot \boldsymbol{e}_j = n_j$; (1.8) can be written as

$$T_i(\boldsymbol{n}) = \tau_{ij} n_j \,. \tag{1.10}$$

The quantity with these nine quantities τ_{ij} as components is known as a **stress tensor**.[5] By definition, τ_{ij} is the x_i-axis component of the force per unit area $\boldsymbol{T}(\boldsymbol{e}_j)$ that acts on the surface perpendicular to the x_j-axis in the direction from the side whose x_j is greater to that whose x_j is smaller. τ_{11}, τ_{22}, τ_{33} are normal stresses, and τ_{ij} $(i \neq j)$ are tangential stresses. The stress tensor is independent of $\boldsymbol{n}$ and is determined by $\boldsymbol{r}$ and t alone. (1.10) shows that the (second-order) tensor serves as a linear operator to make the vector $\boldsymbol{n}$ correspond to the vector $\boldsymbol{T}(\boldsymbol{n})$ (or make the vector $\boldsymbol{n}\delta S$ correspond to the vector $\boldsymbol{T}(\boldsymbol{n})\delta S$ by multiplying both sides of (1.10) by δS). Moreover, the above mentioned definition of τ_{ij} shows that τ_{ij} is the very component of the representation matrix[6] in the Cartesian coordinate system of the linear operator. In reality, the right-hand side of (1.10) is a product of the matrix τ_{ij} and vector n_i. Both vector and tensor are physical quantities and are independent of the coordinate system. Therefore, their components are transformed according to a specific transformation rule for the rotation of the coordinate system. Vector and tensor can also be defined from this point of view (see Appendix A.1).

Note 1.1 Let us provide a little supplement regarding (1.4). $\boldsymbol{T}(\boldsymbol{n})$ in (1.4) is strictly a value at some point in $\Delta P_1P_2P_3$. The same is true for $\boldsymbol{T}(-\boldsymbol{e}_j)$. This can be seen by applying the mean value theorem to the surface integrals: originally each term of (1.4) is a surface integral. If the representative point of the tetrahedron is O, $\boldsymbol{T}(\boldsymbol{n})$ and $\boldsymbol{T}(-\boldsymbol{e}_j)$ are expanded around O, and $\delta S_j = \boldsymbol{n}_j \delta S$ is used, the left-hand side of (1.4) can be written as

$$(\boldsymbol{T}(\boldsymbol{n}, \boldsymbol{r}, t) - \boldsymbol{T}(\boldsymbol{e}_j, \boldsymbol{r}, t) n_j)\, \delta S + O((\delta \ell)^3)\,. \tag{1.11}$$

$\boldsymbol{r}$ is the position vector of the point O. If the tetrahedron is contracted to the point O while maintaining a similar shape, then the first term of (1.11) is $O((\delta\ell)^2)$. This is because the coefficient of δS is constant. The second term represents a difference from the first term and is $O((\delta\ell)^3)$. On the other hand, inertia and external forces are $O((\delta\ell)^3)$; hence, the coefficient of δS in (1.11) has to be 0, and (1.8) is obtained.

---

[5]More precisely, this is a second-order tensor. Moreover, a tensor with τ_{ij} as components may instead be referred to simply as tensor τ_{ij}.

[6]Matrix vertically lining up vectors $\boldsymbol{T}(\boldsymbol{e}_1)$, $\boldsymbol{T}(\boldsymbol{e}_2)$, $\boldsymbol{T}(\boldsymbol{e}_3)$, which are mappings of the basic vectors $\boldsymbol{e}_1$, $\boldsymbol{e}_2$, $\boldsymbol{e}_3$, respectively.

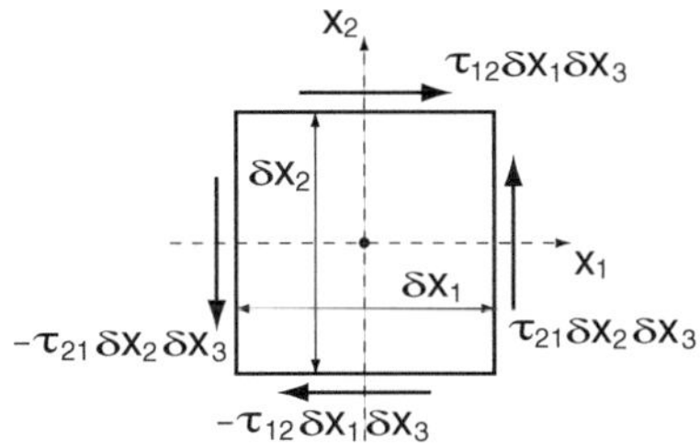

Fig. 1.3 Surface force due to tangential stress acting on a small parallelepiped parallel to the coordinate axes (cross section perpendicular to the x_3-axis)

The balance of the remaining $O((\delta\ell)^3)$ term becomes the fluid equation of motion, which will be discussed later.

A stress tensor is a symmetric tensor; in other words, $\tau_{ij} = \tau_{ji}$. This can be shown from the moment balance of the surface forces around the central axis of the small parallelepiped such as the one in Fig. 1.3, $\delta x_2(\tau_{12}\delta x_1\delta x_3) = \delta x_1(\tau_{21}\delta x_2\delta x_3)$. This is because even in this case, if the size of the parallelepiped is infinitesimally reduced, the angular momentum term and moment term due to external forces can be ignored relative to the moments of the surface forces. In this case, normal stress does not contribute to the moment, thus was omitted in Fig. 1.3.

Now, suppose that the stress acting on the surface with normal vector $\boldsymbol{n}$ satisfies the following relation:

$$\boldsymbol{T}(\boldsymbol{n}) = \lambda\boldsymbol{n}\,. \tag{1.12}$$

In this case, only a normal stress of magnitude $|\lambda|$ acts on its surface. If (1.12) is rewritten using (1.10), it becomes

$$\tau_{ij}n_j = \lambda\,\delta_{ij}n_j\,. \tag{1.13}$$

δ_{ij} is the Kronecker delta and defined as $\delta_{11} = \delta_{22} = \delta_{33} = 1$, $\delta_{ij} = 0\ (i \neq j)$. (1.13) is an eigenvalue problem for the symmetric matrix τ_{ij}, and these are three real eigenvalues $\tau'_1, \tau'_2, \tau'_3$, if multiplicities are counted. The corresponding eigenvectors $\boldsymbol{e}'_1, \boldsymbol{e}'_2, \boldsymbol{e}'_3$ can be selected to be mutually orthogonal to form a right-handed system. If these eigenvectors are selected as the basis of the new Cartesian coordinate system $x'_1x'_2x'_3$, only normal stress acts on each of the surfaces perpendicular to the coordinate axes. Hence, the representation matrix of the stress tensor in the new system becomes

$$\mathrm{diag}\,(\tau'_1,\ \tau'_2,\ \tau'_3)\,. \tag{1.14}$$

Here diag(a, b, c) is the diagonal matrix with diagonal components a, b, c, and $\tau'_1, \tau'_2, \tau'_3$ are called the **principal stresses**. Moreover, these new coordinate axes are called the principal axes of the stress tensor.

As is commonly known, the sum of the diagonal components of the matrix τ_{ij} remains unchanged with respect to the rotation of the Cartesian coordinate system.[7] In other words,

$$\tau_{ii} \equiv \tau_{11} + \tau_{22} + \tau_{33} = \tau_1' + \tau_2' + \tau_3' \tag{1.15}$$

holds true. Furthermore, in the principal axes system, the stress acting through the surface element with normal vectors (n_1', n_2', n_3') is

$$(\tau_1' n_1',\ \tau_2' n_2',\ \tau_3' n_3')\,. \tag{1.16}$$

Exercise 1.1 Show that $\delta S_j = (\boldsymbol{n} \cdot \boldsymbol{e}_j)\delta S$ holds true for the tetrahedron in Fig. 1.2.

1.2.3 Stress Tensors in Stationary Fluids

Now, let us characterize fluids using stress. The characteristic of fluids is that they easily deform relative to solids; however, they resist compression. Hence, a fluid is defined as a substance that cannot resist any type of action trying to deform it without change in volume. The stress tensor (1.14) expressed in the principal axes system will be decomposed into the following two terms (see Appendix A.1):

$$\mathrm{diag}\left(\frac{1}{3}\,\tau_{ii},\ \frac{1}{3}\,\tau_{ii},\ \frac{1}{3}\,\tau_{ii}\right) \tag{1.17}$$

and

$$\mathrm{diag}\left(\tau_1' - \frac{1}{3}\,\tau_{ii},\ \tau_2' - \frac{1}{3}\,\tau_{ii},\ \tau_3' - \frac{1}{3}\,\tau_{ii}\right). \tag{1.18}$$

The contribution of the tensor (1.17) to the stress acting through the surface element with normal vector $\boldsymbol{n}'$ is $(\tau_{ii}/3)\boldsymbol{n}'$. That is, regardless of which way the surface faces, only the normal stress of the same magnitude operates. This type of tensor is called an isotropic tensor. The components of (1.17) can be expressed, using the Kronecker delta, as $(\tau_{ii}/3)\,\delta_{ij}$; actually, any second-order isotropic tensor can be written in the form of a scalar multiple of δ_{ij} (see Appendix A.1). Tensor (1.17) corresponds to states such as that shown in Fig. 1.4. Fluids resist this type of compression (usually $\tau_{ii}/3 < 0$ in fluids) and can maintain a stationary state. On the other hand, (1.18) represents a difference from the isotropic stress tensor, and the sum of the diagonal elements turns out to be 0 from (1.15). Therefore, at least one of the diagonal elements is positive, and one is negative, and is a stress state that tries to purely deform the fluid element without changing volume, as shown in Fig. 1.5. Fluids are unable to resist this stress state. Since body forces are infinitesimal of higher order relative to the surface forces, body force cannot resist this stress state.

[7] Can be obtained from the invariance against the coordinate system rotation of the characteristic polynomial obtained from (1.13).

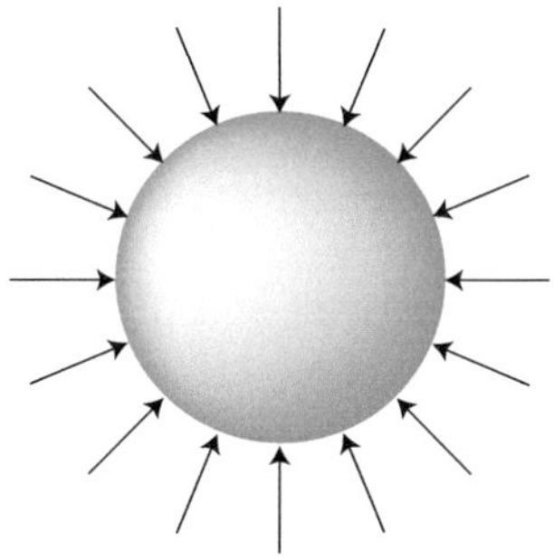

Fig. 1.4 Isotropic compression

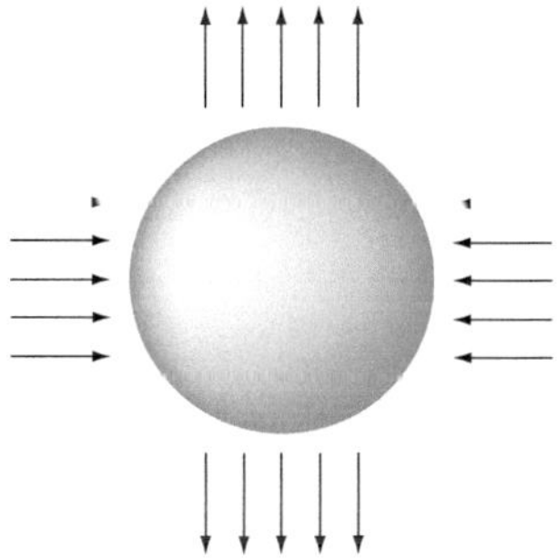

Fig. 1.5 Stress state that tries to deform without changing volume

Therefore, if this sort of stress state exists, fluids cannot maintain the stationary state. Hence, when fluids are stationary,

$$\tau_1' = \tau_2' = \tau_3' = \frac{1}{3}\,\tau_{ii} \tag{1.19}$$

must hold. As a result, in a stationary fluid, the stress tensor can be expressed as

$$\tau_{ij} = -\,p\,\delta_{ij}\,, \tag{1.20}$$

where $p\ (= -\tau_{ii}/3)$ is a function of $\boldsymbol{r}$ only and is called **pressure**.

It is customary, as (1.20), to insert a negative sign on the right-hand side. This is because fluids are usually under a pressured state, and hence, p is positive under the pressured state. (1.20) implies that the force per unit surface area through the surface element with a normal vector $\boldsymbol{n}$ is $-p\boldsymbol{n}$. Moreover, this pressure p in the stationary fluid is thermodynamic pressure (represent this as p_e) itself.

Now, suppose that tangential stress is always zero regardless of whether it is in a stationary state or in motion. Then normal stress is constant regardless of the selection of the surface. In other words, even in this situation, the stress tensor is in the form (1.20). This is because if (1.16) is to be parallel to an arbitrary normal vector $\boldsymbol{n}'$, (1.19) needs to be true.

Generally, a fluid in motion generates tangential stress through viscosity (internal friction). For example, a fluid flowing along a fixed plate exerts tangential stress in the direction of the flow on the plate, and the plate exerts tangential stress on the fluid in the direction opposite the flow. This type of fluid that generates tangential stress through viscosity is called a **viscous fluid**. In contrast, a hypothetical fluid that does not generate tangential stress even in a state of motion is considered and is called an **inviscid** or **perfect fluid**. In a perfect fluid, the stress tensor is in the same form (1.20) as a stationary fluid even when in motion. In other words, in perfect fluids, only isotropic pressure acts as stress. However, generally, the value of pressure p when fluids are in motion is different from that when they are stationary. Usually, in perfect fluids, it is assumed that a locally thermal equilibrium is set up, and hence, the pressure $p(\boldsymbol{r}, t)$ in (1.20) is equivalent to the local thermodynamic pressure $p_e(\boldsymbol{r}, t)$.

Generally, the reason for considering such a hypothetical fluid despite viscosity acting on fluids is because the mathematical treatment of this fluid is far easier than that of viscous fluids; furthermore, flows in the hypothetical fluid frequently express real flows with good approximation.

Note 1.2 When the stress tensor is in the form (1.20), p can be referred to as **hydrostatic pressure**. However, if so, even the pressure of the perfect fluid in relative motion expressed in exactly the same form has to also be referred to as hydrostatic pressure and is inconvenient. Hence, in this book, out of the pressure in (1.20), the words hydrostatic pressure will only be used to refer to the pressure of fluids that are stationary under the influence of external forces (usually gravity).

The (mechanical) pressure on viscous fluids in relative motion is defined by a form that is an extension of (1.20), as shown in Sect. 1.7. This pressure is generally not equal to thermodynamic pressure; however, the difference can be ignored in the phenomenon discussed in this book. Therefore, beyond Chap. 2, all pressure may be considered to be equal to locally thermodynamic pressure.

Exercise 1.2 Water is stationary in a uniform gravitational field. Suppose that atmospheric pressure p_0 is acting on the water surface, obtain the pressure at depth h from the water surface. Here, water density ρ is assumed to be fixed.

1.3 Specifications of Fluid Motions

1.3.1 Lagrangian and Eulerian Specifications

There are two ways of describing fluid motion: the **Lagrangian specification** and the **Eulerian specification**.

In the Lagrangian specification, the movement over time of each point (we will refer to these as **fluid particles**)[8] constructing the fluid as a continuum, as fluid is in

[8]The word fluid particle is often used to indicate an infinitesimal fluid element. Here it is considered to be a "point" that makes up the fluid substance as a continuum.

motion, are investigated. To perform this, it is convenient to mark the fluid particles. The mark can be anything as long as they can be individually identified; however, it is usually the position vector of each fluid particle $\boldsymbol{r}_0 = (x_0, y_0, z_0)$ at initial time $t = 0$.[9] If the fluid particle position vector (position coordinate) at time t is $\boldsymbol{r} = (x, y, z)$, then

$$\boldsymbol{r} = \boldsymbol{r}(\boldsymbol{r}_0, t) \tag{1.21}$$

where $\boldsymbol{r}_0 = \boldsymbol{r}(\boldsymbol{r}_0, 0)$. (1.21) shows that, in this specification, fluid motion can be observed as one of the point transformations with time t as a parameter.

In this specification, the fluid particle's initial position vector $\boldsymbol{r}_0$ and time t are independent variables, and its position vector $\boldsymbol{r}$ is one of the dependent variables. Fixing $\boldsymbol{r}_0$ and partially differentiating $\boldsymbol{r}$ with respect to t means focusing on the fluid particle initially present at $\boldsymbol{r}_0$ and obtaining the rate of change over time of that position vector, thereby calculating the velocity of the fluid particle $\boldsymbol{u} = (u, v, w)$. If this is partially differentiated with respect to t once more, then acceleration is obtained. In other words,

$$\boldsymbol{u} = \frac{\partial \boldsymbol{r}}{\partial t}, \quad \text{or} \quad (u, v, w) = \left(\frac{\partial x}{\partial t}, \frac{\partial y}{\partial t}, \frac{\partial z}{\partial t}\right), \tag{1.22}$$

$$\boldsymbol{a} = \frac{\partial \boldsymbol{u}}{\partial t} = \frac{\partial^2 \boldsymbol{r}}{\partial t^2}. \tag{1.23}$$

Generally, to know the flow, not only the change over time of the fluid particle position but also the density $\rho(\boldsymbol{r}_0, t)$, pressure $p(\boldsymbol{r}_0, t)$, and temperature $T(\boldsymbol{r}_0, t)$ that accompany each particle are required.

In the Eulerian specification, the spatial coordinates $\boldsymbol{r} = (x, y, z)$ and time t are independent variables, and velocity $\boldsymbol{u} = (u, v, w)$ and density ρ, etc. are dependent variables; they can be expressed as

$$\boldsymbol{u} = \boldsymbol{u}(\boldsymbol{r}, t). \tag{1.24}$$

Moreover, if t is fixed, the state of the instantaneous whole flow field becomes apparent, and if $\boldsymbol{r}$ is fixed, the time development of the flow at that point becomes apparent. Generally, one fluid particle after another passes through this one point over time. $\boldsymbol{u}(\boldsymbol{r}, t)$ is called the velocity field, and $\rho(\boldsymbol{r}, t)$ can be called the density field.

Even in the Eulerian specification, there are often situations when, focusing on one fluid particle, time change of the physical quantities of that particle becomes necessary. Time derivative showing the change over time focusing on fluid particles in this way is called the **Lagrangian derivative** or **material derivative** and is denoted as D/Dt.[10] In the Lagrangian specification, this is denoted as $\partial/\partial t$.

---

[9] Sometimes, $\boldsymbol{r}_0 = (x_0, y_0, z_0)$ can be referred to as a material coordinate.

[10] Can also be called a substantial derivative.

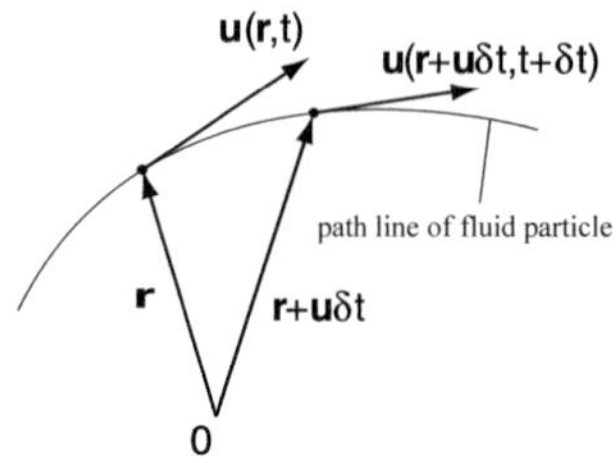

Fig. 1.6 Velocity of fluid particle at time t and $t + \delta t$

Now, let us consider the physical quantity F. In the Eulerian specification, it is a function of $\boldsymbol{r} = (x, y, z)$ and t. If the fluid particle presently under consideration is at position $\boldsymbol{r} = (x, y, z)$ at time t, then after an infinitesimal time δt at time $t + \delta t$, that fluid particle will be at position $\boldsymbol{r} + \boldsymbol{u}\delta t = (x + u\delta t, y + v\delta t, z + w\delta t)$ (Fig. 1.6). Therefore, the increment δF of F in that time is

$$\begin{aligned}\delta F &= F(x + u\delta t, y + v\delta t, z + w\delta t, t + \delta t) - F(x, y, z, t)\\ &= \frac{\partial F}{\partial t}\delta t + \frac{\partial F}{\partial x}u\delta t + \frac{\partial F}{\partial y}v\delta t + \frac{\partial F}{\partial z}w\delta t + O((\delta t)^2).\end{aligned}$$

Hence, the rate of F's change when focusing on specific fluid particle is

$$\frac{DF}{Dt} = \lim_{\delta t \to 0}\frac{\delta F}{\delta t} = \frac{\partial F}{\partial t} + u\frac{\partial F}{\partial x} + v\frac{\partial F}{\partial y} + w\frac{\partial F}{\partial z}. \tag{1.25}$$

The physical quantity F can be arbitrary; hence, if it is expressed as a differential operator, it becomes

$$\frac{D}{Dt} = \frac{\partial}{\partial t} + u\frac{\partial}{\partial x} + v\frac{\partial}{\partial y} + w\frac{\partial}{\partial z} = \frac{\partial}{\partial t} + u_j\frac{\partial}{\partial x_j} = \frac{\partial}{\partial t} + \boldsymbol{u}\cdot\nabla, \tag{1.26}$$

where ∇ expresses the differential operator:

$$\nabla = \boldsymbol{i}\frac{\partial}{\partial x} + \boldsymbol{j}\frac{\partial}{\partial y} + \boldsymbol{k}\frac{\partial}{\partial z} \tag{1.27}$$

and is called the nabla.

In particular, if x, y, z are taken as F, then the following relation holds

$$\frac{Dx}{Dt} = \frac{\partial x}{\partial t} + u\frac{\partial x}{\partial x} + v\frac{\partial x}{\partial y} + w\frac{\partial x}{\partial z} = u, \quad \frac{Dy}{Dt} = v, \quad \frac{Dz}{Dt} = w.$$

Hence, the result can clearly be expressed as expected by

$$\frac{D\boldsymbol{r}}{Dt} = \boldsymbol{u}. \tag{1.28}$$

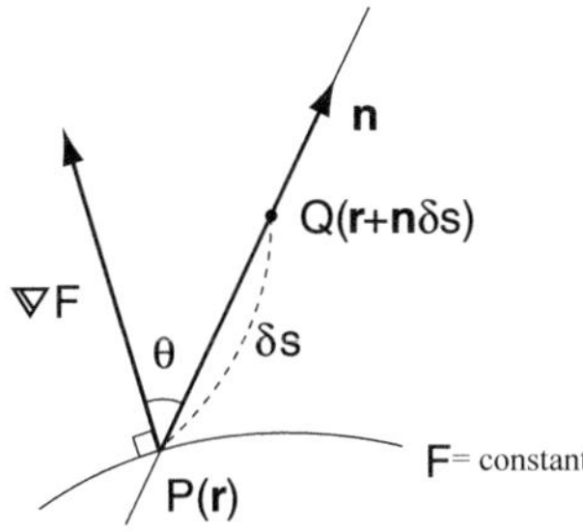

Fig. 1.7 Directional derivative and ∇F

Furthermore, the acceleration of fluid particles can be obtained as a Lagrangian derivative as follows:

$$\boldsymbol{a} = \frac{D\boldsymbol{u}}{Dt} = \frac{\partial \boldsymbol{u}}{\partial t} + u\frac{\partial \boldsymbol{u}}{\partial x} + v\frac{\partial \boldsymbol{u}}{\partial y} + w\frac{\partial \boldsymbol{u}}{\partial z} = \frac{\partial \boldsymbol{u}}{\partial t} + (\boldsymbol{u}\cdot\nabla)\boldsymbol{u}\,. \tag{1.29}$$

When $\boldsymbol{u}$ is independent of time, i.e., when $\partial \boldsymbol{u}/\partial t = 0$, flow is said to be **steady**. From (1.29), it can be observed that velocity field changes along the fluid particles' direction, then acceleration is not zero even if flow is steady.

There are cases in which the Lagrangian specification is particularly useful depending on the problem; however, the Eulerian specification is usually utilized for specifying the flow field. Hereinafter, unless specifically stated otherwise, the Eulerian specification will be used.

Note 1.3 When $\boldsymbol{n}$ is the unit vector and F is any scalar field (arbitrary function of $\boldsymbol{r}$), so that it is easily understood, the following relation is true:

$$\lim_{\delta s\to 0}\frac{F(\boldsymbol{r}+\boldsymbol{n}\delta s) - F(\boldsymbol{r})}{\delta s} = \left(n_1\frac{\partial}{\partial x} + n_2\frac{\partial}{\partial y} + n_3\frac{\partial}{\partial z}\right)F = (\boldsymbol{n}\cdot\nabla)F\,. \tag{1.30}$$

$(\boldsymbol{n}\cdot\nabla)F$ is the rate of change of F at point P in direction $\boldsymbol{n}$ (Fig. 1.7). Therefore, $\boldsymbol{n}\cdot\nabla$ is called a directional derivative in the $\boldsymbol{n}$-direction and is written as $\partial/\partial\boldsymbol{n}$. If $\boldsymbol{u} = |\boldsymbol{u}|\hat{\boldsymbol{u}}$ ($\hat{\boldsymbol{u}}$ is a unit vector), $\boldsymbol{u}\cdot\nabla$ in (1.26) can be written as $|\boldsymbol{u}|(\hat{\boldsymbol{u}}\cdot\nabla)$, that is, the directional derivative in direction $\hat{\boldsymbol{u}}$ multiplied by $|\boldsymbol{u}|$.

Moreover, a vector field called the gradient of F,

$$\nabla F = \frac{\partial F}{\partial x}\boldsymbol{i} + \frac{\partial F}{\partial y}\boldsymbol{j} + \frac{\partial F}{\partial z}\boldsymbol{k}\,, \tag{1.31}$$

accompanies the scalar field F. This can also be written as grad F. If this is used, (1.30) can also be written as

$$\boldsymbol{n}\cdot(\nabla F) = n_1\frac{\partial F}{\partial x} + n_2\frac{\partial F}{\partial y} + n_3\frac{\partial F}{\partial z} = |\nabla F|\cos\theta\,,$$

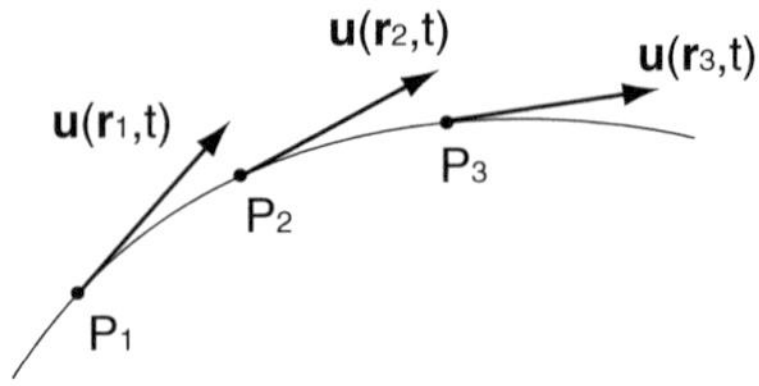

Fig. 1.8 Streamline at time t and its associated velocity vectors, tangential to the streamline

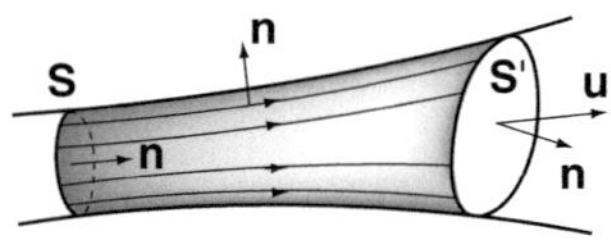

Fig. 1.9 Stream tube. S and S' are two cross sections of the stream tube, and $\boldsymbol{n}$ is the unit vector normal to the cross section and side surface

where, θ is the angle formed by $\boldsymbol{n}$ and ∇F. This considers the maximum value when $\theta = 0$. Therefore, ∇F has the direction that maximizes the rate of increase in F, and $|\nabla F|$ is its maximum rate of increase. Furthermore, this implies that ∇F has a normal vector direction for the surface of F = constant.

Exercise 1.3 Prove that ∇F is a normal vector of a surface F = constant, by considering an arbitrary curve on the surface that passes through point P on the surface.

Exercise 1.4 Prove

$$\frac{D}{Dt}(FG) = \frac{DF}{Dt}G + F\frac{DG}{Dt}.$$

1.3.2 *Streamlines, Path Lines and Streak Lines*

Curves within the flow field whose tangent at each point is parallel to the flow's velocity vector are called **streamlines**. Denoting the line element of a streamline by $d\boldsymbol{r}$, the relation $d\boldsymbol{r} \parallel \boldsymbol{u}$ holds and this is given in components by

$$\frac{dx}{u(x, y, z, t)} = \frac{dy}{v(x, y, z, t)} = \frac{dz}{w(x, y, z, t)}. \tag{1.32}$$

Here t is a parameter (Fig. 1.8). These are the differential equations of streamlines. Solutions of (1.32) for each parameter t take the form $f(x, y, z; t) = C_1$, $g(x, y, z; t) = C_2$ (where C_1 and C_2 are constants of integration), and give a set of streamlines as the intersections of the two surface groups. In a steady flow, the streamlines are time-invariant, whereas in a non-steady flow, they change over time.

The streamlines passing through each points on an arbitrary closed curve in a flow generally form a tube. It is called a **stream tube** (Fig. 1.9). In a steady flow, streamlines and stream tubes are fixed in space. Moreover, then a stream tube determines

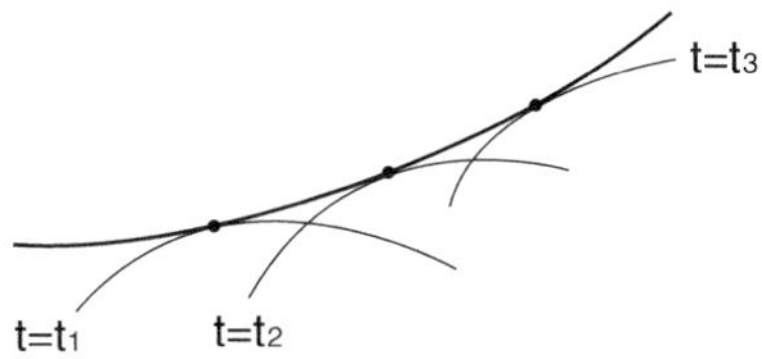

Fig. 1.10 Relationship between path line and streamlines. The particle's path line (*thick line*) contacts one of the streamlines (*thin lines*) at each time t

the mass (the volume if fluid density is constant) of fluid passing through the cross section of the stream tube per unit time (Exercise 1.5).

If a single particle makes a displacement by a distance $d\boldsymbol{r}$ in an infinitesimal time dt, we have

$$d\boldsymbol{r} = \boldsymbol{u}\, dt \tag{1.33}$$

This equation is rewritten in components

$$\frac{dx}{u(x, y, z, t)} = \frac{dy}{v(x, y, z, t)} = \frac{dz}{w(x, y, z, t)} = dt\,. \tag{1.34}$$

Equations (1.34) are the differential equation for the path of a fluid particle. These differential equations differ from (1.32), and x, y, z can be determined as functions of t:

$$x = x(t, C_1),\ \ y = y(t, C_2),\ \ z = z(t, C_3)\,.$$

Here C_1, C_2, and C_3 are constants of integration. This path is called the **path line**.

The velocity vector of a single fluid particle is always tangent to its path line (trajectory) and at each instant to one of streamlines.

Therefore, in non-steady flows, the path line is the envelope of a set of streamlines parameterized by time t (Fig. 1.10).

The flow field in non-steady flows can be visualized by adding a material such as aluminum powder. If this field is photographed with short and long exposure times, the streamline pattern at the moment and flow path lines, respectively, are captured.

The curve traced by the fluid particles passing through given points in space is called a **streak line**. This curve would be drawn if a dye was continuously injected from the given point. The stream, path, and streak lines coincide in steady flows.

1.4 Mass Conservation Law

To understand the flow field, we require not only the velocity $\boldsymbol{u}(\boldsymbol{r}, t)$ but also density $\rho(\boldsymbol{r}, t)$, pressure $p(\boldsymbol{r}, t)$ and temperature $T(\boldsymbol{r}, t)$. In this section, the equation of continuity, one of the fundamental equations which govern these quantities is derived from the mass conservation law.

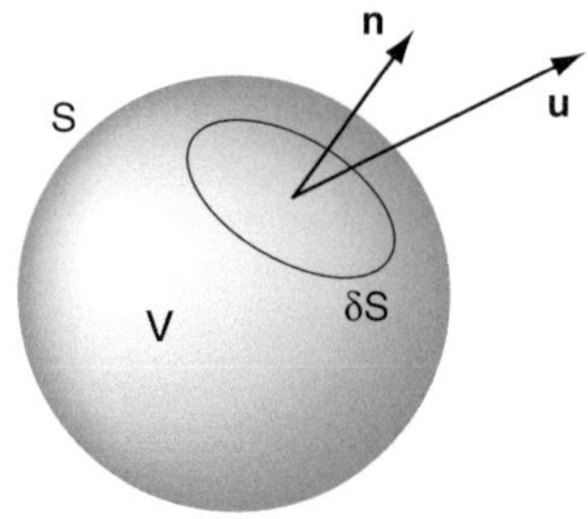

Fig. 1.11 Closed region fixed in space

Consider a closed surface S fixed in space, as shown in Fig. 1.11. Suppose the region surrounded by S to be V. At any time, the mass of the fluid within V is

$$\iiint_V \rho \, dV \, .$$

The net increase of the mass of the fluid within V per unit time is then given by

$$\frac{d}{dt}\iiint_V \rho \, dV = \iiint_V \frac{\partial \rho}{\partial t} \, dV \, .$$

On the other hand, the mass of the fluid flowing out of V per unit time through S is the sum of the masses $\rho \boldsymbol{u} \cdot \boldsymbol{n} \, \delta S$ through infinitesimal surface elements δS per unit time,

$$\iint_S \rho \boldsymbol{u} \cdot \boldsymbol{n} \, dS \, ,$$

where $\boldsymbol{n}$ is the outward unit normal vector of S. Provided that no fluid disappears or is newly created, the increase in fluid mass inside V is derived from fluid material entering through the boundary surface S; that is,

$$\iiint_V \frac{\partial \rho}{\partial t} \, dV = -\iint_S \rho \boldsymbol{u} \cdot \boldsymbol{n} \, dS \, . \tag{1.35}$$

Transforming the right-hand side of (1.35) to a volume integral by Gauss' theorem (see Note 1.4) and rearranging, we obtain

$$\iiint_V \left\{ \frac{\partial \rho}{\partial t} + \nabla \cdot (\rho \boldsymbol{u}) \right\} dV = 0 \, . \tag{1.36}$$

For Equation (1.36) to hold for an arbitrary region V, we require

$$\frac{\partial \rho}{\partial t} + \nabla \cdot (\rho \boldsymbol{u}) = 0 \ \text{ or } \ \frac{\partial \rho}{\partial t} + \frac{\partial}{\partial x_i}(\rho u_i) = 0 \, . \tag{1.37}$$

This expression, derived from the mass conservation law, is called the **equation of continuity**. Moreover, using $\frac{\partial}{\partial x_i}(\rho u_i) = u_i \frac{\partial \rho}{\partial x_i} + \rho \frac{\partial u_i}{\partial x_i}$, or, $\nabla \cdot (\rho \boldsymbol{u}) = (\boldsymbol{u} \cdot \nabla)\rho + \rho \nabla \cdot \boldsymbol{u}$, the equation of continuity can be rewritten as

$$\frac{D\rho}{Dt} + \rho \nabla \cdot \boldsymbol{u} = 0 \quad \text{or} \quad \frac{D\rho}{Dt} + \rho \frac{\partial u_i}{\partial x_i} = 0 . \tag{1.38}$$

When the density of the moving fluid particles is constant, that is, when

$$\frac{D\rho}{Dt} = 0 , \tag{1.39}$$

the flow is said to be **incompressible**. In this case, then, the equation of continuity is reduced to

$$\nabla \cdot \boldsymbol{u} = 0 \quad \text{or} \quad \frac{\partial u_i}{\partial x_i} = 0 . \tag{1.40}$$

According to this definition, any incompressible flow does not imply that the density is constant throughout the flow field. However, if the density is initially constant throughout the flow field, it remains constant. Density can be regarded as a constant in many real-life situations, which significantly simplifies the equations. In particular, a fluid of constant density is called an incompressible fluid. The concepts of incompressible flow and incompressible fluid are different, and should be carefully distinguished,[11] although (1.40) holds in both cases.

Let us consider the meaning of $\nabla \cdot \boldsymbol{u}$. Imagine a closed surface S consisting of the fluid material (not fixed in space) and denote the enclosed fluid portion and its volume by the same symbol, δV. The increase in δV per unit time depends on the expansion of the surface S. Since every point on S moves at velocity $\boldsymbol{u}$ of that point, we have

$$\frac{d\delta V}{dt} = \iint_S \boldsymbol{u} \cdot \boldsymbol{n}\, dS .$$

Rewriting the right-hand side as a volume integral using Gauss' theorem, we obtain

$$\frac{d\delta V}{dt} = \iiint_{\delta V} \nabla \cdot \boldsymbol{u}\, dV = \delta V (\nabla \cdot \boldsymbol{u})(\boldsymbol{r}', t) .$$

Here $\boldsymbol{r}'$ is a point in δV. The final equal sign is based on the mean-value theorem. Therefore, if δV is contracted around the point $\boldsymbol{r}$, resulting in $\boldsymbol{r}' \to \boldsymbol{r}$, we have

$$\lim_{\delta V \to 0} \frac{1}{\delta V} \frac{d\, \delta V}{dt} = \nabla \cdot \boldsymbol{u} . \tag{1.41}$$

Hence, $\nabla \cdot \boldsymbol{u}$ represents the rate of expansion of the infinitesimal fluid element with center $\boldsymbol{r}$. If the volume of an infinitesimal fluid element is δV and its mass is

[11] Not all experts distinguish incompressible flow and incompressible fluid in this way.

$\rho\delta V$, then the following equation is intuitively apparent:

$$\frac{D(\rho\,\delta V)}{Dt} = 0 \tag{1.42}$$

This expression can further be confirmed from (1.38) and (1.41).

Note 1.4 Any vector field $\boldsymbol{u}(\boldsymbol{r})$ is accompanied by a scalar field

$$\nabla\cdot\boldsymbol{u} = \frac{\partial u_i}{\partial x_i} = \frac{\partial u}{\partial x} + \frac{\partial v}{\partial y} + \frac{\partial w}{\partial z}\,.$$

This expression is called the divergence of vector field $\boldsymbol{u}(\boldsymbol{r})$, also written as div$\boldsymbol{u}$. Gauss' theorem holds for the closed surface S with an outward unit normal vector $\boldsymbol{n}$ and the volume V enclosed by the surface S as

$$\iiint_V \nabla\cdot\boldsymbol{u}\,dV = \iint_S \boldsymbol{u}\cdot\boldsymbol{n}\,dS \quad\text{or}\quad \iiint_V \frac{\partial u_i}{\partial x_i}\,dV = \iint_S n_i u_i\,dS\,. \tag{1.43}$$

Formally, when converting a surface integral to a volume integral, the n_i of the former is replaced by a differential operator $\partial/\partial x_i$ in the latter. This replacement is valid in many applications of Gauss' theorem. Note that although the i index notation is adopted in (1.43), implying a summation of three components, the formula (1.43) holds for each case of $i = 1, 2, 3$ without summation. Additionally, then, u_1, u_2, u_3 need not be components of a vector but can specify an arbitrary function.

Note 1.5 In deriving the equation of continuity, mass was assumed to be neither generated in nor vanished from the flow field. If fluid is generated in the flow field, an additional term is added to the right-hand side of (1.35):

$$\iiint_V \sigma(\boldsymbol{r},t)\rho\,dV\,.$$

Here $\sigma(\boldsymbol{r},t)$ is the generation rate of fluid volume per unit time at $\boldsymbol{r}, t$. The equation of continuity is now modified to

$$\frac{\partial\rho}{\partial t} + \nabla\cdot(\rho\boldsymbol{u}) = \rho\sigma(\boldsymbol{r},t) \quad\text{or}\quad \frac{D\rho}{Dt} + \rho\nabla\cdot\boldsymbol{u} = \rho\sigma(\boldsymbol{r},t) \tag{1.44}$$

If the fluid is incompressible, (1.44) is reduced to

$$\nabla\cdot\boldsymbol{u} = \sigma\,. \tag{1.45}$$

This result is discussed in more detail in Sect. 2.5.

Exercise 1.5 Using Gauss'theorem, prove that in a steady flow, the mass (or the volume if the density is uniform) of fluid passing through a cross section of a stream tube per unit time is determined by the stream tube.

1.5 Equations for Perfect Fluids

1.5.1 Euler's Equation

We now derive the equation of motion for fluids. According to d'Alembert's principle, a dynamic problem can be converted to a static problem by including inertia force. We consider the balance between forces acting on the fluid in a spatially fixed region, as shown in Fig. 1.11. Suppose that a fluid is subjected to an external force (body force) of $\boldsymbol{K} = (K_1, K_2, K_3)$ per unit mass. As the mass of the fluid in an infinitesimal volume element δV is $\rho\delta V$, the external force acting on the mass is $K_i\rho\delta V$. Moreover, if the acceleration of that region of fluid is $\boldsymbol{a}$, the inertia force is $-a_i\rho\delta V$. From (1.10), the surface element δS with normal vector $\boldsymbol{n}$ experiences a surface force of $\tau_{ij}n_j\delta S$ imposed by the external fluid. Therefore, the balance between forces throughout the volume V can be written as

$$\iiint_V (K_i - a_i)\rho\, dV + \iint_S \tau_{ij}n_j\, dS = 0\,. \tag{1.46}$$

Applying Gauss' theorem, the surface integral in (1.46) converts to the following volume integral:

$$\iiint_V \left\{(K_i - a_i)\rho + \frac{\partial \tau_{ij}}{\partial x_j}\right\} dV = 0\,. \tag{1.47}$$

Since V is arbitrary, the integrand (if continuous) must be identically 0. Therefore, substituting the acceleration a_i by Du_i/Dt, we obtain

$$\frac{Du_i}{Dt} = K_i + \frac{1}{\rho}\frac{\partial \tau_{ij}}{\partial x_j}\,. \tag{1.48}$$

To complete the derivation, the stress τ_{ij} must be related to the velocity field. Here we assume perfect fluids (viscous fluids will be treated later). As discussed at the end of Sect. 1.2.3, the stress acting on a perfect fluid is given by (1.20). Thus, we have

$$\frac{\partial \tau_{ij}}{\partial x_j} = -\frac{\partial p}{\partial x_j}\delta_{ij} = -\frac{\partial p}{\partial x_i}\,.$$

Expressing (1.48) using the vector notation, we obtain **Euler's equation** of motion for a perfect fluid

$$\frac{D\boldsymbol{u}}{Dt} = \boldsymbol{K} - \frac{1}{\rho}\nabla p\,. \tag{1.49}$$

The components of (1.49) (replacing (K_1, K_2, K_3) with (X, Y, Z)) are

$$\left.\begin{aligned}\frac{\partial u}{\partial t}+u\frac{\partial u}{\partial x}+v\frac{\partial u}{\partial y}+w\frac{\partial u}{\partial z}&=X-\frac{1}{\rho}\frac{\partial p}{\partial x}\\ \frac{\partial v}{\partial t}+u\frac{\partial v}{\partial x}+v\frac{\partial v}{\partial y}+w\frac{\partial v}{\partial z}&=Y-\frac{1}{\rho}\frac{\partial p}{\partial y}\\ \frac{\partial w}{\partial t}+u\frac{\partial w}{\partial x}+v\frac{\partial w}{\partial y}+w\frac{\partial w}{\partial z}&=Z-\frac{1}{\rho}\frac{\partial p}{\partial z}\end{aligned}\right\}. \tag{1.50}$$

More compactly,

$$\frac{\partial u_i}{\partial t}+u_j\frac{\partial u_i}{\partial x_j}=K_i-\frac{1}{\rho}\frac{\partial p}{\partial x_i}. \tag{1.51}$$

This equation indicates that, in the absence of external forces, the flow is driven by the pressure gradient. The most commonly encountered external force is gravity. Considering the z-axis as the vertical axis and denoting the gravitational acceleration by g, the gravitational force is expressed as $\boldsymbol{K} = (0, 0, -g)$.

Multiplying (1.51) by ρ and (1.37) by u_i[12] and adding the results, we get

$$\frac{\partial}{\partial t}(\rho u_i)+\frac{\partial}{\partial x_j}(\rho u_i u_j+p\,\delta_{ij})=\rho K_i\,. \tag{1.52}$$

Equation (1.52) is the momentum conservation form of the equation of motion.

Exercise 1.6 In a stationary fluid ($\boldsymbol{u} = 0$), (1.49) are reduced to

$$\rho\boldsymbol{K}=\nabla p\,.$$

Using this equation, prove that in a uniform gravity field with uniform density, an object floating in a stationary fluid receives buoyancy equal to the weight of the fluid it displaces (Archimedes' principle). For simplicity, assume that the object's surface is entirely surrounded by one fluid.

1.5.2 Equation of State

For perfect fluids we have obtained four equations, that is the equation of continuity (1.37) and Euler's equation (1.49). We assume the fluid is in locally thermal equilibrium. Then, unknown variable are usually six; three are the components of the velocity $\boldsymbol{u}$, and another three are density ρ, pressure p, and temperature T. Therefore, to determine these variables, we need two additional equations. One is the equation

[12]Here the subscript in (1.37) is changed from i to j and the components are summed from 1 to 3. Therefore, notation change does not influence the result.

of energy, which will be derived for the general case in Sect. 1.9. For now, assuming a perfect fluid and adiabatic conditions (in which we can ignore heat conduction and generation), we have

$$\frac{DS}{Dt} = 0\,. \tag{1.53}$$

Here S is the entropy per unit mass. This equation means that the entropy of fluid particles is constant, then the flow is called **isentropic flow**.

The other one is the **equation of state**. Generally, the thermodynamic status of a homogeneous single substance is determined by two thermodynamic variables; for instance,

$$\rho = \rho\,(p, S) \;\text{ or }\; \rho = \rho\,(p, T)\,. \tag{1.54}$$

For ideal gases, such relations are given by

$$p = \rho^{\gamma} \exp\left(\frac{S - S_0}{c_v}\right), \quad p = R\,\rho\,T = (c_p - c_v)\,\rho\,T, \tag{1.55}$$
$$\gamma = \frac{c_p}{c_v}, \quad R = \frac{R_*}{m}.$$

Here c_p and c_v are the specific heat at constant pressure and constant volume per unit mass, respectively, $R_* = 8.314\,\mathrm{J\,K^{-1}mol^{-1}} = 1.986\,\mathrm{cal\,K^{-1}mol^{-1}}$ is the universal gas constant, and m is the molecular weight of the gas. R is the gas constant of the gas, equal to $R = 287.04\,\mathrm{JK^{-1}kg^{-1}}$ in dry air.

When gas undergoes an isothermal (T = constant) or adiabatic (S = constant) change, or when constant density can be assumed, as in most liquids, density is a function of pressure alone and can be expressed as

$$\rho = f\,(p). \tag{1.56}$$

This type of flow is called **barotropic flow**. Barotropic flow is determined by the equation of continuity and Euler's equation, and temperature is determined by the state equation. The equation of energy is not required.

1.5.3 Boundary Conditions

The flow is assumed to be barotropic. Since (1.37) and (1.49) are partial differential equations, initial and boundary conditions are required for solving these. As these equations are first order in time, the initial condition is given by a specified spatial distribution of $\boldsymbol{u}$, p (or ρ) at $t = 0$. That is,

$$\boldsymbol{u}(\boldsymbol{r}, 0) = \boldsymbol{u}_0(\boldsymbol{r}), \quad p\,(\boldsymbol{r}, 0) = p_0(\boldsymbol{r}) \;\; (\text{or } \rho\,(\boldsymbol{r}, 0) = \rho_0(\boldsymbol{r})\,).$$

Initial conditions are not required in steady flow problems.

The boundary conditions must be appropriately specified for the given problem. Typical boundary conditions are listed below.

1.5.3.1 Conditions at Rigid Walls

On the surface of a rigid wall, the normal component of the relative velocity of the fluid and rigid body needs to be zero. Denoting the fluid velocity and normal vector at a point on the rigid body by $\boldsymbol{u}$ and $\boldsymbol{n}$, respectively, this condition is expressed as

$$(\boldsymbol{u} - \boldsymbol{u}_b) \cdot \boldsymbol{n} = 0 \ \ \text{(on the rigid wall surface)}, \tag{1.57}$$

where $\boldsymbol{u}_b$ is the velocity of the rigid body. Under this condition, no gap exists between the fluid and rigid wall, and the fluid will not cave into the rigid wall. Fluid particles on the rigid wall can be considered to flow along the wall's surface. In particular, when the rigid body is stationary, this condition becomes

$$\boldsymbol{u} \cdot \boldsymbol{n} = 0 \ \ \text{(on the rigid wall surface)}. \tag{1.58}$$

Such a restriction does not apply to the tangential component of the flow velocity on the rigid body surface. This is because that the governing equations are also first order in space.

1.5.3.2 Conditions on a Deforming Boundary Surface

Now consider the boundary condition of the case that the fluid boundary deforms in time, as seen on free surface of liquid. Suppose the form of the boundary surface is

$$F(\boldsymbol{r}, t) = 0 \ \text{ or } \ F(x, y, z, t) = 0\,. \tag{1.59}$$

Even in this case, the fluid particles must move along the boundary surface. A fluid particle located at $\boldsymbol{r} = (x, y, z)$ at time t will move to $\boldsymbol{r} + \boldsymbol{u}\delta t$ at time $t + \delta t$; thus, we have

$$F(x + u\delta t, y + v\delta t, z + w\delta t, t + \delta t) = 0\,.$$

Subtracting (1.59) from this equation, dividing the result by δt, and taking the limit $\delta t \to 0$, we obtain the following equation:

$$\frac{DF}{Dt} = \frac{\partial F}{\partial t} + u\frac{\partial F}{\partial x} + v\frac{\partial F}{\partial y} + w\frac{\partial F}{\partial z} = 0. \tag{1.60}$$

Thus, the required boundary condition is that (1.60) holds under the condition of (1.59).

In particular, if (1.59) is independent of t, this condition is reduced to the boundary condition on the rigid wall surface

$$F(x, y, z) = 0\,, \tag{1.61}$$

given by

$$\boldsymbol{u} \cdot \nabla F = u\frac{\partial F}{\partial x} + v\frac{\partial F}{\partial y} + w\frac{\partial F}{\partial z} = 0\,. \tag{1.62}$$

1.6 Local Motion and Deformation of Fluids

A fluid element in a moving fluid not only translates but also rotates and deforms. This local deformation of a fluid is closely connected to stress. Here we focus on a single infinitesimal, spherical fluid element. We denote the position vector of its center point as $\boldsymbol{r}$ and that of another arbitrary point Q in the fluid element as $\boldsymbol{r} + \delta\boldsymbol{r}$. The state of deformation can be learned from the relative velocity $\delta\boldsymbol{u} = \boldsymbol{u}(\boldsymbol{r} + \delta\boldsymbol{r}) - \boldsymbol{u}(\boldsymbol{r})$. Expanding $\boldsymbol{u}(\boldsymbol{r} + \delta\boldsymbol{r})$ as a power series in $\delta\boldsymbol{r}$ and omitting terms beyond the second order of the infinitesimal distance $|\delta\boldsymbol{r}|$, we have

$$\begin{aligned}\boldsymbol{u}(\boldsymbol{r} + \delta\boldsymbol{r}) = \boldsymbol{u}(\boldsymbol{r}) &+ \frac{\partial u}{\partial x}\delta x\boldsymbol{i} + \frac{\partial u}{\partial y}\delta y\boldsymbol{i} + \frac{\partial u}{\partial z}\delta z\boldsymbol{i} \\ &+ \frac{\partial v}{\partial x}\delta x\boldsymbol{j} + \frac{\partial v}{\partial y}\delta y\boldsymbol{j} + \frac{\partial v}{\partial z}\delta z\boldsymbol{j} \\ &+ \frac{\partial w}{\partial x}\delta x\boldsymbol{k} + \frac{\partial w}{\partial y}\delta y\boldsymbol{k} + \frac{\partial w}{\partial z}\delta z\boldsymbol{k}.\end{aligned} \tag{1.63}$$

The differential coefficients such as $\partial u/\partial x$ are specified at the point P. The first term $\boldsymbol{u}(\boldsymbol{r})$ expresses the rigid-body-like translational motion of the fluid element as a whole. The sum of the remaining nine terms describes the relative motion of the point Q for the point P. The motion represented by $(\partial u/\partial x)\delta x\boldsymbol{i}$ is oriented parallel to the x-axis, and its velocity is proportional to the distance δx from the yz plane. In other words, if $\partial u/\partial x$ is positive (negative), this term expresses the uniform stretch (contraction) in the x-direction (Fig. 1.12). Similarly, $(\partial v/\partial y)\delta y\boldsymbol{j}$ and $(\partial w/\partial z)\delta z\boldsymbol{k}$ express the uniform stretch or contraction in the y- and z-directions, respectively.

On the other hand, the term $(\partial u/\partial y)\delta y\boldsymbol{i}$ describes horizontal motion at a velocity whose magnitude is proportional to δy. This expresses shearing motion in the x-direction on a plane parallel to the xz plane, as shown in Fig. 1.13. The remaining terms describe similar motions in the other directions.

In tensor notation, the relative velocity is expressed as

$$\delta u_i = \frac{\partial u_i}{\partial x_j}\delta x_j\,. \tag{1.64}$$

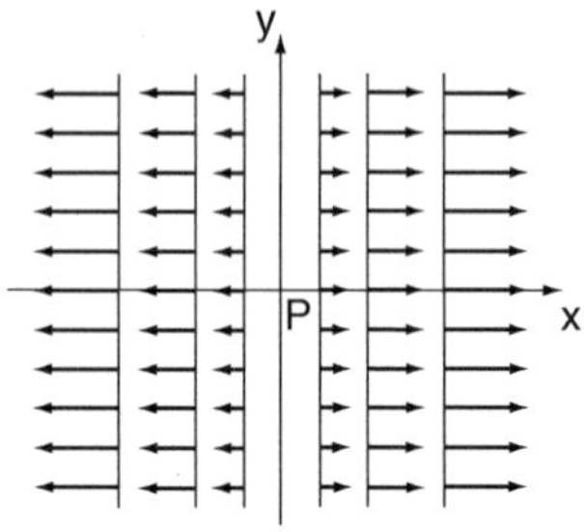

Fig. 1.12 Uniform stretching in the x-direction ($\partial u/\partial x > 0$)

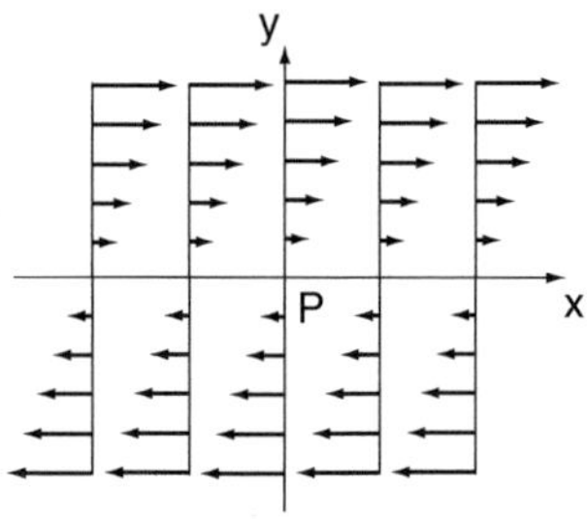

Fig. 1.13 Shearing motion in the x-direction ($\partial u/\partial y > 0$)

The matrix $\partial u_i/\partial x_j$ constitutes a second-order tensor[13] and is called the velocity gradient tensor. This tensor can be split into symmetric and anti-symmetric tensors (see Appendix A.1).

$$e_{ij} = \frac{1}{2}\left(\frac{\partial u_i}{\partial x_j} + \frac{\partial u_j}{\partial x_i}\right), \quad \Omega_{ij} = \frac{1}{2}\left(\frac{\partial u_i}{\partial x_j} - \frac{\partial u_j}{\partial x_i}\right) \tag{1.65}$$

and we have

$$\delta u_i = e_{ij}\delta x_j + \Omega_{ij}\delta x_j\,. \tag{1.66}$$

The symmetric and anti-symmetric tensors make fundamentally different contributions to the relative velocity.

The anti-symmetric tensor Ω_{ij} can be expressed in terms of components (ξ, η, ζ) (or $(\omega_1, \omega_2, \omega_3)$) of the vector field $\boldsymbol{\omega} = \nabla \times \boldsymbol{u}$, known as the **rotation** of the velocity field $\boldsymbol{u}$ (see Note 1.6), as

$$(\Omega_{ij}) = \frac{1}{2}\begin{pmatrix} 0 & -\zeta & \eta \\ \zeta & 0 & -\xi \\ -\eta & \xi & 0 \end{pmatrix}, \tag{1.67}$$

$$\xi = \frac{\partial w}{\partial y} - \frac{\partial v}{\partial z}, \quad \eta = \frac{\partial u}{\partial z} - \frac{\partial w}{\partial x}, \quad \zeta = \frac{\partial v}{\partial x} - \frac{\partial u}{\partial y}, \tag{1.68}$$

[13]This is because the matrix is a linear operator that relates vector $\delta \boldsymbol{r}$ to vector $\delta \boldsymbol{u}$.

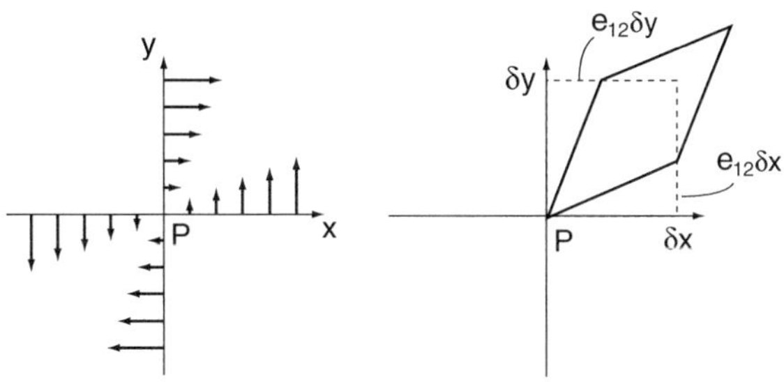

Fig. 1.14 Purely deformational motion described by $e_{12}(\delta x\boldsymbol{j} + \delta y\boldsymbol{i})$

where the vector $\boldsymbol{\omega}$ is called the **vorticity**. Multiplying (1.66) by $\boldsymbol{e}_i$ and summing over i, we obtain

$$\delta\boldsymbol{u} = \delta u_i\boldsymbol{e}_i = e_{ij}\delta x_j\boldsymbol{e}_i + \Omega_{ij}\delta x_j\boldsymbol{e}_i. \tag{1.69}$$

However, as

$$\begin{aligned}\Omega_{ij}\delta x_j\boldsymbol{e}_i &= (\Omega_{12}\delta x_2 + \Omega_{13}\delta x_3)\boldsymbol{e}_1 + (\Omega_{21}\delta x_1 + \Omega_{23}\delta x_3)\boldsymbol{e}_2 \\ &\quad + (\Omega_{31}\delta x_1 + \Omega_{32}\delta x_2)\boldsymbol{e}_3 \\ &= \frac{1}{2}(\eta\delta z - \zeta\delta y)\boldsymbol{i} + \frac{1}{2}(\zeta\delta x - \xi\delta z)\boldsymbol{j} + \frac{1}{2}(\xi\delta y - \eta\delta x)\boldsymbol{k} \\ &= \frac{1}{2}\,\boldsymbol{\omega}\times\delta\boldsymbol{r}\,,\end{aligned} \tag{1.70}$$

we obtain

$$\begin{aligned}\delta\boldsymbol{u} &= e_{11}\delta x\boldsymbol{i} + e_{22}\delta y\boldsymbol{j} + e_{33}\delta z\boldsymbol{k} \\ &\quad + e_{23}(\delta y\boldsymbol{k} + \delta z\boldsymbol{j}) + e_{31}(\delta z\boldsymbol{i} + \delta x\boldsymbol{k}) + e_{12}(\delta x\boldsymbol{j} + \delta y\boldsymbol{i}) \\ &\quad + \frac{1}{2}\,\boldsymbol{\omega}\times\delta\boldsymbol{r}\,.\end{aligned} \tag{1.71}$$

The first line of (1.71) describes the uniform stretch (or contraction) in the direction of the coordinate axes, as explained above. The third term in the second line, $e_{12}(\delta x\boldsymbol{j} + \delta y\boldsymbol{i})$, reduces the angle formed by the x- and y-axes, as shown in Fig. 1.14, by a factor of $2e_{12}$ per unit time. This term describes pure shearing motion of the type that changes a square into a rhombus. The first and second terms in the second line perform similar actions. The third line of (1.71) expresses the velocity field of the **rigid body rotation** with angular velocity $\boldsymbol{\omega}/2$ as seen from its expression.

Since e_{ij} represents the deformation velocity (rate) of the fluid element, it is called the **deformation velocity tensor** or **rate-of-strain tensor**. By transforming the coordinate axes into the principal coordinate of the tensor, this tensor can be expressed as

$$\mathrm{diag}\,(e_1',\ e_2',\ e_3')\,. \tag{1.72}$$

In other words, a fluid element is uniformly stretched or contracted in the direction of the principal axes of the symmetric tensor. As discussed earlier, the expansion rate of the fluid element is the sum of the diagonal elements:

$$e_1' + e_2' + e_3' = e_{ii} = \nabla \cdot \boldsymbol{u}\,. \tag{1.73}$$

Identical to the case of the stress tensor, (1.72) can be decomposed into an isotropic tensor and its difference from the isotropic tensor, respectively, as

$$\mathrm{diag}\left(\frac{1}{3}e_{ii},\ \frac{1}{3}e_{ii},\ \frac{1}{3}e_{ii}\right) \ \text{and}\ \mathrm{diag}\left(e_1' - \frac{1}{3}e_{ii},\ e_2' - \frac{1}{3}e_{ii},\ e_3' - \frac{1}{3}e_{ii}\right). \tag{1.74}$$

The former term denotes isotropic expansion or compression (1.73), and the latter corresponds to purely deformational motion, such as the squeezing of a sphere into an ellipsoid. Summarizing the above, an infinitesimally small fluid element simultaneously undergoes (I) rigid-body translational motion as a whole at flow velocity $\boldsymbol{u}$ at the center, (II) rigid-body rotation of angular velocity $\omega/2$, (III) purely deformational motion without changing its volume, such as spherical to ellipsoid deformations, and (IV) relative isotropic expansion (or contraction) motion at the rate of $\nabla \cdot \boldsymbol{u}$. Motion (II) is contributed by the anti-symmetric tensor Ω_{ij}, whereas the deformational motions (III) and (IV) are embodied in the symmetric tensor e_{ij}. Moreover, (I) and (II) are rigid motions, whereas (III) and (IV) are unique to deformable body. In an irrotational flow in a perfect fluid with constant density (where $\omega = 0$), we have $\nabla \cdot \boldsymbol{u} = 0$. Hence, in that case, the fluid element undergoes only pure deformation (III) as it translates.

Note 1.6 Given a vector field $\boldsymbol{u}(\boldsymbol{r})$, a vector field defined by

$$\begin{aligned} \nabla \times \boldsymbol{u} &= \left(\frac{\partial w}{\partial y} - \frac{\partial v}{\partial z}\right)\boldsymbol{i} + \left(\frac{\partial u}{\partial z} - \frac{\partial w}{\partial x}\right)\boldsymbol{j} + \left(\frac{\partial v}{\partial x} - \frac{\partial u}{\partial y}\right)\boldsymbol{k} \\ &= \begin{vmatrix} \boldsymbol{i} & \boldsymbol{j} & \boldsymbol{k} \\ \dfrac{\partial}{\partial y} & \dfrac{\partial}{\partial y} & \dfrac{\partial}{\partial z} \\ u & v & w \end{vmatrix} \end{aligned} \tag{1.75}$$

is called the rotation of the vector field $\boldsymbol{u}$, expressed as rot $\boldsymbol{u}$ or curl $\boldsymbol{u}$. For this, Stokes' theorem:

$$\oint_C \boldsymbol{u} \cdot d\boldsymbol{r} = \iint_S (\nabla \times \boldsymbol{u}) \cdot \boldsymbol{n}\, dS \tag{1.76}$$

holds. Here, S is an arbitrary surface whose edge is enclosed by a closed curve C (see Fig. 1.15). That is to say, when progressing along C in the direction of the arrow in Fig. 1.15, the front side (which the normal vector $\boldsymbol{n}$ is facing) is seen on the left-hand side.

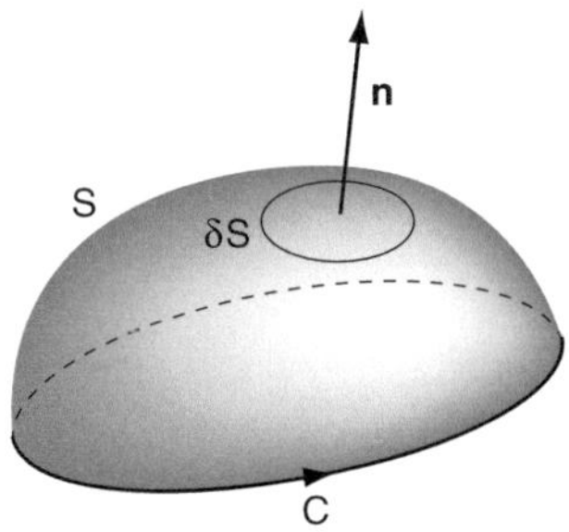

Fig. 1.15 Relationship between the surface and the curve enclosing the surface, in Stokes' theorem

1.7 Relationship Between Rate-of-Strain and Stress

In stationary fluids, stress has normal components only, and its magnitude is independent of the direction of the surface across which the stress act on. In other words, the stress tensor is expressed as

$$\tau_{ij} = -\,p\,\delta_{ij} \tag{1.77}$$

Moreover, if no tangential stress exists (perfect fluid), the stress tensor is still expressed by (1.77) in moving fluids. However, in actual (relative) moving fluids, tangential stress arises owing to viscosity of fluid. In this case, the magnitude of normal stress usually depends on the direction of the surface. Therefore, we define the pressure of fluid in motion as negative average normal stress:

$$p = -\frac{1}{3}\,\tau_{ii} \tag{1.78}$$

As the stresses in stationary fluids and perfect fluids also satisfy this relationship, (1.78) is a generalization of these cases. Moreover, (1.78) is convenient because τ_{ii} remains unchanged by the rotation of the coordinate system. (1.78) is a purely mechanical definition of pressure.

Here it is useful to divide the stress tensor into isotropic pressure and the difference from it;

$$\tau_{ij} = -\,p\,\delta_{ij} + d_{ij}. \tag{1.79}$$

d_{ij} is the anisotropic part of the stress tensor and it brings that the tangential stresses, as well as the sum of its diagonal components, is 0. The tensor d_{ij} is called the **deviatoric stress tensor** and arises entirely from the relative motions of the fluid.

In an ordinary viscous fluid such as air or water for the shear flow, as shown in Fig. 1.16, the side of larger y exerts tangential stress

$$\tau_{12} = d_{12} = \mu\frac{\partial u}{\partial y} \tag{1.80}$$

Fig. 1.16 Shear flow

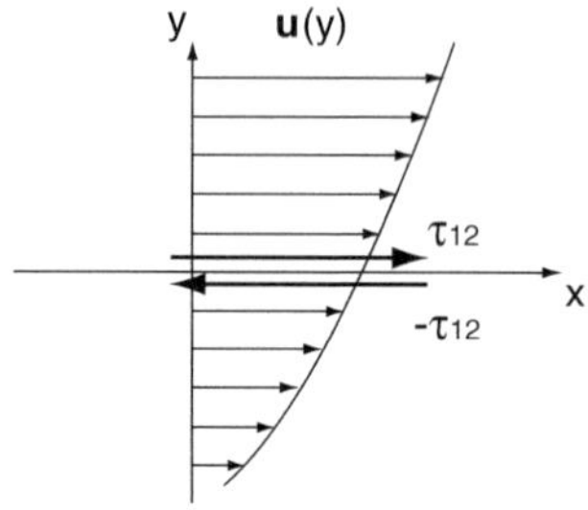

on the side of smaller y through a surface perpendicular to the y-axis.

Of course, the smaller side exerts tangential stress of the same magnitude on the larger side but in the opposite direction. This stress acts in the direction of reducing velocity gap.[14] A fluid producing such tangential stress proportional to the velocity gradient is called a **Newtonian fluid**.[15] This relationship suggests a proportional relationship between the deviatoric stress tensor and the velocity gradient tensor. Since translations and rigid-body rotations of whole fluid elements generate no relative motion, they are unrelated to the viscous force. Therefore, it can be considered that, in the velocity gradient tensor, the rate-of-strain tensor alone relates to the deviatoric stress tensor. We may specify this relationship as $d_{ij} = f(e_{ij})$. Furthermore, if e_{ij} is assumed to be small, and terms beyond second-order are ignored in the Taylor expansion, this relationship can be written as

$$d_{ij} = a_{ijkl}e_{kl} + \cdots \tag{1.81}$$

Here

$$a_{ijkl} = \left(\frac{\partial d_{ij}}{\partial e_{kl}}\right)_{e_{kl}=0}.$$

These terms are material constants independent of e_{ij}. d_{ij} and e_{ij} are second-order tensors; hence, a_{ijkl} is a fourth-order tensor. Moreover, if an isotropic fluid is assumed,[16] the tensors must be isotropic. The fourth-order isotropic tensor is expressed as

$$a_{ijkl} = B\delta_{ij}\delta_{kl} + C\delta_{ik}\delta_{jl} + D\delta_{il}\delta_{jk} \tag{1.82}$$

where B, C, and D are scalars. Substituting (1.82) into (1.81), we get

---

[14] In terms of molecules, this occurs because the molecules on the upper side with large horizontal momentum are transferred to the lower side by molecular thermal motion, whereas those on the lower side with smaller horizontal momentum are transferred to the upper side. In this sense, viscosity can be regarded as the diffusion of momentum.

[15] A fluid in which tangential stress is not proportional to velocity gradient is called a non-Newtonian fluid. This book chiefly focuses on Newtonian fluids.

[16] Materials such as liquid crystals are directional and not isotropic materials.

$$
\begin{aligned}
d_{ij} = a_{ijkl}e_{kl} &= B\delta_{ij}\delta_{kl}e_{kl} + C\delta_{ik}\delta_{jl}e_{kl} + D\delta_{il}\delta_{jk}e_{kl} \\
&= B\delta_{ij}e_{ll} + Ce_{ij} + De_{ji} \\
&= B\Theta\delta_{ij} + (C + D)e_{ij}
\end{aligned}
$$

Here

$$
\Theta = e_{ll} = \frac{\partial u_l}{\partial x_l} = \nabla \cdot \boldsymbol{u}
$$

and we have used $e_{ji} = e_{ij}$. Furthermore, writing $B = \lambda$, $C + D = 2\mu$, we obtain

$$
d_{ij} = 2\mu e_{ij} + \lambda\Theta\delta_{ij} \tag{1.83}
$$

Furthermore, since

$$
d_{ii} = 2\mu e_{ii} + \lambda\Theta\delta_{ii} = (2\mu + 3\lambda)\Theta = 0\,,
$$

we must have

$$
2\mu + 3\lambda = 0\,.
$$

Therefore,

$$
d_{ij} = 2\mu\Big(e_{ij} - \frac{1}{3}\Theta\delta_{ij}\Big)\,. \tag{1.84}
$$

μ (> 0) is a viscosity-related material constant, and is called **viscosity** or coefficient of viscosity. Note that the sum of diagonals of the tensor $e_{ij} - (1/3)\Theta\delta_{ij}$ on the right-hand side of (1.84) is zero, as well as the left-hand side. The form of (1.84) coincides with that of (1.80).

Coefficients of viscosity μ and **kinematic viscosity** ν $(= \mu/\rho)$ of representative fluids such as air and water are listed in Table 1.1.

After all the stress tensor for a Newtonian fluid is given by

$$
\tau_{ij} = -\,p\,\delta_{ij} + 2\mu\Big(e_{ij} - \frac{1}{3}\Theta\delta_{ij}\Big)\,. \tag{1.85}
$$

Table 1.1 Viscosities of representative materials

15 °C, 1 atmospheric pressure	Coefficient of Viscosity μ ($\mathrm{gcm^{-1}s^{-1}} = 10^{-1}\mathrm{kgm^{-1}s^{-1}}$)	Kinematic Viscosity ν ($\mathrm{cm^2s^{-1}} = 10^{-4}\mathrm{m^2/s}$)
Water	0.011	0.011
Air	0.00018	0.15
Mercury	0.016	0.0012
Olive oil	0.99	1.08
Glycerin	23.3	18.5

Here p is defined by (1.78). In particular, in incompressible flows,

$$\tau_{ij} = -\,p\,\delta_{ij} + \mu\Big(\frac{\partial u_i}{\partial x_j} + \frac{\partial u_j}{\partial x_i}\Big)\,. \tag{1.86}$$

1.8 Navier–Stokes Equation

The stress tensor for Newtonian fluids (1.85) developed in the previous section enables us to complete the equation of motion for Newtonian fluids. Substituting (1.85) into (1.48), we obtain

$$\frac{Du_i}{Dt} = K_i - \frac{1}{\rho}\frac{\partial p}{\partial x_i} + \frac{1}{\rho}\frac{\partial}{\partial x_j}\Big[\mu\Big(\frac{\partial u_i}{\partial x_j} + \frac{\partial u_j}{\partial x_i} - \frac{2}{3}\frac{\partial u_k}{\partial x_k}\delta_{ij}\Big)\Big]\,. \tag{1.87}$$

μ is sensitive to temperature, and thus generally depends on both spatial coordinates and time through change in temperature. However, if the temperature change is small throughout the flow field, μ can be regarded as being approximately constant. In this case, (1.87) becomes

$$\frac{Du_i}{Dt} = K_i - \frac{1}{\rho}\frac{\partial p}{\partial x_i} + \frac{\mu}{\rho}\Big(\frac{\partial^2 u_i}{\partial x_j \partial x_j} + \frac{1}{3}\frac{\partial^2 u_j}{\partial x_i \partial x_j}\Big), \tag{1.88}$$

or in vector notation:

$$\frac{D\boldsymbol{u}}{Dt} = \boldsymbol{K} - \frac{1}{\rho}\nabla p + \frac{\mu}{\rho}\nabla^2\boldsymbol{u} + \frac{\mu}{3\rho}\nabla(\nabla\cdot\boldsymbol{u})\,. \tag{1.89}$$

Note that when $\mu = 0$, Eqs. (1.87)–(1.89) are reduced to Euler's equation (1.49).

Furthermore, if the flow is incompressible, (1.89) becomes

$$\frac{D\boldsymbol{u}}{Dt} = \boldsymbol{K} - \frac{1}{\rho}\nabla p + \nu\nabla^2\boldsymbol{u} \tag{1.90}$$

on the basis of (1.40). The quantity $\nu = \mu/\rho$ is the kinematic viscosity, and (1.90) is called the **Navier–Stokes equation**.[17]

In particular, the governing equation for a Newtonian fluid with constant density is

$$\nabla\cdot\boldsymbol{u} = 0\,, \tag{1.91}$$

$$\frac{D\boldsymbol{u}}{Dt} = \boldsymbol{K} - \frac{1}{\rho}\nabla p + \nu\nabla^2\boldsymbol{u}\,. \tag{1.92}$$

[17](1.87) is also sometimes called the Navier–Stokes equation.

Using components, (1.91) and (1.92) are written as

$$\frac{\partial u}{\partial x} + \frac{\partial v}{\partial y} + \frac{\partial w}{\partial z} = 0, \tag{1.93}$$

$$\left.\begin{aligned} \frac{Du}{Dt} &= X - \frac{1}{\rho}\frac{\partial p}{\partial x} + \nu\left(\frac{\partial^2 u}{\partial x^2} + \frac{\partial^2 u}{\partial y^2} + \frac{\partial^2 u}{\partial z^2}\right) \\ \frac{Dv}{Dt} &= Y - \frac{1}{\rho}\frac{\partial p}{\partial y} + \nu\left(\frac{\partial^2 v}{\partial x^2} + \frac{\partial^2 v}{\partial y^2} + \frac{\partial^2 v}{\partial z^2}\right) \\ \frac{Dw}{Dt} &= Z - \frac{1}{\rho}\frac{\partial p}{\partial z} + \nu\left(\frac{\partial^2 w}{\partial x^2} + \frac{\partial^2 w}{\partial y^2} + \frac{\partial^2 w}{\partial z^2}\right) \end{aligned}\right\} . \tag{1.94}$$

Clearly, the parameter governing the flow is ν, not μ. For example, at normal temperature, the μ of air is considerably lower than that of water; however, as the density of air is small, its ν is 15 times larger. This implies that air is stickier than water.

The initial conditions in the above cases are identical to those for perfect fluids; however, the boundary conditions differ. For example, if the fluid is bounded by a solid wall, the fluid velocity on the wall face must match the velocity of the solid wall face. In other words, both tangential and normal components must match on the wall face.[18] This condition is called the viscous or **no-slip condition**.

In particular, if the solid body is at rest, the velocity of fluid on the wall face must also be 0. These conditions apply regardless of the compressibility of the fluid.

1.9 Energy Equations

In this section, we examine an energy equation. As discussed above, energy equations are generally essential for understanding the flow field. Similar to our derivation of the mass and momentum conservation laws, we consider a closed surface S that is fixed in space (Fig. 1.11). If the internal energy of the fluid per unit mass is E, then the energy of the fluid contained in an infinitesimal volume element δV is $(u_i^2/2 + E)\rho\delta V$; hence, the energy per unit time throughout the volume V increases by

$$\frac{d}{dt}\iiint_V \left(\frac{1}{2}u_i^2 + E\right)\rho\, dV = \iiint_V \frac{\partial}{\partial t}\left\{\left(\frac{1}{2}u_i^2 + E\right)\rho\right\} dV . \tag{1.95}$$

Here we consider four components as contributions to this energy increase. The three components of those are energy carried into the volume by the fluid entering through the surface S

[18]Imposing such constraint conditions on the tangential component is appropriate because the differential equations are second order in space in this case.

$$-\iint_S \Big(\frac{1}{2}u_i^2 + E\Big)\rho \boldsymbol{u}\cdot\boldsymbol{n}\, dS = -\iiint_V \frac{\partial}{\partial x_j}\Big\{\rho u_j\Big(\frac{1}{2}u_i^2 + E\Big)\Big\}\, dV\,, \tag{1.96}$$

the work per unit time done by external forces

$$\iiint_V u_i K_i \rho\, dV\,, \tag{1.97}$$

and the work per unit time done by stress on the boundary surface S

$$\iint_S u_i \tau_{ij} n_j\, dS = \iiint_V \frac{\partial}{\partial x_j}(u_i \tau_{ij})\, dV\,. \tag{1.98}$$

As the fourth component we consider the contribution by the heat flowed into through the boundary surface. Many other factors (for example, heat generation absorption during phase transitions or chemical reactions) contribute to real fluid systems but will not be considered here. If we denote the heat flow per unit area per unit time as $\boldsymbol{q}$, the heat energy entering through the boundary surface S per unit time is

$$-\iint_S q_i n_i\, dS\,. \tag{1.99}$$

If the heat flow obeys the Fourier law $\boldsymbol{q} = -k\nabla T$ (where k is the thermal conductivity), (1.99) becomes

$$\iint_S k\frac{\partial T}{\partial x_i} n_i\, dS = \iiint_V \frac{\partial}{\partial x_i}\Big(k\frac{\partial T}{\partial x_i}\Big)\, dV\,. \tag{1.100}$$

(1.95) equals to the sum of contributions from (1.96), (1.97), (1.98), and (1.100). Since this relationship holds in any V, we can write

$$\begin{aligned}\frac{\partial}{\partial t}\Big\{\rho\Big(\frac{1}{2}u_i^2 + E\Big)\Big\} &+ \frac{\partial}{\partial x_j}\Big\{\rho u_j\Big(\frac{1}{2}u_i^2 + E\Big)\Big\} \\ &= \rho u_i K_i + \frac{\partial}{\partial x_j}(u_i\tau_{ij}) + \frac{\partial}{\partial x_i}\Big(k\frac{\partial T}{\partial x_i}\Big)\,.\end{aligned} \tag{1.101}$$

Equation (1.101) is called the equation of energy. With some slight rearrangement, we obtain

$$\begin{aligned}\Big(\frac{\partial\rho}{\partial t} + \frac{\partial(\rho u_j)}{\partial x_j}\Big)\Big(\frac{1}{2}u_i^2 + E\Big) + \rho u_i\Big(\frac{\partial u_i}{\partial t} + u_j\frac{\partial u_i}{\partial x_j}\Big) + \rho\Big(\frac{\partial E}{\partial t} + u_j\frac{\partial E}{\partial x_j}\Big) \\ = \rho u_i K_i + u_i\frac{\partial \tau_{ij}}{\partial x_j} + \frac{\partial u_i}{\partial x_j}\tau_{ij} + \frac{\partial}{\partial x_i}\Big(k\frac{\partial T}{\partial x_i}\Big)\,.\end{aligned} \tag{1.102}$$

By using the equation of continuity

$$\frac{\partial \rho}{\partial t} + \frac{\partial(\rho u_j)}{\partial x_j} = 0$$

and the equation of motion,

$$\frac{\partial u_i}{\partial t} + u_j \frac{\partial u_i}{\partial x_j} = K_i + \frac{1}{\rho}\frac{\partial \tau_{ij}}{\partial x_j} ,$$

the temporal rate of change of the internal energy per unit mass of the fluid element from (1.102) is given as

$$\frac{DE}{Dt} = \frac{\tau_{ij}}{\rho}\frac{\partial u_i}{\partial x_j} + \frac{1}{\rho}\frac{\partial}{\partial x_i}\left(k\,\frac{\partial T}{\partial x_i}\right) = \frac{\tau_{ij} e_{ij}}{\rho} + \frac{1}{\rho}\frac{\partial}{\partial x_i}\left(k\,\frac{\partial T}{\partial x_i}\right).$$

The second equality is true because the symmetric tensor τ_{ij} and the anti-symmetric tensor Ω_{ij} satisfy $\tau_{ij}\Omega_{ij} = 0$ (note that $\tau_{ij}\Omega_{ij} = \tau_{ji}\Omega_{ji} = -\tau_{ij}\Omega_{ij}$). Substituting (1.85) for τ_{ij} on the right-hand side of the above expression, we get

$$\frac{DE}{Dt} = -\frac{p\,\Theta}{\rho} + \frac{2\mu}{\rho}\left(e_{ij}\,e_{ij} - \frac{1}{3}\,\Theta^2\right) + \frac{1}{\rho}\frac{\partial}{\partial x_i}\left(k\,\frac{\partial T}{\partial x_i}\right). \tag{1.103}$$

Here $\Theta = \nabla \cdot \boldsymbol{u}$. This may also be written as

$$\frac{DE}{Dt} = \frac{1}{\rho}(-\,p\,\delta_{ij})\left(\frac{1}{3}\Theta\,\delta_{ij}\right) + \frac{2\mu}{\rho}\left(e_{ij} - \frac{1}{3}\Theta\,\delta_{ij}\right)\left(e_{ij} - \frac{1}{3}\Theta\,\delta_{ij}\right) + \frac{1}{\rho}\frac{\partial}{\partial x_i}\left(k\,\frac{\partial T}{\partial x_i}\right). \tag{1.104}$$

These are alternative forms of the energy equation. The first and second terms on the right-hand side of (1.104) are the contributions from the isotropic and anisotropic parts of the rate-of-strain tensor, respectively. In other words, the second term is the contribution from the purely deformational motion without changing the volume of the fluid element. This term is always non-negative and generates an irreversible conversion of the mechanical energy into internal energy. This term, called the **dissipation function**, is written as

$$\Phi = \frac{2\mu}{\rho}\left(e_{ij}\,e_{ij} - \frac{1}{3}\,\Theta^2\right). \tag{1.105}$$

On the other hand, the first term is positive under compression ($\Theta < 0$) and indicates an increase in internal energy. Under expansion ($\Theta > 0$), the term is negative and indicates decreased internal energy. Thus, the first term can be considered as a reversible interchange between mechanical and internal energies. However, in reality, this holds only approximately. In fluids at rest or perfect fluids, the pressure equals the thermodynamic pressure p_e in a (local) thermal equilibrium state; however, the dynamic pressure (defined as the negative of average normal stress) generally differs from the thermodynamic pressure p_e when a viscous fluid make a relative motion. Here p_e is regarded to be defined from ρ and E by the thermodynamic state equations of fluids, which may be determinable in the non-equilibrium state.[19] The difference between p and p_e is due to the relative fluid motion, and hence can be considered to depend on the velocity gradient, similar to the deviatoric stress tensor. Assuming that the difference is sufficiently small, and considering terms up to first order, the difference is given by

$$p - p_e = b_{ij} \frac{\partial u_i}{\partial x_j} = b_{ij} e_{ij} + b_{ij} \Omega_{ij} \, .$$

The tensor b_{ij} is independent of the velocity distribution but depends on the local state of the fluid. Furthermore, if an isotropic fluid is assumed, it takes the following form:

$$b_{ij} = -\kappa \, \delta_{ij} \, .$$

As the contribution from the anti-symmetric tensor Ω_{ij} vanishes, we obtain

$$p - p_e = -\kappa \, \Theta \, . \tag{1.106}$$

Therefore, the first term on the right-hand side of (1.103) can be written as

$$-\frac{p \, \Theta}{\rho} = -\frac{p_e \Theta}{\rho} + \frac{\kappa \, \Theta^2}{\rho} \, . \tag{1.107}$$

The first term on the right-hand side of (1.107) involves only the equilibrium state pressure and expresses the reversible transformation of energy. In contrast, the second term is definite, and (assuming positive κ) it expresses the dissipation of mechanical energy by the expansion or contraction of the fluid element. Hence, κ is called **bulk viscosity**. Microscopically, ordinary viscosity can be simply interpreted as the diffusion of momentum; however, bulk viscosity is more complicated. If the fluid element is compressed ($\Theta < 0$), the energy in translational motion of molecules will

[19]In relative motions of viscous fluids, the fluid element is generally not in thermal equilibrium. In this situation, the thermodynamic quantities are defined as follows. Density ρ is the mass of the fluid element divided by the volume, and hence can be specified regardless of the equilibrium state. Moreover, the increase in internal energy E caused by the state change is the sum of the work done and the heat added in the process. Thus, the internal energy is also definable in the non-equilibrium state. Other thermodynamic quantities such as temperature and pressure are determined from the ρ and E on the basis of the equation of state. Here the equilibrium state is the state where the fluid element settles eventually when it is suddenly isolated thermally from its surroundings.

likely increase.[20] The energy of the molecule's translational motion determines p. The corresponding p_e is determined by the energy of the translational motion in the equilibrium state, in which energy is evenly distributed to all degrees of freedom. For example, diatomic molecules possess both rotational and oscillatory degrees of freedom as well as the degree of freedom of translational motion, so that the energy is evenly distributed to all those degrees of freedom in the equilibrium state. Therefore, p and p_e are different from each other and generally the difference $p - p_e$ is positive (thus, $\kappa > 0$).

Normally, the second term on the right-hand side of (1.107) is considerably smaller than the first term and can be ignored; however, when the fluid element repeatedly expands or contracts (for example, when sound waves travel), this term can enlarge cumulatively. The dissipation function is then the sum of Φ and $\kappa\Theta^2/\rho$. However, bulk viscosity is effective only in the attenuation of high-frequency sound wave and in the structure of shock wave. Hereafter, we assume that $p = p_e$.

We can rewrite (1.103) in terms of entropy S per unit mass. Regarding S as a function of T and p_e, we have

$$\delta S = \left(\frac{\partial S}{\partial T}\right)_p \delta T + \left(\frac{\partial S}{\partial p_e}\right)_T \delta p_e \,. \tag{1.108}$$

Using[21] $c_p = T\left(\frac{\partial S}{\partial T}\right)_p$, the Maxwell relation $\left(\frac{\partial S}{\partial p_e}\right)_T = -\left(\frac{\partial v}{\partial T}\right)_p$ and the thermal expansion coefficient

$$\alpha = \frac{1}{v}\left(\frac{\partial v}{\partial T}\right)_p = -\frac{1}{\rho}\left(\frac{\partial \rho}{\partial T}\right)_p , \tag{1.109}$$

we obtain

$$\delta S = \frac{c_p}{T}\,\delta T - \left(\frac{\partial v}{\partial T}\right)_p \delta p_e = \frac{c_p}{T}\,\delta T - \alpha v\,\delta p_e = \frac{c_p}{T}\,\delta T - \frac{\alpha}{\rho}\,\delta p_e \,. \tag{1.110}$$

Here $v = 1/\rho$ is the volume per unit mass. Replacing δ by D/Dt[22] in (1.110), and

$$T\delta S = \delta E + p_e\,\delta v = \delta E + p_e\,\delta(1/\rho)\,,$$

the entropy of the fluid element evolves as

$$T\frac{DS}{Dt} = \frac{DE}{Dt} + p_e\frac{D(1/\rho)}{Dt} = c_p\frac{DT}{Dt} - \frac{\alpha\,T}{\rho}\frac{Dp_e}{Dt}\,. \tag{1.111}$$

---

[20] The fluid element moves at macroscopic velocity. Here translational motion refers to its thermal motion relative to its macroscopic velocity. The pressure of the gas is proportional to the average translational energy of its thermal motion.

[21] The subscript of $\partial S/\partial T$ should probably be p_e; however, the subscript e of p_e is omitted.

[22] Each fluid element is regarded as a thermodynamic system. Fluid elements are considered to undergo quasi-static state changes based on the fluid motion.

Adopting (1.103), (1.38), and (1.106), we can derive an alternative form of the energy equation:

$$T\frac{DS}{Dt} = c_p\frac{DT}{Dt} - \frac{\alpha T}{\rho}\frac{Dp_e}{Dt} = \frac{\kappa\Theta^2}{\rho} + \Phi + \frac{1}{\rho}\frac{\partial}{\partial x_i}\left(k\frac{\partial T}{\partial x_i}\right). \tag{1.112}$$

If both viscosity and heat conductivity can be ignored, we have

$$T\frac{DS}{Dt} = c_p\frac{DT}{Dt} - \frac{\alpha T}{\rho}\frac{Dp_e}{Dt} = 0 \tag{1.113}$$

from which (1.53) is obtained.

Finally, when $p = p_e$, as assumed hereafter, the equations above are summarized as

$$\frac{D\rho}{Dt} + \rho\nabla\cdot\boldsymbol{u} = 0, \tag{1.114}$$

$$\frac{Du_i}{Dt} = K_i - \frac{1}{\rho}\frac{\partial p}{\partial x_i} + \frac{1}{\rho}\frac{\partial}{\partial x_j}\left[\mu\left(\frac{\partial u_i}{\partial x_j} + \frac{\partial u_j}{\partial x_i} - \frac{2}{3}\frac{\partial u_k}{\partial x_k}\delta_{ij}\right)\right], \tag{1.115}$$

$$T\frac{DS}{Dt} = c_p\frac{DT}{Dt} - \frac{\alpha T}{\rho}\frac{Dp}{Dt} = \Phi + \frac{1}{\rho}\frac{\partial}{\partial x_i}\left(k\frac{\partial T}{\partial x_i}\right). \tag{1.116}$$

If the state equation

$$f(p, \rho, T) = 0 \tag{1.117}$$

is included, then all the equations are present. However, μ, k, c_p, and α generally depend on the local state of the fluid. In other words, as these parameters are generally functions of ρ and T, these functions need to be known. This complex set of equations is rarely used in the above form; it is usually approximated as appropriate for the specified purpose.

1.10 Vorticity and Vorticity Equation

Vorticity $\boldsymbol{\omega}(\boldsymbol{r}, t) = \nabla\times\boldsymbol{u}$ was introduced in Sect. 1.6 as a rotation of the velocity field $\boldsymbol{u}(\boldsymbol{r}, t)$. If vorticity $\boldsymbol{\omega}$ exists at a point $\boldsymbol{r}$, a fluid element centered at $\boldsymbol{r}$ is rigidly rotating at an angular velocity of $\boldsymbol{\omega}/2$.

When describing the velocity field, we introduced the concept of streamlines. Similarly, we introduce vortex lines in the vorticity field. A space curve in the flow, whose tangent is parallel to the vorticity vector $\boldsymbol{\omega}$ at each point, is called a **vortex line**. Similar to streamlines, if $\boldsymbol{\omega} = (\xi, \eta, \zeta)$, the equation of the vortex line is $d\boldsymbol{r} \parallel \boldsymbol{\omega}$, that is,

$$\frac{dx}{\xi(x,y,z,t)} = \frac{dy}{\eta(x,y,z,t)} = \frac{dz}{\zeta(x,y,z,t)}. \tag{1.118}$$

Moreover, the surface created by vortex lines passing through each points on a closed curve is called a **vortex tube**.[23] The strength of the tube will be defined in Sect. 2.2. A very thin vortex tube whose immediate outside is surrounded by a fluid region of vorticity 0 is called a **vortex filament**.[24] Denoting the area of the orthogonal cross section of the vortex filament as σ and the vorticity magnitude as ω, the strength of the vortex filament is defined as $\Gamma = \omega\sigma$. As shown in Sect. 2.2, Γ is invariant along the vortex filament.

We now derive the equation governing the vorticity $\boldsymbol{\omega} = \nabla \times \boldsymbol{u}$; that is, the vorticity equation. Let us start from the equation of motion when μ is constant (1.89).

$$\frac{\partial \boldsymbol{u}}{\partial t} + (\boldsymbol{u}\cdot\nabla)\boldsymbol{u} = \boldsymbol{K} - \frac{1}{\rho}\nabla p + \nu\nabla^2\boldsymbol{u} + \frac{\nu}{3}\nabla(\nabla\cdot\boldsymbol{u}). \tag{1.119}$$

The density ρ can change but we assume to be able to approximate $\nu = \mu/\rho$ by a fixed value. Moreover, we assume the external force $\boldsymbol{K}$ can be derived from the potential; in other words, $\boldsymbol{K} = -\nabla\Pi$. Using the formula

$$(\boldsymbol{u}\cdot\nabla)\boldsymbol{u} = \frac{1}{2}\nabla(\boldsymbol{u}\cdot\boldsymbol{u}) - \boldsymbol{u}\times(\nabla\times\boldsymbol{u}), \tag{1.120}$$

we rewrite (1.119) as

$$\frac{\partial \boldsymbol{u}}{\partial t} + \nabla\left(\frac{1}{2}|\boldsymbol{u}|^2 + \Pi - \frac{\nu}{3}\nabla\cdot\boldsymbol{u}\right) = \boldsymbol{u}\times\boldsymbol{\omega} - \frac{1}{\rho}\nabla p + \nu\nabla^2\boldsymbol{u}. \tag{1.121}$$

Applying $\nabla\times$ to both sides of this equation, and noting that for any scalar field ϕ

$$\nabla\times\nabla\phi = 0, \tag{1.122}$$

and also that

$$\nabla\times(\phi\boldsymbol{u}) = \nabla\phi\times\boldsymbol{u} + \phi\nabla\times\boldsymbol{u}, \tag{1.123}$$

we derive the **vorticity equation** as

[23] Generally, an arbitrary closed curve does not always produce a tube. Thus, we confine our discussion to simple closed curves that can be reduced to a point without passing outside the fluid.

[24] Vortex filaments can be defined in several ways. In one definition, a vortex filament is simply a vortex tube with an infinitesimally small cross section. Here we adopt the definition in [6]. As recently recognized, an important requirement of a vortex filament is that the neighboring exterior of the filament has zero vorticity. Moreover, [6] called the limit of $\sigma \to 0, \omega \to \infty$ with keeping Γ constant as the line vortex and distinguished it from a vortex filament. Here this limit is also termed a vortex filament. Intuitively, a vortex filament is a solitary vortex region of filament form.

$$\frac{\partial \omega}{\partial t} = \nabla \times \big(\boldsymbol{u} \times \omega \big) + \frac{1}{\rho^2} \nabla \rho \times \nabla p + \nu \nabla^2 \omega \,. \tag{1.124}$$

The second term on the right-hand side of (1.124) generates vorticity in the fluid and is sometimes called the **baroclinic vector**. The third term on the right-hand side represents the diffusion of vorticity.

The vorticity equation can be expressed in a slightly different form. Using identities

$$\nabla \times (\boldsymbol{u} \times \boldsymbol{v}) = (\boldsymbol{v} \cdot \nabla)\boldsymbol{u} - (\boldsymbol{u} \cdot \nabla)\boldsymbol{v} + (\nabla \cdot \boldsymbol{v})\boldsymbol{u} - (\nabla \cdot \boldsymbol{u})\boldsymbol{v} \tag{1.125}$$

and

$$\nabla \cdot (\nabla \times \boldsymbol{u}) = 0 \,, \tag{1.126}$$

(1.124) can be written as

$$\frac{D\omega}{Dt} = \big(\omega \cdot \nabla\big)\, \boldsymbol{u} - \big(\nabla \cdot \boldsymbol{u}\big)\, \omega + \frac{1}{\rho^2}\, \nabla \rho \times \nabla p + \nu \nabla^2 \omega \,. \tag{1.127}$$

Dividing (1.127) by ρ and applying the equation of continuity, we obtain

$$\frac{D}{Dt}\Big(\frac{\omega}{\rho}\Big) = \Big(\frac{\omega}{\rho} \cdot \nabla\Big)\boldsymbol{u} + \frac{1}{\rho^3} \nabla \rho \times \nabla p + \frac{\nu}{\rho} \nabla^2 \omega \,. \tag{1.128}$$

If conditions such as barotropic flow ($\nabla p \parallel \nabla \rho$), incompressibility ($\nabla \cdot \boldsymbol{u} = 0$), and inviscid fluid ($\nu = 0$) are specified in these equations, we can state the vorticity equation appropriate for the situation.

In particular, if the fluid density ρ is fixed, the vorticity equation (1.127) is reduced to

$$\frac{D\omega}{Dt} = \big(\omega \cdot \nabla\big)\boldsymbol{u} + \nu \nabla^2 \omega \,. \tag{1.129}$$

Moreover, if the fluid is inviscid, this is further reduced to

$$\frac{D\omega}{Dt} = \big(\omega \cdot \nabla\big)\boldsymbol{u} \,. \tag{1.130}$$

In a two-dimensional flow, as ω is orthogonal to the flow surface, $(\omega \cdot \nabla)\boldsymbol{u} = 0$. Therefore, (1.129) simplifies to

$$\frac{D\omega}{Dt} = \nu \nabla^2 \omega \,. \tag{1.131}$$

In this case, the vorticity of the fluid particles simply diffuses due to the viscosity. In particular, for an inviscid fluid, (1.131) becomes

$$\frac{D\boldsymbol{\omega}}{Dt} = 0 \tag{1.132}$$

and the vorticity of the fluid particles is constant.

The first terms on the right-hand sides of (1.129) and (1.130), $(\boldsymbol{\omega} \cdot \nabla)\boldsymbol{u}$, appear only in three-dimensional flow fields. Let us consider the physical meaning of this term. First, we consider the equation satisfied by an infinitesimal line element consist of fluid particles joining a point $\boldsymbol{r}$ to a point $\boldsymbol{r}' = \boldsymbol{r} + \delta\boldsymbol{r}$ apart from the point $\boldsymbol{r}$ by $\delta\boldsymbol{r}$, The difference between the flow velocity $\boldsymbol{u}(\boldsymbol{r} + \delta\boldsymbol{r})$ and $\boldsymbol{u}(\boldsymbol{r})$ at the terminal and initial points of $\delta\boldsymbol{r}$ is $D\delta\boldsymbol{r}/Dt$. By ignoring terms of or exceeding the second order in $|\delta\boldsymbol{r}|$, we have

$$\frac{D\delta\boldsymbol{r}}{Dt} = (\delta\boldsymbol{r} \cdot \nabla)\boldsymbol{u} \,. \tag{1.133}$$

The form of (1.133) is identical to that of (1.130). Now, suppose that at time $t = 0$, $\delta\boldsymbol{r}$ is a part of the vortex line at point $\boldsymbol{r}$ and that $\delta\boldsymbol{r} = \epsilon\boldsymbol{\omega}$ (where ϵ is a constant). Subtracting a ϵ-multiple of (1.130) from (1.133), we obtain

$$\frac{D}{Dt}(\delta\boldsymbol{r} - \epsilon\,\boldsymbol{\omega}) = (\delta\boldsymbol{r} - \epsilon\,\boldsymbol{\omega}) \cdot \nabla\boldsymbol{u} \,. \tag{1.134}$$

If $\boldsymbol{u}$ is known, (1.134) may be viewed as a system of ordinary differential equations describing the time evolution of the components of the vector $\delta\boldsymbol{r} - \epsilon\boldsymbol{\omega}$ associated with a fluid particle initially located at a point $\boldsymbol{r}$ when the fluid particle is traced. Clearly, $\delta\boldsymbol{r} - \epsilon\boldsymbol{\omega} = 0$ is a solution that satisfies both (1.134) and the initial conditions. Based on the uniqueness theorem for systems of ordinary differential equations,[25] this is also the only solution to the initial value problem. Therefore, $\delta\boldsymbol{r} - \epsilon\,\boldsymbol{\omega}$ is always true. The fluid particles initially forming the vortex line form the vortex line at all times; in other words, the vortex line moves with the fluid, and the vorticity is proportional to the length of the fluid line element$|\delta\boldsymbol{r}|$. As vortex tubes are constructed by vortex lines, they are characterized by the same behavior. Furthermore, even if the fluid density varies, provided that the flow is barotropic, $\boldsymbol{\omega}$ can be replaced with $\boldsymbol{\omega}/\rho$ in (1.130), and it becomes apparent that vortex lines and vortex tubes move with the fluid.[26] It should be noted that vortex lines and vortex tubes are not merely geometric objects but have substance. This will be demonstrated from a different perspective in Chap. 2.

If $\boldsymbol{e}$ is the unit vector in the direction of $\boldsymbol{\omega}$, we can state $\boldsymbol{\omega} = \omega\boldsymbol{e}$, $\delta\boldsymbol{r} = \delta r\boldsymbol{e}$. Then using (1.133), we obtain

$$(\boldsymbol{\omega} \cdot \nabla)\boldsymbol{u} = \omega(\boldsymbol{e} \cdot \nabla)\boldsymbol{u} = \frac{\omega}{\delta r}\frac{D\delta\boldsymbol{r}}{Dt} = \omega\frac{D\boldsymbol{e}}{Dt} + \frac{\omega}{\delta r}\frac{D\delta r}{Dt} \tag{1.135}$$

---

[25] For example, if $\partial u_i/\partial x_j$ are bounded, this holds.

[26] This situation is sometimes said that vorticity is frozen in the fluid. In a perfectly conducting fluid (see Note 2.1), the magnetic flux density $\boldsymbol{B}$ satisfies (1.130); hence, the lines of magnetic force also move with the fluid. This is referred for magnetic field to be frozen in the fluid.

The first term in the last part of (1.135) indicates the effect of directional changes in $\boldsymbol{\omega}$ (as well as the fluid line element), and the second term is the increased vorticity caused by the extension of the fluid line element.

In this way, $(\boldsymbol{\omega} \cdot \nabla)\boldsymbol{u}$ or $((\boldsymbol{\omega}/\rho) \cdot \nabla)\, \boldsymbol{u}$ either strengthens vorticity or changes its direction; however, vorticity cannot be generated from a zero vorticity state. Therefore, in barotropic flows, vorticity is never generated inside the fluid but can be generated only at the boundaries.

Exercise 1.7 Prove Eqs. (1.122), (1.126), (1.123), (1.120), and (1.125).

Note 1.7 Let us consider the relationship between angular momentum and vorticity. Denote the surface formed by the fluid (not fixed in space) by S and the fluid volume enclosed by it by V. The equation of angular momentum of this portion of the fluid is given by

$$\frac{d}{dt}\iiint_V (\boldsymbol{r} \times \boldsymbol{u})\rho\, dV = \iiint_V (\boldsymbol{r} \times \boldsymbol{K})\rho\, dV + \iint_S (\boldsymbol{r} \times \boldsymbol{T}(\boldsymbol{n}))\, dS. \qquad (1.136)$$

Here $\boldsymbol{r}$ is a position vector relative to an appropriately located origin O. The left-hand side of (1.136) is the temporal rate of change of the angular momentum of the fluid within V relative to the origin. The first and second terms on the right-hand side are the moments of the external forces and surface forces, respectively.[27] From (1.42), we have $D(\rho dV)/Dt = 0$; hence, the left-hand side of (1.136) can also be written as

$$\frac{d}{dt}\iiint_V (\boldsymbol{r} \times \boldsymbol{u})\rho\, dV = \iiint_V \frac{D}{Dt}(\boldsymbol{r} \times \boldsymbol{u})\rho\, dV = \iiint_V \left(\boldsymbol{r} \times \frac{D\boldsymbol{u}}{Dt}\right)\rho\, dV\, . \qquad (1.137)$$

Expressing (1.136) in tensor notation using the third-order alternating tensor ϵ_{ijk} (see Appendix A.1.5), we get

$$\frac{d}{dt}\iiint_V \epsilon_{ijk} x_j u_k \rho\, dV = \iiint_V \epsilon_{ijk} x_j K_k \rho\, dV + \iint_S \epsilon_{ijk} x_j \tau_{kl} n_l\, dS. \qquad (1.138)$$

Applying Gauss' theorem to the surface integral on the right-hand side of (1.138), we obtain

$$\begin{aligned}\iint_S \epsilon_{ijk} x_j \tau_{kl} n_l\, dS &= \iiint_V \epsilon_{ijk} \frac{\partial}{\partial x_l}(x_j \tau_{kl})\, dV \\ &= \iiint_V \epsilon_{ijk} \tau_{kj}\, dV + \iiint_V \epsilon_{ijk} x_j \frac{\partial \tau_{kl}}{\partial x_l}\, dV \\ &= \iiint_V \epsilon_{ijk} x_j \frac{\partial \tau_{kl}}{\partial x_l}\, dV\, . \end{aligned} \qquad (1.139)$$

[27] Here it is assumed that the fluid under consideration has not the intrinsic couples, as in ordinary fluids.

Here we used $\partial x_j/\partial x_l = \delta_{jl}$ and $\epsilon_{ijk}\tau_{kj} = 0$. The latter statement is true because ϵ_{ijk} and τ_{kj} are anti-symmetric and symmetric, respectively, with regard to j, k. Using (1.139) and the equation of motion (1.48), we immediately see that the equation of angular momentum (1.136) or (1.138) actually holds.

Let us now derive the equation of vorticity from (1.138) and (1.139). For simplicity, we assume the fluid density is constant. V is assumed as a spherical fluid element centered at the origin O. Furthermore, we suppose $K_i = -\partial \Pi/\partial x_i$. Expanding u_k about the origin up to first-order terms in x_l (this approximation is exact if the fluid element is contracted to the point O), the angular momentum of the fluid element around O at some time t is

$$\begin{aligned}\iiint_V \epsilon_{ijk} x_j u_k \rho\, dV &\cong \iiint_V \epsilon_{ijk} x_j \left(u_k + x_l \frac{\partial u_k}{\partial x_l}\right) \rho\, dV \\ &= \epsilon_{ijk} u_k \left(\iiint_V x_j \rho\, dV\right) + \epsilon_{ijk} \frac{\partial u_k}{\partial x_l} \iiint_V x_j x_l \rho\, dV \\ &- \epsilon_{ijk} \frac{\partial u_k}{\partial x_l} \frac{1}{2} I\, \delta_{jl} = \frac{1}{2} I\, \epsilon_{ijk} \frac{\partial u_k}{\partial x_j} = \frac{1}{2} I\, \omega_l. \end{aligned} \tag{1.140}$$

Here u_k and $\partial u_k/\partial x_l$ are evaluated at O, and hence are unrelated to the integral. Since the sphere is centered at O, which is also the center of gravity, we have $\iiint_V x_j \rho\, dV = 0$. Moreover, I is the moment of inertia of the sphere around an arbitrary axis passing through O, and $\epsilon_{ijk}\partial u_k/\partial x_j = \omega_i$ is the vorticity. The right-hand side of this equation is the angular momentum of a sphere rigidly rotating at an angular velocity of $\boldsymbol{\omega}/2$. This simple form of the angular momentum is applicable to a spherical fluid element. If the inertial tensor around the fluid element's center of gravity O is anisotropic, the angular momentum will include not only the vorticity ω_i but also the rate-of-strain tensor e_{ij}.

Furthermore, the moment of external forces is

$$\begin{aligned}\iiint_V \epsilon_{ijk} x_j K_k\, \rho\, dV &\cong -\epsilon_{ijk} \frac{\partial \Pi}{\partial x_k} \iiint_V x_j \rho\, dV - \epsilon_{ijk} \frac{\partial^2 \Pi}{\partial x_k \partial x_l} \iiint_V x_j x_l \rho\, dV \\ &= -\frac{1}{2} I\, \epsilon_{ijk} \frac{\partial^2 \Pi}{\partial x_k \partial x_l} \delta_{jl} = -\frac{1}{2} I\, \epsilon_{ijk} \frac{\partial^2 \Pi}{\partial x_k \partial x_j} = 0.\end{aligned}$$

Similarly, from (1.139) and (1.86), the moment of the surface force is

$$\begin{aligned}\iint_S \epsilon_{ijk} x_j \tau_{kl} n_l\, dS &= \iiint_V \epsilon_{ijk} x_j \left(-\frac{\partial p}{\partial x_k} + \mu \frac{\partial^2 u_k}{\partial x_l^2}\right) dV \\ &\cong \epsilon_{ijk} \left(-\frac{\partial p}{\partial x_k} + \mu \frac{\partial^2 u_k}{\partial x_l^2}\right) \iiint_V x_j\, dV \\ &+ \epsilon_{ijk} \left(-\frac{1}{\rho} \frac{\partial^2 p}{\partial x_k \partial x_m} + \nu \frac{\partial^2}{\partial x_l^2} \frac{\partial u_k}{\partial x_m}\right) \iiint_V x_j x_m \rho\, dV\end{aligned}$$

$$= \epsilon_{ijk}\Big(-\frac{1}{\rho}\frac{\partial^2 p}{\partial x_k \partial x_m} + \nu \frac{\partial^2}{\partial x_l^2}\frac{\partial u_k}{\partial x_m}\Big)\frac{1}{2}\, I\,\delta_{jm} = \frac{1}{2}\, I\,\nu\,\nabla^2 \omega_i .$$

Therefore, (1.138) becomes

$$\frac{D}{Dt}\Big(\frac{1}{2}\, I\,\omega_i\Big) = \frac{1}{2}\, I\,\nu\,\nabla^2\,\omega_i\,. \tag{1.141}$$

This equation shows that if $\nu = 0$, the angular momentum of the spherical fluid element is conserved. However, as mentioned above, this simple form of the angular momentum, $I\omega_i/2$, applies only for the spherical fluid element at time t. In the next instant, the fluid element is generally deformed by relative motion; thus, the left-hand side of this equation is not easily calculated. Therefore, it is convenient to replace this term with the right-hand side of (1.137).

$$\begin{aligned}
\iiint_V \epsilon_{ijk} x_j\,\frac{Du_k}{Dt}\rho\, dV &= \iiint_V \epsilon_{ijk} x_j \Big(\frac{\partial u_k}{\partial t} + u_m \frac{\partial u_k}{\partial x_m}\Big)\rho\, dV \\
&\cong \iiint_V \epsilon_{ijk} x_j x_l \Big[\frac{\partial}{\partial t}\Big(\frac{\partial u_k}{\partial x_l}\Big) + u_m \frac{\partial}{\partial x_m}\Big(\frac{\partial u_k}{\partial x_l}\Big) + \frac{\partial u_m}{\partial x_l}\frac{\partial u_k}{\partial x_m}\Big]\rho\, dV \\
&= \epsilon_{ijk}\Big[\frac{D}{Dt}\Big(\frac{\partial u_k}{\partial x_l}\Big) + \frac{\partial u_m}{\partial x_l}\frac{\partial u_k}{\partial x_m}\Big]\frac{1}{2}\, I\,\delta_{jl} \\
&= \frac{1}{2}\, I\,\frac{D\omega_i}{Dt} + \frac{1}{2}\, I\,\epsilon_{ijk}\frac{\partial u_m}{\partial x_j}\frac{\partial u_k}{\partial x_m}\,.
\end{aligned}$$

Here, since we can confirm

$$\epsilon_{ijk}\frac{\partial u_m}{\partial x_j}\frac{\partial u_k}{\partial x_m} = -\omega_j \frac{\partial u_i}{\partial x_j}\,,$$

we obtain

$$\iiint_V \epsilon_{ijk} x_j \frac{Du_k}{Dt}\rho\, dV \cong \frac{1}{2}\, I\,\frac{D\omega_i}{Dt} - \frac{1}{2}\, I\,\omega_j \frac{\partial u_i}{\partial x_j}\,, \tag{1.142}$$

from which we arrive at (1.129). The meaning of $(\boldsymbol{\omega}\cdot\nabla)\boldsymbol{u}$ has been previously discussed; however, here we observe that this term arises from the deformation of the fluid element.

Note 1.8 (*Literature for further study*)
For further studies, textbooks [1, 2, 4, 5, 7] may be referred. The advanced readers may also refer some other textbooks e.g. [3], although written in Japanese.

References

1. Aris, R.: Vectors, Tensors, and the Basic Equations of Fluid Mechanics. Prentice-Hall (1962)
2. Batchelor, G.K.: An Introduction to Fluid Dynamics. Cambridge University Press, Cambridge (1967)

3. Imai, I.: Fluid Dynamics, vol. I. Shokabo (1973)
4. Jeffreys, H.: Cartesian Tensors. Cambridge University Press, Cambridge (1969)
5. Landau, L.D., Lifshitz, E.M.: Fluid Mechanics. Butterworth-Heinemann, Oxford (1987)
6. Saffman, P.G.: Vortex Dynamics. Cambridge University Press, Cambridge (1992)
7. Whitham, G.B.: The Navier-Stokes Equations of Motion. In: Rosenhead, L. (ed.) Laminar Boundary Layers, Part I. Oxford University Press, Oxford (1963)

Chapter 2
Dynamics of Perfect Fluids

As discussed in the previous chapter, the viscosity of fluids induces tangential stresses in relatively moving fluids. A familiar example is water being poured into a rotating cylindrical container. Although the container is circular, water is pulled toward the container wall as it begins its movement. This example highlights that real fluids are viscous in nature.

In contrast, inviscid fluids are called perfect or ideal fluids. Apart from extremely special cases, such as the superfluid phenomenon, perfect fluids do not exist in reality; however, they are abstract entities that simplify the theoretical development of fluid dynamics. Indeed, the inviscid assumption is extremely useful to theoretical analysis, and fortunately, water and air encountered in daily use have low viscosity. Thus, the actual flow characteristics are explained well by the dynamics of perfect fluids. Conversely, by investigating the motion characteristics of perfect fluids, we can accurately understand the effect of viscosity and other properties of real fluids. Hence, this chapter discusses the dynamics and analytical method of perfect fluids.

2.1 Lagrange's Vortex Theorem

In Sect. 1.10, the vorticity equation for general viscous fluids was derived. Vorticity ($\boldsymbol{\omega} = \nabla \times \boldsymbol{u}$) relates to the angular velocity of the rigid rotation of fluid particles; thus, it is expected that the laws analogous to the law of angular momentum conservation in the dynamics of particle systems should exist in those of perfect fluids. Actually, an extremely useful vortex theorem can be derived from the assumption of perfect fluids; thus, we will discuss this subject in this chapter.

First, we assume Euler's equation of perfect fluids with barotropic flow; that is, density ρ and pressure p are related through $\rho = f(p)$ and external forces are assumed as conservative forces. Rearranging, we obtain

D.E.S.S.T. Kyushu Univ., *Fluid Dynamics for Global Environmental Studies*,
DOI 10.1007/978-4-431-56499-7_2

$$\left.\begin{aligned}\frac{\partial \boldsymbol{u}}{\partial t} &= -\nabla\left(\frac{1}{2}q^2 + P + \Pi\right) + \boldsymbol{u} \times \boldsymbol{\omega}\,, \\ q = |\boldsymbol{u}|\,, \quad P &= \int^p \frac{dp'}{\rho} = \int^p \frac{dp'}{f(p')}\end{aligned}\right\} \tag{2.1}$$

Here Π is the potential of external forces. For incompressible fluids moving in a uniform gravitational field, P and Π in (2.1) are, respectively, reduced to

$$P = \frac{p}{\rho}\,, \quad \Pi = gz\,, \tag{2.2}$$

where g is the gravitational acceleration, and z is the vertical coordinate (assuming upwards as the positive direction).

Applying the $\nabla\times$ operator to both sides of (2.1), we obtain $\nabla \times \boldsymbol{u} = \boldsymbol{\omega}$, and given that $\nabla \times \nabla F = 0$ for an arbitrary scalar function F, we immediately obtain

$$\frac{\partial \boldsymbol{\omega}}{\partial t} = \nabla \times (\boldsymbol{u} \times \boldsymbol{\omega})\,. \tag{2.3}$$

Alternatively, by rearranging the equation in Sect. 1.10 (Exercise 2.1), we obtain

$$\frac{D}{Dt}\left(\frac{\boldsymbol{\omega}}{\rho}\right) = \left(\frac{\boldsymbol{\omega}}{\rho}\cdot\nabla\right)\boldsymbol{u}\,. \tag{2.4}$$

Equation (2.3) or (2.4) is called the vorticity equation for perfect fluids.

According to this equation, the temporal evolution of vorticity is unrelated to external force and pressure. This can be understood as follows: considering an infinitesimally small spherical fluid element, the external force operates on the center of mass of the sphere, while the pressure operates normal to the surface of the sphere. Thus, no rotation moment is generated around the center of mass, and the angular momentum of an infinitesimally small spherical fluid element remains unchanged during motion. This implies that if an infinitesimally small sphere is not rotating at a given instant, it will never rotate; conversely, if it is rotating initially, it will rotate forever. Since the angular velocity of a minute fluid element is half its vorticity, vortices can neither arise nor cease in a perfect fluid. This concept underlies **Lagrange's vortex theorem**. More simply, Lagrange's vortex theorem is called the theory of non-arising and non-ceasing vortices.

Using (2.4), let us explain Lagrange's vortex theorem from the perspective of a differential equation. If $\boldsymbol{\omega} = 0$ at $t = 0$, the material derivative of the fluid particle, from (2.4), is

$$\left[\frac{D}{Dt}\left(\frac{\boldsymbol{\omega}}{\rho}\right)\right]_{t=0} = 0\,. \tag{2.5}$$

An infinitesimally small time instant later ($t = \delta t$), $\boldsymbol{\omega}$ of the fluid particle remains 0 because we have

$$\left(\frac{\omega}{\rho}\right)_{t=\delta t} = \left(\frac{\omega}{\rho}\right)_{t=0} + \left[\frac{D}{Dt}\left(\frac{\omega}{\rho}\right)\right]_{t=0} \cdot \delta t = 0 \,. \tag{2.6}$$

Considering that time progresses in infinitesimal increments, $\omega = 0$ at any arbitrary time.

In contrast, if $\omega \neq 0$ at $t = 0$, the above reasoning stipulates that $\omega \neq 0$ forever. Suppose that we can progress backward from some time at which $\omega = 0$. Following the above argument, $\omega = 0$ at $t = 0$, contradicting our assumption that $\omega \neq 0$ at $t = 0$. This implies that when an inviscid barotropic fluid moves under conservative forces, no vortices are generated or dissipated.

However, as discussed in the opening paragraph, real fluids are viscous in nature. If viscosity is significant, vortices are generated in the vicinity (boundary layer) of a body surface. This is because even if $\omega = 0$ in an area far from the body, by virtue of non-slip boundary conditions at the body surface, the $\omega = 0$ solution cannot be established. However, if the boundary layer breaks away from the body, the vortex forming the boundary layer will enter the main flow without disappearing, although new vortices rarely form within the fluid. This indicates that the perfect fluid assumption is fulfilled with satisfactory accuracy in regions slightly distant from body surfaces.

Note 2.1 Equation (2.3) was described as a vorticity equation; however, an analogous relationship is observed in magnetohydrodynamics. Specifically, when a fluid is a perfect conductor (with zero electrical resistance), and there is a magnetic field in this fluid, the magnetic flux density B satisfies the equation where ω in (2.3) is replaced by B. Therefore, the magnetic flux density has properties similar to vorticity in perfect fluids, and the magnetic field is neither generated nor dissipated inside the fluid of a perfect electrical conductor.

Now, when $\omega = \nabla \times \boldsymbol{u} = 0$ in a certain region of the flow, the flow is said to be irrotational in that region. As an identical equation in vector analysis, $\nabla \times \nabla \Phi = 0$ is established at all times relative to an arbitrary scalar function Φ. Thus, in an irrotational flow, there exists a scalar function Φ such that

$$\boldsymbol{u} = \nabla \Phi \,. \tag{2.7}$$

As the gradient of this function is velocity, Φ is called the **velocity potential**. In three-dimensional problems, a vector $\boldsymbol{u}$ with three unknowns can be calculated from a single scalar function Φ, which significantly simplifies the mathematical treatment. Generally, a flow that can be described in terms of velocity potential is called a potential flow, and an irrotational flow is a permanent potential flow. Velocity potential will be explained in detail in Sect. 2.5.

Exercise 2.1 Derive (2.1) and (2.4).

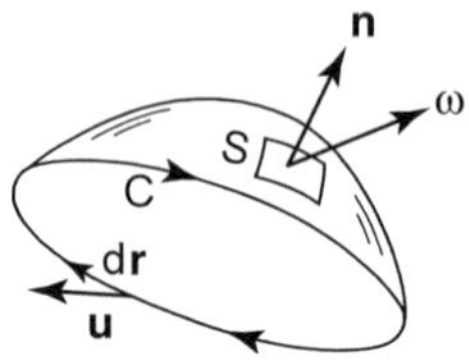

Fig. 2.1 Relationship between circulation and vorticity

2.2 Circulation and Vorticity

Consider any closed curve C in a flow and integrate $\boldsymbol{u}$ along C.

$$\Gamma(C) = \oint_C \boldsymbol{u} \cdot d\boldsymbol{r} = \oint_C u_s \, ds \,. \tag{2.8}$$

Here u_s is the tangential component of C intercepting flow velocity vector $\boldsymbol{u}$, and $\Gamma(C)$ is called the **circulation** along C.

Using Stokes' theorem of vector analysis (see Fig. 2.1 and Note 1.6), the above equation can be rearranged as

$$\Gamma(C) = \iint_S (\nabla \times \boldsymbol{u}) \cdot \boldsymbol{n} \, dS = \iint_S \boldsymbol{\omega} \cdot \boldsymbol{n} \, dS = \iint_S \omega_n \, dS \,. \tag{2.9}$$

Equation (2.9) relates circulation to vorticity. Specifically, circulation $\Gamma(C)$ is equal to the integral of the normal component $\boldsymbol{\omega}_n$ of the vorticity $\boldsymbol{\omega}$ on a curved surface S surrounded by a closed curve C.

Consider an extremely thin vortex tube with a cross-sectional area σ. Taking the orthogonal section of the vortex tube as the integral surface in (2.9), $\omega_n = \omega$ can be considered fixed within the cross-sectional surface, and the circulation becomes

$$\Gamma(C) = \iint_S \omega_n \, dS = \omega \iint_S dS = \omega \, \sigma \,. \tag{2.10}$$

This result is independent of the position along the vortex tube (i.e., does not depend on the choice of C), and is hence called the strength of the vortex tube.

Note that if a single vortex tube is considered, the strength of the vortex tube is constant regardless of the position; the thinner the tube, the larger the vorticity ω, and thus the larger the rotational angular velocity of the fluid particles.

A wide vortex tube can be treated as a collection of many thin vortex tubes and expressed as

$$\Gamma(C) = \iint_S \omega_n \, dS = \sum_{i=1}^{N} \omega_i \, \sigma_i \,. \tag{2.11}$$

According to (2.11), the circulation along an arbitrary closed curve C equals the sum of the strengths of the vortex tubes passing through C.

We now relate the circulation $\Gamma(C)$ to the velocity potential Φ. Substituting (2.7) and (2.8) into (2.11) gives

$$\Gamma(C) = \oint_C \nabla\Phi \cdot d\boldsymbol{r} = \oint_C \frac{\partial \Phi}{\partial x_j} dx_j = \big[\,\Phi\,\big]_C . \tag{2.12}$$

In other words, the circulation along an arbitrary closed curve C inside a fluid region is the change in Φ along one circuit of C. Since the circulation is non-zero, the velocity potential is multi-valued. However, the flow velocity is a single-valued function (that is, at a given location, the flow velocity settles to a fixed value). Hence, when obtaining the flow velocity from $\boldsymbol{u} = \nabla\Phi$, the multi-valued property disappears.

2.3 Circulation Theorem and Vortex Theorem

2.3.1 Kelvin's Circulation Theorem

Section 2.2 only describes the properties of the vortex tube at a given time. Here we consider how the vortex tube changes over time. Hence, we evaluate the temporal change in circulation along a closed curve C, which moves with the flow. In terms of the material derivative (2.8), we have

$$\frac{D}{Dt}\Gamma(C) = \oint_C \frac{D}{Dt}\big(\boldsymbol{u}\cdot d\boldsymbol{r}\big) = \oint_C \frac{D\boldsymbol{u}}{Dt}\cdot d\boldsymbol{r} + \oint_C \boldsymbol{u}\cdot \frac{D}{Dt} d\boldsymbol{r} . \tag{2.13}$$

Here, from the Euler equation and (2.1),

$$\left.\begin{aligned} \frac{D\boldsymbol{u}}{Dt} &= -\nabla\big(P + \Pi\big), \\ \boldsymbol{u}\cdot\frac{D}{Dt} d\boldsymbol{r} &= \boldsymbol{u}\cdot d\boldsymbol{u} = d\Big(\frac{1}{2}q^2\Big). \end{aligned}\right\} \tag{2.14}$$

Therefore, we have

$$\frac{D}{Dt}\Gamma(C) = \left[\frac{1}{2}q^2 - P - \Pi\right]_C . \tag{2.15}$$

Here $[\cdots]_C$ indicates the change along one circuit of the closed curve C. As the flow velocity q and pressure function P are physical quantities, (2.15) is a single-valued function of location. In general, the potential Π of external forces is also a single-valued function; hence, the right-hand side of (2.15) is 0. In other words, when an inviscid barotropic fluid moves under conservative forces, the circulation along a closed curve C formed by the fluid particles is constant in time. This concept is called **Kelvin's circulation theorem**.

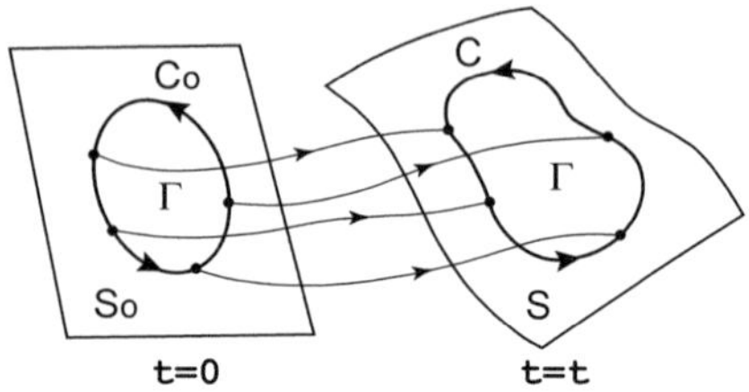

Fig. 2.2 A vortex surface (tube) moves as a vortex surface (tube), and its strength is constant

It should be noted that, as circulation and vorticity are related by (2.10), Lagrange's vortex theorem can be explained in terms of Kelvin's circulation theorem.

2.3.2 Helmholtz Vortex Theorem

A curved surface created from vortex lines in a flow is called a vortex surface. Here we focus on one such vortex surface S_0 at $t = 0$ (see Fig. 2.2). Considering a closed curve C_0 on S_0, no vortex line passes through C_0; hence, $\Gamma(C_0) = 0$.

Now suppose that C_0 at a certain time t forms a closed curve C because of the movement of fluid particles. By Kelvin's circulation theorem, we have $\Gamma(C) = \Gamma(C_0) = 0$. However, according to the concept of non-arising and non-ceasing vortices, any existing vortex surface should never disappear. Thus, $\Gamma(C) = 0$ indicates that a closed curve C can be defined on the vortex surface. It now becomes evident that "through fluid motion, a vortex surface remains a vortex surface." The vortex tube discussed in Sect. 2.2 is one such vortex surface; moreover, as the strength of a given vortex tube is fixed throughout space, the strength of the vortex tube is invariably maintained in space and time. This is called the **Helmholtz vortex theorem**.

Now consider an extremely thin vortex tube with a cross-sectional area σ. Vortex tubes move with fluid; hence, if we consider an infinitesimally small length δs of the vortex tube, $\rho\sigma\delta s$ is fixed in time by the mass conservation law. Moreover, the strength $\Gamma = \omega\sigma$ of the vortex tube defined by (2.10) is constant by the Helmholtz circulation theorem; thus, $\rho\delta s/\omega =$ constant must be true. In particular, as ρ is constant in an incompressible fluid, $\omega \propto \delta s$.

From this discussion, it follows that if a vortex tube is extended or contracted owing to temporal changes in the fluid, the vorticity will proportionately increase or decrease, respectively.

2.3.3 Flow and Circulation Around a Wing Section

To enhance our understanding of the abovementioned circulation and vortex theorems, we demonstrate the theorems on flow around a wing section. Consider that

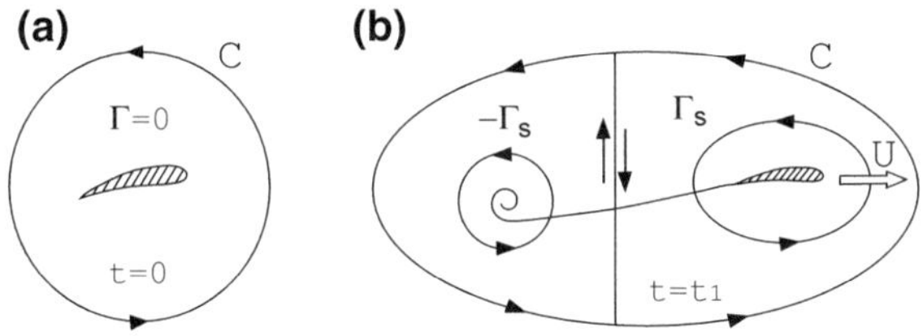

Fig. 2.3 Explanation of Kelvin's circulation theorem part 1: **a** $\Gamma = 0$ for a wing section in a static fluid, **b** Even when the wing section moves with a velocity U, $\Gamma = 0$ as a whole

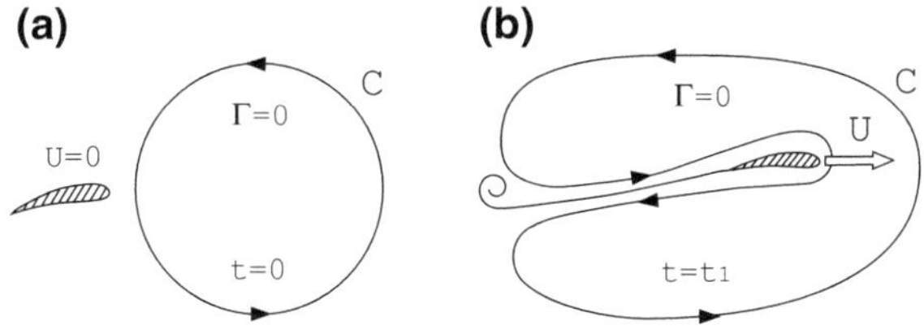

Fig. 2.4 Demonstration of Kelvin's circulation theorem part 2: **a** $\Gamma = 0$ in the fluid region in front of the wing section, **b** $\Gamma = 0$ as long as the same fluid region is considered

the wing is placed in a static fluid. Most areas of the fluid are irrotational; thus, the circulation is zero (see Fig. 2.3a). If the wing starts to move with velocity U in a real fluid, a thin boundary layer develops around the wing surface because of viscosity. However, the boundary layer formed beneath the wing section cannot reach the upper surface by fully traveling around the trailing edge and instead separates from the trailing edge to form a vortex. A short time later, a separation point on the wing section's upper surface moves to the trailing edge. When the flow becomes smooth, this vortex eventually flows toward the rear of the wing section, where it becomes a clockwise free vortex, as shown in Fig. 2.3. Suppose that the circulation of this vortex (called a starting vortex) is $-\Gamma_s$. According to Kelvin's circulation theorem, the circulation is initially 0; hence, the circulation along the closed curve surrounding both the starting vortex and wing section is also 0. In other words, a circulation Γ_s of the same strength as the starting vortex exists in the opposite direction around the wing section (see Fig. 2.3b). This circulation generates a **lift**, known as the **Kutta–Joukowski theorem**:

$$L = \rho U \Gamma_s \,. \tag{2.16}$$

The forces operating on bodies will be explained in Sect. 2.10.

Kelvin's circulation theorem can be understood from the closed curves shown in Fig. 2.4. We first focus on the fluid region in front of the wing section at $t = 0$. The circulation around the closed curve C surrounding this region is 0. Once the wing starts to move, after a short time, the wing section will intercept the curve C, deforming it (Fig. 2.4b). In Kelvin's circulation theorem, the curve encloses the same fluid region at all subsequent times; thus, the circulation along the closed curve shown in Fig. 2.4b certainly remains at 0.

2.4 Bernoulli's Theorem

We now introduce Bernoulli's theorem, one of the most important theorems in fluid dynamics. As we shall see, Bernoulli's theorem is based on the law of conservation of energy in fluid dynamics.

2.4.1 *Irrotational Flows*

As discussed above, the velocity potential Φ exists in an irrotational flow with $\omega = 0$. Substituting (2.7) into Euler's equation (2.1), we have

$$\nabla\left(\frac{\partial \Phi}{\partial t} + \frac{1}{2}q^2 + P + \Pi\right) = 0\,. \tag{2.17}$$

Therefore,

$$\frac{\partial \Phi}{\partial t} + \frac{1}{2}q^2 + P + \Pi = f(t)\,. \tag{2.18}$$

Here $f(t)$ is a function of t only. For functions of time, $\nabla f(t) = 0$; thus, $f(t)$ can be included in the definition of velocity potential, and the right-hand side of (2.18) can be set to 0. However, in applications, it is more convenient to leave $f(t)$ as a pressure function in (2.18).

In any case, (2.18) is used to determine the pressure from the velocity potential, and it is known as **Bernoulli's pressure equation**. If the flow is irrotational, (2.18) can be applied to the whole flow field. Of course, the flow can be non-steady. The reader should note that Bernoulli's pressure equation is derived from the law of conservation of momentum (Euler's equation), as is clearly shown in the derivation.

2.4.2 *Steady Flows*

In steady flows, the following is immediately obtained from (2.1):

$$\nabla\left(\frac{1}{2}q^2 + P + \Pi\right) \equiv \nabla H = \boldsymbol{u} \times \boldsymbol{\omega}\,, \tag{2.19}$$

$$H = \frac{1}{2}q^2 + P + \Pi\,. \tag{2.20}$$

A surface with $H =$ constant is called a **Bernoulli surface** (see Fig. 2.5).

The vector $\boldsymbol{u} \times \boldsymbol{\omega}$ is perpendicular to both $\boldsymbol{u}$ and $\boldsymbol{\omega}$ and (as seen in (2.19)) parallel to ∇H. Put another way, both $\boldsymbol{u}$ and $\boldsymbol{\omega}$ are parallel to a Bernoulli surface, and both streamlines and vortex lines reside on Bernoulli surfaces. Thus, it can be concluded that, on one streamline, $H =$ const. This is called **Bernoulli's theorem**, and flow can be established even in the presence of vortices. However, it is different from the pressure equation (2.18), and the flow must be steady.

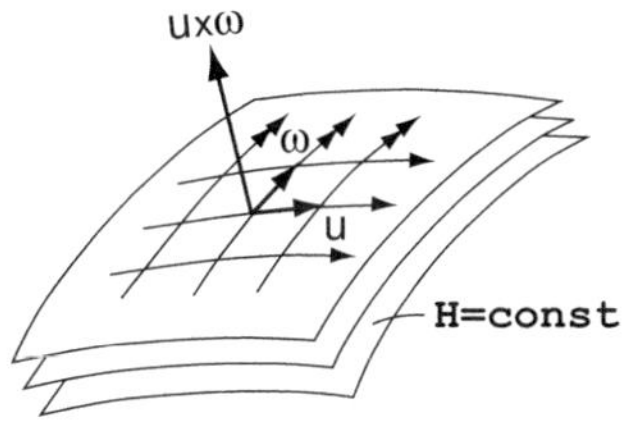

Fig. 2.5 Bernoulli surfaces

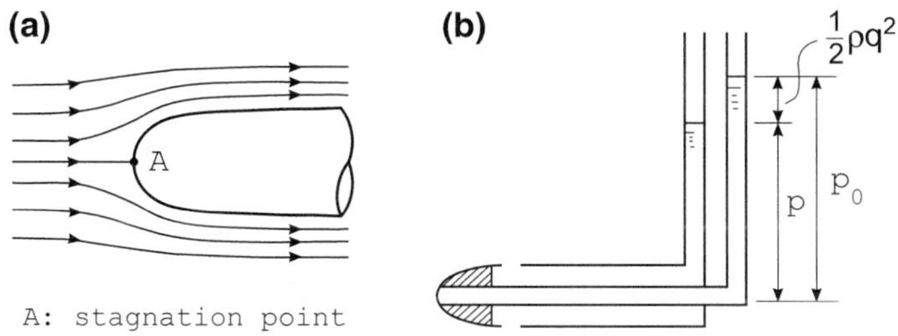

Fig. 2.6 Principle of the Pitot tube; **a** Pressure at the stagnation point A is p_0, **b** Dynamic pressure is the difference between p_0 and the pressure at a point parallel to the flow (denoted p)

Bernoulli's theorem was obtained by integrating Euler's equation, which is based on the law of conservation of momentum. However, (2.20) is remarkably similar to the energy conservation law in a system of particles. The first term on the right-hand side of (2.20) can be interpreted as the kinetic energy per unit mass, the third term can be interpreted as the potential energy due to external forces, and the second term can be interpreted as the potential energy reserved as work done by the pressure. As H is fixed on streamlines, Bernoulli's theorem can be interpreted as an energy conservation law of fluid dynamics.

There are many fluid phenomena that can be explained by Bernoulli's theorem, and there are also many measurement devices that use Bernoulli's theorem. Here we introduce the **Pitot tube**, a device that measures flow velocities.

If a body with a rounded front surface is placed in the flow, a point forms in front of the body where the fluid velocity $\boldsymbol{u} = 0$. This point (indicated as point A in Fig. 2.6a) is called a **stagnation point**. Let the pressure at the stagnation point (called the stagnation pressure) be p_0. Applying Bernoulli's theorem to the streamlines passing through the stagnation point and ignoring the height difference (corresponding to the potential of external forces), we have

$$p + \frac{1}{2}\rho q^2 = \text{const} = p_0 \,. \qquad (2.21)$$

Consider $(1/2)\rho q^2$ as a type of pressure and call it **dynamic pressure**. Consider pressure p as **static pressure** in relation to this; then, the total pressure is the sum

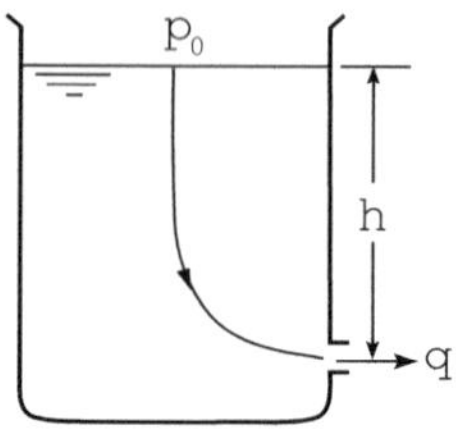

Fig. 2.7 Torricelli's theorem and its application

of $(1/2)\rho q^2$ and p. Along the streamline, the total pressure is fixed and equal to the stagnation pressure p_0.

If a tube such as that shown in Fig. 2.6b faces the direction of the flow, and the total and static pressures p_0 and p are measured through a hole in the front and side of the tube, respectively, the flow velocity q can be given by rearranging (2.21) to

$$q = \sqrt{\frac{2(p_0 - p)}{\rho}} . \tag{2.22}$$

A tube to measure the flow velocity based on this principle is called a Pitot tube.

Exercise 2.2 (*Torricelli's theorem*)
Consider the velocity when fluid inside a large container flows out through a hole made in the wall. In Fig. 2.7, suppose that the hole is made at $z = 0$, and the water surface is at $z = h$. Assuming that the surface area of the water is large relative to the cross-sectional area of the hole, the flow velocity at the water surface is approximately 0. Moreover, the pressure at the water surface equals the atmospheric pressure p_0. Using Bernoulli's theorem, determine the velocity of the fluid flowing through the hole. The flow velocity can be computed from Torricelli's equation $q = \sqrt{2gh}$.

Exercise 2.3 A cylindrical water tank with water surface area A is filled with fluid to height H. The fluid is then released through a small hole of area a at the bottom of the tank. Assuming that $A/a \geq 1$, and that atmospheric pressure acts on the liquid surface and the fluid that has flowed out, calculate

1. The time for the depth H to decrease to $H/2$, $T_{1/2}$.
2. The ratio of $T_{1/2}$ and the time taken for H to become 0, T_1.

2.5 Velocity Potential

Here we present a more detailed analysis of irrotational flow in incompressible perfect fluids. As shown in (2.7), irrotational flow is associated with a velocity potential such that

$$\boldsymbol{u} = \nabla \Phi . \tag{2.23}$$

2.5.1 The Laplace Equation

In Sect. 1.4, we derived the following continuity equation for an incompressible fluid:

$$\nabla \cdot \boldsymbol{u} = 0\,. \tag{2.24}$$

From (2.23) and (2.24), we have

$$\nabla^2 \Phi = 0\,. \tag{2.25}$$

In other words, velocity potential is governed by the **Laplace equation**.

The boundary condition of the velocity potential Φ can be obtained by substituting (2.23) into the boundary condition of the velocity $\boldsymbol{u}$ shown in Sect. 1.5. When the boundary surface is a rigid wall,

$$\boldsymbol{n} \cdot \nabla \Phi - \frac{\partial \Phi}{\partial n} - \boldsymbol{n} \cdot \boldsymbol{u}_b \equiv V_n\,. \tag{2.26}$$

Here $\boldsymbol{n}$ is a vector normal to the boundary surface, and V_n is the normal component of the velocity of the boundary surface $\boldsymbol{u}_b$.

In this way, problems seeking the velocity field in an incompressible irrotational flow result in problems that seek solutions to the Laplace equation that fulfills the given boundary conditions (harmonic functions). The harmonic function theory is mathematically complete, and its results are extensively used in fluid dynamics. (An analysis in spherical coordinates is shown below). If we can define the velocity potential Φ, the flow velocity and pressure function ($P = p/\rho$) can be computed from (2.23) and Bernoulli's pressure equation (2.18), respectively.

Note that the Laplace equation determining the flow velocity field is the same in both steady and non-steady motions. This indicates that the velocity field is determined at each moment and is unrelated to the history of the fluid motion. Therefore, regardless of whether the flow is steady or unsteady, the flow pattern will be identical provided that the boundary conditions remain the same. However, as shown by Bernoulli's pressure equation, pressure differs between steady and non-steady flows, and the fluid forces acting on bodies (obtained by integrating the pressure) depend on the history of the fluid motion.

Note 2.2 The Laplace equation expresses the mass conservation law; hence, the surface integral in

$$\iiint_V \nabla^2 \Phi \, dV = \iint_S \frac{\partial \Phi}{\partial n} \, dS = 0 \tag{2.27}$$

is a generalized expression that is valid in any coordinate system. Applying (2.27) to an infinitely small volume element in an arbitrary Cartesian coordinate system, we can express the Laplace equation in that coordinate system without requiring calculations for complicated coordinate transformations. For example, consider the

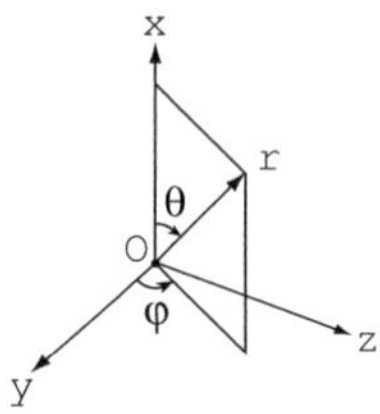

Fig. 2.8 Spherical coordinate

spherical coordinate system (Fig. 2.8).

$$\left.\begin{aligned} x &= r\,\cos\theta\,,\\ y &= r\,\sin\theta\,\cos\varphi\,,\\ z &= r\,\sin\theta\,\sin\varphi\,. \end{aligned}\right\} \tag{2.28}$$

An infinitesimally small line element in three-dimensional Cartesian coordinates is described by $(\delta x)^2 + (\delta y)^2 + (\delta z)^2$. Using (2.28), its representation in spherical coordinates is

$$\delta s_1 = \delta r,\ \delta s_2 = r\,\delta\theta,\ \delta s_3 = r\,\sin\theta\,\delta\varphi\,. \tag{2.29}$$

In terms of these coordinates, the surface integral of (2.27) can be thus rewritten:

$$\begin{aligned} \frac{\partial}{\partial s_1}\left(\frac{\partial\Phi}{\partial s_1}\,\delta s_2\,\delta s_3\right)\delta s_1 + \frac{\partial}{\partial s_2}\left(\frac{\partial\Phi}{\partial s_2}\,\delta s_3\,\delta s_1\right)\delta s_2 \\ + \frac{\partial}{\partial s_3}\left(\frac{\partial\Phi}{\partial s_3}\,\delta s_1\,\delta s_2\right)\delta s_3 = 0\,. \end{aligned} \tag{2.30}$$

Substituting (2.29) into (2.30), dividing the whole result by $\delta r\ \delta\theta\ \delta\varphi$, and rearranging, we obtain

$$\nabla^2\Phi = \frac{\partial}{\partial r}\left(r^2\frac{\partial\Phi}{\partial r}\right) + \frac{1}{\sin\theta}\frac{\partial}{\partial\theta}\left(\sin\theta\frac{\partial\Phi}{\partial\theta}\right) + \frac{1}{\sin^2\theta}\frac{\partial^2\Phi}{\partial\varphi^2} = 0\,. \tag{2.31}$$

2.5.2 *Sources and Sinks*

Imagine that uniform flow is output in all directions by a "source point" within an infinite fluid region. This flow is spherically symmetric with the source point at the center, and the flow velocity is zero at infinity.

This type of spherically symmetric flow can be solved by the Laplace equation. In spherical coordinates (r, θ, φ) as shown in Fig. 2.8, the velocity potential Φ becomes a function of r alone, and (2.31) is reduced to

$$\nabla^2 \Phi = \frac{d}{dr}\left(r^2 \frac{d\Phi}{dr}\right) = 0\,. \tag{2.32}$$

This general solution can be easily obtained by integration. With m and C as integration constants, it is given as

$$\Phi = -\frac{m}{r} + C\,. \tag{2.33}$$

The flow velocity is also a function of r alone, and it is given by

$$u_r = \frac{\partial \Phi}{\partial r} = \frac{m}{r^2}\,. \tag{2.34}$$

If $m > 0$, $u_r > 0$, and hence, the flow radiates from the origin. In fluid dynamics, this flow is described as originating from a **source**[1] at the origin. In contrast, if $m < 0$, the flow converges to the origin. This is called flow caused by the **sink**.

We now calculate the amount of flow Q released from the source point. Because flow is conserved, the amount of flow passing through an arbitrary closed curved surface S containing the origin is independent of the choice of the curved surface S. Assuming a spherical surface S with a radius r, Q is calculated as

$$Q = \iint_S \frac{\partial \Phi}{\partial n}\, dS = \iint_S u_r\, dS = \int_0^{2\pi} d\varphi \int_0^{\pi} \frac{m}{r^2}\, r^2 \sin\theta\, d\theta = 4\pi m\,. \tag{2.35}$$

Therefore, the constant m in (2.33) is related to the amount of flow released from the source point, known as the strength[2] of the source (If the origin is a sink, its strength is given by $-m = |m|$).

The constant C in (2.33) makes no contribution to the spatial differential value; hence, it can be arbitrarily selected. Setting $C = 0$ and substituting the result from (2.35), the velocity potential due to the source becomes

$$\Phi = -\frac{m}{r} = -\frac{Q}{4\pi}\left(\frac{1}{r}\right). \tag{2.36}$$

Equation (2.36) satisfies the Laplace equation at points other than the source ($r = 0$) but is singular at the origin. Let us consider the mass conservation law in fluid regions containing singular source points that release flow into the region. From (2.27), the amount of flow is given by

$$Q = \iint_S \frac{\partial \Phi}{\partial n}\, dS = \iiint_V \nabla^2 \Phi\, dV\,. \tag{2.37}$$

[1]A source can also be referred to as an outlet.

[2]Sometimes, it is appropriate to define the flow volume Q, instead of m, as the strength of the source.

However, to satisfy the mass conservation law (expressed by a continuum equation), we must have

$$\nabla^2 \Phi = Q\,\delta(\boldsymbol{r}) \tag{2.38}$$

(see Note 2.3). Here $\delta(r)$ is the Dirac delta function that becomes 0 everywhere except at $r = 0$.

In summary, the velocity potential due to a source at the origin (2.36) satisfies (2.38). Mathematically, $Q = 1$ in (2.36) is called the fundamental solution[3] of the Laplace equation.

If several (N) sources of different strengths exist at locations other than the origin, the total velocity potential is the sum of the N individual potentials. Furthermore, in the limiting case of a continuous distribution of sources within a region V, the summed velocity potential becomes an integral. Mathematically, the velocity potential is then given by

$$\Phi_{\mathrm{P}} = -\frac{1}{4\pi}\sum_{k=1}^{N}\frac{Q_k}{r_k} \to -\frac{1}{4\pi}\iiint_V \frac{\sigma_{\mathrm{Q}}}{r}\,dV\;;\quad r = |\boldsymbol{r}_{\mathrm{P}} - \boldsymbol{r}_{\mathrm{Q}}|\,. \tag{2.39}$$

In (2.39), the subscripts P and Q represent a point and the integral point of interest in the fluid, respectively. In a continuous distribution, (2.38) becomes

$$\nabla\cdot\boldsymbol{u} = \nabla^2\Phi = \sigma(\boldsymbol{r}) \tag{2.40}$$

where $\sigma(\boldsymbol{r})$ is called the density of the source distribution. Equation (2.40) is known as **Poisson's equation**.

Note 2.3 Dirac's delta function $\delta(x)$ can be defined in several ways; however, it is defined here as a generalized function that extracts the value at a given point.

If $F(x)$ is a continuous test function that disappears beyond a certain finite interval, the delta function possesses properties where the following is true:

$$\int_{-\infty}^{\infty} F(x)\,\delta(x - x_0)\,dx = \int_{-\infty}^{\infty} F(x + x_0)\,\delta(x)\,dx = F(x_0)\,. \tag{2.41}$$

In other words, if an integral contains the delta function, the argument of the delta function is a generalized function (symbol function) that extracts the value of the test function $F(x_0)$ at $x = 0$.

Therefore, expanding in three dimensions as $x_0 = 0$, for any continuous function $F(r)$, we have

$$\iiint_V F(\boldsymbol{r})\,\delta(\boldsymbol{r})\,dV = F(0)\,. \tag{2.42}$$

[3]Also known as the principal solution of the Laplace equation.

Let us reconsider (2.37). Since $\nabla^2 \Phi = 0$ if the origin ($r = 0$) is outside the integral region V, the integral of (2.37) is 0. Conversely, if the integral region encloses the origin, the integral of (2.37) gives the amount of flow Q. Comparing (2.37) with $F(\boldsymbol{r}) = Q$ (constant) and (2.42),

$$Q = \iiint_V Q\,\delta(\boldsymbol{r})\,dV \tag{2.43}$$

we can obtain (2.38), that is, $\nabla^2 \Phi = Q\,\delta(\boldsymbol{r})$.

Note 2.4 We use the variable separation method to solve (2.31), assuming symmetry about the x-axis (i.e., no φ dependence). As Φ is a function of r and θ, we substitute $\Phi = R(r)\,\Theta(\theta)$ into (2.31) to obtain

$$\frac{d}{dr}\left(r^2\frac{dR}{dr}\right)\frac{1}{R} = -\frac{1}{\sin\theta}\frac{d}{d\theta}\left(\sin\theta\frac{d\Theta}{d\theta}\right)\frac{1}{\Theta}\,. \tag{2.44}$$

The left-hand side of (2.44) is a function of r only, while the right-hand side is a function of θ only. For both sides to be equal, they must each equal a constant. Writing $n(n+1)$ for convenience, we get

$$\frac{d}{dr}\left(r^2\frac{dR}{dr}\right) - n(n+1)\,R = 0\,, \tag{2.45}$$

$$\frac{1}{\sin\theta}\frac{d}{d\theta}\left(\sin\theta\frac{d\Theta}{d\theta}\right) + n(n+1)\,\Theta = 0\,. \tag{2.46}$$

Since (2.45) is a differential equation of the same order, its solution is easily obtained. Denoting $\cos\theta$ by a new variable, z, (2.46) becomes a Legendre differential equation. Therefore, we obtain

$$\left.\begin{aligned} R(r) &= A\,r^n + \frac{B}{r^{n+1}}\,,\\ \Theta(\theta) &= C\,P_n(\cos\theta) + D\,Q_n(\cos\theta)\,. \end{aligned}\right\} \tag{2.47}$$

However, A, B, C, and D are arbitrary constants, and P_n and Q_n are Legendre functions. Assuming that no singularities exist at $\cos\theta = \pm 1$, $D = 0$ and n must be integer values, $n = 0, 1, 2\ldots$. Furthermore, by adding another condition where $\nabla\Phi \to 0$ at infinity, $A = 0$ must be true, and we obtain

$$\Phi = \sum_{n=0}^{\infty} B_n \frac{P_n(\cos\theta)}{r^{n+1}}\,, \tag{2.48}$$

which is the solution to the Laplace equation.

Here $P_n(z)$ is called **Legendre's polynomial**. If no singularities exist at any θ of $z = \cos\theta$, $P_n(z)$ is given by

$$P_n(z) = \frac{1}{2^n\, n!} \frac{d^n}{dz^n} \left(z^2 - 1 \right)^n , \tag{2.49}$$

$$P_0(\cos\theta) = 1 , \quad P_1(\cos\theta) = \cos\theta , \quad \ldots . \tag{2.50}$$

When $n = 0$ in (2.48), the velocity potential reduces to (2.36) already derived from the source. When $n = 1$, (2.48) gives the velocity potential due to a doublet, as shown later.

Note 2.5 (2.48) can also be expressed as follows:

$$\Phi = \sum_{n=0}^{\infty} B_n \frac{(-1)^n}{n!} \frac{\partial^n}{\partial x^n} \left(\frac{1}{r} \right) . \tag{2.51}$$

To prove this mathematically, we use the relation equation of the Legendre function:

$$\frac{1}{\sqrt{(x-\xi)^2 + y^2 + z^2}} = \frac{1}{\sqrt{r^2 - 2r\,\xi\,\cos\theta + \xi^2}} = \sum_{n=0}^{\infty} \frac{P_n(\cos\theta)}{r^{n+1}}\, \xi^n . \tag{2.52}$$

Setting the left-hand side to $f(\xi)$, we obtain the following series expansion in ξ

$$f^{(n)}(0) = \left. \frac{\partial^n f(\xi)}{\partial \xi^n} \right|_{\xi=0} = (-1)^n \left. \frac{\partial^n f(\xi)}{\partial x^n} \right|_{\xi=0} = (-1)^n \frac{\partial^n}{\partial x^n} \left(\frac{1}{r} \right) ,$$

from which we find that

$$f(\xi) = \sum_{n=0}^{\infty} \frac{f^{(n)}(0)}{n!}\, \xi^n = \sum_{n=0}^{\infty} \frac{(-1)^n}{n!} \frac{\partial^n}{\partial x^n} \left(\frac{1}{r} \right) \xi^n . \tag{2.53}$$

Comparing (2.52) with (2.53), it becomes apparent that (2.48) and (2.51) are equal.

2.5.3 *Doublet*

Let us consider a situation where a source and a sink of the same strength are infinitely close to each other. In other words, consider a sink of strength m at origin O and a source of strength m at a point Q at a distance δs away (Fig. 2.9). From (2.36), we have

$$\Phi = \frac{m}{r} - \frac{m}{r_1} , \tag{2.54}$$

where $r_1 = \sqrt{r^2 - 2r\delta s \cos\theta + (\delta s)^2}$.

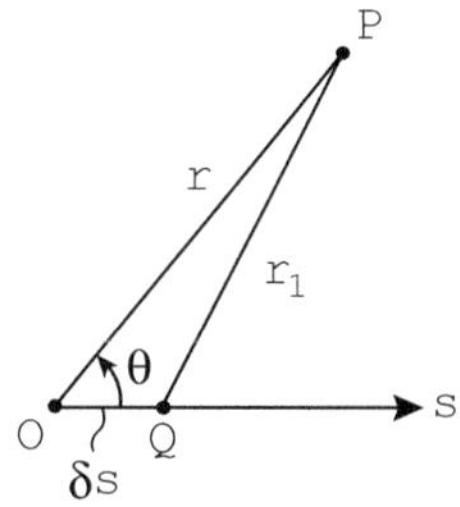

Fig. 2.9 Limiting values of source and sink

As $\delta s \to 0$, the limit approaches a fixed value $m\delta s \to \mu$. Here r_1 is equal to the left-hand side of (2.52), with $\xi = \delta s$. Alternatively, we can Taylor expand around $1/r_1$. As $\delta s \to 0$, both approaches give

$$\frac{1}{r_1} \sim \frac{1}{r} + \frac{\cos\theta}{r^2}\delta s + O((\delta s)^2)\,. \tag{2.55}$$

Substituting (2.55) into (2.54), and denoting $m\delta s = \mu$, we obtain

$$\Phi = -\mu\frac{\cos\theta}{r^2}\,. \tag{2.56}$$

Equation (2.56) specifies the velocity potential of a **doublet** with the axis oriented along $\overrightarrow{\mathrm{OQ}}$, and μ denotes its strength. This is none other than the solution to $n = 1$ in (2.48) shown in Note 2.4 (with $-\mu$ equivalent to B_1 in (2.48)). Equation (2.56) can be rewritten as

$$\Phi = -\frac{\mu}{r^2}\frac{\boldsymbol{r}}{r}\cdot\boldsymbol{e} = \mu\,\boldsymbol{e}\cdot\nabla\left(\frac{1}{r}\right) = -\mu\frac{\partial}{\partial s}\left(\frac{-1}{r}\right). \tag{2.57}$$

Here $\boldsymbol{e}$ is a unit vector in the direction $\overrightarrow{\mathrm{OQ}}$. The vector $\mu\boldsymbol{e}$ is called the moment of the doublet. If $\boldsymbol{e}$ is oriented along the x-axis, (2.57) gives the solution to $n = 1$ in (2.51) shown in Note 2.5.

Equation (2.57) shows that the velocity potential of a doublet is the negative of the velocity potential $(-1/r)$ of a unit strength source differentiated along the direction of the doublet axis (in this case, differentiated with respect to s).

Differentiating (2.57) with respect to s, we obtain the velocity potential of two infinitely close doublets, i.e., the potential of a **quadruple** source. Iterating this procedure, (2.51) can be physically interpreted as the velocity potential of 2^n overlapping sources (**multipoles**) oriented in the direction of the x-axis.

The above relationship between the multipole and unit strength source velocity potentials also holds in two-dimensional theory, as will be shown in Sect. 2.9.

Exercise 2.4 Consider a flow generated by the overlap of a uniform flow and a source at the origin O. This flow occurs when a uniform flow meets a cylindrical

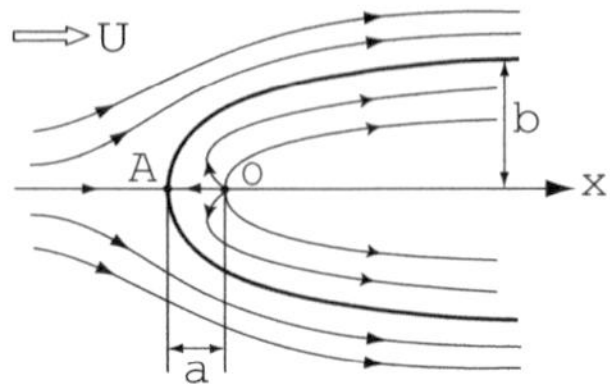

Fig. 2.10 Flow around a semi-infinite body

body of semi-infinite length with a round end (semi-infinite body; see Fig. 2.10). Point A is a stagnation point.

The uniform flow has velocity U and moves in the positive direction of the x-axis. Considering the strength of the source as m, the velocity potential is

$$\Phi = Ux - \frac{m}{r} . \tag{2.58}$$

Obtain the distance a between the origin O and the stagnation point A and the cylindrical radius b under the infinite downstream produced by the branching streamlines from point A.

2.5.4 *Flow Around a Sphere*

Now consider a sphere of radius a moving through a static fluid with velocity U. The center of the sphere O is taken as the origin, and the propagation direction is the x-axis (see Fig. 2.11). In spherical coordinates (r, θ, φ), considering the polar axis as the x-axis, the boundary condition equation for the velocity potential at the sphere's surface is (from 2.26)

$$\frac{\partial \Phi}{\partial r} = U \cos\theta \qquad \text{at } r = a . \tag{2.59}$$

Given that $\nabla\Phi \to 0$ at $r = \infty$, the general solution satisfying this condition is (2.48) or (2.51), as shown above. Thus, the coefficient B_n that satisfies (2.59) is

$$B_0 = 0, \quad B_1 = -\frac{1}{2} U a^3, \quad B_n = 0 \quad (n \geq 2) . \tag{2.60}$$

Essentially,

$$\Phi = -\frac{Ua^3}{2} \frac{\cos\theta}{r^2} \tag{2.61}$$

is the velocity potential of the flow field around the sphere. As evident from (2.56), this represents the flow when there is a doublet of strength $Ua^3/2$ in the x-axis direction. In other words, the flow induced by a sphere moving through a static fluid

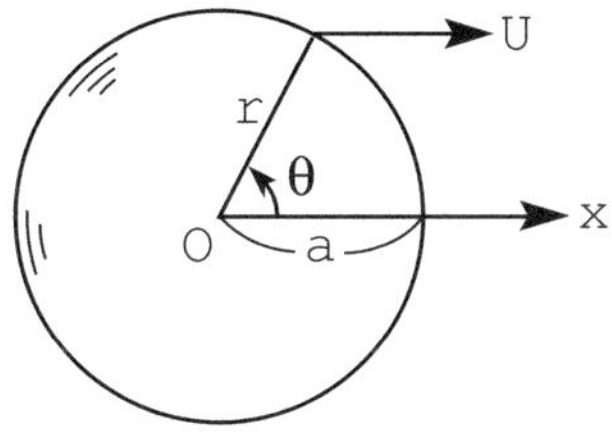

Fig. 2.11 Sphere moving parallel to the x-axis with velocity U

is equal to the flow due to a doublet whose axis aligns in the direction of motion of the center of the sphere.

Next, we consider a sphere of radius a fixed in a uniform flow with velocity U. In this case, a uniform flow with flow velocity U in the negative x-axis direction can be superimposed onto the flow field disturbed by the sphere, as shown in Fig. 2.11. The velocity potential of this uniform flow is

$$\Phi = -Ux = -Ur\cos\theta\,. \tag{2.62}$$

Hence, the velocity potential for the whole is given by the sum of (2.61) and (2.62):

$$\Phi = -U\left(r + \frac{a^3}{2r^2}\right)\cos\theta\,, \tag{2.63}$$

and the corresponding flow velocity is

$$\left.\begin{aligned} u_r &= \frac{\partial\Phi}{\partial r} = -U\left(1 - \frac{a^3}{r^3}\right)\cos\theta\,,\\ u_\theta &= \frac{\partial\Phi}{r\,\partial\theta} = U\left(1 + \frac{a^3}{2r^3}\right)\sin\theta\,. \end{aligned}\right\} \tag{2.64}$$

Clearly, $u_r = 0$ at the spherical surface ($r = a$). Therefore, the magnitude of the flow velocity at this surface is given by

$$q = |u_\theta| = \frac{3}{2}\,U|\sin\theta|\,. \tag{2.65}$$

The flow velocity is maximized at $\theta = \pi/2$, with magnitude $q_{\max} = 1.5U$. Moreover, at the stagnations points, which should lie on the x-axis (at $\theta = 0, \pi$), $q = 0$ from (2.65).

Exercise 2.5 Consider a uniform flow with velocity U with an overlapping source of strength m at $x = -\ell$ and a sink of strength m at $x = +\ell$. There are two stagnation points, A and B, as shown in Fig. 2.12. The curved surface created by diverging streamlines passing through these points is called the Rankine ovoid. Note that, in this way, when the sum of the strengths of the source and sink is 0, the diverging

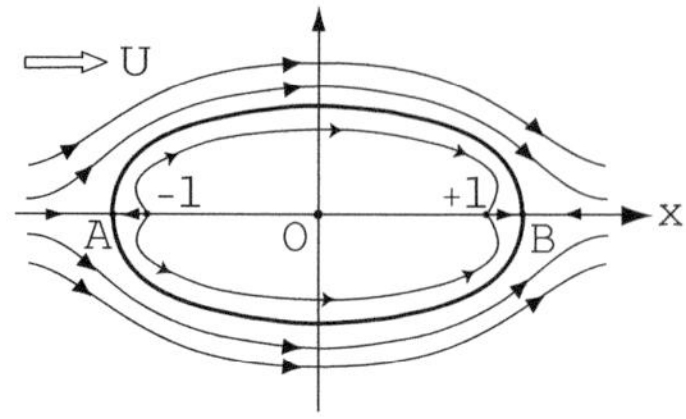

Fig. 2.12 Flow around the Rankine ovoid

streamlines create a closed curved surface equivalent to a body in a uniform flow. The doublet expressing the flow around the sphere is an example of this.

Assume that the coordinates a of stagnation points due to the Rankine ovoid are functions of m and ℓ. Obtain the maximum radius b and the maximum flow velocity $u_{\max}$ of the diverging streamlines as functions of m, ℓ, and U.

2.6 Vector Potential

The previous section considered irrotational flows in the presence of source distribution within the flow field (including isolated sources). In other words,

$$\nabla \cdot \boldsymbol{u}_1 = \sigma\,, \quad \nabla \times \boldsymbol{u}_1 = 0\,. \tag{2.66}$$

This section focuses on rotational flow created by vorticity $\boldsymbol{\omega}$. In other words, consider a velocity field $\boldsymbol{u}_2$ satisfying

$$\nabla \cdot \boldsymbol{u}_2 = 0\,, \quad \nabla \times \boldsymbol{u}_2 = \boldsymbol{\omega}\,. \tag{2.67}$$

Once $\boldsymbol{u}_1$ and $\boldsymbol{u}_2$ are obtained for (2.66) and (2.67), in a general case containing sources and vortices, it can be obtained from overlapping $\boldsymbol{u} = \boldsymbol{u}_1 + \boldsymbol{u}_2$.

As previously discussed, the velocity potential is derived from the condition $\nabla \times \boldsymbol{u} = 0$ on the basis of the identity equation $\nabla \times \nabla \Phi = 0$ of vector analysis. Similarly, another identity equation of vector analysis, $\nabla \cdot (\nabla \times \boldsymbol{A}) = 0$, can be established; thus, from the condition $\nabla \cdot \boldsymbol{u} = 0$ in (2.67), we can derive a **vector potential** $\boldsymbol{A}$ that satisfies

$$\boldsymbol{u} = \nabla \times \boldsymbol{A}\,. \tag{2.68}$$

Since the condition $\nabla \cdot \boldsymbol{u} = 0$ holds for incompressible fluids, the vector potential can also be defined for viscous fluids.

Now, substituting (2.68) into the second equation of (2.67), we obtain

$$\nabla \times (\nabla \times \boldsymbol{A}) = \nabla(\nabla \cdot \boldsymbol{A}) - \nabla^2 \boldsymbol{A} = \boldsymbol{\omega}\,. \tag{2.69}$$

For convenience, we specify another condition for $\boldsymbol{A}$:

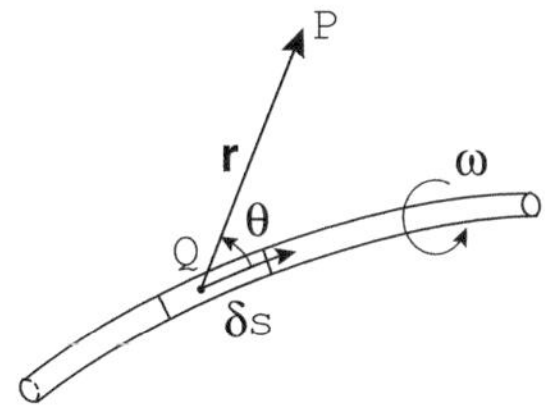

Fig. 2.13 Velocity at point P induced by a vortex filament

$$\nabla \cdot \boldsymbol{A} = 0 \tag{2.70}$$

(Whether the obtained solution actually satisfies this condition will be confirmed later (see Note 2.6).)

From (2.69) and (2.70), we have

$$\nabla^2 \boldsymbol{A} = -\boldsymbol{\omega} \tag{2.71}$$

as the governing equation on A. Equation (2.71) is a Poisson equation of the same shape as (2.40); thus, its solution can be found in the same way as for (2.39):

$$\boldsymbol{A}_{\mathrm{P}} = \frac{1}{4\pi}\iiint_V \frac{\boldsymbol{\omega}_{\mathrm{Q}}}{r}\,dV\ ; \quad r = |\boldsymbol{r}_{\mathrm{P}} - \boldsymbol{r}_{\mathrm{Q}}|\,. \tag{2.72}$$

Substituting (2.72) into (2.68), we get

$$\boldsymbol{u} = \frac{1}{4\pi}\iiint_V \nabla_{\mathrm{P}} \times \frac{\boldsymbol{\omega}_{\mathrm{Q}}}{r}\,dV = \frac{1}{4\pi}\iiint_V \boldsymbol{\omega}_{\mathrm{Q}} \times \frac{\boldsymbol{r}}{r^3}\,dV\,, \tag{2.73}$$

which can be used to calculate the velocity induced by the vortex.

To investigate the meaning of (2.73), we consider the contribution from an infinitesimally small part of the vortex filament (Sect. 1.10) with the length of δs and cross-sectional area of σ. This filament passes through point Q (see Fig. 2.13). Since $dV = \sigma\delta s$ in this case, the flow velocity induced by point P based on (2.73) is

$$\delta\boldsymbol{u} = \frac{\boldsymbol{\omega}_{\mathrm{Q}} \times \boldsymbol{r}}{4\pi r^3}\,\sigma\,\delta s = \frac{\omega_{\mathrm{Q}}\,\sigma}{4\pi r^2}\left(\frac{\boldsymbol{\omega}_{\mathrm{Q}}}{\omega_{\mathrm{Q}}}\delta s \times \frac{\boldsymbol{r}}{r}\right) = \frac{\Gamma}{4\pi}\frac{\delta\boldsymbol{s} \times \boldsymbol{r}}{r^3}\,. \tag{2.74}$$

Here $\Gamma = \omega_{\mathrm{Q}}\sigma$ is the strength of the vortex filament (the circulation indicated in (2.10)), and $\delta\boldsymbol{s} = (\boldsymbol{\omega}_{\mathrm{Q}}/\omega_{\mathrm{Q}})\,\delta s$ is the linear element vector along the axis of the vortex filament.

In (2.74), Γ (strength of the current) and $\boldsymbol{u}$ (magnetic field vector) are the same as those in the **Biot–Savart principle** of electromagnetism. In this regard, the vorticity-contributed velocity field behaves similar to an electromagnetic field.

The vector potential of rotating flows is comparable with the velocity potential of irrotational flows. However, the vector potential is still a vector; thus, it is the

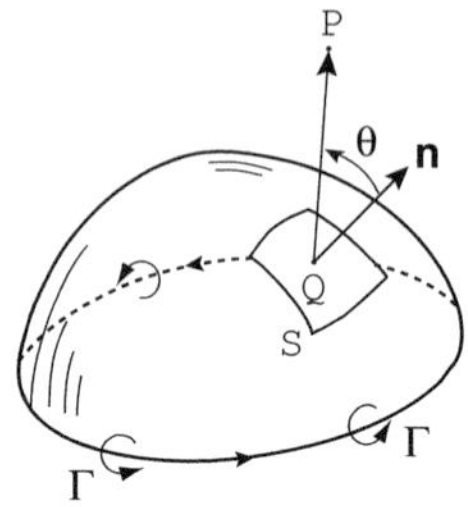

Fig. 2.14 Relationship between vortex filament and doublet

same as the original velocity component as an unknown value. Therefore, the vector potential is not used as generally as the velocity potential.

Note 2.6 For (2.72), it is important to show that $\nabla \cdot \boldsymbol{A} = 0$:

$$\begin{aligned}\nabla \cdot \boldsymbol{A} &= \frac{1}{4\pi} \iiint_V \omega_Q \cdot \nabla_P\left(\frac{1}{r}\right) dV = -\frac{1}{4\pi} \iiint_V \omega_Q \cdot \nabla_Q\left(\frac{1}{r}\right) dV \\ &= -\frac{1}{4\pi} \iiint_V \nabla_Q \cdot \left(\omega_Q \frac{1}{r}\right) dV \quad (\leftarrow \nabla \cdot \boldsymbol{\omega} = 0) \\ &= -\frac{1}{4\pi} \iint_S \frac{\boldsymbol{n} \cdot \boldsymbol{\omega}}{r} dS = 0 \qquad (\leftarrow \boldsymbol{n} \cdot \boldsymbol{\omega} = 0)\end{aligned}$$

Note 2.7 When a vortex filament of strength $\Gamma = \omega\,\sigma$ creates a closed curve C, it induces a velocity field equivalent to that of a doublet distribution on a curved surface S bounded by perimeter C.

The velocity field induced by a vortex filament of strength Γ, as shown in Fig. 2.14, is determined from (2.74) as

$$\boldsymbol{u}_P = \frac{\Gamma}{4\pi} \oint_C \frac{d\boldsymbol{s}_Q \times \boldsymbol{r}}{r^3} = \frac{\Gamma}{4\pi} \oint_C d\boldsymbol{s}_Q \times \nabla_Q\left(\frac{1}{r}\right). \tag{2.75}$$

Applying Stokes' theorem to (2.75), and rearranging while paying careful attention to $\nabla_Q(1/r) = -\nabla_P(1/r)$, we obtain

$$\begin{aligned}\boldsymbol{u}_P &= -\frac{\Gamma}{4\pi} \iint_S (\boldsymbol{n} \times \nabla_Q) \times \nabla_P\left(\frac{1}{r}\right) dS \\ &= \nabla_P\left[-\frac{\Gamma}{4\pi} \iint_S \frac{\partial}{\partial n_Q}\left(\frac{1}{r}\right) dS\right].\end{aligned} \tag{2.76}$$

This is identical to the velocity field in

$$\left.\begin{aligned}\boldsymbol{u} &= \nabla \Phi \\ \Phi &= -\frac{\Gamma}{4\pi} \iint_S \frac{\partial}{\partial n_Q}\left(\frac{1}{r}\right) dS = -\frac{\Gamma}{4\pi} \iint_S \frac{\cos\theta}{r^2} dS\end{aligned}\right\} \tag{2.77}$$

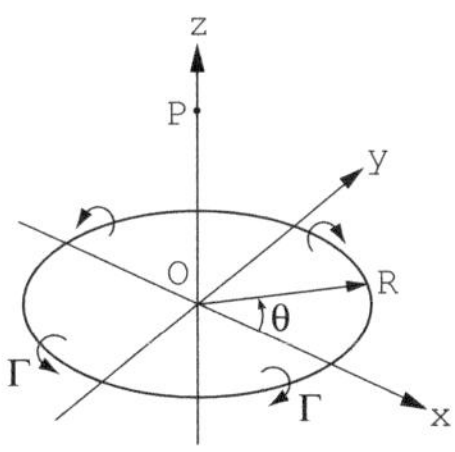

Fig. 2.15 Velocity induced by a circular vortex

This Φ can be interpreted as the velocity potential due to the uniform distribution of a doublet with the axis pointing normal to the curved surface S calculated by (2.56) or (2.57). Here the surface area S is introduced through Stokes' theorem; therefore, (2.77) holds for any curved surface bounded by a closed curve with perimeter C.

Exercise 2.6 Using the Biot–Savart principle, show that the velocity induced by an infinitely long and straight vortex filament is given by

$$u = \frac{\Gamma}{2\pi h} \tag{2.78}$$

where h is the perpendicular distance from point P to the straight vortex filament.

Exercise 2.7 Consider a circular vortex filament of radius R and strength Γ as shown in Fig. 2.15. Obtain the velocity w induced in the z-axis direction at point P(0, 0, z) as a function of Γ, R, and z using the following two methods:

1. Biot–Savart analogy (2.74)
2. Velocity potential of the doublet distribution (2.77)

Then, show that the solutions from both methods are the same.

2.7 Stream Function

As explained in Sect. 2.6, the vector potential is a less useful concept than the velocity potential. However, in two- or three-dimensional problems with axial symmetry, it becomes a scalar function and can then be used analogously to the velocity potential.

2.7.1 Two-Dimensional Flow

We first consider a two-dimensional problem. In two-dimensional problems, $\boldsymbol{u} = (u, v, 0)$ is a function of (x, y) alone. The vector potential then takes the form $\boldsymbol{A} = (0, 0, \Psi)$. Here

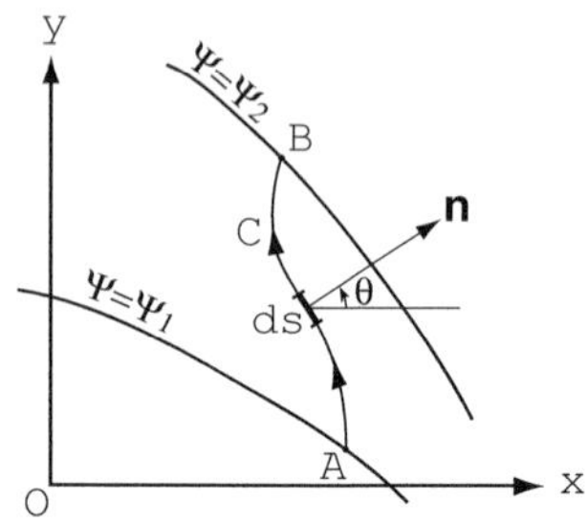

Fig. 2.16 Relationship between streamline, flow amount, and flow velocity

$$\boldsymbol{u} = \nabla \times \boldsymbol{A} = \left(\frac{\partial \Psi}{\partial y}, -\frac{\partial \Psi}{\partial x}, 0 \right). \tag{2.79}$$

Note that Ψ is none other than the **stream function**.[4] This is because the curve, $\Psi = \text{const}$, equals the streamline. For example, from (2.79), we have

$$d\Psi = \frac{\partial \Psi}{\partial x} dx + \frac{\partial \Psi}{\partial y} dy = -v\,dx + u\,dy = 0. \tag{2.80}$$

Therefore,

$$\frac{dx}{u} = \frac{dy}{v}. \tag{2.81}$$

Clearly, when $\Psi = \text{const}$, the streamline line element $d\boldsymbol{s}$ is parallel to the flow velocity $\boldsymbol{u}$.

Note that the only condition for the stream function is $\nabla \cdot \boldsymbol{u} = 0$; thus, it can be defined for incompressible viscous fluids. Actually, it is easy to verify that $\nabla \cdot \boldsymbol{u} = 0$ is automatically satisfied by (2.79).

Now consider two streamlines $\Psi = \Psi_1$ and $\Psi = \Psi_2$ as shown in Fig. 2.16. Take an arbitrary curve C joining point A on Ψ_1 and point B on Ψ_2 and specify a normal vector $\boldsymbol{n} = (n_x, n_y)$ whose positive direction is left to right. The line element $d\boldsymbol{s} = (dx, dy)$ on C is described by

$$dx + i\,dy = ds\, e^{i\,(\theta + \pi/2)} = i\,ds\,(n_x + i\,n_y). \tag{2.82}$$

From (2.79) and (2.82), the rate of change in Ψ along ds is given by

$$\begin{aligned} \frac{\partial \Psi}{\partial s} &= \frac{\partial \Psi}{\partial x}\frac{\partial x}{\partial s} + \frac{\partial \Psi}{\partial y}\frac{\partial y}{\partial s} = -\frac{\partial \Psi}{\partial x} n_y + \frac{\partial \Psi}{\partial y} n_x \\ &= v\,n_y + u\,n_x = \boldsymbol{u} \cdot \boldsymbol{n} = u_n\,. \end{aligned} \tag{2.83}$$

[4]Often called a flow function or streamline function.

Therefore,

$$\left[\Psi\right]_{\mathrm{A}}^{\mathrm{B}} = \int_C d\Psi = \int_C u_n\, ds = Q\,. \tag{2.84}$$

In other words, $\Psi_2 - \Psi_1$ specifies the amount of flow between the two streamlines. Moreover, (2.83) indicates that when seeking a directional component of velocity $\boldsymbol{u}$, the derivative of Ψ in the counterclockwise right-angled direction should be calculated. In two-dimensional polar coordinates, the velocity components are given by

$$u_r = \frac{\partial \Psi}{r\partial \theta}\,, \qquad u_\theta = -\frac{\partial \Psi}{\partial r}\,. \tag{2.85}$$

We now seek the stream function (vector potential) induced by two-dimensional vortices. The vorticity $\boldsymbol{\omega} = \nabla \times \boldsymbol{u}$ only has a z-component (i.e., perpendicular to the x-y plane), and

$$\boldsymbol{\omega} = (0,\, 0,\, \omega)\,, \quad \omega = \frac{\partial v}{\partial x} - \frac{\partial u}{\partial y}\,. \tag{2.86}$$

Thus, substituting (2.79), we can see that

$$\nabla^2 \Psi = -\,\omega \tag{2.87}$$

is a governing equation and is identical to (2.71). Now, by paying attention to the fact that the solution in the two-dimensional problem that is equivalent to the solution, $1/4\pi r$, in the three-dimensional problem is $-(1/2\pi)\log r$ (see (2.120) below), similar to (2.72), we can obtain

$$\Psi_{\mathrm{P}} = -\frac{1}{2\pi}\iint_S \omega_{\mathrm{Q}} \log r\, dS\,; \quad r = |\boldsymbol{r}_{\mathrm{P}} - \boldsymbol{r}_{\mathrm{Q}}|\,. \tag{2.88}$$

Moreover, if vortices are concentrated at the origin, and equivalently, if the flow contains vortex filaments (with cross-sectional area σ), then similar to (2.38), we can replace ω_{Q} with $\Gamma\delta(\boldsymbol{r}_{\mathrm{Q}})$. From the property of the delta function shown in Note 2.3, the following expressions apply:

$$\Psi_{\mathrm{P}} = -\frac{\Gamma}{2\pi}\log r\,, \quad \Gamma = \omega_{\mathrm{Q}}\,\sigma\,. \tag{2.89}$$

Later, this result will be derived again from the complex velocity potential (see (2.122)).

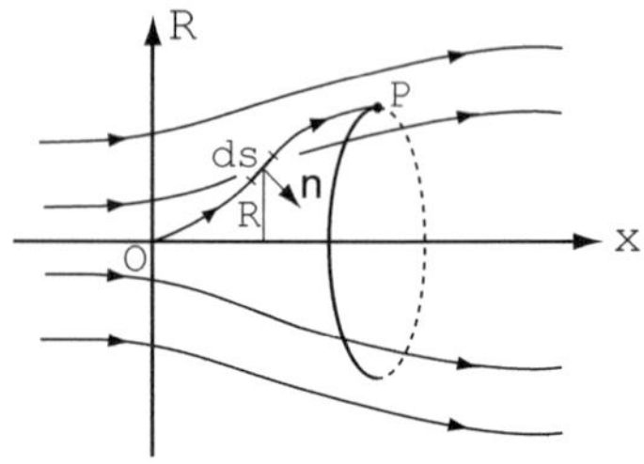

Fig. 2.17 Three-dimentional axisymmetric flow

2.7.2 Three-Dimensional Axisymmetric Flow

Next, let us consider a three-dimensional **axisymmetric** flow.[5] In a cylindrical coordinate system (x, R, θ) with symmetry about the x-axis, the flow field is independent of θ. Therefore, in a plane center containing the axis of symmetry (see Fig. 2.17), axisymmetric flow is seen to be similar to a two-dimensional flow. However, a real flow would rotate Fig. 2.17 around the x-axis; consequently, the vortex lines would circle in the plane perpendicular to the x-axis.

If a point P in the flow is connected to the origin O by an arbitrary planar curve C, then similar to the derivation of (2.84), the amount of flow passing from left to right of the rotation surface, obtained by rotating C around the x-axis, is given by

$$Q = [\,\Psi\,]_{\mathrm{O}}^{\mathrm{P}} = \int_C d\Psi = \int_C u_n\, 2\pi R\, ds\,. \tag{2.90}$$

For axisymmetric flows, the effect of rotation around the x-axis is considered, and it can be redefined as $\Psi = 2\pi\,\Psi_a$. Then, the velocity in the normal direction can be obtained with

$$u_n = \frac{1}{R}\frac{\partial \Psi_a}{\partial s}\,. \tag{2.91}$$

The velocity components along the x- and R-axes, (u_x, u_R), are calculated from

$$u_x = \frac{1}{R}\frac{\partial \Psi_a}{\partial R}\,, \quad u_R = -\frac{1}{R}\frac{\partial \Psi_a}{\partial x}\,, \tag{2.92}$$

as the δs components of u_x and u_R are δR and $-\delta x$, respectively.

$\Psi_a = \Psi/2\pi$ defined in (2.91) is equivalent to the stream function in a two-dimensional flow and is called **Stokes' stream function**. It is worth reiterating that, in three-dimensional problems, such scalar functions are limited to axisymmetric flows.

[5]Here axisymmetric flow only refers to flow with a zero velocity component in the θ-direction.

Exercise 2.8 The above-described flows around semi-infinite bodies (see Fig. 2.10), spheres (see Fig. 2.11), and the Rankine ovoid (see Fig. 2.12) are all three-dimensional axisymmetric flows. Obtain the Stokes' stream function Ψ_a in each of these flows.

2.8 Complex Velocity Potential

In this section, we restrict our discussion to incompressible, irrotational flows in two-dimensional problems. As discussed in Sect. 2.5, a velocity potential exists since it is irrotational flow, and it can be expressed from (2.23) and (2.40):

$$(u,\ v) = \left(\frac{\partial \Phi}{\partial x},\ \frac{\partial \Phi}{\partial y}\right), \tag{2.93}$$

$$\nabla^2 \Phi = \sigma(x,\ y)\,. \tag{2.94}$$

Here, $\sigma(x, y)$ is the source distribution density within the fluid.

On the other hand, as mentioned in the Sect. 2.7, the continuity equation gives rise to a stream function (two-dimensional vector potential) Ψ, and from (2.79) and (2.87), we can obtain

$$(u,\ v) = \left(\frac{\partial \Psi}{\partial y},\ -\frac{\partial \Psi}{\partial x}\right), \tag{2.95}$$

$$\nabla^2 \Psi = -\,\omega(x,\ y)\,. \tag{2.96}$$

Here $\omega(x, y)$ is the component of the vorticity vector perpendicular to the x-y plane.

In the fields in which the source distribution $\sigma(x, y)$ and vorticity distribution $\omega(x, y)$ are isolated on a certain point, σ and ω behave as delta functions, and in fluid regions outside those point, we have

$$\nabla^2 \Phi = 0\,, \quad \nabla^2 \Psi = 0\,. \tag{2.97}$$

In other words, Φ and Ψ are both harmonic functions. Now, from (2.93) and (2.95), we have

$$\left.\begin{aligned} u &= \frac{\partial \Phi}{\partial x} = \frac{\partial \Psi}{\partial y}\,, \\ v &= \frac{\partial \Phi}{\partial y} = -\frac{\partial \Psi}{\partial x}\,. \end{aligned}\right\} \tag{2.98}$$

Equation (2.98) is the **Cauchy–Riemann relation** derived from the complex function theory. That is,

$$f(z) = \Phi(x,\ y) + i\,\Psi(x,\ y)\,, \quad z = x + i\,y\,. \tag{2.99}$$

Equation (2.98) expresses the necessary and sufficient conditions for $f(z)$ to be an analytic function of z. This $f(z)$ is called a **complex velocity potential**.

Note 2.8 If $f(z)$ is analytic (holomorphic) at $z = z_0$, $f(z)$ is by definition differentiable at $z = z_0$ and its vicinity. In the case of real functions, in order for the derivative $f'(x)$ to exist at $x = x_0$, the right- and left-hand limits must be the same. In complex functions, this type of one-dimensional approach cannot work, and it has to be unrelated to the direction of approach. Instead, the required conditions are the Cauchy–Riemann relations, as elaborated below.

First, the differential of a complex function $f(z)$ is defined as

$$f'(z) = \lim_{h \to 0} \frac{f(z+h) - f(z)}{h}, \quad h = \delta x + i\,\delta y\,. \tag{2.100}$$

Here from (2.99),

$$f(z+h) - f(z) = \frac{\partial \Phi}{\partial x}\,\delta x + \frac{\partial \Phi}{\partial y}\,\delta y + i\left(\frac{\partial \Psi}{\partial x}\,\delta x + \frac{\partial \Psi}{\partial y}\,\delta y\right) + O(h^2)\,.$$

Therefore,

$$f'(z) = \lim_{h \to 0} \frac{(\Phi_x + i\,\Psi_x)\delta x + (\Phi_y + i\,\Psi_y)\delta y}{\delta x + i\,\delta y}\,. \tag{2.101}$$

As the method of $h \to 0$ is arbitrarily obtained, when $\delta y = 0$ or $\delta x = 0$, we have

$$f'(z) = \Phi_x + i\,\Psi_x = \Psi_y - i\,\Phi_y\,. \tag{2.102}$$

We have stated (2.102) without proof, but the equation is clearly satisfied by (2.98). In other words, a derivative function $f'(z)$ exists if (2.98) holds.

Next, if (2.98) is assumed in (2.101), we have

$$\begin{aligned} f'(z) &= \lim_{h \to 0} \frac{(\Phi_x + i\,\Psi_x)\delta x + (-\Psi_x + i\,\Phi_x)\delta y}{\delta x + i\,\delta y} \\ &= \lim_{h \to 0} \frac{(\Phi_x + i\,\Psi_x)(\delta x + i\,\delta y)}{\delta x + i\,\delta y}\,. \end{aligned} \tag{2.103}$$

Regardless of the value of $h = \delta x + i\delta y$, the limit approaches a fixed value $f'(z) = \Phi x + i\Psi x$ as $h \to 0$. In other words, (2.98) is a sufficient condition for differentiability of $f(z)$.

Equations (2.98) and (2.99) can be rearranged similar to (2.103) to give

$$\frac{df}{dz} = \frac{\partial \Phi}{\partial x} + i\,\frac{\partial \Psi}{\partial x} = u - i\,v = q\,e^{-i\,\theta} \tag{2.104}$$

(see Fig. 2.18.) This quantity is called **complex velocity**.

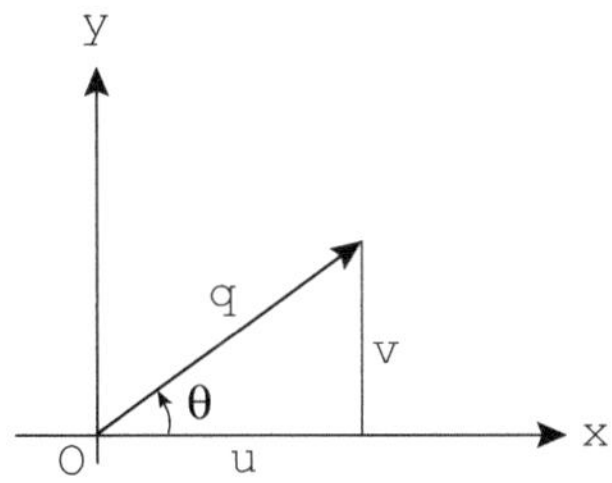

Fig. 2.18 Relationship between complex velocity, and q and θ

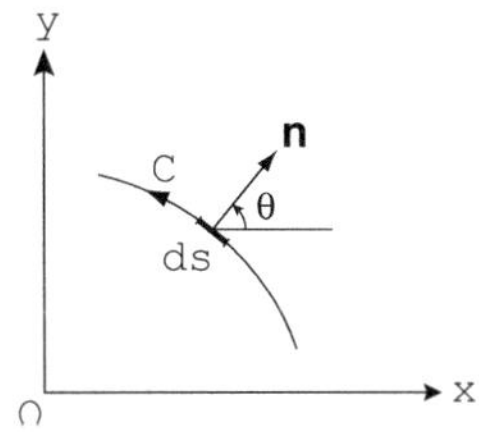

Fig. 2.19 Relationship between the normal line, and dz and ds

Now consider a contour integral of the complex velocity along an arbitrary closed curve C on the z-plane.

$$\oint_C \frac{df}{dz} dz = \oint_C df = [\, f \,]_C = [\, \Phi \,]_C + i\, [\, \Psi \,]_C \,. \tag{2.105}$$

As shown with (2.82), we have $dz = dx + i\, dy = i\, ds(n_x + i\, n_y)$ (see Fig. 2.19). Together with (2.104), this gives

$$\begin{aligned} \oint_C \frac{df}{dz} dz &= \oint_C (u\, dx + v\, dy) + i \oint_C (u\, n_x + v\, n_y)\, ds \\ &= \oint_C \boldsymbol{u} \cdot d\boldsymbol{r} + i \oint_C u_n\, ds = \Gamma(C) + i\, Q(C)\,. \end{aligned} \tag{2.106}$$

The relationship equations obtained from (2.105) and (2.106) are same as those described in (2.12) and (2.84). In other words, $[\Phi]_C$ is the same as the circulation along C, $\Gamma(C)$, and $[\Psi]_C$ is the same as the amount of flow passing from left to right through C, $Q(C)$.

As already noted in (2.94) and (2.96), vortex filaments and isolated sources concentrated in the flow field are permitted in incompressible, irrotational flows, and we considered fluid region excluding these features. Therefore, if such concentrated activities occur within a closed curve C, neither $\Gamma(C)$ nor $Q(C)$ is 0. In other words, $f(z) = \Phi + i\, \Psi$ must be a multi-valued function. However, as this polyvalence is a constant, it is lost when differentiating the function and the complex velocity becomes a single-valued function.

2.9 Simple Two-Dimensional Potential Flows

In Sect. 2.5, for three-dimensional problems, we discussed the velocity potential induced by a source or a doublet and the velocity potential that expresses the flows around a sphere. Here we discuss the two-dimensional flows corresponding to these complex velocity potentials and evaluate their characteristics.

2.9.1 Uniform Flow

We denote the uniform flow velocity by U (> 0). If the direction of the flow and the x-axis form the angle, α, from (2.104), we obtain

$$f(z) = U\,e^{-i\,\alpha}\,z\,. \tag{2.107}$$

Equation (2.107) can be interpreted as the rotation of the coordinate axis. In other words, as shown in Fig. 2.20, if the O-xy coordinate system ($z = x + i\,y$) is rotated counterclockwise through an angle α to yield the O-XY coordinate system ($Z = X + i\,Y$), the systems are related by

$$Z = z\,e^{-i\alpha}\,. \tag{2.108}$$

The complex velocity potential of the uniform flow in the positive direction of the x-axis in the O-XY coordinate system is $f(Z) = U Z$. When this is expressed with the O-xy coordinate system, it is clear that (2.108) can be substituted to obtain (2.107).

2.9.2 Flow Around a Corner

Consider

$$f(z) = A\,z^n \qquad (A > 0,\ n > 0) \tag{2.109}$$

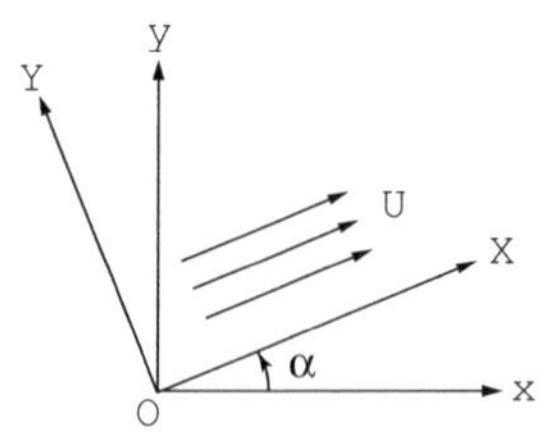

Fig. 2.20 Uniform flow parallel to the x-axis

As already discussed, when $n = 1$, the flow is uniform. Substituting $z = r\exp(i\theta)$ in (2.109), we obtain

$$f(z) = A\,r^n\,e^{in\theta} = A\,r^n(\cos n\theta + i\,\sin n\theta)\,, \tag{2.110}$$

which can be separated into real and complex parts:

$$\Phi = A\,r^n\cos n\theta\,, \quad \Psi = A\,r^n\sin n\theta\,. \tag{2.111}$$

$\Psi =$ constant represents the streamlines as shown in (2.81); thus, setting $\Psi = C$ $(C > 0)$ in (2.111) gives

$$r = \left(\frac{C}{A}\right)^{\frac{1}{n}}\frac{1}{(\sin n\theta)^{1/n}}\,. \tag{2.112}$$

The special case of $C = 0$ is an asymptote of these streamlines. As $\sin n\theta = 0$, from (2.111), we have

$$\theta = k\,\frac{\pi}{n}\,, \quad (k = 0,\ \pm 1,\ \pm 2, \ldots)\,. \tag{2.113}$$

These terms describe radial lines passing through the origin, obtained by successively rotating the positive component of the x-axis $(k = 0)$ through π/n in the positive and negative directions. A group of straight lines can be regarded as a fixed wall.

From (2.109), the complex velocity is given as follows:

$$\frac{df}{dz} = A\,n z^{n-1} = A\,n\,r^{n-1}\,e^{i(n-1)\theta}\,. \tag{2.114}$$

Thus, the magnitude of velocity is

$$q = \left|\frac{df}{dz}\right| = A\,n\,r^{n-1}\,. \tag{2.115}$$

If $n > 1$ in (2.113), the angle encountered by the flow is less than π, as shown in Fig. 2.21, and the origin $(r = 0)$ is a stagnation point. However, at infinity $(r \to \infty)$, the flow velocity is infinitely large. On the other hand, if $1/2 < n < 1$, the flow passes around a convex corner (see Fig. 2.22) and its velocity at the origin is infinite. According to Bernoulli's theorem (2.18), an infinitely large flow velocity indicates that the pressure is infinitely negative $(p = -\infty)$, clearly contradicting real phenomena. Therefore, fluid flows around convex angles cannot be assumed as inviscid and incompressible.

The special case where $n = 1/2$ indicates flow along a plate of semi-infinite length extending from the origin in the x-axis direction. Flow around a plate will be discussed in Sect. 2.11.

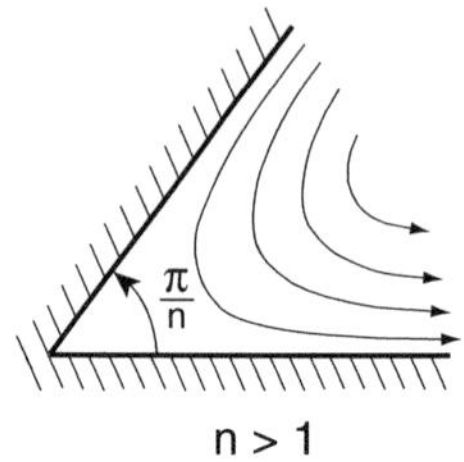

Fig. 2.21 Flow around a corner formed by two walls ($n > 1$)

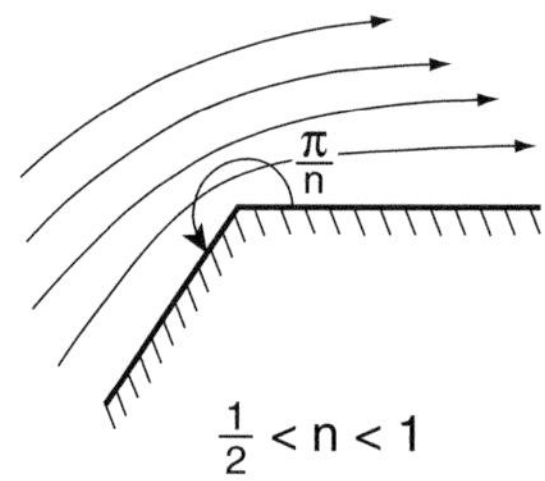

Fig. 2.22 Flow going at a convex angle ($1/2 < n < 1$)

2.9.3 Sources and Sinks

Consider the following

$$f(z) = m \log z \qquad (m\text{: real number}). \tag{2.116}$$

Setting $z = r\exp(i\theta)$ in (2.116) gives

$$f(z) = m \log(r\, e^{i\,\theta}) = m(\log r + i\,\theta)\,.$$

Splitting this expression into real and complex components, we obtain

$$\Phi = m \log r\,, \quad \Psi = m\,\theta\,. \tag{2.117}$$

Since $\Psi =$ constant represents a streamline, setting $\theta =$ constant specifies a group of lines radiating from the origin.

Thus, the flow velocity in the radial direction is given by

$$u_r = \frac{\partial \Phi}{\partial r} = \frac{\partial \Psi}{r \partial \theta} = \frac{m}{r}\,, \tag{2.118}$$

from which we see that if $m > 0$, then $u_r > 0$ (see Fig. 2.23). Therefore, similar to the three-dimensional problem, the flow when $m > 0$ is called a flow by a two-dimensional source of the strength m. If $m < 0$, the flow of strength m is drawn into a sink.

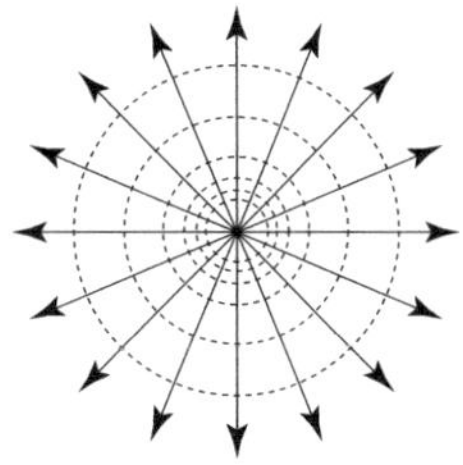

Fig. 2.23 Streamlines emitted by a source

The change in Ψ throughout a single cycle about the origin calculated from (2.105) and (2.106) is

$$Q(C) = \left[\Psi\right]_C = m\left[\theta\right]_C = 2\,\pi\,m\,. \tag{2.119}$$

In other words, the flow amount Q is a 2π multiple of the strength of the source[6] m.

Therefore, the velocity potential due to a two-dimensional source, given by (2.117), is

$$\Phi = m\log r = \frac{Q}{2\pi}\log r = -\frac{Q}{2\pi}\log\left(\frac{1}{r}\right). \tag{2.120}$$

It should be noted that (2.36) is the corresponding three-dimensional representation of (2.120). In other words, setting $Q = 1$ in (2.120) yields the basic solution of the two-dimensional Laplace equation (this argument was used in the explanation of (2.88)).

Exercise 2.9 Assume that the x-axis is a rigid wall surface (floor). Consider a two-dimensional source (strength is m) above the origin O at a distance a, and an uniform flow passing through this system from right to left at a speed U (Fig. 2.24). In this situation,

1. Obtain the complex velocity potential.
2. Obtain the pressure distribution along the x-axis, and calculate the point where the pressure will be the lowest.

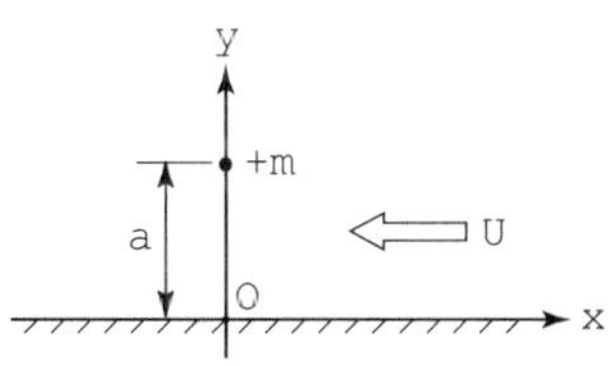

Fig. 2.24 Source and uniform flow on a rigid wall surface

[6]Similar to the three-dimensional case, Q is a measure of the strength of the source.

2.9.4 Vortex Filaments

Consider the following:

$$f(z) = -i\kappa \log z \qquad \text{(where } \kappa \text{ is a real number).} \tag{2.121}$$

Similar to (2.117), we have

$$f(z) = -i\,\kappa(\log r + i\theta)\,.$$

Therefore,

$$\Phi = \kappa\theta\,, \quad \Psi = -\kappa \log r\,. \tag{2.122}$$

As Ψ = constant is defined by r = const, the streamlines of this flow form a group of concentric circles centered at the origin (Fig. 2.25).

The circumferential component of the velocity is given by (2.122), (2.85), and (2.98) as

$$u_\theta = \frac{\partial \Phi}{r\,\partial\theta} = -\frac{\partial \Psi}{\partial r} = \frac{\kappa}{r} = \frac{\Gamma}{2\pi r}\,. \tag{2.123}$$

If $\kappa > 0$, the flow is counterclockwise (see Fig. 2.24). Moreover, (2.123) is identical to (2.78) calculated from the Biot–Savart principle.

Performing a contour integration along an arbitrary closed curve C enclosing the origin, the circulation along C is determined from (2.105) and (2.106) as follows:

$$\Gamma(C) = \left[\Phi\right]_C = \kappa\left[\theta\right]_C = 2\,\pi\,\kappa\,. \tag{2.124}$$

Equation (2.124) describes the flow when the vorticity is focused at the origin (a vortex filament). The stream function described by (2.122) and (2.124) is consistent with (2.88), previously obtained by solving the Poisson equation.

Exercise 2.10 Figure 2.26 illustrates a vortex filament with a counterclockwise circulation Γ at distance a above the origin O in place of an outlet. The conditions are similar to those in Exercise 2.9. In this situation,

1. Obtain the complex velocity potential.

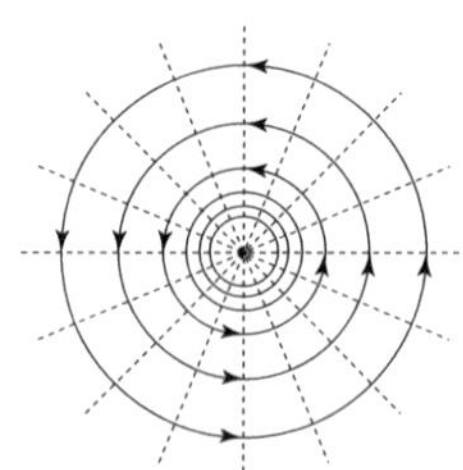

Fig. 2.25 Streamlines induced by a vortex filament

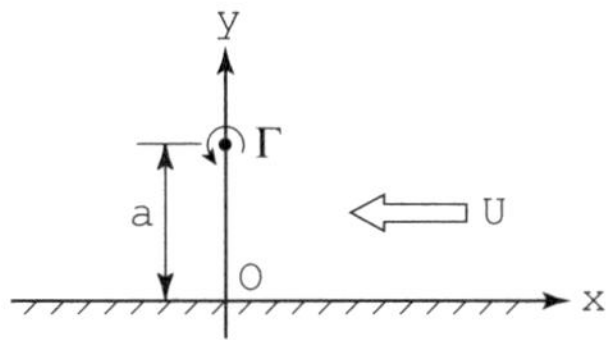

Fig. 2.26 Vortex filament and uniform flow on rigid surface

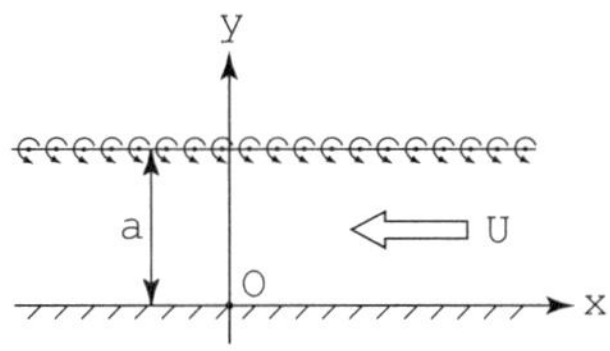

Fig. 2.27 Continuous vortex layer and uniform flow on a rigid surface

2. Obtain the condition (value of Γ) under which the vortex filament settles at the original position $(0, a)$.

Exercise 2.11 Consider an infinitely long, continuous vortex layer (strength per unit length is γ) parallel to and at a distance a above the x-axis (Fig. 2.27). Furthermore, assume that the x-axis is a rigid wall surface. If uniform flows passing through this system from right to left overlap, obtain the velocity distribution along the y-axis and the condition (value of γ) under which the velocity at the wall is 0.

2.9.5 Doublet

Similar to the three-dimensional problem, a doublet is formed when a source becomes infinitely close to a sink of identical strength m.

When a sink of strength m at the origin O is separated from a source of strength m by distance δs (Fig. 2.28), the complex velocity potential is given by

$$
\begin{aligned}
f(z) &= m \log \frac{z - z_0}{z} = m \log\left(1 - \frac{z_0}{z}\right) \\
&= -m \sum_{n=1}^{\infty} \frac{1}{n}\left(\frac{z_0}{z}\right)^n .
\end{aligned}
\tag{2.125}
$$

Here $z_0 = \delta s \exp(i\,\alpha)$, and $\delta s \to 0$.

Instead, if we consider the limiting case $m\,\delta s \to \overline{\mu}$, the velocity potential becomes

$$
f(z) = -\overline{\mu}\,\frac{e^{i\,\alpha}}{z} \tag{2.126}
$$

$$
= -\mu \frac{d}{dz}\big(\log z\big), \tag{2.127}
$$

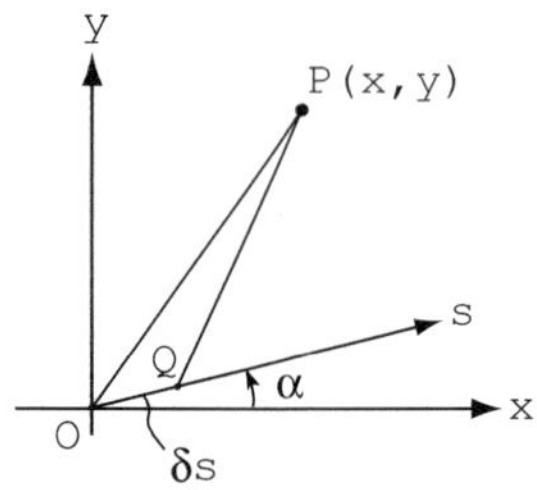

Fig. 2.28 Limiting value of the two-dimensional source and sink

where $\mu = \overline{\mu}\, e^{i\,\alpha}$. Equation (2.127) is equivalent to (2.57) in the three-dimensional problem. That is, the complex velocity potential of a doublet is obtained by differentiating the complex velocity potential of a unit strength source with respect to z and reversing its sign.

Again, analogous to the three-dimensional problem, differentiating (2.126) with respect to z yields the multipole complex velocity potential, which is the general solution to the two-dimensional Laplace equation. In other words, the solutions corresponding to (2.51) or (2.48) in the three-dimensional problem are given by

$$f(z) = k_0 \log z + \sum_{n=1}^{\infty} \frac{k_n}{z^n} \tag{2.128}$$

$$= k_0 \log z + \sum_{n=1}^{\infty} k_n \frac{(-1)^{n+1}}{(n-1)!} \frac{d^n}{dz^n} (\log z) \tag{2.129}$$

The negative power series term in (2.128) is infinitely holomorphic and is known as Laurent's series in complex function theory. (A positive power series represents flow around the corner of two walls as shown in (2.109) and infinite flow velocity at infinite distance.)

Note 2.9 Let us solve the two-dimensional Laplace equation by the method of separation of variables. The Laplace equation in polar coordinates (r, θ) is

$$\nabla^2 \Phi = \frac{1}{r}\frac{\partial}{\partial r}\left(r\frac{\partial \Phi}{\partial r}\right) + \frac{1}{r^2}\frac{\partial^2 \Phi}{\partial \theta^2} = 0\,. \tag{2.130}$$

When assuming $\Phi = R(r)\Theta(\theta)$ and substituting into (2.130), we can rearrange as follows:

$$r\frac{d}{dr}\left(r\frac{dR}{dr}\right)\frac{1}{R} = -\frac{d^2\Theta}{d\theta^2}\frac{1}{\Theta} = n^2 \text{ (a constant)}\,. \tag{2.131}$$

Thus, these can be rewritten as two ordinary differential equations

$$r\frac{d}{dr}\left(r\frac{dR}{dr}\right) - n^2 R = 0\,, \tag{2.132}$$

$$\frac{d^2\Theta}{d\theta^2} + n^2\Theta = 0\,. \tag{2.133}$$

Separately considering $n = 0$ and $n \neq 0$, the solution for which $\nabla\Phi \to 0$ at infinity is given by

$$\Phi = A_0 \log r + B_0\theta + \sum_{n=1}^{\infty} \frac{1}{r^n}\left(A_n \cos n\theta - B_n \sin n\theta \right). \tag{2.134}$$

This is the same expression as the one obtained by substituting $z = r\exp(i\theta)$ and $k_n = A_n - iB_n$ into the Laurent series (2.128) and considering the real part ($\Phi = \mathrm{Re} f$).

Exercise 2.12 Similar to the three-dimensional problem, the flow around a two-dimensional semi-infinite body can be obtained by overlapping a uniform flow with a source at origin O. Obtain the complex velocity potential of such a flow. Calculate the distance a between the origin and stagnation point and the height b of a semi-infinite body placed infinitely downstream (see Fig. 2.10). Compare with the results of the three-dimensional problems and discuss the three-dimensional effect.

2.9.6 *Flow Around a Cylinder*

Equation (2.128) is the generalized two-dimensional complex velocity potential for which $\nabla\Phi \to 0$ at infinity. We now use this to determine uniform flows around a cylinder.

Suppose that a cylinder of radius a is placed in a uniform flow along the x-axis in the negative direction (Fig. 2.29). Overlapping the velocity potential of a uniform flow on (2.128), we obtain the complex velocity potential:

$$f(z) = -Uz + k_0 \log z + \sum_{n=1}^{\infty} \frac{k_n}{z^n}\,. \tag{2.135}$$

By (2.26), the body surface condition is $V_n = 0$. This condition is equivalent to $\Psi = \mathrm{Im} f = \text{const}$ when setting $z = a\exp(i\theta)$, that is, the cylinder surface is a streamline, we thus have

$$\Psi = -Ua\sin\theta + \kappa \log a + m\,\theta - \sum_{n=1}^{\infty} \frac{1}{a^n}\left(B_n \cos n\theta + A_n \sin n\theta \right), \tag{2.136}$$

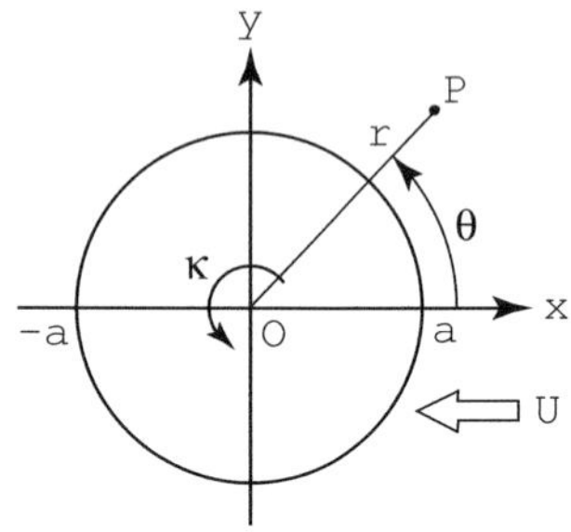

Fig. 2.29 Two-dimensional flow around a cylinder

where $k_0 = m - i\kappa$ and setting $k_n = A_n - i B_n$ have been substituted. The condition of $\Psi =$ constant gives the following conditions:

$$m = 0, \quad A_1 = -Ua^2, \quad B_1 = 0, \quad k_n = 0\ (n \geq 2) \tag{2.137}$$

Substituting these conditions into (2.135), the complex velocity potential is given by

$$f(z) = -U\left(z + \frac{a^2}{z}\right) - i\,\kappa \log z\,. \tag{2.138}$$

Same as in the three-dimensional problem, it indicates the flow introduced by a cylinder because of a doublet of strength Ua^2 whose axis opposes the flow direction (i.e., whose axis is oriented along the positive x-axis). This solution differs from that of the three-dimensional problem in the presence of the counterclockwise vortex filament $-i\kappa \log z$ at the origin. It should be noted that, the strength κ of this vortex filament is unknown at this moment. In other words, to determine κ, aside from the dynamic surface condition of a body such as normal velocity $= 0$, another condition is necessary. This extra condition will likely be imposed by viscous effects, such as a specific stagnation point on the body surface. Kelvin's circulation theorem states that the circulation around a body is related to the initial vortices shed from boundary layers close to the surface of the body. The magnitude of this circulation is presumably determined by viscous effects inside the boundary layer. Later, we will demonstrate that the circulation around a two-dimensional wing moves the stagnation point toward the back edge of the wing, a condition known as **Kutta's condition**.

We now determine the flow velocity at the cylinder surface. Substituting $z = a\exp(i\theta)$ into (2.138), we obtain

$$\Phi = -2Ua\cos\theta + \kappa\,\theta\,, \tag{2.139}$$

$$u_\theta = \frac{\partial \Phi}{a\,\partial\theta} = 2U\sin\theta + \frac{\kappa}{a}\,. \tag{2.140}$$

In the absence of circulation ($\kappa = 0$), the flow velocity is given by

$$q = |u_\theta| = 2U|\sin\theta|\,, \tag{2.141}$$

whose maximum value is $q_{\max} = 2U$. On a three-dimensional spherical surface, the maximum flow velocity is 1.5 times the uniform velocity U (see (2.65)). As the flow is restricted to the x-y plane in two-dimensional problems, the increase in speed around the body is greater than that in a three-dimensional flow field. To obtain the pressure distribution, we substitute (2.140) into Bernoulli's pressure equation to get

$$\begin{aligned} p &= p_0 - \frac{1}{2}\rho q^2 = p_0 - \frac{1}{2}\rho u_\theta^2 \\ &= \rho\left(U^2\cos 2\theta - 2U\frac{\kappa}{a}\sin\theta\right) + \text{const}\,. \end{aligned} \tag{2.142}$$

The fluid force acting on the cylinder is obtained by integrating the pressure. As the positive direction of the normal is outward from the cylinder, it is given by $n_x + in_y = \exp(i\theta)$, and we obtain

$$\left.\begin{aligned} X &= -\int_0^{2\pi} p\cos\theta\, ad\theta = 0\,, \\ Y &= -\int_0^{2\pi} p\sin\theta\, ad\theta = \rho U\Gamma\,, \end{aligned}\right\} \tag{2.143}$$

where $\Gamma = 2\pi\kappa$ is the (counterclockwise) circulation around the cylinder.

The force components acting parallel and perpendicular to the flow are called **drag** and **lift**, respectively. According to (2.143), the drag is 0, which contradicts observed phenomena. This situation, called **d'Alembert's paradox**, arises because viscosity generates a large wake behind the cylinder, whereas the streamlines of inviscid flows are right-left symmetric; hence, the pressure is also right-left symmetric.

On the other hand, the lift force according to the Kutta–Joukowski theorem (introduced in (2.16)) is $\rho U\Gamma$. If a counterclockwise circulation coexists with uniform right-left flow, the flow is accelerated more on the upper side of the cylinder than on the lower side, and the pressure is reduced by Bernoulli's theorem. This pressure difference generates lift. The same principle, called the **Magnus effect**, operates when rotation is added to a baseball curve or drop throw.

Exercise 2.13 Obtain the complex velocity potential of a uniform flow around a cylinder of radius a, tilted by angle α above against the x-axis (Fig. 2.30). Assume that the strength of the clockwise vortex filament is κ. Consider two stagnation points (S_1 and S_2) on the surface of the cylinder and show what happens when κ is determined under the condition that point S_2 matches the back end (point B).

Exercise 2.14 A cylinder of radius a is intercepted by a uniform flow in the positive x-axis direction (flow velocity U). Show that the point at which the flow velocity is cU lies on the curve $(c^2 - 1)r^4 + 2a^2r^2\cos 2\theta = a^4$. Moreover, show that the point at which the flow velocity forms an angle α with the x-axis lies on the curve $r^2\sin\alpha = a^2\sin(\alpha + 2\theta)$.

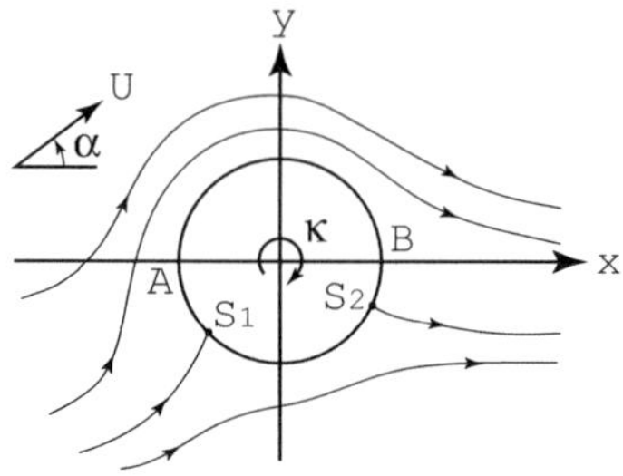

Fig. 2.30 Calculation of circulation flow around a cylinder

2.10 Forces Acting on a Body

In the previous section, (2.143) was solved for flows around a cylinder. However, it can be shown that (2.143) holds for more general, arbitrary bodies in two dimensions. The force acting on a body in a flow $\boldsymbol{F} = (X, Y)$ is obtained by integrating the pressure along the body surface; hence,

$$\boldsymbol{F} = -\oint_{C_B} p\,\boldsymbol{n}\,ds\,. \tag{2.144}$$

Here C_B is a closed curve representing the body surface, and the positive direction of the normal points away from the body. The pressure p is

$$p = -\rho\left(\frac{\partial \Phi}{\partial t} + \frac{1}{2}q^2\right). \tag{2.145}$$

As shown in (2.82), the boundary surface satisfies

$$dz = dx + i\,dy = i\,ds\,(n_x + i\,n_y)\,,$$

i.e.,

$$(n_x + i\,n_y)\,ds = -i\,dz\,. \tag{2.146}$$

From the above expressions, we obtain the following complex equation:

$$X + i\,Y = -i\,\frac{\rho}{2}\oint_{C_B}\left\{\frac{\partial}{\partial t}(f + f^*) + \left|\frac{df}{dz}\right|^2\right\}dz\,, \tag{2.147}$$

where f^* is the complex conjugate of f. The moment acting on a body can be calculated similarly. Denoting the moment around the origin as M, from (2.146), $n_x\,ds = dy$, $n_y\,ds = -dx$; hence,

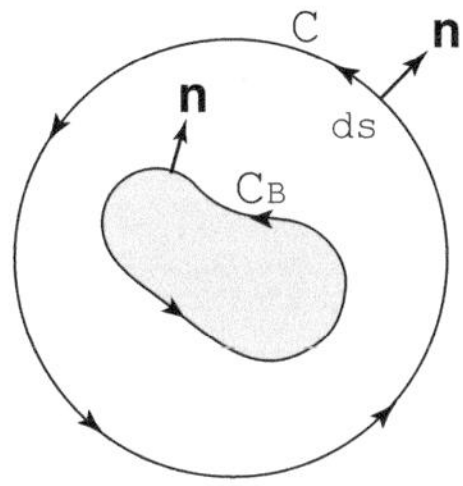

Fig. 2.31 Replacement of the integral path

$$M = -\oint_{C_B} p\,(x\,n_y - y\,n_x\,)\,ds$$
$$= \oint_{C_B} p\,(\,x\,dx + y\,dy\,) = \frac{1}{2}\oint_{C_B} p\,d(zz^*)\,. \tag{2.148}$$

Substituting (2.145) into (2.148), we obtain

$$M = -\frac{\rho}{4}\oint_{C_B}\left\{\frac{\partial}{\partial t}(\,f + f^*) + \left|\frac{df}{dz}\right|^2\right\} d(zz^*)\,. \tag{2.149}$$

2.10.1 The Case of Steady Motion

In steady flows, Ψ = constant at the surface of the body, and we have $df = d\Phi = df^*$. Additionally,

$$\left|\frac{df}{dz}\right|^2 dz^* = \frac{df}{dz}\frac{df^*}{dz^*}\,dz^* = \frac{df}{dz}\frac{df}{dz}\,dz = \left(\frac{df}{dz}\right)^2 dz \tag{2.150}$$

Thus, the complex conjugate of (2.147) is

$$X - i\,Y = i\,\frac{\rho}{2}\oint_{C_B}\left(\frac{df}{dz}\right)^2 dz\,. \tag{2.151}$$

Furthermore, since the derivative function df/dz is holomorphic in the fluid region, the integral path on the body surface C_B can be replaced by an arbitrary closed curve C surrounding the body (see Fig. 2.31). Thus, we have

$$X - i\,Y = i\,\frac{\rho}{2}\oint_{C}\left(\frac{df}{dz}\right)^2 dz \tag{2.152}$$

Equation (2.152) is called **Blasius' first formula**.

By an arrangement similar to that of (2.150), we obtain the following expression for the moment:

$$\left|\frac{df}{dz}\right|^2 d(zz^*) = \frac{df}{dz}\frac{df^*}{dz^*}(z\,dz^* + z^*dz) = 2\,\mathrm{Re}\left\{\left(\frac{df}{dz}\right)^2 z\,dz\right\}. \tag{2.153}$$

Moreover, $(df/dz)^2 z$ is holomorphic with respect to z; hence, the line integral along C_B can be replaced by an integral along an arbitrary closed curve C. Therefore,

$$M = -\frac{\rho}{2}\,\mathrm{Re}\oint_C \left(\frac{df}{dz}\right)^2 z\,dz. \tag{2.154}$$

Equation (2.154) is called Blasius' second formula.

In determining the contour integrals in (2.152) and (2.154), it is convenient to introduce the following Laurent series. In other words, we can use the following formula

$$\left.\begin{aligned} &\text{When } F(z) = \sum_{n=-\infty}^{\infty} a_n(z - z_0)^n\,, \\ &\oint_C F(z)\,dz = 2\pi\, i\, a_{-1}\,. \end{aligned}\right\} \tag{2.155}$$

Therefore, we need to only retain the $1/z$ terms in the integrands of (2.152) and (2.154).

Extracting the $1/z$ terms contributing to (2.152) from the general solution (2.135), we can write

$$\left(\frac{df}{dz}\right)^2 = -2U\,\frac{(m - i\,\kappa)}{z} + \cdots, \tag{2.156}$$

which, together with (2.152), (2.155), and (2.156), gives

$$\begin{aligned} X - i\,Y &= -i\,\frac{\rho}{2}\,2U\,(m - i\,\kappa)\,2\pi\, i \\ &= \rho U Q - i\,\rho U \Gamma\,. \end{aligned} \tag{2.157}$$

Here $Q = 2\pi m$ and $\Gamma = 2\pi\kappa$.

This result is independent of the body shape, as evidenced from the integral path in (2.152). Moreover, as observed in the Rankine ovoid, a body with closed streamlines implies that the sum of the sources and sinks is 0, that is, $Q = 0$. Under these circumstances, we encounter d'Alembert's paradox, that is, the drag acting on a body with ordinary rigid boundary surfaces is always zero in a perfect fluid. Of course, if the body itself ejects fluid, it exerts a thrust of $X = \rho U Q$.

2.10.2 *The Case of Non-steady Motion*

Equation (2.157) provides a general solution to flows intercepted by steadily moving bodies. In this section, we consider a cylinder of radius a moving non-steadily through

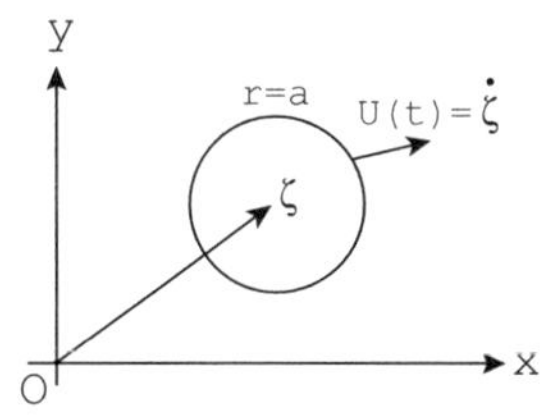

Fig. 2.32 Non-steady motion around a cylinder. $U(t)$ is time dependent

a static fluid with velocity $U(t)$. The coordinates at the center of the cylinder are given by $z = \zeta$ in a spatially fixed coordinate system (O, x, y).

In this case, the complex velocity potential is given by (2.138) with the uniform flow component being removed; that is,

$$f(z) = -\frac{Ua^2}{z-\zeta} - i\kappa \log(z-\zeta)\,. \tag{2.158}$$

The first term Ua^2, in terms of fluid dynamics, is the strength of a doublet. As in (2.127), here we regard U in (2.158) as a complex number specifying the direction of progression.

The velocity of this cylinder is given by

$$\frac{d\zeta}{dt} = \dot{\zeta} = U(t) \tag{2.159}$$

(see Fig. 2.32).

Now, from (2.158), we obtain

$$\frac{\partial f(z)}{\partial t} = -\frac{a^2\dot{U}}{z-\zeta} - \frac{a^2U^2}{(z-\zeta)^2} + i\kappa\frac{U}{z-\zeta}\,, \tag{2.160}$$

$$\frac{df(z)}{dz} = \frac{Ua^2}{(z-\zeta)^2} - \frac{i\kappa}{z-\zeta}\,. \tag{2.161}$$

Equations (2.160) and (2.161) can be used to calculate (2.147). The contour integral of (2.147) assumes the center of the cylinder $z = \zeta$ as the origin. Moreover, z and its complex conjugate z^* are considered as independent variables. Thus, by the residue theorem of (2.155), we have

$$\oint_{C_B} \frac{\partial}{\partial t}\{f + f^*\}\, d(z-\zeta) = \left(-a^2\dot{U} + i\kappa U\right) 2\pi i\,, \tag{2.162}$$

$$\oint_{C_B} \left|\frac{df}{dz}\right|^2 d(z-\zeta) = \oint_{C_B} \left(\frac{Ua^2}{z^2} - \frac{i\kappa}{z}\right)\left(\frac{Ua^2}{z^{*2}} + \frac{i\kappa}{z^*}\right) dz$$
$$= \oint_{C_B} \left(\frac{Ua^2}{z^2} - \frac{i\kappa}{z}\right)\left(\frac{U}{a^2}z^2 + \frac{i\kappa}{a^2}z\right) dz$$
$$= i\kappa U\, 2\pi i \tag{2.163}$$

where we use $zz^* = a^2$. Substituting these results into (2.147) and $2\pi\kappa = \Gamma$ as the result, we obtain

$$X + iY = i\frac{\rho}{2}\left\{i\,2\pi a^2\dot{U} + 2U\Gamma\right\} = -\rho\pi a^2\dot{U} + i\rho U\Gamma\,. \tag{2.164}$$

Please note that this equation considers U as a complex number. Thus, it is not necessarily true that the first and second items on the right side correspond to the X and Y of fluid force, respectively.

The second term on the right side, $\rho U\Gamma$, is the same as that in (2.143) in its shape, and is the lift in the Kutta–Joukowski theorem. It is indicated that this lift moves in a vertical direction (iU) relative to the direction of the movement U.

On the other hand, the first term on the right side is inertial flow strength proportional to acceleration $\dot{U}$. The fact that its magnitude is negative indicates that it is working as inertial resistance in the direction opposite to that of the movement. This magnitude is called **added mass**, and for a cylinder, it is expressed as follows:

$$m = \rho\pi a^2\,. \tag{2.165}$$

In other words, it is equal to the mass of the fluid displaced by the cylinder.

Generally, when a body moves through fluid, it must displace the fluid by a certain amount, which increases the inertia of the body. The inertial increase is the added mass. Added mass can be defined as the amount of fluid particles accelerated by the motion of the body. However, the range of fluid particles accelerated by the motion of the body must extend infinitely. Therefore, to calculate the added mass, all fluid particles accelerated by the motion should be proportionally weighted and integrated over the displaced mass.

For an arbitrarily shaped body, the number of accelerated fluid particles, and hence the added mass, will depend on the direction of motion of the body. This aspect of fluid dynamics starkly contrasts with the inertia of a body in space, which is unrelated to its direction of motion. To improve our understanding of added mass, let us consider the kinetic energy of irrotational flow. To generalize the discussion, we discuss the three-dimensional problem. The kinetic energy within a region V surrounded by a closed surface S is given by

$$E = \frac{1}{2}\iiint_V \rho q^2\, dV = \frac{\rho}{2}\iiint_V \nabla\Phi\cdot\nabla\Phi\, dV\,. \tag{2.166}$$

Since Φ satisfies the Laplace equation, by applying Gauss' theorem, we obtain

$$E = \frac{\rho}{2}\iiint_V \nabla\cdot(\Phi\nabla\Phi)\,dV = -\frac{\rho}{2}\iint_S \boldsymbol{n}\cdot(\Phi\nabla\Phi)\,dS = -\frac{\rho}{2}\iint_S \Phi\frac{\partial\Phi}{\partial n}\,dS \tag{2.167}$$

The negative sign of the surface integral indicates the positive direction of the inward normal of the fluid from the boundary surface.

Equation (2.167) shows that kinetic energy can be solely calculated from Φ at the boundary surface and normal flow velocity $\partial\Phi/\partial n$. From the boundary condition (2.26), $\partial\Phi/\partial n$ on the surface of a body in non-steady motion equals normal velocity V_n and thus can be expressed as follows:

$$\frac{\partial\Phi}{\partial n} = V_n = \boldsymbol{V}(t)\cdot\boldsymbol{n}(\boldsymbol{r})\,. \tag{2.168}$$

Here the velocity potential Φ can be expressed as an overlap of components linearly separated in the direction of motion, as

$$\Phi(\boldsymbol{r},\, t) = \boldsymbol{V}(t)\cdot\boldsymbol{\phi}(\boldsymbol{r}) = V_j(t)\,\phi_j(\boldsymbol{r})\,, \tag{2.169}$$

where $\phi_j(\boldsymbol{r})$ is the velocity potential induced by a body moving at unit velocity in the jth direction j ($j = 1 \sim 6$). The function $\phi_j(\boldsymbol{r})$ depends only on the body's shape. For example, for a sphere of radius a moving in the x-axis direction,

$$\phi_1 = -\frac{a^3}{2}\frac{\cos\theta}{r^2} = -\frac{a^3}{2}\frac{x}{r^3}\,, \tag{2.170}$$

as can be understood from (2.61). From (2.158), a cylinder moving in the same direction induces a potential

$$\phi_1 = -a^2\frac{\cos\theta}{r} = -a^2\frac{x}{r^2}\,. \tag{2.171}$$

Note that (r, θ) of (2.170) is expressed in spherical coordinates, whereas (r, θ) of (2.171) is expressed in two-dimensional polar coordinates.

Substituting (2.169) into (2.167), we obtain

$$E = \frac{1}{2}\left[-\rho\iint_S \phi_i\frac{\partial\phi_j}{\partial n}\,dS\right]V_i^2 \equiv \frac{1}{2}m_{jj}V_i^2\,. \tag{2.172}$$

However, the kinetic energy of the fluid can be expressed as follows:

$$m_{jj} = -\rho\iint_S \phi_j\frac{\partial\phi_j}{\partial n}\,dS = -\rho\iint_S \phi_j n_j\,dS\,. \tag{2.173}$$

Equation (2.173) gives the energy of the fluid particles accelerated by the non-steady motion of the body. Comparing this expression to the kinetic energy of the material at some point in space, m_{jj} of (2.173) can be defined as the added mass in direction j of the motion. In the case of a cylinder, from (2.171) we have

$$m_{11} = \rho a \int_0^{2\pi} \cos\theta \, a \cos\theta \, d\theta = \rho a^2 \int_0^{2\pi} \cos^2\theta \, d\theta = \rho\pi a^2 , \tag{2.174}$$

which equals the result from (2.165).

On the other hand, using (2.170), the mass added by a moving sphere is given as

$$m_{11} = \frac{\rho a}{2} \iint_S \cos^2\theta \, dS = \frac{\rho a}{2} \int_0^{2\pi} d\varphi \int_0^{\pi} \cos^2\theta \, a^2 \sin\theta \, d\theta = \rho\frac{2}{3}\pi a^3 \tag{2.175}$$

In other words, in terms of the proportion of the mass of the fluid replaced by the body of interest (coefficient of added mass), it is 1.0 for a cylinder and 0.5 for a sphere. As seen, flows disturbed by bodies in a three-dimensional system will spread in three orthogonal directions; hence, the coefficient of the added mass is generally smaller than that for two-dimensional bodies.

2.11 Flow Around a Flat Plate

If a mapping function is used, flows around a two-dimensional flat plate can be analyzed relatively easily using our previously gained knowledge of flow fields around a cylinder. The analytical method is described below.

Consider a uniform flow of velocity U parallel to the x-axis on the z-plane ($z = x + iy$). The complex velocity potential of this flow is

$$f(z) = Uz . \tag{2.176}$$

This situation is equivalent to the plate of zero thickness aligned parallel to the x-axis.

Now consider a cylinder of radius a placed in the uniform flow with the velocity U oriented parallel to the ξ-axis in the ζ-plane ($\zeta = \xi + i\eta$) separate from the z-plane (in which circulation is absent). Using (2.138), the complex velocity potential can be expressed as follows:

$$f(\zeta) = U\left(\zeta + \frac{a^2}{\zeta}\right) . \tag{2.177}$$

Defining

$$z = \zeta + \frac{a^2}{\zeta} \qquad (a > 0) \tag{2.178}$$

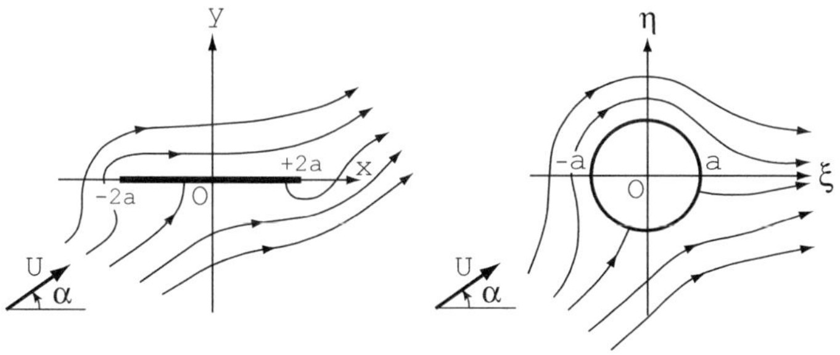

Fig. 2.33 Flow around a flat plate (*left*) and flow around a cylinder (*right*) are related by the Joukowski transformation (2.178)

as an analytical mapping function relating the z-plane to the ζ-plane, we can establish the relationship between the flow around a cylinder and the flow around a flat plate through a mapping function. Equation (2.178) is called the **Joukowski transformation**. Setting $\zeta = a\exp(i\theta)$ in (2.178), we get

$$z = x + iy = 2a\cos\theta\,, \tag{2.179}$$

which maps the cylindrical surface in the z-plane to a linear section (flat plate) of length 4a on the x-axis.

On the other hand, we can express ζ in (2.178) in terms of z:

$$\zeta = \frac{1}{2}\left(z + \sqrt{z^2 - 4a^2}\right). \tag{2.180}$$

We have only selected the positive sign because the outside region of the cylinder is considered to correspond to the z-plane (in other words, $\zeta \to \infty$ and $z \to \infty$ correspond to one another, and in the long distance limit, $\zeta \simeq z$).

We now consider a uniform flow tilted by an angle α from the real axis (ξ-axis) on the ζ-plane (Fig. 2.33). Rotating the coordinate axis does not change the fact that it is a flow around a cylinder. Thus, according to (2.180), we can replace ζ by $\zeta\, e^{-i\alpha}$, and from (2.177),

$$f(\zeta) = U\left(\zeta\, e^{-i\alpha} + \frac{a^2}{\zeta}\, e^{i\alpha}\right). \tag{2.181}$$

Inserting (2.180) into (2.181), we immediately obtain the flow when a uniform flow in the direction that forms an angle α with the x-axis on a flat plate is placed on the x-axis:

$$\begin{aligned} f(z) &= U\left(\frac{z + \sqrt{z^2 - 4a^2}}{2}\, e^{-i\alpha} + \frac{z - \sqrt{z^2 - 4a^2}}{2}\, e^{i\alpha}\right) \\ &= U\left(z\cos\alpha - i\sqrt{z^2 - 4a^2}\,\sin\alpha\right). \end{aligned} \tag{2.182}$$

As a special case of (2.182), if $\alpha = -\pi/2$, a uniform flow of velocity U from the positive y-direction is perpendicular to the flat plate. In this case, the flow is described by a complex velocity potential, given by

$$f(z) = iU\sqrt{z^2 - 4a^2}\,. \tag{2.183}$$

The second term on the right-hand side of (2.182) represents the flow introduced by the flat plate. Around the edge of the flat plate ($z = \pm 2a$), the flow becomes equivalent to that of $n = 1/2$ in (2.109), which describes flow around a corner.

The velocity of flow around the flat plate can be directly determined by differentiating (2.182) with respect to z. Alternatively, we can consider ζ as a parameter, and using the calculation on the ζ-plane and a mapping function gives

$$\frac{df}{dz} = \frac{df}{d\zeta}\frac{d\zeta}{dz} = \frac{df}{d\zeta}\Big/\frac{dz}{d\zeta}\,. \tag{2.184}$$

In any case, when discussing flows around a convex angle, we established that the flow velocity $q = \infty$ at $z = \pm 2a$ (i.e., at the ends of the flat plate).

Now consider that the flow described by (2.181) is overlapped with a clockwise circulation $\Gamma\ (= 2\pi\kappa)$, representing a flow around a cylinder (This flow is considered in Exercise 2.13). This complex velocity potential is

$$f = U\left(\zeta\, e^{-i\alpha} + \frac{a^2}{\zeta}\, e^{i\alpha}\right) + i\kappa\log\zeta\,. \tag{2.185}$$

Using (2.184), the flow velocity around the flat plate is given by

$$\frac{df}{dz} = \frac{U\left(e^{-i\alpha} - a^2\, e^{i\alpha}/\zeta^2\right) + i\kappa/\zeta}{1 - a^2/\zeta^2}\,. \tag{2.186}$$

If we impose a limit on the flow velocity at $\zeta = a$, corresponding to the rear end of the flat plate ($z = +2a$), we can set the numerator in (2.186) to 0:

$$U\left(e^{-i\alpha} - e^{i\alpha}\right) + i\,\frac{\kappa}{a} = 0\,.$$

Therefore,

$$\kappa = 2a\,U\sin\alpha \quad (\Gamma = 4\pi a\,U\sin\alpha) \tag{2.187}$$

As noted in (2.138), the circulation could not be solely determined from the kinetic boundary conditions at the body surface. However, in deriving (2.187), we imposed an additional condition that the flow velocity is limited at the end of the flat plate. This condition, called **Kutta's condition**, enables the determination of the circulation.

The condition (2.187) can also be directly calculated on the z-plane after substituting (2.180) into (2.185). That is,

$$f(z) = U\left(z\cos\alpha - i\sqrt{z^2-4a^2}\,\sin\alpha\right) + i\kappa\log\frac{z+\sqrt{z^2-4a^2}}{2}\,, \tag{2.188}$$

$$\frac{df}{dz} = U\cos\alpha + \frac{i}{\sqrt{z^2-4a^2}}\left(\kappa - z\,U\sin\alpha\right). \tag{2.189}$$

It shows that (2.187) is the necessary condition for the flow to be limited at $z = +2a$.

Next, we consider the force acting on a flat plate. Equation (2.157) is applicable to an arbitrarily shaped body; hence, the lift on a flat plate is given by

$$L = \rho\,U\,\Gamma = 4\pi\rho\,a\,U^2\sin\alpha \tag{2.190}$$

and is oriented perpendicular to the uniform flow. This result is known as the Kutta–Joukowski theorem.

Let us obtain this result from calculation on the z-plane. Although the flow velocity and pressure need to be obtained for above and below the flat plate, note that $\sqrt{z^2-4a^2}$ is a two-valued function and equals $\pm i\sqrt{4a^2-x^2}$ on the top (positive) and bottom (negative) surfaces of the flat plate. Hence, from (2.189), the flow velocity and pressure are given by

$$u_\pm = U\cos\alpha \pm \frac{(\kappa - xU\sin\alpha)}{\sqrt{4a^2-x^2}}\,, \quad v = 0 \tag{2.191}$$

$$p_\pm = -\frac{1}{2}\rho\left[U^2\cos^2\alpha + \frac{(\kappa - xU\sin\alpha)^2}{4a^2-x^2} \pm 2U\cos\alpha\frac{(\kappa - xU\sin\alpha)}{\sqrt{4a^2-x^2}}\right] \tag{2.192}$$

where the subscripts + and − indicate the upper and lower sides of the flat plate, respectively. From (2.192), the fluid force acting in the y-axis direction becomes

$$\begin{aligned} Y &= -\oint_{S_B} p\,n_y\,ds = \int_{-2a}^{2a}(p_- - p_+)\,dx \\ &= 2\rho\,U\cos\alpha\int_{-2a}^{2a}\frac{(\kappa - xU\sin\alpha)}{\sqrt{4a^2-x^2}}\,dx \\ &= 2\pi\rho\,U\,\kappa\cos\alpha \\ &= 4\pi\rho\,a\,U^2\sin\alpha\,\cos\alpha \end{aligned} \tag{2.193}$$

Equation (2.193) is the $\cos\alpha$-component of the lift L; that is, it specifies the component of the lift in the direction of the y-axis. Moreover, the moment around the origin (counterclockwise) is calculated as

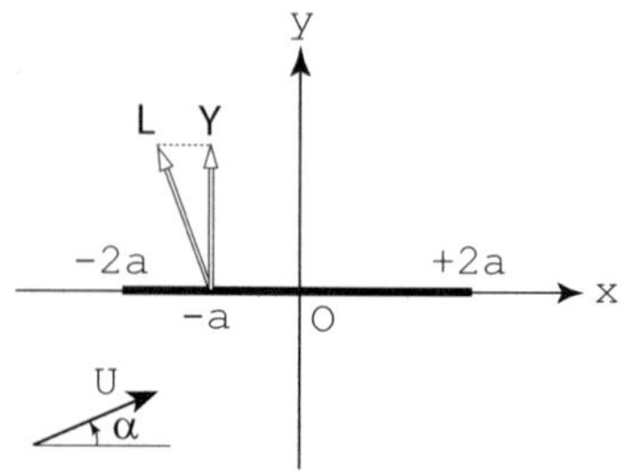

Fig. 2.34 Lift force acts at a point 1/4 unit distant from the front end of the flat plate

$$M = -\oint_{S_B} p\,x\,n_y\,ds = \int_{-2a}^{2a} (p_- - p_+)\,x\,dx$$
$$= 2\rho U \cos\alpha \int_{-2a}^{2a} \frac{(\kappa - xU\sin\alpha)}{\sqrt{4a^2 - x^2}}\,x\,dx = -4\pi\rho a^2 U^2 \sin\alpha\cos\alpha\,. \tag{2.194}$$

Therefore, the lift is applied at $M/Y = -a$, i.e., at $1/4$ of the plate's full length ($4a$) from the front end of the flat plate (see Fig. 2.34).

Note 2.10 Let us consider the force acting on the flat plate in the x-axis direction. Since the lift L acts perpendicular to the uniform flow, the x-axis component of the lift is given by

$$X = -L\sin\alpha = -4\pi\rho a\,U^2\sin^2\alpha \tag{2.195}$$

This result can be obtained by directly integrating the pressure on the flat plate. Although the integration on the flat plate is difficult, it can be considerably simplified by applying Blasius' first formula (2.152) to the distant flow field. First, from (2.189), we have

$$\left(\frac{df}{dz}\right)^2 = U^2\cos^2\alpha - \frac{(\kappa - zU\sin\alpha)^2}{z^2 - 4a^2} + 2iU\cos\alpha\frac{(\kappa - zU\sin\alpha)}{\sqrt{z^2 - 4a^2}}\,. \tag{2.196}$$

In the far field, we can make the following approximation:

$$\frac{1}{\sqrt{z^2 - 4a^2}} \sim \frac{1}{z}\left(1 + \frac{2a^2}{z^2}\right), \quad \frac{1}{z^2 - 4a^2} \sim \frac{1}{z^2}\left(1 + \frac{4a^2}{z^2}\right). \tag{2.197}$$

Removing the $1/z$ terms, we get

$$\left(\frac{df}{dz}\right)^2 \sim \frac{2\kappa U}{z}(\sin\alpha + i\cos\alpha) + \cdots\,. \tag{2.198}$$

Applying (2.144), we obtain

$$X - i\,Y = i\,\frac{\rho}{2}\,2\kappa U\,(\sin\alpha + i\,\cos\alpha)\,2\pi i = -4\pi\rho\,a\,U^2(\sin\alpha + i\,\cos\alpha)\,\sin\alpha\,. \tag{2.199}$$

From (2.199), it is easily seen that

$$\left.\begin{aligned} X &= -4\pi\rho\,a\,U^2\sin\alpha\sin\alpha = -L\,\sin\alpha \\ Y &= 4\pi\rho\,a\,U^2\sin\alpha\cos\alpha = L\,\cos\alpha \end{aligned}\right\}\,. \tag{2.200}$$

While on the topic, let us also calculate the moment around the origin using Blasius' second formula (2.154). From (2.196), removing the $1/z$ terms in $(df/dz)^2 z$, we obtain

$$\left(\frac{df}{dz}\right)^2 z \sim \frac{(2aU\,\sin\alpha)^2 - \kappa^2}{z} - i\,\frac{(2aU)^2}{z}\cos\alpha\sin\alpha + \cdots\,. \tag{2.201}$$

Substituting (2.201) into (2.154) and applying (2.155), the moment is given by

$$M = -\frac{\rho}{2}\left(-i\,4\,a^2U^2\cos\alpha\sin\alpha\right)2\pi i = -4\pi\rho\,a^2U^2\cos\alpha\sin\alpha\,. \tag{2.202}$$

Equation (2.202) is identical to (2.194) obtained by integrating the pressure.

Exercise 2.15 As in Fig. 2.35, consider a uniform flow of velocity $U = 1$ from the positive direction of the y-axis intercepting a flat plate of length $2a$ (moving in the positive direction of the y-axis at unit speed).

From (2.83), the complex velocity potential is given by

$$f(z) = i\sqrt{z^2 - a^2}.$$

To extract the flow alone, when a flat plate moves in the direction of y-axis, the complex velocity potential $f(z) = -i\,z$ of a uniform flow in the positive direction of the y-axix ($U = 1$) should be overlapped. Thus,

$$f(z) = i\,(\sqrt{z^2 - a^2} - z)\,. \tag{2.203}$$

Obtain the velocity potential in this situation at the flat plate's upper surface ($y = 0_+$) and the lower surface ($y = 0_-$), then use (2.173) to calculate the added mass and hence show that

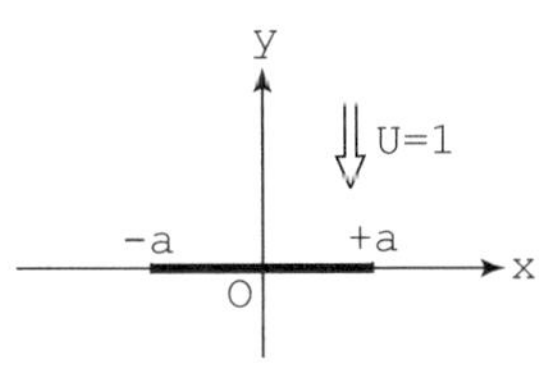

Fig. 2.35 A flat plate in a uniform flow

$$m = \rho \pi a^2 \, . \tag{2.204}$$

Note that this result is the same as the added mass (2.174) for a cylinder of radius a (a circle circumscribed by a flat plate of length $2a$).

Exercise 2.16 Consider Joukowski's transformation as a mapping function between the physical surface $z = x + iy$ and a mapped plane $\zeta = \xi + i\eta$:

$$z = \zeta + \frac{a^2}{\zeta} \quad (a > 0) \, .$$

Here are a circle of radius R ($R > a$) with a center at the origin of the ζ-plane is mapped to an ellipse with a longer axis $A = R + a^2/R$ and a shorter axis $B = R - a^2/R$. Obtain added mass when this elliptic cylinder is moving in the direction of the long axis (x-axis) and the short axis (y-axis).

Note 2.11 (*Literature for further study*)
For further studies, textbooks [2, 3, 6, 8–10] may be referred. The advanced readers may also refer some other textbooks [1, 4, 5, 7, 11, 12], although they are written in Japanese.

References

1. Adachi, C.: Vectors and Tensors. Baifukan (1957)
2. Aris, R.: Vectors, Tensors, and the Basic Equations of Fluid Mechanics. Prentice-Hall, Englewood Cliffs (1962)
3. Batchelor, G.K.: An Introduction to Fluid Dynamics. Cambridge University Press, Cambridge (1967)
4. Hino, M.: Fluid Dynamics. Asakura Publishing, Japan (1992)
5. Imai, I.: Fluid Dynamics. Shokabo, Tokyo (1973)
6. Jeffreys, H.: Cartesian Tensors. Cambridge University Press, Cambridge (1969)
7. Kambe, T. (ed.): Fluid Dynamics. Shokabo, Tokyo (1995)
8. Landau, L.D., Lifshitz, E.M.: Fluid Mechanics. Butterworth-Heinemann, Oxford (1987)
9. Newman, J.N.: Marine Hydrodynamics. The MIT Press, Cambridge (1977)
10. Prandtl, L., Tietjens, O.G.: Fundamentals of Hydro- and Aeromechanics (translated by Rosenhead, L.). Dover, New York (1957)
11. Tatsumi, T.: Fluid Dynamics. Baifukan, Tokyo (1982)
12. Tomita, Y.: Introduction to Fluid Dynamics. Yokendo, Hongo (1975)

Chapter 3
Theory of Free Surface Waves

Various phenomena related to waves on the sea surface present several interesting research themes in the fields of physical oceanography and ocean engineering. The term "waves" includes not only surface waves generated at the boundary between the atmosphere and ocean but also internal waves generated within different densities of stratified fluid. The frequencies (or wavelengths) of ocean waves are highly diverse, ranging from high-frequency ripples to ultra-low-frequency waves associated with the Coriolis force of the Earth's rotation. In this chapter, we restrict ourselves to wavelengths on the order of several tens of centimeters to several tens of meters, which typify **free surface waves** observed on the general sea surface.

In such free surface waves, the effect of viscosity is negligibly small; thus, we can regard typical ocean fluids as incompressible perfect fluids, as discussed in the previous chapter. Moreover, if the fluid motion is considered to start from a static or irrotational (non-vortex) state, then the subsequent fluid motion is irrotational by Lagrange's vortex theorem. In other words, as flow fields can be described by their velocity potential, how to find the velocity potential is explained below.

3.1 Boundary Condition Equations on a Free Surface

As shown in (2.25) of Sect. 2.5, the governing equation of velocity potential Φ is a Laplace equation. To solve this equation, we must assign boundary conditions, as when determining the flow field around a sphere or cylinder. However, the free surfaces of ocean waves are complicated because of gravitational field effects, and the boundary condition equation (referred to as the free surface condition equation) is nonlinear in Φ, as shown below. Nonetheless, ocean wave phenomena can be largely explained by the linear theory. When solving the linearized form, we must also linearize the boundary value problem; hence, this procedure will also be explained.

Now let the static water surface lie in a Cartesian coordinate plane (x, y) with the z-axis directed vertically upwards. If the boundary surface in the spatial coordinate system is specified as $F(x, y, z, t) = 0$, the boundary condition can always be

D.E.S.S.T. Kyushu Univ., *Fluid Dynamics for Global Environmental Studies*,
DOI 10.1007/978-4-431-56499-7_3

obtained by differentiating the boundary surface equation. This is because the fluid particles on the boundary surface must move with the boundary surface. Thus, the fluid boundary satisfies

$$\frac{DF}{Dt} = \left(\frac{\partial}{\partial t} + \nabla\Phi \cdot \nabla\right) F = 0 \qquad \text{on } F = 0\,. \tag{3.1}$$

The normal vector on the boundary surface is determined as $\boldsymbol{n} = \nabla F/|\nabla F|$. Dividing (3.1) by $|\nabla F|$, we obtain

$$\boldsymbol{n} \cdot \nabla\Phi = \frac{\partial\Phi}{\partial n} = -\frac{1}{|\nabla F|}\frac{\partial F}{\partial t} \equiv V_n \qquad \text{on } F = 0\,. \tag{3.2}$$

Equation (3.2) gives Φ at the boundary; furthermore, in principle, it is applicable to any boundary surface. The condition equation (3.1) is called the **kinematic condition**.

In water wave problems, the free surface is also one of the boundary surfaces. Expressing this surface as $z = \zeta(x, y, t)$, F in (3.1) is given by

$$F = z - \zeta\,(x,\ y,\ t) = 0\,. \tag{3.3}$$

Therefore, the boundary condition equation of the velocity potential can be derived from either (3.1) or (3.2). However, in (3.3), $\zeta(x, y, t)$ must also be sought, and it is not usually known until a solution has been obtained. This implies that an additional boundary condition equation must be specified for a free surface.

Fortunately, this additional condition is available because we can assume that the pressure on the free surface always equals atmospheric pressure. Thus, from Bernoulli's pressure equation (2.18), we can specify

$$\frac{\partial\Phi}{\partial t} + \frac{1}{2}\nabla\Phi \cdot \nabla\Phi + g\zeta = 0 \quad \text{on } z = \zeta\,(x,\ y,\ t)\,. \tag{3.4}$$

In principle, if ζ is eliminated from (3.1) and (3.4), we obtain a boundary condition equation in terms of Φ. However, the resulting equation is nonlinear. Furthermore, as evident from (3.4), the position to which the boundary condition equation should be applied is unknown. Hence, to simplify the analytical treatment, we linearize the boundary condition equation.

Assuming that the wave amplitude (ζ) is small, Φ and ζ in (3.4) can be regarded as having the same order; thus, we can eliminate third- and higher-order terms in Φ and ζ. Then, (3.4) and (3.1) are, respectively, expressed as

$$\zeta = -\frac{1}{g}\frac{\partial\Phi}{\partial t} + O(\Phi^2)\,, \tag{3.5}$$

$$-\frac{\partial\zeta}{\partial t} + \frac{\partial\Phi}{\partial z} + O(\zeta\Phi) = 0\,. \tag{3.6}$$

Eliminating ζ from both equations, we obtain

$$\frac{\partial^2 \Phi}{\partial t^2} + g\frac{\partial \Phi}{\partial z} + O(\Phi^2) = 0 \qquad \text{on } z = \zeta\,. \tag{3.7}$$

However, this boundary condition equation should be satisfied on the water surface $z = \zeta$. Since ζ is infinitesimally small, we expand the water surface around the still water surface $z = 0$ as a Taylor series:

$$\Phi(x,\ y,\ z,\ t) = \Phi(x,\ y,\ 0,\ t) + \zeta\left(\frac{\partial \Phi}{\partial z}\right)_{z=0} + \cdots\,. \tag{3.8}$$

Clearly, the error generated by applying (3.7) at $z = 0$ is greater than $O(\Phi^2)$. Therefore, the **linearized free surface** can be expressed as follows:

$$\frac{\partial^2 \Phi}{\partial t^2} + g\frac{\partial \Phi}{\partial z} = 0 \qquad \text{on } z = 0\,. \tag{3.9}$$

Once Φ is determined, the water surface displacement $z = \zeta$ under the linear theory can be obtained from (3.5).

The above procedure is appropriate when deriving the linearized boundary condition equations; however, it is complicated if higher-order boundary condition equations are sought. We now introduce a slightly more useful method that yields the same results.

On the free surface, along with fluid particle movement, pressure is permanently fixed at atmospheric pressure; hence, pressure differential can be assumed to be 0. That is,

$$\frac{D}{Dt}\left(\frac{p}{\rho}\right) = \left(\frac{\partial}{\partial t} + \nabla\Phi\cdot\nabla\right)\left[\frac{\partial \Phi}{\partial t} + \frac{1}{2}\nabla\Phi\cdot\nabla\Phi + gz\right] = 0 \quad \text{on } z = \zeta\,. \tag{3.10}$$

Therefore,

$$\frac{\partial^2 \Phi}{\partial t^2} + g\frac{\partial \Phi}{\partial z} + 2\nabla\Phi\cdot\nabla\left(\frac{\partial \Phi}{\partial t}\right) + \frac{1}{2}\nabla\Phi\cdot\nabla\left(\nabla\Phi\cdot\nabla\Phi\right) = 0 \quad \text{on } z = \zeta\,. \tag{3.11}$$

Eliminating terms having order beyond $O(\Phi^2)$ from this equation, we readily obtain (3.9).

If ζ is determined from (3.5), and we perform Taylor expansion (3.8), the high-order free surface condition equation at $z = 0$ can also be relatively easily derived from (3.11). For example, the free surface condition equation up to $O(\Phi^2)$ is

$$\frac{\partial^2 \Phi}{\partial t^2} + g\frac{\partial \Phi}{\partial z} = -2\nabla\Phi\cdot\nabla\frac{\partial \Phi}{\partial t} + \frac{1}{g}\frac{\partial \Phi}{\partial t}\frac{\partial}{\partial z}\left(\frac{\partial^2 \Phi}{\partial t^2} + g\frac{\partial \Phi}{\partial z}\right) \quad \text{on } z = 0\,. \tag{3.12}$$

Exercise 3.1 In obtaining (3.2), if the boundary surface is given by $F(x, y, z) = 0$, the normal vector can be calculated from $\boldsymbol{n} = \nabla F/|\nabla F|$. On the basis of this equation, calculate the normal vector of a three-dimensional rotating ellipsoid indicated by the following equation:

$$x = a\cos\theta,\ \ y = b\sin\theta\cos\varphi,\ \ z = b\sin\theta\sin\varphi$$

The result should be as follows:

$$n_1 = \epsilon\cos\theta/\Delta,\ n_2 = \sin\theta\cos\varphi/\Delta,\ n_3 = \sin\theta\sin\varphi/\Delta\,.$$

where

$$\Delta = \sqrt{\sin^2\theta + \epsilon^2\cos^2\theta},\quad \epsilon = b/a\,.$$

Exercise 3.2 By eliminating ζ from (3.1) and (3.4) and applying Taylor expansion as shown in (3.8), derive the second-order free surface condition equation shown in (3.12).

3.2 Small-Amplitude Progressive Waves

In this section, we discuss the simplest theoretical water waves—small-amplitude two-dimensional progressive waves. For simplicity, we assume fixed water depth (at $z = -h$) and that the waves propagate in the positive x-direction (see Fig. 3.1). If the amplitude of the progressive wave is a, and the circular frequency is ω, the wave on the water surface behaves as a sine wave:

$$z = \zeta(x, t) = a\ \cos(\omega t - kx)\,. \tag{3.13}$$

Here k is called the **wavenumber**. If the wavelength of the progressive wave is λ, we have $k = 2\pi/\lambda$.

The phase of the sine function (3.13) is $\omega t - kx$. The reader should note that (3.13) expresses a wave progressing in the positive direction along the x-axis. To observe this, consider an infinitesimally subsequent time, $t + \delta t$. Since the wave propagates to the positive x-axis, the wave form should be the same at position $x + \delta x$. Therefore, we have

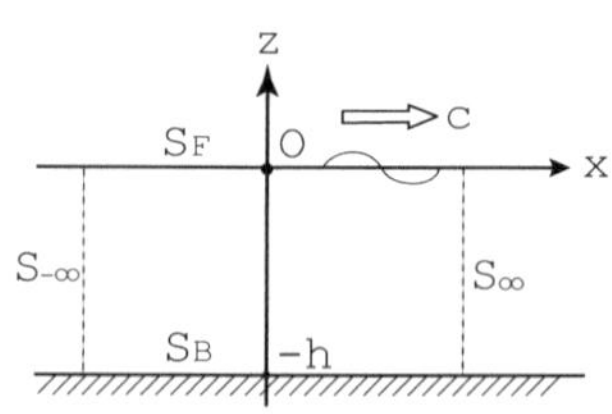

Fig. 3.1 Coordinate system for analyzing small-amplitude waves

$$\omega t - kx = \omega(t + \delta t) - k(x + \delta x)\,,$$

$$\frac{\delta x}{\delta t} = \frac{\omega}{k} \equiv c > 0 \tag{3.14}$$

from which the propagation velocity is seen to be positive. c in (3.14) is called the **phase velocity**. The phase function of a wave progressing in the negative direction along the x-axis is clearly $\omega t + kx$. When expressing these quantities in a generalized form $f(\omega t - kx)$, we note that

$$\frac{\partial f}{\partial t} + c\,\frac{\partial f}{\partial x} = \left(\omega - c\,k\right) f' = 0\,. \tag{3.15}$$

Next, we seek the velocity potential Φ expressing two-dimensional progressive waves. The governing equation of Φ is a two-dimensional Laplace equation. The boundary surfaces surrounding the fluid region (see Fig. 3.1) are the free surface S_F, the bottom of the water S_B, and two virtual boundary surfaces at $x \to \pm\infty$, $S_{\pm\infty}$. Φ is then determined to satisfy the following boundary condition equations.

Continuity equation $[L]$
$$\frac{\partial^2 \Phi}{\partial x^2} + \frac{\partial^2 \Phi}{\partial z^2} = 0 \quad \text{for } z \le 0\,, \tag{3.16}$$

Free surface condition $[F]$
$$\frac{\partial^2 \Phi}{\partial t^2} + g\frac{\partial \Phi}{\partial z} = 0 \quad \text{on } z = 0\,, \tag{3.17}$$

Water depth condition $[B]$
$$\frac{\partial \Phi}{\partial z} = 0 \quad \text{on } z = -h\,. \tag{3.18}$$

The boundary conditions at $S_{\pm\infty}$ are not explicitly stated; however, since $S_{\pm\infty}$ is a virtual surface, we use a condition that provides a physically plausible solution. In the current problem, we understand that the wave propagates in the positive direction along the x-axis. As mentioned above, this condition is satisfied for phase functions of the form $f(\omega t - kx)$. Hence, an appropriate form of the velocity potential is

$$\Phi(x,\ z,\ t) = Z(z)\sin(\omega t - kx)\,. \tag{3.19}$$

Substituting this expression into (3.16), the differential equation in $Z(z)$ becomes

$$\frac{d^2 Z}{dz^2} - k^2\,Z = 0\,, \tag{3.20}$$

which has the general solution

$$Z(z) = D_1\,e^{kz} + D_2\,e^{-kz}\,, \tag{3.21}$$

where D_1 and D_2 are arbitrary constants.

To determine these constants, we substitute them into the free surface condition $[F]$ and the bottom condition $[B]$, yielding two equations for two unknowns:

$$\left.\begin{aligned} D_1(\omega^2 - gk) + D_2(\omega^2 + gk) = 0\,, \\ D_1\, e^{-kh} - D_2\, e^{kh} = 0\,. \end{aligned}\right\} \tag{3.22}$$

Solutions other than the trivial solution $D_1 = D_2 = 0$ exist if

$$\begin{vmatrix} \omega^2 - gk & \omega^2 + gk \\ e^{-kh} & -e^{kh} \end{vmatrix} = 0\,, \tag{3.23}$$

That is,

$$k \tanh kh = \frac{\omega^2}{g} \equiv K\,. \tag{3.24}$$

Equation (3.24) is the characteristic equation that must be satisfied by k, ω, and g. Let the terms in the second equation of (3.22) denote

$$D_1\, e^{-kh} = D_2\, e^{kh} \equiv \frac{1}{2} D\,, \tag{3.25}$$

then the following expression contains a single unknown

$$\Phi(x, z, t) = D \cosh k(z + h) \sin(\omega t - kx) \tag{3.26}$$

As $[F]$ and $[B]$ both describe homogeneous boundary conditions, D cannot be known. Therefore, we seek a description of the wave on the free surface. This wave is expressed by (3.13), and hence using (3.5), we obtain

$$\begin{aligned} \zeta = -\frac{1}{g}\left(\frac{\partial \Phi}{\partial t}\right)_{z=0} &= -\,D\,\frac{\omega}{g} \cosh(kh) \cos(\omega t - kx) \\ &= a\, \cos(\omega t - kx) \end{aligned} \tag{3.27}$$

from which D can be determined.

$$D = -\,\frac{ga}{\omega \cosh kh} \tag{3.28}$$

Finally, the sought velocity potential can be expressed as follows.

$$\Phi = -\frac{ga}{\omega} \frac{\cosh k(z + h)}{\cosh kh} \sin(\omega t - kx)\,. \tag{3.29}$$

Alternatively, Φ can be expressed in a complex notation as

$$\Phi(x, z, t) = \mathrm{Re}\left[\, \phi(x, z)\, e^{i\omega t} \,\right], \tag{3.30}$$

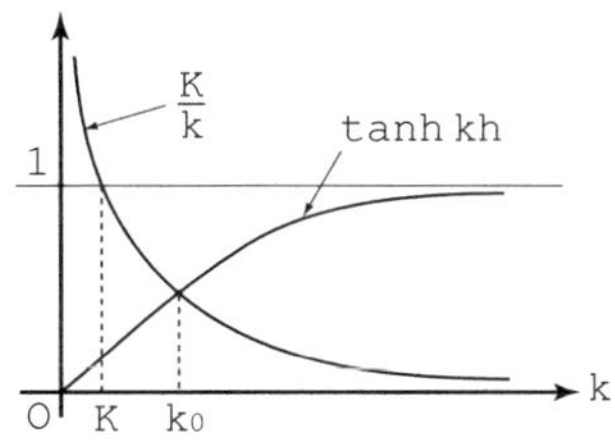

Fig. 3.2 Wavenumber k_0 at limited water depth is greater than the value K at infinite depth

$$\phi(x,z) = \frac{iga}{\omega} \frac{\cosh k(z+h)}{\cosh kh} e^{-ikx} . \tag{3.31}$$

In other words, the temporal term can be separately expressed as $e^{i\omega t}$, and the components of Φ are pure real parts of $[\phi(x,z)\, e^{i\omega t}]$. (Here we have introduced the complex number notation for convenience in later calculations; this notation should not be confused with the complex velocity potential $f(z)$ discussed in Chap. 2.)

We now ponder on the physical meaning of (3.24). From (3.14) and (3.24), the phase velocity c is given by

$$c = \frac{\omega}{k} = \sqrt{\frac{g}{k} \tanh kh} = \sqrt{\frac{g\lambda}{2\pi} \tanh \frac{2\pi h}{\lambda}} . \tag{3.32}$$

Equation (3.32) clearly indicates that the phase velocity changes with the wavelength λ. A practical wave consists of an infinite number of sine waves with different wavelengths. Thus, from (3.32), we see that the waveform created from an infinite number of sine waves with different wavelength changes in time; this phenomenon is called wave dispersion. Equations (3.24) and (3.32), relating the phase velocity to the wavelength, are called **dispersion relations**.

Generally, k cannot be obtained explicitly from (3.24); however, as $y = \tanh kh$ is a monotonically increasing function, we can always find a value of k that satisfies (3.24). The situation is illustrated in Fig. 3.2 (where k is expressed as k_0). When the water is infinitely deep ($h \to \infty$), $\tanh kh \to 1$ and $k_0 = K$. Since $K < k_0$ at limited water depth, as the water becomes shallower, the wavelength gradually decreases from its value at infinite depth.

If the water is infinitely deep ($h \to \infty$) or infinitely shallow ($h \to 0$), (3.32) reduces, respectively, to

$$c = \sqrt{\frac{g\lambda}{2\pi}} \qquad (h \to \infty) , \tag{3.33}$$

or

$$c = \sqrt{gh} \qquad (h \to 0) . \tag{3.34}$$

When the water depth becomes shallow (alternatively, when the wavelength becomes long because both results in $h/\lambda \to 0$), the phase velocity need not be related to the wavelength; thus, the wave is not dispersive. Moreover, (3.33) is a good approximation when $\tanh kh \sim 1$. When $kh = 2\pi h/\lambda \geq 2.65$, or $\lambda \leq 2.4\,\mathrm{h}$, the error is less than 1.0%. Therefore, in practice, provided that the water depth exceeds half of the wavelength, the water depth can be approximately treated as infinity for those waves.

3.3 Water Particle Trajectory and Mass Flux

We now seek the trajectory of water particles caused by a wave. Here because we focus on the motion of a specific fluid particle, the Lagrangian approach is appropriate. If the position of the water particle is $\boldsymbol{r}_0(t) = (\,x_0(t), z_0(t)\,)$, then

$$\frac{d\boldsymbol{r}_0}{dt} = \boldsymbol{u}(\boldsymbol{r}_0,\, t)\,. \tag{3.35}$$

$\boldsymbol{r}_0$ is temporally dynamic, and if it moves a small distance of $O(a)$ in Euler coordinates $\boldsymbol{r} = (x, z)$, then by Taylor expansion around $\boldsymbol{r} = (x, z)$, we obtain

$$\frac{d\boldsymbol{r}_0}{dt} = \boldsymbol{u}(\boldsymbol{r}, t) + (\boldsymbol{r}_0 - \boldsymbol{r}) \cdot \nabla \boldsymbol{u} + O(a^3)\,. \tag{3.36}$$

Here $\boldsymbol{u}(\boldsymbol{r}, t) = \nabla \Phi$, and hence, from (3.29), the trajectory of the water particle (to $O(a)$) is

$$\left.\begin{aligned} (x_0 - x) &= \int \frac{\partial \Phi}{\partial x}\, dt = a\, \frac{\cosh k(z+h)}{\sinh kh} \sin(\omega t - kx)\,, \\ (z_0 - z) &= \int \frac{\partial \Phi}{\partial z}\, dt = a\, \frac{\sinh k(z+h)}{\sinh kh} \cos(\omega t - kx)\,. \end{aligned}\right\} \tag{3.37}$$

From this, we understand that a water particle at shallow depth traces an elliptical orbit. Moreover, if the wave propagates in the positive direction, the water particle follows a clockwise trajectory. Regarding the flow velocity, the crest of the wave travels at the same velocity and in the same direction as that of the wave itself, whereas the trough travels at the same velocity in the reverse direction.

Substituting (3.37) into (3.36) as a correction term, from (3.24) and (3.29), we have

$$\begin{aligned} \frac{dx_0}{dt} &= a\,\omega\, \frac{\cosh k(z+h)}{\sinh kh} \cos(\omega t - kx) \\ &\quad + \frac{1}{2}\omega k a^2\, \frac{\cosh 2k(z+h) - \cos 2(\omega t - kx)}{\sinh^2 kh} + O(a^3)\,, \end{aligned} \tag{3.38}$$

$$\frac{dz_0}{dt} = -a\omega\, \frac{\sinh k(z+h)}{\sinh kh} \sin(\omega t - kx) + O(a^3)\,. \tag{3.39}$$

That is, as the horizontal velocity component contains a time-independent term of $O(a^2)$, the trajectory of the water particle is not fixed, and on average, it moves in a horizontal direction. This movement is called Stokes drift.

The average mass transport within one period is calculated from (3.38). The time average of E can be calculated by

$$\overline{E} \equiv \frac{1}{T}\int_0^T E\,dt\,, \tag{3.40}$$

therefore the time average of mass flux is

$$\overline{M} = \rho\int_{-h}^{0}\overline{\frac{dx_0}{dt}}\,dz = \frac{1}{2}\,\frac{\rho\omega a^2}{\tanh kh} = \frac{1}{2}\,\rho g a^2\,\frac{1}{c}\,, \tag{3.41}$$

where T is the period, and c is the phase velocity given by (3.32).

Stokes drift can be explained by the nonlinear effect of $O(a^2)$ described above; it is also easily observed by watching the movement of a small floating object on a sea surface.

How would this phenomenon appear from Euler's perspective? Even by considering a velocity potential of order $O(a^2)$, the components below the wave trough move in a periodic manner (see Note 3.1). Therefore, the mass flux below the wave trough is zero; however, if mass flux does exist, we must consider the $O(a^2)$ term of the contribution from the component above the wave trough. The time average value of this term is,

$$\overline{M} = \overline{\int_{-a}^{\zeta}\rho\frac{\partial\Phi}{\partial x}\,dz} \approx \overline{\rho\,(\zeta + a)\frac{\partial\Phi}{\partial x}\bigg|_{z=0}} = \frac{1}{2}\,\rho g a^2\,\frac{k}{\omega}\,, \tag{3.42}$$

which is identical to (3.41).

Note 3.1 Consider the second-order velocity potential $\Phi^{(2)}$ proportional to $O(a^2)$. $\Phi^{(2)}$ should satisfy the free surface condition equation (3.12). Substituting (3.30) and (3.31) into the right-hand side of (3.12), we have

$$\frac{\partial^2\Phi^{(2)}}{\partial t^2} + g\,\frac{\partial\Phi^{(2)}}{\partial z} = Q \qquad \text{on } z = 0\,, \tag{3.43}$$

$$\begin{aligned} Q &\equiv \left[-2\nabla\Phi\cdot\nabla\frac{\partial\Phi}{\partial t} + \frac{1}{g}\,\frac{\partial\Phi}{\partial t}\,\frac{\partial}{\partial z}\left(\frac{\partial^2\Phi}{\partial t^2} + g\,\frac{\partial\Phi}{\partial z}\right)\right]_{z=0} \\ &= \mathrm{Re}\left[-i\,\frac{3}{2}\,\frac{g^2a^2}{\omega}\,k^2\left(1 - \tanh^2 kh\right)e^{i\,2(\omega t - kx)}\right] \end{aligned} \tag{3.44}$$

$\Phi^{(2)}$ should satisfy both the Laplace equation and the bottom condition at $z = -h$, and if the phase function takes the form $2(\omega t - kx)$, then

$$\left.\begin{aligned} \Phi^{(2)} &= \mathrm{Re}\Big[\,\phi^{(2)}(x,z)\,e^{i\,2\omega t}\,\Big],\\ \phi^{(2)}(x,z) &= D\,\frac{\cosh 2k(z+h)}{\cosh 2kh}\,e^{-i\,2kx} \end{aligned}\right\} \tag{3.45}$$

where D is undetermined and can be determined from the free surface condition (3.43). The result is

$$D = i\,\frac{3}{4}\,\frac{g^2a^2}{\omega}\,\frac{k(1-\tanh^2 kh)}{(2\tanh kh - \tanh 2kh)}\,. \tag{3.46}$$

Substituting (3.46) into (3.45) and rearranging using the dispersion relation, we eventually obtain

$$\phi^{(2)}(x,z) = i\,\frac{3}{8}\,\omega a^2\,\frac{\cosh 2k(z+h)}{\sinh^4 kh}\,e^{-i\,2kx}\,. \tag{3.47}$$

This is called the second approximation of the Stokes wave based on the perturbation method. In the case of infinite water depth ($h \to \infty$), as $Q = 0$ from (3.44), (we can also let $h \to \infty$ on the basis of (3.47)) $\Phi^{(2)}$ does not exist.

Exercise 3.3 Obtain the velocity potential for a progressive wave at infinite water depth. Using this, calculate the free surface form up to $O(a^2)$. Using Taylor expansion such as (3.8), the calculation equation can be expressed, from (3.4), as follows:

$$\begin{aligned} \zeta &= \Big(1 + \zeta\,\frac{\partial}{\partial z} + \cdots\Big)\Big\{-\frac{1}{g}\Big(\frac{\partial\Phi}{\partial t} + \frac{1}{2}\nabla\Phi\cdot\nabla\Phi\Big)\Big\}_{z=0}\\ &= -\frac{1}{g}\Big(\frac{\partial\Phi}{\partial t} + \frac{1}{2}\nabla\Phi\cdot\nabla\Phi - \frac{1}{g}\frac{\partial\Phi}{\partial t}\frac{\partial^2\Phi}{\partial z\,\partial t}\Big)_{z=0} + O(\Phi^3) \end{aligned}$$

Moreover, explain the characteristic of the obtained free surface form in words.

3.4 Group Velocity

In Sect. 3.2, it was made apparent that a wave with a single circular frequency ω and wavenumber k propagates with phase velocity given by (3.32). We now consider a "group" of waves, each with slightly different ω and k. If we overlap two waves that differ by

$$\delta\omega = \omega_2 - \omega_1\,, \qquad \delta k = k_2 - k_1\,, \tag{3.48}$$

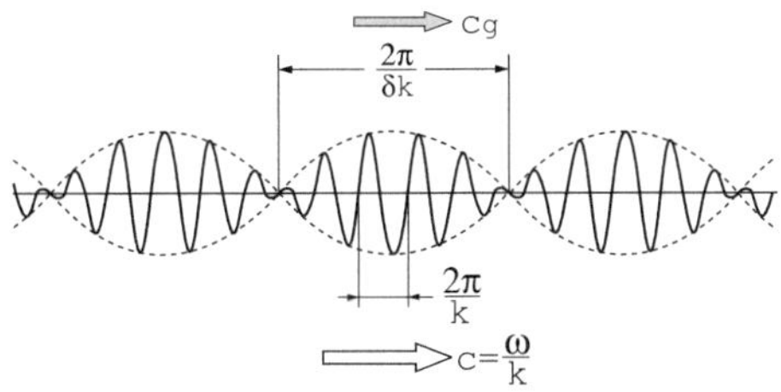

Fig. 3.3 Amplitude modulation establishes an envelope around the amplitude of a basic sine wave (carrier wave), and its propagation velocity (group velocity) is $\delta\omega/\delta k$

we obtain

$$\begin{aligned}\zeta &= \mathrm{Re}\Big[\, A_1\, e^{i\,(\omega_1 t - k_1 x)} + A_2\, e^{i\,(\omega_2 t - k_2 x)} \Big] \\ &= \mathrm{Re}\Big[\, A_1 \Big\{ 1 + \frac{A_2}{A_1}\, e^{i\,(\delta\omega\cdot t - \delta k\cdot x)} \Big\}\, e^{i\,(\omega_1 t - k_1 x)} \Big]. \end{aligned} \tag{3.49}$$

In (3.49), the term in curly brackets expresses amplitude modulation. As δk and $\delta\omega$ are both small, the amplitude changes are slow. This term represents a group of waves. The group propagates at a velocity known as the group velocity (Fig. 3.3). By similar consideration, as used to derive (3.14), the **group velocity** is given by

$$c_g = \frac{\delta\omega}{\delta k}\,. \tag{3.50}$$

Here we have assumed that $\delta\omega \to 0$ and $\delta k \to 0$; however, if t and x are large such that the limits of $\delta\omega \cdot t$ and $\delta k \cdot x$ are non-zero, then the amplitude modulation of (3.49) will persist. In this scenario, the limit of the group velocity c_g is given by

$$c_g = \frac{d\omega}{dk} = \frac{d(kc)}{dk} = c + k\,\frac{dc}{dk} = c - \lambda\,\frac{dc}{d\lambda}\,. \tag{3.51}$$

Using (3.24), and assuming finite water depth, c_g is given by

$$c_g = \frac{1}{2}\,c\left[1 + \frac{2kh}{\sinh 2kh}\right]. \tag{3.52}$$

where c is the phase velocity.

Considering the deep water wave limit ($h \to \infty$) or shallow water wave limit ($h \to 0$), (3.52) respectively reduces to

$$c_g = \frac{1}{2}\,c \qquad (h \to \infty) \tag{3.53}$$

or

$$c_g = c \qquad (h \to 0) \tag{3.54}$$

That is, if the water is infinitely deep, the group velocity is half of the phase velocity, whereas, for non-dispersing shallow water waves, the group velocity equals the phase velocity.

In the above commentary, we considered only two components (ω_1, k_1) and (ω_2, k_2). How should we handle more than three components or more general cases? When answering this question, we should consider an amplitude spectrum of continuous wavenumbers clustered around a central wavenumber k.

If the amplitude spectrum is assumed to be normally distributed, the overall wave form becomes

$$\zeta = \mathrm{Re}\Big[A\sqrt{\frac{\alpha}{\pi}} \int_{-\infty}^{\infty} e^{-\alpha(k'-k)^2}\, e^{i\,\{\omega(k')t-k'x\}}\, dk' \Big]. \tag{3.55}$$

The parameter α expresses the concentration of the spectral components. As $\alpha \to \infty$, it becomes a delta function:

$$\lim_{\alpha\to\infty} \sqrt{\frac{\alpha}{\pi}}\, e^{-\alpha k^2} = \delta(k)\,. \tag{3.56}$$

Therefore, if $\alpha \gg 1$, non-zero amplitudes are observed only in the vicinity of $k' = k$ and the following approximation holds:

$$\omega(k') = \omega(k) + (k'-k)\,\frac{d\omega}{dk}\,. \tag{3.57}$$

The exponent of (3.55) is

$$\begin{aligned} &- \alpha(k'-k)^2 + i\,\{\omega(k')t - k'x\} \\ &= i\,(\omega t - kx) - \frac{1}{4\alpha}\Big(x - \frac{d\omega}{dk}t\Big)^2 - \alpha\Big\{ (k'-k) + \frac{i}{2\alpha}\Big(x - \frac{d\omega}{dk}t\Big)\Big\}^2 \end{aligned} \tag{3.58}$$

Thus, integrating (3.55) with respect to k', we get

$$\zeta \approx \mathrm{Re}\Big[A\exp\Big\{-\frac{1}{4\alpha}\Big(x - \frac{d\omega}{dk}t\Big)^2\Big\}\, e^{i\,(\omega t - kx)} \Big]. \tag{3.59}$$

As in Fig. 3.4, (3.59) describes the envelope around the wave group that propagates at velocity $d\omega/dk$ with a peak at $x = (d\omega/dk)t$. This is of course identical to (3.51), the group velocity.

Exercise 3.4 If the phase velocity of progressive waves in infinite water depth is expressed as c, then the following can be obtained from (3.32) and (3.52):

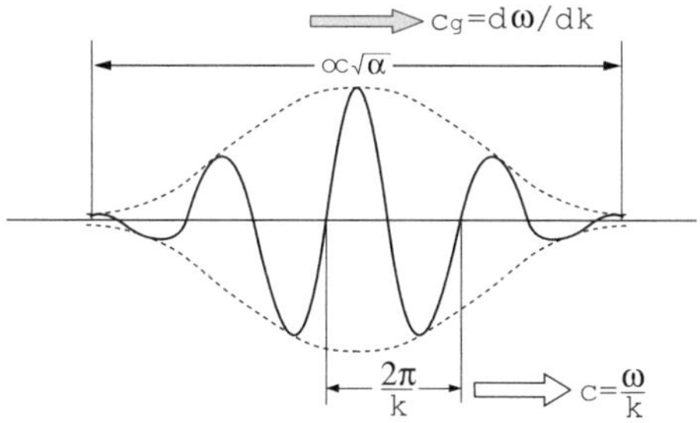

Fig. 3.4 (3.59) expresses the unique wave group with the maximum amplitude at $x = (d\omega/dk)t$ propagating at velocity $d\omega/dk$

$$\frac{c}{c_\infty} = \sqrt{\tanh kh}\,, \tag{3.60}$$

$$\frac{c_g}{c_\infty} = \frac{c}{c_\infty}\frac{1}{2}\Big[1 + \frac{2kh}{\sinh 2kh}\Big]. \tag{3.61}$$

Plot these curves within the range $0 \le kh \le 3$ with horizontal axis kh. (3.60) is a monotonically increasing function with respect to kh but (3.61) takes a maximum value at some value of kh. Seek the equation that provides this kh by differentiating (3.61).

3.5 Principle of Energy Conservation

This section is devoted to the principle of energy conservation for a perfect fluid. This principle is used to investigate the characteristics of wave motion and calculate fluid forces exerted on objects.

Drawing from our knowledge of general dynamics, the energy within a certain fluid region V is the sum of the kinetic and potential energies, and can be expressed as

$$E = \iiint_{V(t)} \rho\Big[\frac{1}{2}q^2 + gz\Big]dV = \iiint_{V(t)} \rho\Big[\frac{1}{2}\nabla\Phi\cdot\nabla\Phi + gz\Big]dV\,. \tag{3.62}$$

We now consider the temporal rate of change of this energy. When the boundary surface S of a fluid region V in (3.62) moves with an outward facing normal velocity U_n, we need to express the derivative in the following Lagrangian form:

$$\begin{aligned}\frac{dE}{dt} &= \rho\frac{d}{dt}\iiint_{V(t)}\Big[\frac{1}{2}q^2 + gz\Big]dV \\ &= \rho\iiint_V \frac{\partial}{\partial t}\Big[\frac{1}{2}q^2 + gz\Big]dV + \rho\iint_S\Big[\frac{1}{2}q^2 + gz\Big]U_n\,dS\,.\end{aligned} \tag{3.63}$$

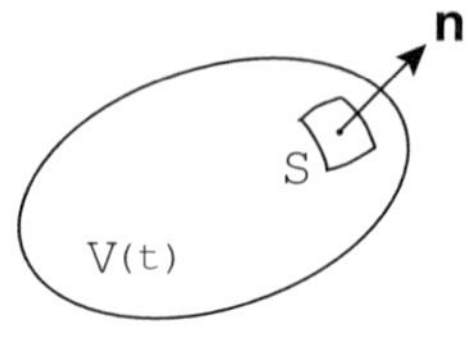

Fig. 3.5 Application of Gauss' theorem

Assuming that the pressure is the atmospheric pressure, from Bernoulli's pressure equation, we have

$$\frac{1}{2}q^2 + gz = -\left(\frac{p}{\rho} + \frac{\partial \Phi}{\partial t}\right). \tag{3.64}$$

Furthermore,

$$\frac{\partial}{\partial t}\left[\frac{1}{2}q^2\right] = \nabla \cdot \left(\frac{\partial \Phi}{\partial t}\nabla \Phi\right). \tag{3.65}$$

Applying Gauss' theorem (Fig. 3.5), the temporal rate of energy change is given as

$$\frac{dE}{dt} = \rho \iint_S \left[\frac{\partial \Phi}{\partial t}\frac{\partial \Phi}{\partial n} - \left(\frac{p}{\rho} + \frac{\partial \Phi}{\partial t}\right)U_n\right] dS\,. \tag{3.66}$$

Here the boundary surface S consists of the free surface S_F, the object surface S_H, and the boundary surface lying at infinity S_∞. Regarding S_∞ as a non-moving boundary surface, at each of the boundary surfaces, we should specify

$$\left.\begin{array}{ll} \text{on } S_\infty & U_n = 0\,, \\ \text{on } S_H & \dfrac{\partial \Phi}{\partial n} = U_n = V_n\,, \\ \text{on } S_F & \dfrac{\partial \Phi}{\partial n} = U_n\,, \quad p = 0 \end{array}\right\} \tag{3.67}$$

Since V_n is the normal velocity of a body, it can be given by (3.2). Substituting (3.67) into (3.66), we obtain

$$\frac{dE}{dt} = -\iint_{S_H} p\,V_n\,dS + \rho\iint_{S_\infty}\frac{\partial \Phi}{\partial t}\frac{\partial \Phi}{\partial n}\,dS\,. \tag{3.68}$$

Now let us consider the time average of (3.68) over one period. Given that the positive direction of the normal points away from the fluid region, the first term on the right-hand side of (3.68) equals the work exerted on the body by the fluid, but has an opposite sign. That is, it expresses the work exerted by the body on the fluid on the basis of the action-reaction law, W_D. The temporal rate of change of the total energy within the fluid region of interest, dE/dt, is 0 in the time-averaged case; therefore, the left-hand side of (3.68) vanishes, and the energy conservation law is expressed as

$$W_D \equiv -\overline{\iint_{S_H} p\, V_n\, dS} = -\rho \iint_{S_\infty} \overline{\frac{\partial \Phi}{\partial t} \frac{\partial \Phi}{\partial n}}\, dS\,. \tag{3.69}$$

From this equation, we can understand the relationship between the wave exerting a damping force on a body and the energy held in the progressive wave generated by movement of the body. When the body is fixed ($V_n = 0$), the left-hand side of (3.69) is of course 0. Even if the body is not fixed, provided it is not forcibly moved by external driving devices, it performs no work against the fluid. In such cases, the left-hand side of (3.69) is also 0.

3.6 Progressive Wave Energy and Propagation Velocity

We now use the result of the previous section to calculate the energy held by a two-dimensional progressive wave. In this analysis, we consider a fluid region of unit length and width along the y-axis. We calculate the one-period time average of the energy in the unit length along the x-axis (Fig. 3.6). From (3.62), we have

$$\overline{E} = \rho \overline{\int_{-h}^{\zeta} \Big[\frac{1}{2} \nabla\Phi \cdot \nabla\Phi + gz \Big] dz}\,, \tag{3.70}$$

where,

$$\zeta = -\frac{1}{g} \frac{\partial \Phi}{\partial t} \Big|_{z=0} \tag{3.71}$$

Here we consider Φ up to second order, ignoring third- and higher-order terms. Moreover, because the static water surface contains potential energy regardless of wave motion, it is excluded from the calculation of progressive wave energy. Under these conditions, (3.70) becomes

$$\overline{E} = \frac{1}{2}\rho \int_{-h}^{0} \overline{\Big[\Big(\frac{\partial \Phi}{\partial x}\Big)^2 + \Big(\frac{\partial \Phi}{\partial z}\Big)^2 \Big]} dz + \frac{1}{2}\rho g \overline{\zeta^2} + O(\Phi^3)\,. \tag{3.72}$$

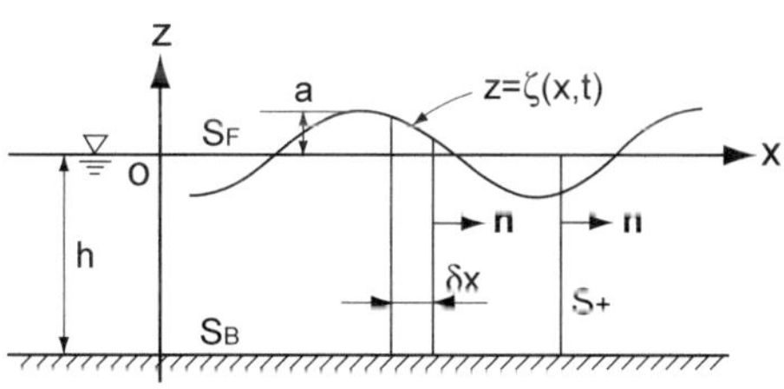

Fig. 3.6 Inspection surface for the calculation of progressive wave energy and propagation velocity

We can now substitute (3.30) into this equation and compute the time average; however, for this calculation, the following equation is useful:

$$\overline{\mathrm{Re}\big[A\, e^{i\,\omega t}\big]\,\mathrm{Re}\big[B\, e^{i\,\omega t}\big]} = \frac{1}{2}\,\mathrm{Re}\big[\,A\,B^{*}\big], \tag{3.73}$$

where B^* is the complex conjugate of B.

We first calculate the time average, then substitute (3.31) and make integration over z. The following equation is obtained :

$$\begin{aligned} \overline{E} &= \frac{1}{4}\rho\Big(\frac{gak}{\omega}\Big)^2 \frac{1}{\cosh^2 kh}\int_{-h}^{0} \cosh 2k(z+h)\,dz + \frac{1}{4}\,\rho g a^2 \\ &= \frac{1}{4}\,\rho g a^2 + \frac{1}{4}\,\rho g a^2 = \frac{1}{2}\,\rho g a^2\,. \end{aligned} \tag{3.74}$$

In other words, the kinetic and potential energies are both equal to $\rho g a^2/4$. Furthermore, it should be noted that these are dependent only on wave amplitude and are unrelated to water depth.

By comparing (3.74) with the previously obtained (3.41), the following relationship clearly holds:

$$\overline{E} = \overline{M}\,c\,. \tag{3.75}$$

That is, the energy per unit length of the progressive wave is the product of the time-averaged mass flux M and the phase velocity c.

We now consider the temporal rate of energy change of the progressive wave. The formula is given by (3.68), but no bodies are present in this case; thus, the first term on the right-hand side of (3.68) vanishes. Clearly, the energy changes only through the inspection surface parallel to the still water surface. If the unit width is parallel to the y-axis, and we consider two vertical sections separated by δx along the x-axis ($S(x)$ and $S(x+\delta x)$), the energy change at a point is

$$\begin{aligned} \frac{\partial E}{\partial t}\,\delta x &= \rho\Big(\int_{S(x+\delta x)} - \int_{S(x)}\Big)\frac{\partial \Phi}{\partial t}\frac{\partial \Phi}{\partial x}\,dz \\ &= \rho\frac{\partial}{\partial x}\int_{-h}^{0}\frac{\partial \Phi}{\partial t}\frac{\partial \Phi}{\partial x}\,dz\cdot\delta x + O(\Phi^3)\,. \end{aligned} \tag{3.76}$$

Inserting (3.30) and (3.31) as Φ in (3.76) and computing the time average by equation (3.73), we obtain

$$\begin{aligned} \frac{\partial \overline{E}}{\partial t} &= \frac{1}{2}\rho\frac{\partial}{\partial x}\Big[\mathrm{Re}\int_{-h}^{0}(i\,\omega\phi)\frac{\partial \phi^*}{\partial x}\,dz\Big] \\ &= -\frac{\partial}{\partial x}\Big[\frac{1}{4}\,\rho g a^2\,\frac{\omega}{k}\Big\{1+\frac{2kh}{\sinh 2kh}\Big\}\Big]. \end{aligned} \tag{3.77}$$

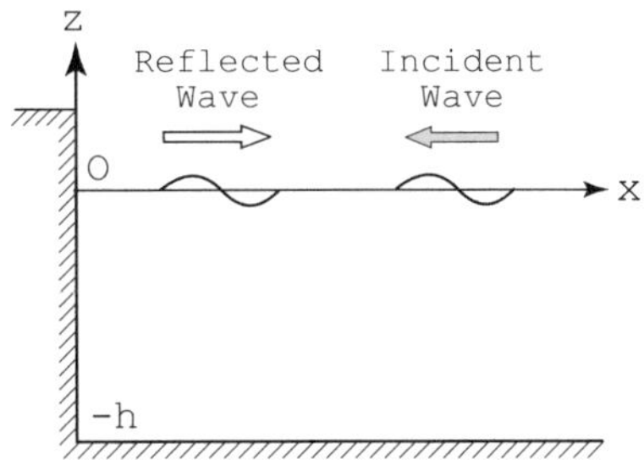

Fig. 3.7 Incident waves reflected by the wall surface create standing waves

From (3.52) and (3.74), we also observe that

$$\frac{\partial \overline{E}}{\partial t} + \frac{\partial}{\partial x}\left(c_g \overline{E}\right) = 0\,. \tag{3.78}$$

Equation (3.78) is expressed in the same form as (3.15), which is derived for progressive waves, and it indicates that total energy E held by progressive waves is propagated by group velocity c_g.

Exercise 3.5 Prove (3.73). Then, show the equation transformation to obtain (3.77).

3.7 Standing Waves

Thus far, we have considered only progressive waves in the positive x-direction. We now consider the reflection of waves from a fixed wall in a vertical plane at $x = 0$ (Fig. 3.7). The boundary condition equation in this situation is

$$\frac{\partial \Phi}{\partial x} = 0 \qquad \text{at } x = 0\,. \tag{3.79}$$

This condition can be satisfied by the superposition of two waves progressing in opposite directions along the x-axis. From (3.29), we obtain

$$\begin{aligned} \Phi_s &= -\frac{ga}{\omega}\,\frac{\cosh k(z+h)}{\cosh kh}\left\{\sin(\omega t + kx) + \sin(\omega t - kx)\right\} \\ &= -\frac{2ga}{\omega}\,\frac{\cosh k(z+h)}{\cosh kh}\cos kx\ \sin \omega t\,, \end{aligned} \tag{3.80}$$

which definitely satisfies (3.79). The waves take the form

$$\zeta_s = -\frac{1}{g}\left(\frac{\partial \Phi_s}{\partial t}\right)_{z=0} = 2a \cos kx\ \cos \omega t\,. \tag{3.81}$$

As can be understood from (3.80) or (3.81), waves generated by reflections from a vertical wall are not actually progressive waves but merely vibrate up and down at a fixed position. Such waves are called **standing waves**.

The wave described by (3.81) has its maximum amplitude at points $x = m\pi/k$ $(m = 0, \pm 1, \pm 2, \ldots)$, and its amplitude is 0 at $x = (m + \frac{1}{2})\pi/k$. In other words, the wall $(x = 0)$ is also the point of maximum amplitude. The points of maximum and zero amplitudes are called **loops** and **nodes**, respectively.

As shown in (3.79), water particles occupying loop positions do not move horizontally; however, they vibrate in the vertical direction. In contrast, water particles at node positions, at which the amplitude is permanently 0, do not move vertically but propagate in the horizontal direction.

As $\partial\Phi/\partial x = 0$ at $x = m\pi/k$, (3.80) can also be considered to express the oscillation of water inside a container partitioned by two vertical fixed walls at positions $x = m\pi/k$. The horizontal width ℓ of this container must be a natural number multiple of the half wavelength $\lambda/2$. Set up in this way, the natural oscillation of the water is a standing wave of wavelength $\lambda = 2\ell/\mathrm{m}$ $(m = 1, 2, \ldots)$. From (3.24), the natural frequency ω_m is given as

$$\omega_m = \sqrt{g\frac{m\pi}{\ell}\tanh\left(\frac{m\pi h}{\ell}\right)}. \tag{3.82}$$

If the water is shallow (permitting longwave approximation), (3.82) gives the natural period T_m as

$$T_m = \frac{2\pi}{\omega_m} = \frac{2\ell}{m}\frac{1}{\sqrt{gh}}. \qquad (m = 1,\ 2,\ \ldots) \tag{3.83}$$

This period (particularly, when $m = 1$) is consistent with the oscillation period of **seiche** waves, the natural frequency in real lakes and bays.

Incidentally, standing waves are not limited to the one-dimensional waves described above but can also exist as two-dimensional waves in regions surrounded by arbitrarily closed curves in the x-y plane. The general solution to this situation requires numerical calculation; however, it is readily found if the longwave approximation applies.

In the longwave approximation, $h \ll \lambda$. Integrating the three-dimensional Laplace equation over $-h \leq z \leq 0$, the terms not containing the z-differential can be regarded as fixed; thus, we have

$$\left(\frac{\partial\Phi}{\partial z}\right)_{z=0} = -h\left(\frac{\partial^2\Phi}{\partial x^2} + \frac{\partial^2\Phi}{\partial y^2}\right), \tag{3.84}$$

considering the bottom condition at $z = -h$.

Substituting (3.84) into the free surface condition and adopting the phase velocity in the longwave approximation, namely, $c = \sqrt{gh}$, we obtain

$$\frac{\partial^2 \Phi}{\partial t^2} = c^2\left(\frac{\partial^2 \Phi}{\partial x^2} + \frac{\partial^2 \Phi}{\partial y^2}\right), \qquad c = \sqrt{gh}\,. \tag{3.85}$$

In this equation, Φ is not considered to be a function of z, but rather as $\Phi = \Phi(x, y, t)$. Equation (3.85) is a two-dimensional wave equation and requires a boundary condition on the boundary curve C_B on the x-y plane. This boundary condition is given as

$$\frac{\partial \Phi}{\partial n} = 0 \qquad \text{on } C_B\,. \tag{3.86}$$

Now, when Φ expressing the standing waves is given by

$$\Phi(x, y, t) = \phi(x, y)\sin \omega t\,, \tag{3.87}$$

Equation (3.85) becomes

$$\frac{\partial^2 \phi}{\partial x^2} + \frac{\partial^2 \phi}{\partial y^2} + k^2\phi = 0\,, \qquad k = \frac{\omega}{c}\,. \tag{3.88}$$

Equation (3.88) is known as Helmholtz's equation.

Let us consider (3.88) in a simple rectangular boundary (Fig. 3.8). Specifying the boundary line C_B by $x = 0$, $x = a$, and $y = 0$, $y = b$, (3.86) can be written as follows:

$$\left.\begin{aligned} \frac{\partial \phi}{\partial x} &= 0\,, \quad \text{at } x = 0,\ a \\ \frac{\partial \phi}{\partial y} &= 0\,, \quad \text{at } y = 0,\ b \end{aligned}\right\} \tag{3.89}$$

Equation (3.88) can be solved by the variable separation method. Expressing ϕ as the product of functions of two independent variables $\phi(x, y) = X(x)Y(y)$ and substituting into (3.88), we obtain the following differential equation:

$$X'' + p^2 X = 0\,, \quad Y'' + q^2 Y = 0\,, \tag{3.90}$$

where $k^2 = p^2 + q^2$. If the solutions to both differential equations must satisfy (3.89), they are given by

$$\left.\begin{aligned} X &= C\cos(p\,x)\,, \ p = \frac{m\pi}{a} \\ Y &= D\cos(q\,y)\,, \ q = \frac{n\pi}{b} \end{aligned}\right\} \tag{3.91}$$

where C and D are arbitrary constants. Substituting these results into $\phi(x, y) = X(x)Y(y)$, we get

$$\phi(x, y) = A_{mn}\cos\frac{m\pi}{a}x\,\cos\frac{n\pi}{b}y \tag{3.92}$$

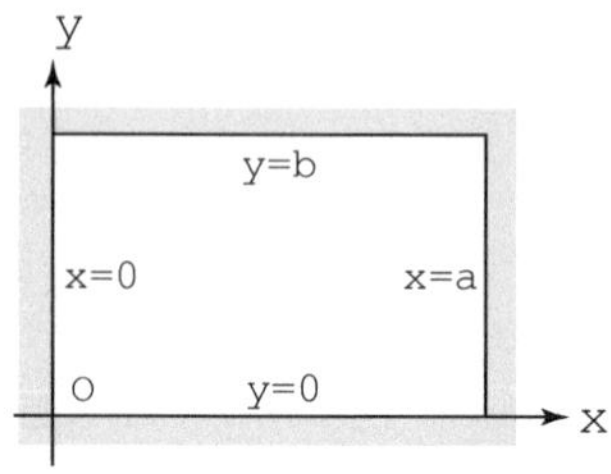

Fig. 3.8 Long wave approximation enclosed by a rectangular boundary

$$k^2 = \pi^2 \left\{ \left(\frac{m}{a}\right)^2 + \left(\frac{n}{b}\right)^2 \right\} \tag{3.93}$$

Here m and n are natural numbers, and A_{mn} is an arbitrary constant. The wavenumber k in (3.93) is the eigenvalue in the longwave approximation. If this equation is considered independent of y, it reduces to that of a one-dimensional wave, and it is clearly consistent with (3.83). A general two-dimensional long standing wave inside a rectangular container overlaps all natural oscillations of the form (3.92).

Exercise 3.6 The calculation in (3.80) was formulated assuming the boundary condition (3.79), with waves fully reflected at $x = 0$. Suppose, instead, that the wall at $x = 0$ is extremely soft, with a zero-pressure boundary condition ($p = 0$). Find the velocity potential of the reflected waves in this situation.

Exercise 3.7 Next, consider that the wall at $x = 0$ is elastic and displaces in the horizontal direction by an extent proportional to the pressure at that point. Let the constant of proportionality be α, and find the boundary condition equation at $x = 0$. Additionally, find the velocity potential of the reflected wave satisfying this condition, and confirm that the limits as $\alpha \to 0$ and $\alpha \to \infty$ are the limits previously obtained in (3.80) and Exercise 3.6.

Exercise 3.8 From the above results, find the wave reflection coefficient (reflected wave amplitude/incident wave amplitude). Furthermore, calculate the work exerted on the fluid by the wall at $x = 0$. Use the following expression from (3.69) to calculate the work:

$$W_D = -\rho \int_{-h}^{0} \overline{\frac{\partial \Phi}{\partial t} \frac{\partial \Phi}{\partial x}} \bigg|_{x=0} dz = \frac{\rho}{2} \mathrm{Re} \int_{-h}^{0} i\,\omega\,\phi^* \frac{\partial \phi}{\partial x} \bigg|_{x=0} dz\,.$$

Explain the physical meaning of the reflection coefficient and the calculated work.

Note 3.2 (*Literature for further study*)

For further studies, textbooks [2–4, 8–10] may be referred. The advanced readers may also refer some other textbooks [1, 5–7, 11–13], although they are written in Japanese.

References

1. Adachi, C.: Vectors and Tensors. Baifukan (1957)
2. Aris, R.: Vectors, Tensors, and the Basic Equations of Fluid Mechanics. Prentice-Hall, Upper Saddle River (1962)
3. Batchelor, G.K.: An Introduction to Fluid Dynamics. Cambridge University Press, Cambridge (1967)
4. Gill, A.E.: Atmosphere-Ocean Dynamics. Academic Press, Cambridge (1982)
5. Hino, M.: Fluid Dynamics. Asakura Publishing (1992)
6. Imai, I.: Fluid Dynamics, vol. I. Shokabo (1973)
7. Kambe, T. (ed.): Fluid Dynamics. Shokabo (1995)
8. Landau, L.D., Lifshitz, E.M.: Fluid Mechanics. Butterworth-Heinemann, Oxford (1987)
9. Newman, J.N.: Marine Hydrodynamics. The MIT Press, Cambridge (1977)
10. Prandtl, L., Tietjens, O.G.: Fundamentals of Hydro- and Aeromechanics (translated by Rosenhead, L.). Dover (1957)
11. Tatsumi, T.: Fluid Dynamics. Baifukan (1982)
12. Tomita, Y.: Introduction to Fluid Dynamics. Yokendo (1975)
13. Unoki, S., Kubota, M.: Waves and Currents in the Ocean. Tokai University Press, Tokai (1996)

Chapter 4
Dynamics of Viscous Fluids

Chapter 2 investigated the dynamics of perfect fluids. Since there is no tangential stress within perfect fluids, in parallel fluid flows (for example), the fluid particles on neighboring streamlines flow independently. However, in real fluids, if a portion of the fluid is fast-flowing, the fluid on adjacent streamlines is likely accelerated by tangential stress occurring within the fluid. Such fluids developing tangential stress are called viscous fluids.

Here we focus on incompressible viscous fluids of uniform density. As derived in Sect. 1.8, the motions of such fluids are governed by the equation of continuity and the Navier–Stokes equation, expressed as

$$\nabla \cdot \boldsymbol{u} = 0\,, \tag{4.1}$$

$$\frac{D\boldsymbol{u}}{Dt} = \boldsymbol{K} - \frac{1}{\rho}\nabla p + \nu\nabla^2\boldsymbol{u}\,. \tag{4.2}$$

In this chapter, (4.1) and (4.2) form the basis of all subsequent discussions.

4.1 Reynolds Number and the Law of Similarity

When the external force $\boldsymbol{K}$ in the Navier–Stokes equation (4.2) is conservative, $\boldsymbol{K}$ can be expressed in terms of potential Π as $\boldsymbol{K} = -\nabla\Pi$. Writing $p^* \equiv p + \rho\Pi$, (4.2) can be rewritten as

$$\frac{D\boldsymbol{u}}{Dt} = -\frac{1}{\rho}\nabla p^* + \nu\nabla^2\boldsymbol{u}\,. \tag{4.3}$$

D.E.S.S.T. Kyushu Univ., *Fluid Dynamics for Global Environmental Studies*,
DOI 10.1007/978-4-431-56499-7_4

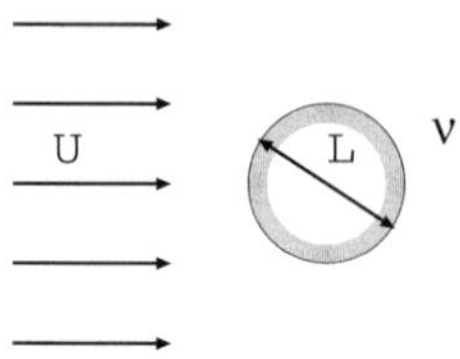

Fig. 4.1 Representative scales of flow. U is the velocity scale, L is the length scale and ν is the coefficient of kinematic viscosity

Dissecting the terms in the Navier–Stokes equation (4.3), $D\boldsymbol{u}/Dt$ is called the **inertial term**,[1] $-(1/\rho)\nabla p^*$ is the **pressure term**, and $\nu\nabla^2\boldsymbol{u}$ is the **viscous term**.

Now, with some representative length scale characterizing fluid motion as L and velocity scale as U (Fig. 4.1), the quantities appearing in (4.3) are non-dimensionalized as

$$\left.\begin{aligned} \tilde{\boldsymbol{x}} &\equiv \frac{\boldsymbol{x}}{L}, \quad \tilde{t} \equiv \frac{t}{L/U}, \\ \tilde{\boldsymbol{u}} &\equiv \frac{\boldsymbol{u}}{U}, \quad \tilde{p}^* \equiv \frac{p^*}{\rho U^2} = \frac{p+\rho\Pi}{\rho U^2}. \end{aligned}\right\} \tag{4.4}$$

In (4.4), "$\tilde{\ }$" denotes that the quantity is dimensionless. Using these terms to non-dimensionalize the motion equation (4.3), we get

$$\frac{D\tilde{\boldsymbol{u}}}{D\tilde{t}} = -\tilde{\nabla}\tilde{p}^* + \frac{1}{Re}\tilde{\nabla}^2\tilde{\boldsymbol{u}}, \tag{4.5}$$

$$Re \equiv \frac{UL}{\nu} \tag{4.6}$$

where Re is a dimensionless number known as the **Reynolds number** and is given by

$$\frac{\text{Inertial term}}{\text{Viscous term}} \cong \frac{U \cdot U/L}{\nu U/L^2} = \frac{UL}{\nu} = Re. \tag{4.7}$$

As shown in (4.7), the Reynolds number specifies the ratio of inertial term to viscous term. When $Re \gg 1$, the inertia nearly balances the pressure gradient in a flow. In contrast, when $Re \ll 1$, the pressure gradient is balanced by the viscous force.

Consider two flows around geometrically similar bodies. According to the non-dimensionalized equation (4.5), even if the scales of the bodies, the fluid velocity magnitudes, or the coefficients of viscosity differ, if two fluids have the same Reynolds number, they will yield the same solutions in terms of the non-dimensionalized variables $\tilde{\boldsymbol{u}}(\tilde{\boldsymbol{r}}, \tilde{t})$ and $\tilde{p}^*(\tilde{\boldsymbol{r}}, \tilde{t})$. Consequently, their flow fields are similar (Fig. 4.2). This is called **Reynolds' law of similarity**. The law of similarity implies that even when the boundary conditions are the same, flows with different Reynolds numbers are

[1] In particular, $(\boldsymbol{u} \cdot \nabla)\boldsymbol{u}$ is called the **advection term** (or **convection term**). It can also be called the nonlinear term, because of the product of independent variables $\boldsymbol{u}$.

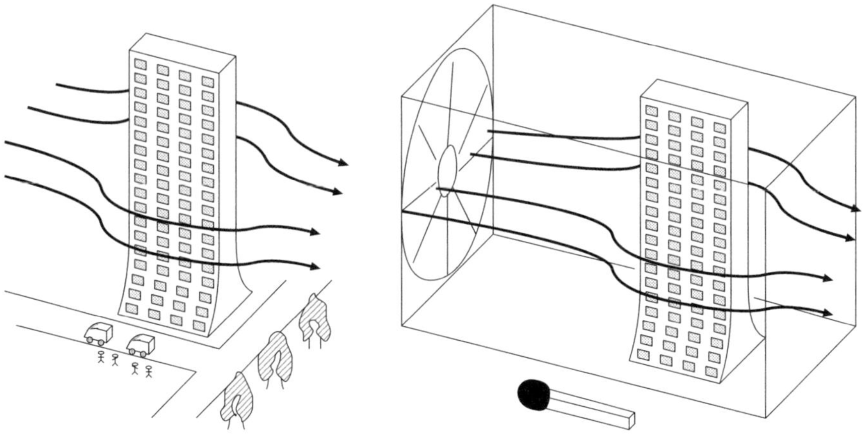

Fig. 4.2 Two flows with the same Reynolds number are similar regardless of scale

inherently different. Flows with a low Reynolds number are simple and dominated by the viscous term, whereas those with a high Reynolds number are complex.

4.2 Exact Solutions of the Navier–Stokes Equation

Because the Navier–Stokes equation is non-linear, in general, it is difficult to obtain exact solutions for arbitrary flows. However, exact solutions are easily found in some special cases, for example, when a flow is unidirectional (called **parallel flow**) or when it traces concentric circles around an axis. In these cases, the nonlinear term becomes identically 0. Several representative exact solutions of the Navier–Stokes equation are considered in this section.

4.2.1 Couette Flow

Consider a fluid with uniform density surrounded by two parallel plates, as shown in Fig. 4.3. Assuming that the lower plate is stationary and the upper plate moves parallel to the lower plate with a uniform velocity U, we contemplate the fluid's movement between the plates. In this case, the flow is seen to be two-dimensional in the x-y plane. Here let us take x-axis in the upper plate movement direction and y-axis in the vertically upward direction from the lower plate. Then, the continuity and Navier–Stokes equations are given by

$$\frac{\partial u}{\partial x} + \frac{\partial v}{\partial y} = 0\,, \tag{4.8}$$

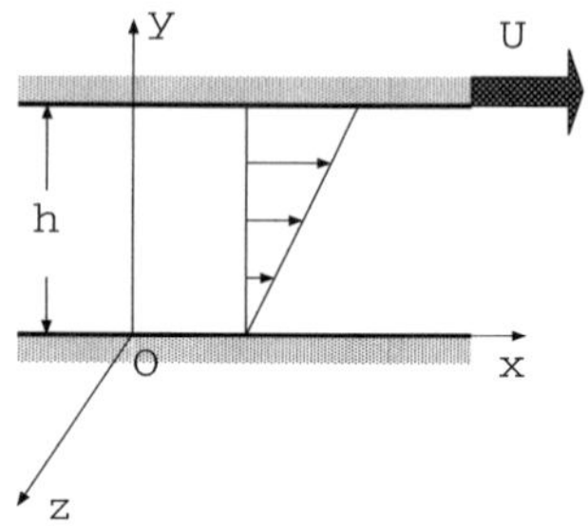

Fig. 4.3 Couette flow

$$\frac{\partial u}{\partial t} + u\frac{\partial u}{\partial x} + v\frac{\partial u}{\partial y} = -\frac{1}{\rho}\frac{\partial p}{\partial x} + \nu\left(\frac{\partial^2 u}{\partial x^2} + \frac{\partial^2 u}{\partial y^2}\right), \tag{4.9}$$

$$\frac{\partial v}{\partial t} + u\frac{\partial v}{\partial x} + v\frac{\partial v}{\partial y} = -\frac{1}{\rho}\frac{\partial p}{\partial y} + \nu\left(\frac{\partial^2 v}{\partial x^2} + \frac{\partial^2 v}{\partial y^2}\right). \tag{4.10}$$

The boundary conditions are

$$\left.\begin{aligned} u = 0\,,\ v = 0 \text{ at } y = 0\,, \\ u = U\,,\ v = 0 \text{ at } y = h\,. \end{aligned}\right\} \tag{4.11}$$

Assuming that the flow is steady ($\partial/\partial t = 0$) and uniform in the x-direction ($\partial/\partial x = 0$), the equation of continuity (4.8) gives $\partial v/\partial y = 0$. However, to satisfy the boundary condition $v = 0$ at $y = 0$, $v = 0$ in the whole region, and the Navier–Stokes equations (4.9) and (4.10) are reduced to

$$0 = -\frac{1}{\rho}\frac{\partial p}{\partial x} + \nu\frac{\partial^2 u}{\partial y^2}, \tag{4.12}$$

$$0 = -\frac{1}{\rho}\frac{\partial p}{\partial y}. \tag{4.13}$$

Supposing that no pressure gradient develops even in the direction of the flow ($\partial p/\partial x = 0$) and the boundary conditions are

$$\left.\begin{aligned} u = 0 \text{ at } y = 0\,, \\ u = U \text{ at } y = h \end{aligned}\right\} \tag{4.14}$$

Equation (4.12) is simply solved as

$$u = U\frac{y}{h} \tag{4.15}$$

and the flow velocity of the fluid between both plates becomes linearly distributed. This type of flow is known as **Couette flow**.

4.2.2 *Plane Poiseuille Flow*

Assuming that a fluid surrounded by two stationary flat parallel plates flows under a fixed pressure gradient ($\partial p/\partial x \neq 0$) as shown in Fig. 4.4, the flow velocity is parabolically distributed as

$$u(y) = \frac{1}{2\rho\nu}\frac{dp}{dx}(y-h)y\,. \tag{4.16}$$

This flow is called a **plane Poiseuille flow**. In this type of flow, the maximum velocity $U_{\rm max}$, flow amount Q, and mean velocity $U_{\rm mean}$ are, respectively, given by

$$U_{\rm max} = -\frac{1}{8\rho\nu}\frac{dp}{dx}h^2, \tag{4.17}$$

$$Q = -\frac{1}{12\rho\nu}\frac{dp}{dx}h^3, \tag{4.18}$$

$$U_{\rm mean} = \frac{2}{3}U_{\rm max}\,. \tag{4.19}$$

Exercise 4.1 Solve the plane Poiseuille flow equation and derive the velocity distribution (4.16). Furthermore, derive the maximum velocity (4.17), flow amount (4.18), and mean velocity (4.19).

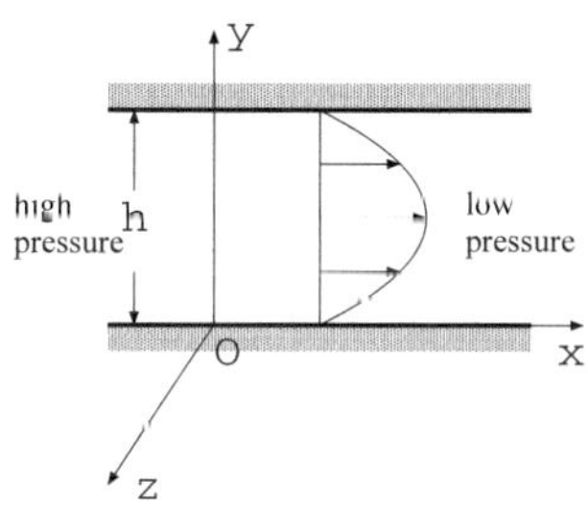

Fig. 4.4 Plane Poiseuille flow

4.2.3 Hagen–Poiseuille Flow

Now, consider a uniform flow in an infinitely long tube with a circular cross section of radius a (Fig. 4.5). The continuity and Navier–Stokes equations in cylindrical coordinates are given by

$$\frac{\partial u_x}{\partial x} + \frac{1}{r}\frac{\partial}{\partial r}(r u_r) + \frac{1}{r}\frac{\partial u_\phi}{\partial \phi} = 0, \tag{4.20}$$

$$\frac{\partial u_x}{\partial t} + u_x\frac{\partial u_x}{\partial x} + u_r\frac{\partial u_x}{\partial r} + \frac{u_\phi}{r}\frac{\partial u_x}{\partial \phi} = -\frac{1}{\rho}\frac{\partial p}{\partial x} + \nu\left[\frac{\partial^2 u_x}{\partial x^2} + \frac{1}{r}\frac{\partial}{\partial r}\left(r\frac{\partial u_x}{\partial r}\right) + \frac{1}{r^2}\frac{\partial^2 u_x}{\partial \phi^2}\right], \tag{4.21}$$

$$\frac{\partial u_r}{\partial t} + u_x\frac{\partial u_r}{\partial x} + u_r\frac{\partial u_r}{\partial r} + \frac{u_\phi}{r}\frac{\partial u_r}{\partial \phi} - \frac{u_\phi^2}{r} = -\frac{1}{\rho}\frac{\partial p}{\partial r} + \nu\left[\frac{\partial^2 u_r}{\partial x^2} + \frac{1}{r}\frac{\partial}{\partial r}\left(r\frac{\partial u_r}{\partial r}\right) + \frac{1}{r^2}\frac{\partial^2 u_r}{\partial \phi^2} - \frac{u_r}{r^2} - \frac{2}{r^2}\frac{\partial u_\phi}{\partial \phi}\right], \tag{4.22}$$

$$\frac{\partial u_\phi}{\partial t} + u_x\frac{\partial u_\phi}{\partial x} + u_r\frac{\partial u_\phi}{\partial r} + \frac{u_\phi}{r}\frac{\partial u_\phi}{\partial \phi} + \frac{u_r u_\phi}{r} = -\frac{1}{\rho r}\frac{\partial p}{\partial \phi} + \nu\left[\frac{\partial^2 u_\phi}{\partial x^2} + \frac{1}{r}\frac{\partial}{\partial r}\left(r\frac{\partial u_\phi}{\partial r}\right) + \frac{1}{r^2}\frac{\partial^2 u_\phi}{\partial \phi^2} + \frac{2}{r^2}\frac{\partial u_r}{\partial \phi} - \frac{u_\phi}{r^2}\right] \tag{4.23}$$

(Cylindrical coordinates are expressed as (x, r, ϕ); see Appendix A.2 (2)). The boundary condition is

$$u_x = u_r = u_\phi = 0 \quad \text{at } r = a\,. \tag{4.24}$$

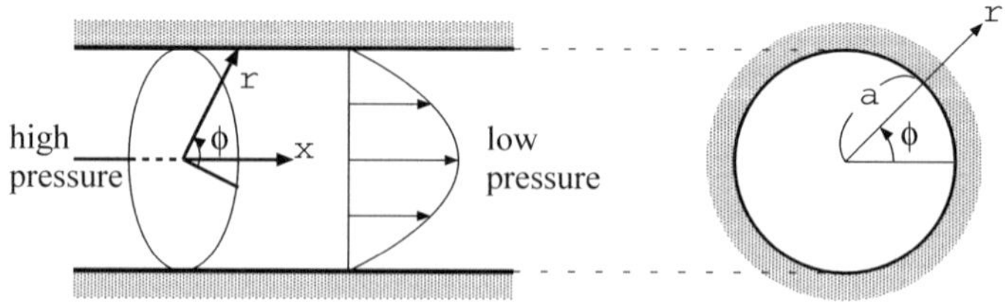

Fig. 4.5 Hagen–Poiseuille flow

Assuming that the flow is steady ($\partial/\partial t = 0$) and uniform in the x- and ϕ-directions ($\partial/\partial x = 0, \partial/\partial\phi = 0$), pressure can also be assumed to be uniform in the ϕ-direction ($\partial p/\partial\phi = 0$) by the symmetry of the system. From the continuity equation (4.20) and the boundary conditions, we have $u_r = 0$. Furthermore, assuming that $u_\phi = 0$, (4.23) certainly holds, Eqs. (4.21) and (4.22) become

$$0 = -\frac{1}{\rho}\frac{\partial p}{\partial x} + \frac{\nu}{r}\frac{\partial}{\partial r}\left(r\frac{\partial u_x}{\partial r}\right), \tag{4.25}$$

$$0 = -\frac{1}{\rho}\frac{\partial p}{\partial r}. \tag{4.26}$$

From Eq. (4.26) and the symmetric assumption of $\partial p/\partial\phi = 0$, the pressure p becomes a function of x only. Writing $\partial p/\partial x$ in (4.25) as dp/dx, we have

$$\frac{\nu}{r}\frac{\partial}{\partial r}\left(r\frac{\partial u_x}{\partial r}\right) = \frac{1}{\rho}\frac{dp}{dx}. \tag{4.27}$$

The right- and left-hand sides of (4.27) are functions of x and r only, respectively; hence both are constant. Multiplying both sides by r/ν and integrating with respect to r, we obtain

$$r\frac{\partial u_x}{\partial r} = \frac{1}{2\rho\nu}\frac{dp}{dx}r^2 + C_1.$$

Dividing both sides of this expression by r and integrating give

$$u_x = \frac{1}{4\rho\nu}\frac{dp}{dx}r^2 + C_1\log r + C_2. \tag{4.28}$$

Applying the boundary conditions

$$\left.\begin{array}{r} u_x = 0 \quad \text{at } r = a, \\ u_x \text{ should be finite at } r = 0, \end{array}\right\} \tag{4.29}$$

C_1 and C_2 are determined as

$$C_1 = 0, \quad C_2 = -\frac{a^2}{4\rho\nu}\frac{dp}{dx} \tag{4.30}$$

and the flow velocity distribution is

$$u_x = \frac{1}{4\rho\nu}\frac{dp}{dx}(r^2 - a^2). \tag{4.31}$$

Using this result, the maximum velocity $U_{\max}$, flow amount Q, and mean velocity U_{mean} are obtained as

$$U_{\max} = \frac{-1}{4\rho\nu}\frac{dp}{dx}a^2, \tag{4.32}$$

$$Q \equiv \int_0^a u_x 2\pi r\, dr = \frac{1}{4\rho\nu}\frac{dp}{dx}\int_0^a (r^2 - a^2)2\pi r\, dr = -\frac{\pi}{8\rho\nu}\frac{dp}{dx}a^4, \tag{4.33}$$

$$U_{\text{mean}} \equiv \frac{Q}{\pi a^2} = -\frac{1}{8\rho\nu}\frac{dp}{dx}a^2 = \frac{1}{2}U_{\max}. \tag{4.34}$$

The flow velocity within the tube develops a parabolic distribution. The amount of flow is proportional to the fourth power of the tube's radius and pressure gradient, and inversely proportional to the coefficient of viscosity. This relationship was experimentally discovered by German waterworks engineer Hagen and French doctor Poiseuille; hence, such flows inside a tube are called **Hagen–Poiseuille flows** (or simply **Poiseuille flows**).

Note 4.1 In the preceding discussion, $u_\phi = 0$ clearly holds from the problem setting but can be formally deduced as follows.

Assuming that u_ϕ is non-zero, from (4.23), we have

$$0 = \nu\left[\frac{1}{r}\frac{\partial}{\partial r}\left(r\frac{\partial u_\phi}{\partial r}\right) - \frac{u_\phi}{r^2}\right]. \tag{4.35}$$

Equation (4.35) is in the same form as (4.54), which appears in the discussions of Couette flow in coaxial tubes (see Sect. 4.2.5). A similar integration gives

$$u_\phi = C_1 r + \frac{C_2}{r}. \tag{4.36}$$

However, under the boundary conditions,

$$\left.\begin{array}{r} u_\phi = 0 \quad \text{at } r = a\,, \\ u_\phi \text{ should be finite at } r = 0\,, \end{array}\right\} \tag{4.37}$$

C_1 and C_2 both become 0, and we have

$$u_\phi = 0\,. \tag{4.38}$$

4.2.4 Rayleigh Flow

Imagine a flow expanding semi-infinitely upward but bounded by an infinitely wide plane at the bottom. Initially, it remains at rest. Let us discuss the fluid motion induced by the flat plate that starts to move along itself at a fixed velocity U at a certain moment (Fig. 4.6). Suppose that the fluid motion is uniform in the x-direction, i.e., $\partial/\partial x = 0$

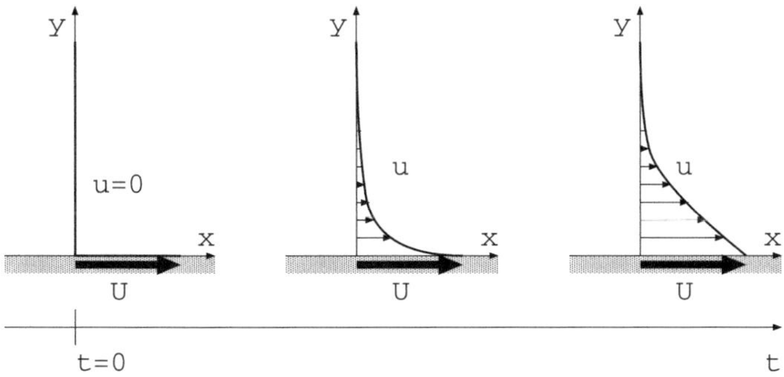

Fig. 4.6 Rayleigh flow

in the continuity equation (4.8) and Navier–Stokes equations (4.9) and (4.10). Since $v = 0$ in the same way as the case of Couette flow, (4.9) becomes

$$\frac{\partial u}{\partial t} = \nu \frac{\partial^2 u}{\partial y^2} . \tag{4.39}$$

The boundary and initial conditions are

$$\left. \begin{array}{l} u = U \text{ at } y = 0 , \\ u \to 0 \text{ as } y \to \infty , \end{array} \right\} \tag{4.40}$$

$$u = 0 \quad \text{at } t = 0 . \tag{4.41}$$

Note that the system in this problem is not characterized by a representative length or time scale. If the variables y, t, and u are non-dimensionalized as $y = L\tilde{y}$, $t = T\tilde{t}$, and $u = U\tilde{u}$, then (4.39)–(4.41) become

$$\frac{\partial \tilde{u}}{\partial \tilde{t}} = \frac{\nu T}{L^2} \frac{\partial^2 \tilde{u}}{\partial \tilde{y}^2} , \tag{4.42}$$

$$\left. \begin{array}{l} \tilde{u} = 1 \text{ at } \tilde{y} = 0 , \\ \tilde{u} \to 0 \text{ as } \tilde{y} \to \infty , \end{array} \right\} \tag{4.43}$$

$$u = 0 \quad \text{at } \tilde{t} = 0 . \tag{4.44}$$

Furthermore, the boundary conditions (4.43) and initial conditions (4.44) described with non-dimensional independent variables become independent of L and T. Even if the time scale T is changed to CT, provided that the length scale is simultaneously changed from L to $\sqrt{C}L$, the dimensionless form of the Navier–Stokes equation

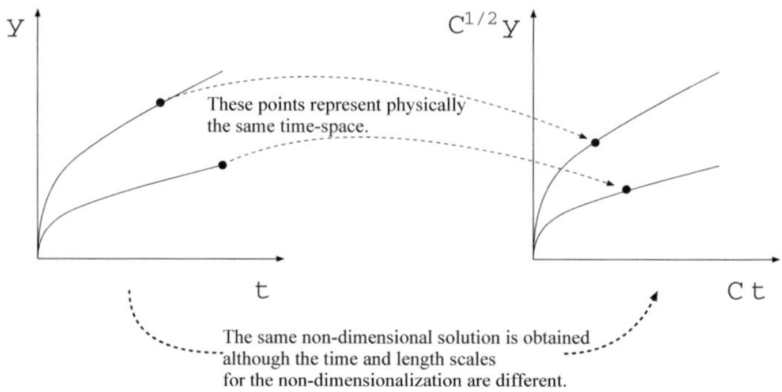

Fig. 4.7 Similar solution

(4.42) remains unchanged. Consequently, the dimensionless Navier–Stokes equation yields the same solution. The solution when y and t are non-dimensionalized by L and T is $\tilde{u}(\tilde{y}, \tilde{t})$, i.e., $u(y/L, t/T)/U$. When y and t are non-dimensionalized by $\sqrt{C}L$ and CT, the solution is $\tilde{u}(\tilde{y}, \tilde{t})$, i.e., $u(y/(\sqrt{C}L), t/(CT))/U$. As these solutions are equal, it suggests that any combination of y and t will yield the same velocity u, provided that the ratio $y/\sqrt{t}$ remains constant. (See Fig. 4.7) Phrased differently, although u is intrinsically a function of both y and t, it can be expressed as a function of a single variable $y/\sqrt{t}$.

Here the solution $u(y, t)$ is expressed in the form

$$u = U\tilde{u}(\eta), \quad \eta \equiv \frac{y}{2\sqrt{\nu t}}. \tag{4.45}$$

(Here, the coefficient of kinematic viscosity ν is introduced to make the equation dimensionless.) Such a solution, which retains its shape after a scale change, is called a **self-similar solution** or a **similar solution**.

Substituting (4.45) into (4.39), we get

$$\left.\begin{aligned} \frac{\partial u}{\partial t} &= U\frac{d\tilde{u}}{d\eta}\frac{\partial \eta}{\partial t} = U\frac{d\tilde{u}}{d\eta}\frac{-y}{4t\sqrt{\nu t}} = -U\frac{d\tilde{u}}{d\eta}\frac{\eta}{2t}, \\ \frac{\partial^2 u}{\partial y^2} &= U\left(\frac{1}{2\sqrt{\nu t}}\frac{d}{d\eta}\right)^2\tilde{u} = \frac{U}{4\nu t}\frac{d^2\tilde{u}}{d\eta^2}. \end{aligned}\right\} \tag{4.46}$$

Hence, (4.39) becomes

$$\frac{d^2\tilde{u}}{d\eta^2} + 2\eta\frac{d\tilde{u}}{d\eta} = 0 \tag{4.47}$$

and the boundary condition and initial conditions become

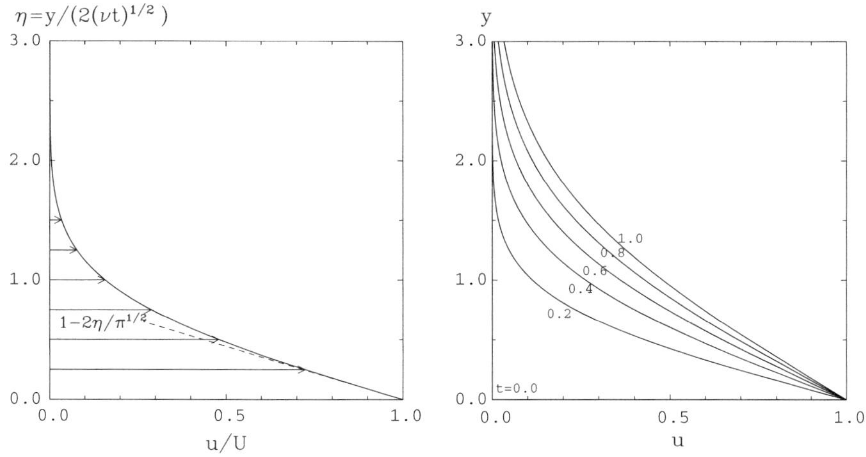

Fig. 4.8 Distribution of Rayleigh flow u: dimensionless graph (*left*) and change over time in the case $U = 1$ and $\nu = 1$ (*right*)

$$\left.\begin{aligned} \tilde{u} &= 1 \ \text{ at } \eta = 0, \\ \tilde{u} &\to 0 \text{ as } \eta \to \infty \,. \end{aligned}\right\} \tag{4.48}$$

Solving (4.47) for $d\tilde{u}/d\eta$

$$\frac{d\tilde{u}}{d\eta} = C_1 e^{-\eta^2} \,,$$

and integrating again gives

$$\tilde{u} = C_1 \frac{\sqrt{\pi}}{2} \mathrm{erf}(\eta) + C_2 \quad \text{where} \quad \mathrm{erf}(\eta) \equiv \frac{2}{\sqrt{\pi}} \int_0^{\eta} e^{-\xi^2}\, d\xi \,. \tag{4.49}$$

The integration constant is determined from the initial and boundary conditions (4.48). Also note that the function $\mathrm{erf}(\eta)$ satisfies $\mathrm{erf}(0) = 0$ and $\mathrm{erf}(\eta) \to 1$ as $\eta \to \infty$. Thus, we obtain $C_1 = -2/\sqrt{\pi}$ and $C_2 = 1$, and u is finally given by

$$u = U\big\{1 - \mathrm{erf}(\eta)\big\}. \tag{4.50}$$

The argument in the velocity solution (4.50), η, is defined in (4.45) as $y/(2\sqrt{\nu t})$. Hence, the range over which the flow is generated by the plate movement is proportional to $\sqrt{\nu t}$ (Fig. 4.8). This flow is called **Rayleigh flow**.

Exercise 4.2 Obtain the flow formed by an infinitely long flat plate with an oscillating surface. The motion of the flat plate is given by $u = U \cos \omega t$, where ω is the angular frequency (assume that the plate steadily oscillates for a sufficiently long time, and the flow has a regular oscillation). The solution should take the form $u = \mathrm{Re}[f(y)\exp(i\omega t)]$.

4.2.5 *Couette Flow in Coaxial Cylinders*

In addition to parallel flow, the Navier–Stokes equation can be exactly solved for two-dimensional flow in concentric circles. In analyzing this type of flow, it is convenient to use (4.20)–(4.23), which express the continuity and Navier–Stokes equations in cylindrical coordinates. The flow is uniform in the ϕ-direction ($\partial/\partial\phi = 0$), and forms concentric circles on the r-ϕ plane ($u_r = 0$). It is a two-dimensional flow ($u_x = 0$). Since the system is symmetric, the pressure can also be assumed to be uniform in the ϕ-direction ($\partial p/\partial\phi = 0$). Under these conditions, the continuity equation (4.20) always holds, and (4.21)–(4.23) are expressed, respectively, as

$$0 = -\frac{1}{\rho}\frac{\partial p}{\partial x}, \tag{4.51}$$

$$-\frac{u_\phi^2}{r} = -\frac{1}{\rho}\frac{\partial p}{\partial r}, \tag{4.52}$$

$$\frac{\partial u_\phi}{\partial t} = \nu\left[\frac{1}{r}\frac{\partial}{\partial r}\left(r\frac{\partial u_\phi}{\partial r}\right) - \frac{u_\phi}{r^2}\right]. \tag{4.53}$$

As an example of such cylindrical flows, we investigate the steady flow of a fluid between two coaxial cylinders (Fig. 4.9). Let us assume that the radii of inner and outer cylinders are a_1 and a_2, respectively. The inner and outer cylinders rotate at fixed angular velocities of Ω_1 and Ω_2, respectively. As steady flow is assumed, the problem equates to solving (4.53) with the left-hand side set to 0:

$$0 = \nu\left[\frac{1}{r}\frac{\partial}{\partial r}\left(r\frac{\partial u_\phi}{\partial r}\right) - \frac{u_\phi}{r^2}\right]. \tag{4.54}$$

Summarizing the terms inside the brackets [ ], we obtain

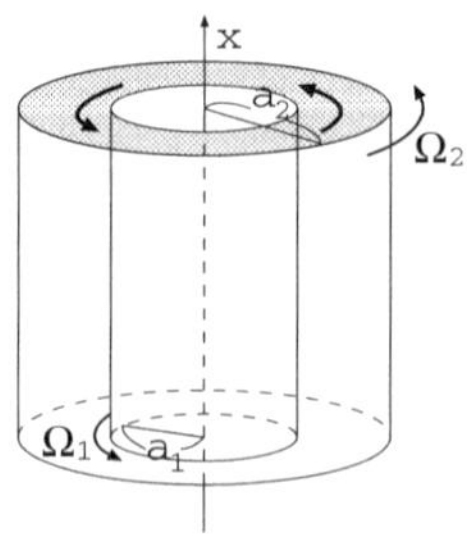

Fig. 4.9 Couette flow between coaxial cylinders

$$\nu \frac{\partial}{\partial r}\left[\frac{1}{r}\frac{\partial}{\partial r}\left(r u_\phi\right)\right] = 0\,. \tag{4.55}$$

The boundary conditions are given by

$$\left.\begin{aligned} u_\phi &= \Omega_1 a_1 \text{ at } r = a_1\,, \\ u_\phi &= \Omega_2 a_2 \text{ at } r = a_2\,. \end{aligned}\right\} \tag{4.56}$$

Dividing both sides of (4.55) by ν and integrating, we get

$$\frac{1}{r}\frac{\partial}{\partial r}\left(r u_\phi\right) = 2C_1\,.$$

Furthermore, multiplying both sides by r, integrating, and then dividing by r, we obtain

$$u_\phi = C_1 r + \frac{C_2}{r}\,. \tag{4.57}$$

From the boundary conditions

$$\Omega_1 a_1 = C_1 a_1 + \frac{C_2}{a_1}\,, \quad \Omega_2 a_2 = C_1 a_2 + \frac{C_2}{a_2}\,,$$

the constants of integration, C_1 and C_2, are obtained as

$$C_1 = \frac{\Omega_2 a_2^2 - \Omega_1 a_1^2}{a_2^2 - a_1^2}\,, \quad C_2 = \frac{(\Omega_1 - \Omega_2) a_1^2 a_2^2}{a_2^2 - a_1^2}\,, \tag{4.58}$$

and the velocity distribution is given by

$$u_\phi = \frac{(\Omega_2 a_2^2 - \Omega_1 a_1^2) r^2 + (\Omega_1 - \Omega_2) a_1^2 a_2^2}{r(a_2^2 - a_1^2)}\,. \tag{4.59}$$

Substituting this u_ϕ into (4.51) and (4.52) and integrating, we obtain the pressure distribution in the flow.

4.2.6 Attenuated Vortex

Vortex filaments with circulation Γ were investigated in Sect. 2.9.4. Such filaments persist in perfect fluids but gradually attenuate in viscous fluids. Let us investigate this situation. Assuming cylindrical coordinates with the x-axis oriented along the direction of the vortex filament (Fig. 4.10), the basic equations become (4.51)–(4.53), and the initial condition is

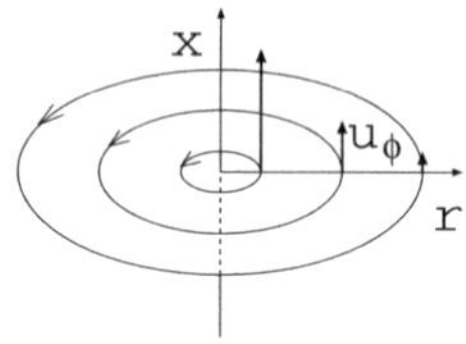

Fig. 4.10 Attenuation of vortex filament

$$u_\phi = \frac{\Gamma}{2\pi r} \quad \text{at } t = 0 \tag{4.60}$$

(see (2.123)). Moreover, the deformation velocity of the flow induced by this initial vortex becomes infinitely large along the x-axis. In a viscous fluid, an infinitely large deformation velocity generates an infinitely large stress. Such singularities are implausible, and must be eliminated. To ensure that no singularity exists at $r = 0$, we impose the following condition:

$$u_\phi = 0 \quad \text{at } r = 0\,. \tag{4.61}$$

Similar to the treatment of Rayleigh flow in Sect. 4.2.4, the solution is obtained in the form of a similar solution:

$$u_\phi = \frac{\Gamma}{2\pi r}\tilde{u}_\phi(\eta) \left(= \frac{\Gamma}{4\pi\sqrt{\nu t}}\frac{\tilde{u}_\phi(\eta)}{\eta} \right), \quad \eta \equiv \frac{r}{2\sqrt{\nu t}}\,. \tag{4.62}$$

Consequently, both sides of (4.53) transform to

$$\left.\begin{aligned} \frac{\partial u_\phi}{\partial t} &= -\frac{\Gamma}{2\pi r}\frac{d\tilde{u}_\phi}{d\eta}\frac{\eta}{2t} = -\frac{\Gamma}{2\pi}\frac{1}{4t\sqrt{\nu t}}\frac{d\tilde{u}_\phi}{d\eta}\,, \\ \nu\left[\frac{1}{r}\frac{\partial}{\partial r}\left(r\frac{\partial u_\phi}{\partial r}\right) - \frac{u_\phi}{r^2}\right] &= \frac{\Gamma}{2\pi}\frac{1}{8t\sqrt{\nu t}}\frac{d}{d\eta}\left(\frac{1}{\eta}\frac{d\tilde{u}_\phi}{d\eta}\right) \end{aligned}\right\} \tag{4.63}$$

and (4.53) becomes

$$2\frac{d\tilde{u}_\phi}{d\eta} + \frac{d}{d\eta}\left(\frac{1}{\eta}\frac{d\tilde{u}_\phi}{d\eta}\right) = 0\,. \tag{4.64}$$

The initial and boundary conditions become

$$\left.\begin{aligned} &\tilde{u}_\phi \to 1 \text{ as } \eta \to \infty\,, \\ &\tilde{u}_\phi = 0 \text{ at } \eta = 0\,. \end{aligned}\right\} \tag{4.65}$$

Integrating (4.64), we obtain

$$2\tilde{u}_\phi + \frac{1}{\eta}\frac{d\tilde{u}_\phi}{d\eta} = 2C_1\,.$$

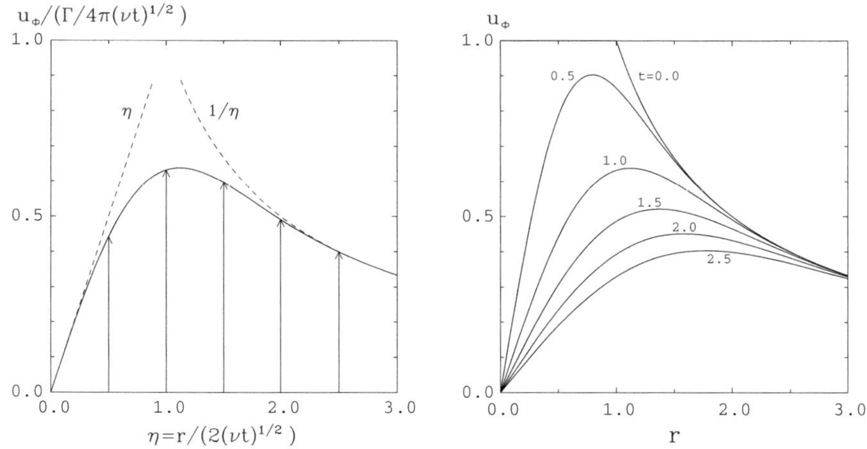

Fig. 4.11 Distribution of u_ϕ in attenuated vortex: (*left*) non-dimensionalized graph and (*right*) temporal evolution of velocity distribution for $\Gamma = 2\pi$, $\nu = 0.25$

Therefore,

$$2\eta(\tilde{u}_\phi - C_1) + \frac{d}{d\eta}(\tilde{u}_\phi - C_1) = 0\,. \tag{4.66}$$

Further integration gives

$$\tilde{u}_\phi - C_1 = C_2 \exp(-\eta^2)\,. \tag{4.67}$$

The integration constants are determined from the initial and boundary conditions (4.65) as $C_1 = 1$ and $C_2 = -1$. Therefore, u_ϕ is finally solved as

$$u_\phi = \frac{\Gamma}{2\pi r}\left[1 - \exp\left(-\frac{r^2}{4\nu t}\right)\right]. \tag{4.68}$$

This result is shown in Fig. 4.11.

In this system, we can also approximate the solutions far from and near the cylindrical axis. In the region far from the cylindrical axis, the term $\exp(-r^2/4\nu t)$ becomes negligible, and therefore,

$$u_\phi \sim \frac{\Gamma}{2\pi r} \quad \text{for } r \gg 2\sqrt{\nu t}\,. \tag{4.69}$$

Sufficiently far from the axis, viscosity exerts a negligible effect and the original velocity distribution of the vortex filament is maintained. By contrast, near the axis, $\exp x \sim 1 + x$ (for $x \ll 1$), and the approximate solution becomes

$$u_\phi \sim \frac{\Gamma}{2\pi r}\left[1 - \left(1 - \frac{r^2}{4\nu t}\right)\right] = \frac{\Gamma}{8\pi\nu t} r \qquad \text{for } r \ll 2\sqrt{\nu t}\,. \tag{4.70}$$

Since u_ϕ is proportional to r, the flow rigidly rotates with a constant angular velocity. The approximate solution (4.70) indicates that viscosity induces rigid rotation of the fluid around the cylindrical axis. The area influenced by the viscosity expands in proportion to $\sqrt{t}$. Since the angular velocity of rigidly rotating flow is inversely proportional to t, the fluid rotation gradually slows down toward motionless state.

4.3 Flows at Low Reynolds Number

4.3.1 Stokes Approximation

In the previous section, exact solutions were found for nonadvective flows, in which the nonlinear term in the Navier–Stokes equation is identically 0. Generally, finding exact solutions to the Navier–Stokes equation in any other case is difficult, but the nonlinear term, which is difficult to handle, becomes small when the Reynolds number is small. This section focuses on flows with at a low Reynolds number. In such flows, the flow velocity is extremely small and the inertial term can be ignored compared with the viscous term as an approximation of Navier–Stokes equation.

The dimensionless Navier–Stokes equation is given by (4.5):

$$\frac{\partial \tilde{\boldsymbol{u}}}{\partial \tilde{t}} + (\tilde{\boldsymbol{u}} \cdot \tilde{\nabla})\tilde{\boldsymbol{u}} = -\tilde{\nabla}\tilde{p}^* + \frac{1}{Re}\tilde{\nabla}^2\tilde{\boldsymbol{u}}\,. \tag{4.71}$$

In flows at low Reynolds numbers, where $Re \ll 1$, the inertial term, particularly the nonlinear (advection) term, can be ignored compared with other terms. The continuity equation and the Navier–Stokes equation neglecting the advection term are written in the dimensional form as

$$\nabla \cdot \boldsymbol{u} = 0, \tag{4.72}$$

$$\frac{\partial \boldsymbol{u}}{\partial t} = \boldsymbol{K} - \frac{1}{\rho}\nabla p + \nu\nabla^2\boldsymbol{u}\,. \tag{4.73}$$

This linear approximation to the Navier–Stokes equation at a low Reynolds number is called **Stokes' approximation**. Flows satisfying this approximation equation are called **Stokes flows**.

Applying $\nabla\times$ to (4.73), we obtain the vorticity equation in Stokes flow

$$\frac{\partial \boldsymbol{\omega}}{\partial t} = \nu\nabla^2\boldsymbol{\omega}\,. \tag{4.74}$$

Clearly, (4.74) becomes the vorticity equation (1.129) derived from the full Navier–Stokes equation along with excluding the nonlinear term. Moreover, applying $\nabla\cdot$ to both sides of (4.73) and using the continuity equation, we obtain

$$\nabla^2 p = 0\,. \tag{4.75}$$

When the boundary condition is given as a condition for the pressure, the solution to (4.75) is the pressure field of Stokes flow.

4.3.2 Slow Flow Around a Sphere Placed in a Uniform Flow

If a sphere of radius a is placed in a uniform flow of velocity U, the spherical surface is subjected to adhesive boundary conditions, and the flows can be solved from the continuity and Navier–Stokes equations. Slow flows are appropriately solved by the Stokes approximations (4.72) and (4.73). In practice, however, the vorticity equation (4.74) is more easily solved than the equation of motion (4.73). Thus, this problem is appropriately solved by (4.72) and (4.74).

For convenience, the problem is set up in spherical coordinates. In this coordinate system, the equations of continuity (4.72) and vorticity (4.73) are written as

$$\frac{1}{r^2}\frac{\partial}{\partial r}\left(r^2 u_r\right) + \frac{1}{r\sin\theta}\frac{\partial}{\partial\theta}(\sin\theta u_\theta) + \frac{1}{r\sin\theta}\frac{\partial u_\phi}{\partial\phi} = 0\,, \tag{4.76}$$

$$\frac{\partial\omega_r}{\partial t} = \nu\left[\nabla^2\omega_r - \frac{2\omega_r}{r^2} - \frac{2}{r^2\sin\theta}\frac{\partial}{\partial\theta}(\omega_\theta\sin\theta) - \frac{2}{r^2\sin^2\theta}\frac{\partial\omega_\phi}{\partial\phi}\right], \tag{4.77}$$

$$\frac{\partial\omega_\theta}{\partial t} = \nu\left[\nabla^2\omega_\theta + \frac{2}{r^2}\frac{\partial\omega_r}{\partial\theta} - \frac{\omega_\theta}{r^2\sin^2\theta} - \frac{2\cos\theta}{r^2\sin^2\theta}\frac{\partial\omega_\phi}{\partial\phi}\right], \tag{4.78}$$

$$\frac{\partial\omega_\phi}{\partial t} = \nu\left[\nabla^2\omega_\phi + \frac{2}{r^2\sin^2\theta}\frac{\partial\omega_r}{\partial\phi} + \frac{2\cos\theta}{r^2\sin^2\theta}\frac{\partial\omega_\theta}{\partial\phi} - \frac{\omega_\phi}{r^2\sin^2\theta}\right], \tag{4.79}$$

where

$$\nabla^2 \equiv \frac{1}{r^2}\frac{\partial}{\partial r}\left(r^2\frac{\partial}{\partial r}\right) + \frac{1}{r^2\sin\theta}\frac{\partial}{\partial\theta}\left(\sin\theta\frac{\partial}{\partial\theta}\right) + \frac{1}{r^2\sin\theta}\frac{\partial^2}{\partial\phi^2}\,. \tag{4.80}$$

($\nabla\cdot A$ and $\nabla^2 A$ are expressed in spherical coordinates (r, θ, ϕ) in Appendix A.2 (3)).

As shown in Fig. 4.12, the direction of uniform flow U at infinity is chosen as the polar axis. From symmetry considerations, the flow is clearly uniform in the ϕ-direction ($\partial/\partial\phi = 0$) and there is no ϕ-velocity component ($u_\phi = 0$). In such axisymmetric flows, it is convenient to introduce the stream function Ψ_a defined by

$$\left.\begin{aligned} u_r &= \frac{1}{r^2\sin\theta}\frac{\partial\Psi_a}{\partial\theta}\,, \\ u_\theta &= -\frac{1}{r\sin\theta}\frac{\partial\Psi_a}{\partial r}\,. \end{aligned}\right\} \tag{4.81}$$

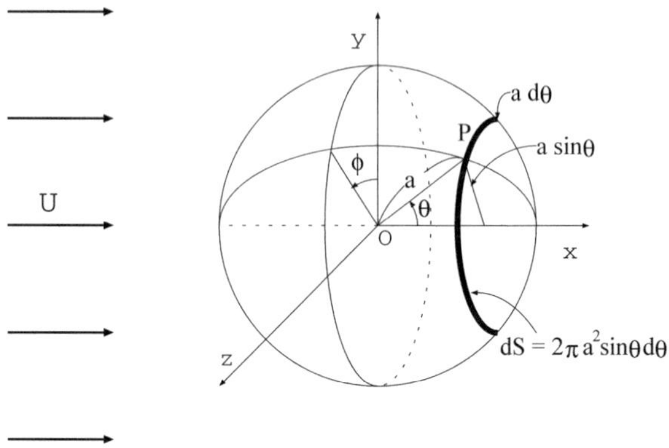

Fig. 4.12 Flow around a sphere and setting of spherical coordinates

(See Sect. 2.7.2. However, since this function was expressed in cylindrical coordinates in Sect. 2.7.2, to adopt (2.92) to (4.81), coordinate transformations must be done). By introducing Eq. (4.81), the equation of continuity (4.76) is satisfied. Using the stream function Ψ_a, we solve the ϕ-component (4.79) of the vorticity equation corresponding to steady Stokes' flow (as we set $\partial/\partial\phi = 0$ and $u_\phi = 0$, the r-component equation (4.77) and θ-component equation (4.78) are automatically satisfied). Hence, setting (4.79) to a steady state ($\partial/\partial t = 0$), we obtain

$$\begin{aligned}(\nabla^2\boldsymbol{\omega})_\phi &= \nabla^2\omega_\phi + \frac{2}{r^2\sin\theta}\underbrace{\frac{\partial\omega_r}{\partial\phi}}_{0} + \frac{2\cos\theta}{r^2\sin^2\theta}\underbrace{\frac{\partial\omega_\theta}{\partial\phi}}_{0} - \frac{\omega_\phi}{r^2\sin^2\theta}\\ &= \frac{1}{r^2}\frac{\partial}{\partial r}\left(r^2\frac{\partial\omega_\phi}{\partial r}\right) + \frac{1}{r^2\sin\theta}\frac{\partial}{\partial\theta}\left(\sin\theta\frac{\partial\omega_\phi}{\partial\theta}\right) + \frac{1}{r^2\sin^2\theta}\underbrace{\frac{\partial^2\omega_\phi}{\partial\phi^2}}_{0} - \frac{\omega_\phi}{r^2\sin^2\theta}\\ &= \left[\frac{1}{r^2}\frac{\partial}{\partial r}\left(r^2\frac{\partial}{\partial r}\right) + \frac{1}{r^2\sin\theta}\frac{\partial}{\partial\theta}\left(\sin\theta\frac{\partial}{\partial\theta}\right) - \frac{1}{r^2\sin^2\theta}\right]\omega_\phi\\ &= 0\,. \end{aligned} \tag{4.82}$$

In polar coordinates, ω_ϕ becomes

$$\omega_\phi \equiv (\nabla\times\boldsymbol{u})_\phi = \frac{1}{r}\left(\frac{\partial(ru_\theta)}{\partial r} - \frac{\partial u_r}{\partial\theta}\right). \tag{4.83}$$

Substituting (4.81) in (4.83) (see spherical coordinate (r, θ, ϕ) form of $\nabla\times\boldsymbol{A}$ in Appendix A.2 (3)), in terms of Ψ_a, we get

$$\omega_\phi = -\frac{1}{r}\left[\frac{\partial}{\partial r}\left(\frac{1}{\sin\theta}\frac{\partial \Psi_a}{\partial r}\right)+\frac{\partial}{\partial\theta}\left(\frac{1}{r^2\sin\theta}\frac{\partial \Psi_a}{\partial\theta}\right)\right]$$
$$= -\frac{1}{r\sin\theta}\underbrace{\left[\frac{\partial^2\Psi_a}{\partial r^2}+\frac{\sin\theta}{r^2}\frac{\partial}{\partial\theta}\left(\frac{1}{\sin\theta}\frac{\partial\Psi_a}{\partial\theta}\right)\right]}_{X}. \tag{4.84}$$

Substituting (4.84) into the polar coordinate expressions of $(\nabla^2\omega)_\phi$ and expressing with Ψ_a, we get

$$(\nabla^2\omega)_\phi = -\frac{1}{r^2}\frac{\partial}{\partial r}\left[r^2\frac{\partial}{\partial r}\left(\frac{X}{r\sin\theta}\right)\right]-\frac{1}{r^2\sin\theta}\frac{\partial}{\partial\theta}\left[\sin\theta\frac{\partial}{\partial\theta}\left(\frac{X}{r\sin\theta}\right)\right]+\frac{X}{r^3\sin^3\theta}$$
$$= -\frac{1}{r^2\sin\theta}\frac{\partial}{\partial r}\left(r\frac{\partial X}{\partial r}-X\right)-\frac{1}{r^3\sin\theta}\frac{\partial}{\partial\theta}\left(\frac{\partial X}{\partial\theta}-\cot\theta\, X\right)+\frac{X}{r^3\sin^3\theta}$$
$$= -\frac{1}{r\sin\theta}\frac{\partial^2 X}{\partial r^2}-\frac{1}{r^3\sin\theta}\frac{\partial^2 X}{\partial\theta^2}+\frac{\cos\theta}{r^3\sin^2\theta}\frac{\partial X}{\partial\theta}\underbrace{-\frac{X}{r^3\sin^3\theta}+\frac{X}{r^3\sin^3\theta}}_{0}$$
$$= -\frac{1}{r\sin\theta}\frac{\partial^2 X}{\partial r^2}-\frac{1}{r^3}\frac{\partial}{\partial\theta}\left(\frac{1}{\sin\theta}\frac{\partial X}{\partial\theta}\right)$$
$$= -\frac{1}{r\sin\theta}\left[\frac{\partial^2}{\partial r^2}+\frac{\sin\theta}{r^2}\frac{\partial}{\partial\theta}\left(\frac{1}{\sin\theta}\frac{\partial}{\partial\theta}\right)\right]X. \tag{4.85}$$

Finally, Eq. (4.82) becomes

$$\left[\frac{\partial^2}{\partial r^2}+\frac{\sin\theta}{r^2}\frac{\partial}{\partial\theta}\left(\frac{1}{\sin\theta}\frac{\partial}{\partial\theta}\right)\right]^2\Psi_a = 0. \tag{4.86}$$

In polar coordinates, the non-slip condition at $r = a$ and the boundary condition of the uniform flow U at infinity become

$$\left.\begin{array}{lll} u_r = u_\theta = 0 & \text{at} & r = a,\\ u_r \to U\cos\theta,\ \ u_\theta \to -U\sin\theta & \text{as} & r\to\infty. \end{array}\right\} \tag{4.87}$$

Together with the boundary condition (4.87), the differential equation (4.86) is solved to obtain Ψ_a as follows:

$$\Psi_a = \frac{1}{4}Ua^2\sin^2\theta\left(\frac{2r^2}{a^2}-\frac{3r}{a}+\frac{a}{r}\right). \tag{4.88}$$

Exercise 4.3 Assuming that solutions to the differential equation (4.86) are of the form $\Psi_a^{(n)} \equiv r^n\sin^2\theta$, show that there are four values of n that can satisfy (4.86). Express the solution to (4.86) in terms of these four linear combinations and determine their coefficients from the boundary condition (4.87). Hence, derive (4.88).

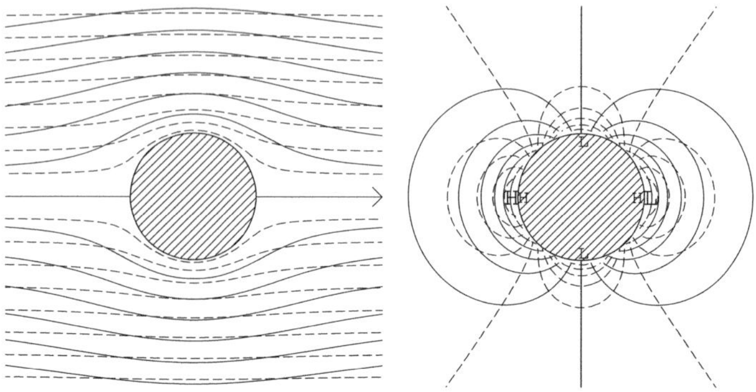

Fig. 4.13 Stokes flow (*solid lines*) and potential flow (*broken lines*) around a sphere: streamline (*left*) and pressure distribution (*right*)

Since the obtained stream function Ψ_a is symmetric about the plane $\theta = \pi/2$, the flow is symmetric ahead and behind the uniform flow by virtue of ignoring the advection term in the Stokes approximation (Fig. 4.13). The flow velocity components u_r and u_θ are obtained from Ψ_a as

$$u_r = U\cos\theta\left(1 - \frac{3a}{2r} + \frac{a^3}{2r^3}\right), \tag{4.89}$$

$$u_\theta = U\sin\theta\left(-1 + \frac{3a}{4r} + \frac{a^3}{4r^3}\right). \tag{4.90}$$

We now calculate the pressure in the flow. Since the flow is steady, from (4.73), we have

$$\nabla p = \rho\nu\nabla^2 \boldsymbol{u}\,. \tag{4.91}$$

From (4.89) and (4.90) (and also from $\nabla^2 \boldsymbol{A}$ in spherical coordinates (r, θ, ϕ); see Appendix A.2 (3)), the r-component on the right-hand side of (4.91) can be written as

$$(\nabla^2 \boldsymbol{u})_r = \nabla^2 u_r - \frac{2u_r}{r^2} - \frac{2}{r^2\sin\theta}\frac{\partial}{\partial\theta}(u_\theta\sin\theta) - \frac{2}{r^2\sin^2\theta}\underbrace{\frac{\partial u_\phi}{\partial\phi}}_{0}$$

$$= \underbrace{\frac{1}{r^2}\frac{\partial}{\partial r}\left(r^2\frac{\partial u_r}{\partial r}\right)}_{3U\cos\theta\frac{a^3}{r^5}} + \underbrace{\frac{1}{r^2\sin\theta}\frac{\partial}{\partial\theta}\left(\sin\theta\frac{\partial u_r}{\partial\theta}\right)}_{-2U\cos\theta\frac{1}{r^2}(1-\frac{3a}{2r}+\frac{a^3}{2r^3})}$$

$$
+\frac{1}{r^2\sin^2\theta}\underbrace{\frac{\partial^2 u_r}{\partial\phi^2}}_{0}-\frac{2u_r}{r^2}\underbrace{-\frac{2}{r^2\sin\theta}\frac{\partial}{\partial\theta}(u_\theta\sin\theta)}_{-4U\cos\theta\frac{1}{r^2}(-1+\frac{3a}{4r}+\frac{a^3}{4r^3})}
$$

$$
=3U\cos\theta\frac{a}{r^3}\,. \tag{4.92}
$$

Therefore, the r-component of (4.91) becomes

$$
\frac{\partial p}{\partial r}=3\rho\nu U\cos\theta\frac{a}{r^3}\,. \tag{4.93}
$$

Integrating both sides of (4.93) from infinity to r, p is obtained as

$$
p=p_\infty-\frac{3\mu a U}{2r^2}\cos\theta \tag{4.94}
$$

where p_∞ is the pressure at $r\to\infty$.

Using (4.89) and (4.90), we can determine the deformation velocity $e_{r\theta}$ (see deformation velocity tensor in spherical coordinates (r,θ,ϕ) in Appendix A.2 (3)):

$$
e_{r\theta}=\frac{1}{2}\Bigg[\underbrace{\frac{1}{r}\frac{\partial u_r}{\partial\theta}}_{-U\sin\theta(1-\frac{3a}{2r}+\frac{a^3}{2r^3})}+r\underbrace{\frac{\partial}{\partial r}\left(\frac{u_\theta}{r}\right)}_{U\sin\theta(\frac{1}{r^2}-\frac{3a}{2r^3}-\frac{a^3}{r^5})}\Bigg]=-\frac{3}{4}U\sin\theta\frac{a^3}{r^4}. \tag{4.95}
$$

Hence, the shear stress $\tau_{r\theta}$ is given by

$$
\tau_{r\theta}=2\mu e_{r\theta}=-\frac{3}{2}\frac{U\mu\sin\theta}{r}\frac{a^3}{r^3}\,. \tag{4.96}
$$

Exercise 4.4 Obtain the deformation velocity tensor $(e_{rr}, e_{\theta\theta}, e_{\phi\phi}, e_{r\theta}, e_{\theta\phi}, e_{\phi r})$ in a Stokes flow past a sphere, and show that the shear stress operating on the surface of the sphere is all zero except for $\tau_{r\theta}$.

Integrating the stress and shear stress applied to the surface of the sphere, the total drag in the flow direction, F_x, is obtained as follows (Fig. 4.14).

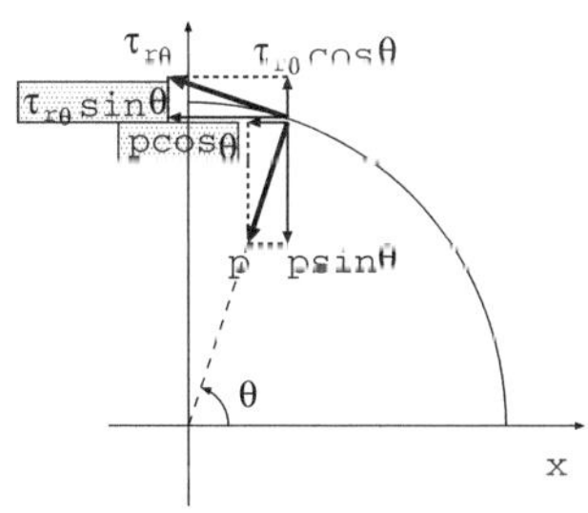

Fig. 4.14 Calculation of drag force acting on the sphere

$$
\begin{aligned}
F_x &= -\int_S \tau_{r\theta}|_{r=a} \sin\theta\, dS - \int_S p|_{r=a} \cos\theta\, dS \\
&= \int_0^{\pi} \frac{3}{2}\frac{U\mu\sin\theta}{a} 2\pi a^2 \sin^2\theta\, d\theta - \int_0^{\pi}\left(p_\infty - \frac{3\mu U}{2a}\cos\theta\right) 2\pi a^2 \cos\theta\sin\theta\, d\theta \\
&= 6\pi\mu U a\,.
\end{aligned}
\tag{4.97}
$$

This result shows that in flows with a low Reynolds number, the resistive force acting on the sphere is proportional to the velocity U. This relationship is known as Stokes' law of resistance.

Defining the drag coefficient C_D of an object with cross-sectional area A in a uniform flow by $C_D \equiv F/(\frac{1}{2}\rho U^2 A)$, the drag coefficient of a sphere in a Stokes flow is given by

$$
C_D = \frac{6\pi\mu U a}{\frac{\rho}{2}U^2\pi a^2} = \frac{24}{Re}\,. \tag{4.98}
$$

4.4 Boundary Layer Theory

4.4.1 Flows with High Reynolds Number

We now consider the opposite situation for the previous section: flows with a high Reynolds number. In such flows, the viscous term in the Navier–Stokes equation is small, and if that term is ignored, the Navier–Stokes equation becomes the same as Euler's equation, which describes the motion of a perfect fluid. As estimated, essentially, as the Reynolds number increases (viscosity decreases), the fluid increasingly resembles a perfect fluid. However, fluids of any viscosity, no matter how small, behave fundamentally differently from perfect fluids.

Note 4.2 To exemplify the difference between a viscous fluid with sufficiently low viscosity and a non-viscous fluid (perfect fluid flow), let us consider a two-dimensional steady flow with closed streamlines.

In a steady flow, where $\partial/\partial t = 0$, the two-dimensional vorticity equation (1.131) is reduced to

$$
(\boldsymbol{u}\cdot\nabla)\omega = \nu\nabla^2\omega \tag{4.99}
$$

where $\omega \equiv \partial v/\partial x - \partial u/\partial y$. Introducing the following stream function Ψ

$$
u = \frac{\partial\Psi}{\partial y}\,, \quad v = -\frac{\partial\Psi}{\partial x}\,, \tag{4.100}
$$

Equation (4.99) can be written as[2]

[2]Left-hand side of (4.101) is in the Jacobian form: $-J(\Psi,\omega)$. Equation (4.105) is deduced from a mathematically well-known property, in which f becomes a function of g if $J(f,g) = 0$.

$$\frac{\partial \Psi}{\partial y}\frac{\partial \omega}{\partial x} - \frac{\partial \Psi}{\partial x}\frac{\partial \omega}{\partial y} = \nu \nabla^2 \omega \,. \tag{4.101}$$

Because the right-hand side of (4.101) can be ignored at sufficiently small viscosity, we can write

$$\frac{\partial \Psi}{\partial y}\frac{\partial \omega}{\partial x} - \frac{\partial \Psi}{\partial x}\frac{\partial \omega}{\partial y} = 0 \,. \tag{4.102}$$

In other words,

$$\frac{\partial \omega}{\partial x}\bigg/\frac{\partial \omega}{\partial y} = \frac{\partial \Psi}{\partial x}\bigg/\frac{\partial \Psi}{\partial y} \,. \tag{4.103}$$

Therefore,

$$\nabla \omega \parallel \nabla \Psi \,. \tag{4.104}$$

This result indicates that the isolines of ω and Ψ are parallel (Fig. 4.15). Therefore, ω can be expressed as a function of Ψ:

$$\omega = \omega(\Psi). \tag{4.105}$$

Integrating (4.99) over the area $S(\Psi)$ surrounded by a closed streamline $C(\Psi)$ (Fig. 4.16)

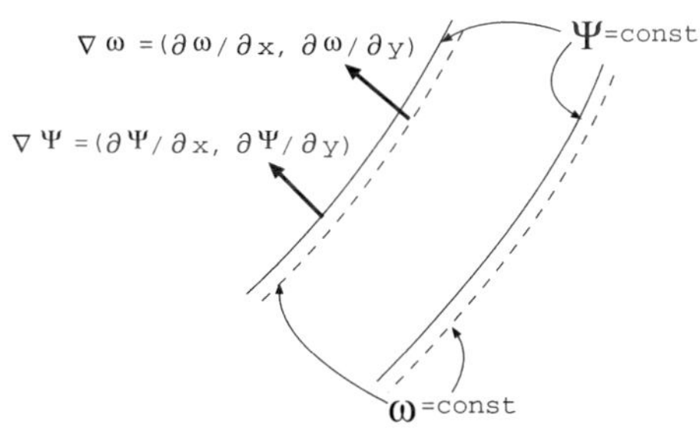

Fig. 4.15 Contour line of vorticity and streamline are parallel

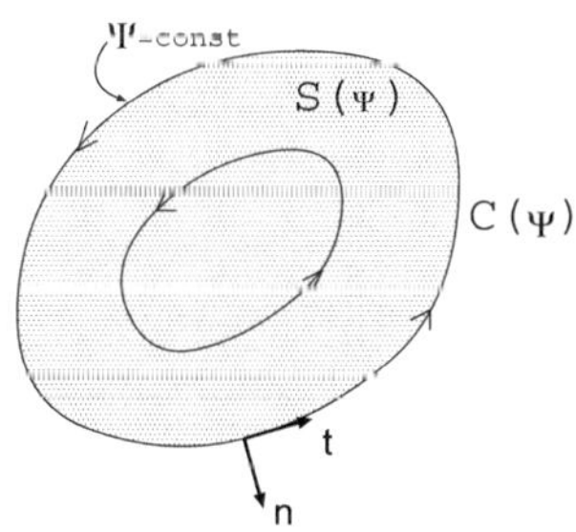

Fig. 4.16 Closed streamline $C(\Psi)$ and area $S(\Psi)$ surrounded by it

$$\int_{S(\Psi)} (\boldsymbol{u} \cdot \nabla)\omega \, dS = \nu \int_{S(\Psi)} \nabla^2 \omega \, dS. \tag{4.106}$$

Both sides of (4.106) are

$$\text{(Left-hand side)} \underset{\nabla\cdot\boldsymbol{u}=0}{\overset{}{=}} \int_{S(\Psi)} \nabla \cdot (\omega \boldsymbol{u}) \, dS \underset{\text{Gauss' theorem}}{=} \int_{C(\Psi)} \omega \boldsymbol{u} \cdot \boldsymbol{n} \, d\ell \underset{\boldsymbol{u}\cdot\boldsymbol{n}=0}{=} 0,$$

$$\text{(Right-hand side)} = \nu \int_{S(\Psi)} \nabla \cdot (\nabla\omega) \, dS \underset{\text{Gauss' theorem}}{=} \nu \int_{C(\Psi)} (\nabla\omega) \cdot \boldsymbol{n} \, d\ell = \nu \frac{d\omega}{d\Psi} \int_{C(\Psi)} (\nabla\Psi) \cdot \boldsymbol{n} \, d\ell$$

$$= -\nu \frac{d\omega}{d\Psi} \int_{C(\Psi)} \boldsymbol{u} \cdot \boldsymbol{t} \, d\ell = -\nu \frac{d\omega}{d\Psi} \Gamma(\Psi)$$

where $\boldsymbol{n}$ and $\boldsymbol{t}$ are unit normal and tangential vectors of the streamline $C(\Psi)$, respectively, and $\Gamma(\Psi)$ is the circulation along $C(\Psi)$. Hence,

$$\nu \frac{d\omega}{d\Psi} \Gamma(\Psi) = 0. \tag{4.107}$$

As long as it is not motionless, circulation $\Gamma(\Psi)$ along the streamline is not zero. Therefore, if $\nu \neq 0$,

$$\frac{d\omega}{d\Psi} = 0 \tag{4.108}$$

and in the area where the streamline is closed,

$$\omega = \text{const}\,. \tag{4.109}$$

(4.109) indicates that vorticity within the area of a closed streamline becomes uniform in two-dimensional steady flow with sufficiently small viscosity. This is called the **Prandtl–Batchelor theorem**.

Vorticity is transported by advection $-(\boldsymbol{u} \cdot \nabla)\omega$ and viscosity $\nabla^2\omega$. Comparing the role of these two effects at each point shows that advection dominates when viscosity is small. Therefore, essentially, contour lines of vorticity follow the streamline so that there is no temporal change in vorticity. This is what (4.103) and (4.104) imply. However, advective transport only occurs along the streamline, and once it travels around the streamline, it returns. Therefore, regardless of the amount of advective transport at each point, if it is integrated within a closed streamline, the effect of the advection vanishes, and vorticity transport over the streamline is only conducted by viscosity. To have a steady flow, vorticity transport in or out of a streamline cannot occur. Thus, vorticity gradient cannot exist, and the conclusion of (4.109) is derived. (See Fig. 4.17.)

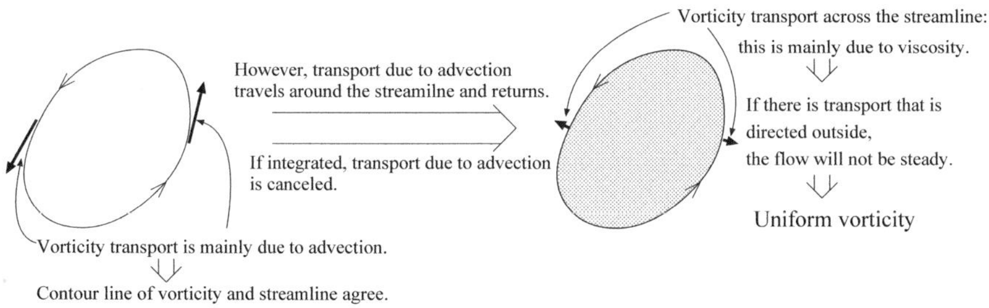

Fig. 4.17 Prandtl–Batchelor theorem seen from vorticity balance

In the above argument, we now replace "a viscous fluid of sufficiently low viscosity" with "a perfect fluid of zero viscosity." The derivation of (4.105) is equally applicable to the perfect fluid, and ω is similarly a function of Ψ. However, in the argument in the latter half, when $\nu = 0$, (4.108) cannot be derived from (4.107). In a perfect fluid, even if vorticity is not uniform, (4.107) can be true. Thus, as long as $\omega = \omega(\Psi)$, a steady flow of various vorticity distributions can exist, and it becomes a completely different flow from viscous fluid with sufficiently small viscosity.

Moreover, as can be understood from the derivation of the theorem, when the flow is two-dimensional and some physical quantity is transferred only through advection and weak viscosity or diffusion and there are no sources or suctions, a similar conclusion is attained. The large-scale movement of the atmosphere and oceans is a semi-two-dimensional flow where movement in the vertical direction is suppressed. Potential vorticity (details are explained in Sect. 7.2) is a physical quantity that satisfies such conditions; thus, in a closed streamline with small external force, the potential vorticity is mainly uniform.

The difference between a low-viscosity fluid and a perfect fluid manifests most strikingly near object boundaries. Even when the viscosity is sufficiently small, viscous fluid is a viscous fluid, and the required boundary conditions are fundamentally different from those of a perfect fluid. In the boundary conditions of a stationary rigid wall, only the normal velocity component at the surface of an object in perfect fluids must be 0 (see Sect. 1.5.3). In contrast, velocity must be 0, along with the tangential component for viscous fluids (see final part of Sect. 1.8). If flows with a high Reynolds number behave almost identical to perfect fluids in the flow interior, the flow along the boundary retains large values until it is close to the solid wall (tangential velocity component does not have to be 0 at the solid wall in a perfect fluid). However, the velocity of viscous fluids must be 0 at the boundary of a solid. Consequently, even if the velocity itself is relatively small, there exist layers with large shear deformation velocity near the body surface. Since the viscous term is a product of the velocity's (second order) derivative and the coefficient of viscosity, within this layer, the viscous term cannot be ignored because of the large shear deformation velocity, even though the coefficient is small. Such a thin layer along a fixed surface is called a **boundary layer**.

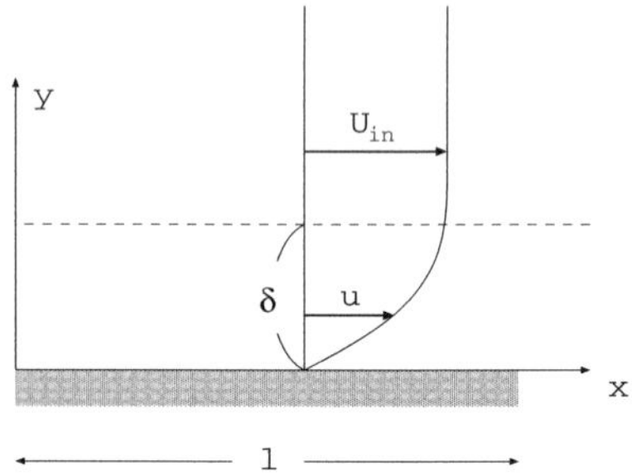

Fig. 4.18 Boundary layer

4.4.2 Boundary Layer Equation

Consider a two-dimensional flow with a boundary plate on one side (Fig. 4.18). The flow field with a sufficiently large Reynolds number can be divided into two areas: area where the viscous term can be ignored, and a boundary layer where large velocity gradients formed near the wall and viscous term cannot be ignored. The thickness of the boundary layer, δ, is considerably smaller than the representative length of the flow, l. Thus,

$$\delta \ll l\,. \tag{4.110}$$

Now evaluate the order of each term in the equation that describes the flow inside the boundary layer.

First, the equation of continuity for a two-dimensional flow is given by

$$\frac{\partial u}{\partial x} + \frac{\partial v}{\partial y} = 0\,. \tag{4.111}$$

However, if we assume $u \sim O(U)$, $x \sim O(l)$, and $y \sim O(\delta)$, the equation of continuity requires the following condition to be satisfied:

$$v \sim O\left(\frac{\delta}{l}U\right)\,. \tag{4.112}$$

To account for the order specified above, we introduce the following non-dimensional variables:

$$t = \frac{l}{U}\tilde{t},\ x = l\tilde{x},\ y = \delta\tilde{y},\ u = U\tilde{u},\ v = \frac{\delta}{l}U\tilde{v},\ p = \rho U^2\tilde{p}. \tag{4.113}$$

In (4.113), dimensionless quantities are indicated by $\tilde{}$.

Using these scales to non-dimensionalize the Navier–Stokes equation, we get

$$\frac{\partial \tilde{u}}{\partial \tilde{t}} + \tilde{u}\frac{\partial \tilde{u}}{\partial \tilde{x}} + \tilde{v}\frac{\partial \tilde{u}}{\partial \tilde{y}} = -\frac{\partial \tilde{p}}{\partial \tilde{x}} + \frac{1}{Re}\left[\frac{\partial^2 \tilde{u}}{\partial \tilde{x}^2} + \left(\frac{l}{\delta}\right)^2 \frac{\partial^2 \tilde{u}}{\partial \tilde{y}^2}\right], \tag{4.114}$$

$$\frac{\partial \tilde{v}}{\partial \tilde{t}} + \tilde{u}\frac{\partial \tilde{v}}{\partial \tilde{x}} + \tilde{v}\frac{\partial \tilde{v}}{\partial \tilde{y}} = -\left(\frac{l}{\delta}\right)^2 \frac{\partial \tilde{p}}{\partial \tilde{y}} + \frac{1}{Re}\left[\frac{\partial^2 \tilde{v}}{\partial \tilde{x}^2} + \left(\frac{l}{\delta}\right)^2 \frac{\partial^2 \tilde{v}}{\partial \tilde{y}^2}\right]. \tag{4.115}$$

Within the boundary layer, the viscosity and inertial terms are of the same order even for small $1/Re$. However, in (4.114) and (4.115), the viscous and inertial terms must be of the same order even when $Re \gg 1$. To satisfy this condition, we specify

$$\frac{1}{Re}\left(\frac{l}{\delta}\right)^2 \sim O(1). \tag{4.116}$$

Hence, the thickness of the boundary layer is given by

$$\frac{\delta}{l} \sim \frac{1}{\sqrt{Re}}. \tag{4.117}$$

As (4.117) is inversely proportionally to $\sqrt{Re}$, the boundary layer becomes increasingly thin as the Reynolds number increases. Extracting the terms of largest orders only from (4.114) and (4.115), we get

$$\frac{\partial \tilde{u}}{\partial \tilde{t}} + \tilde{u}\frac{\partial \tilde{u}}{\partial \tilde{x}} + \tilde{v}\frac{\partial \tilde{u}}{\partial \tilde{y}} = -\frac{\partial \tilde{p}}{\partial \tilde{x}} + \frac{1}{Re}\left(\frac{l}{\delta}\right)^2 \frac{\partial^2 \tilde{u}}{\partial \tilde{y}^2}, \tag{4.118}$$

$$\frac{\partial \tilde{p}}{\partial \tilde{y}} = 0. \tag{4.119}$$

According to (4.119), no pressure gradient exists along the y-direction in the boundary layer, and the boundary layer is penetrated by adjacent exterior pressure. If the flow outside the boundary layer is assumed to be parallel to the flat plate and given by $U_{\text{in}}(x, t)$, the following relation is established:

$$\frac{\partial U_{\text{in}}}{\partial t} + U_{\text{in}}\frac{\partial U_{\text{in}}}{\partial x} = -\frac{1}{\rho}\frac{\partial p}{\partial x}. \tag{4.120}$$

From the above scaling argument, the boundary layer equation for a two-dimensional flow along a flat plate is given by

$$\frac{\partial u}{\partial x} + \frac{\partial v}{\partial y} = 0, \tag{4.121}$$

$$\frac{\partial u}{\partial t} + u\frac{\partial u}{\partial x} + v\frac{\partial u}{\partial y} = -\frac{1}{\rho}\frac{\partial p}{\partial x} + \nu\frac{\partial^2 u}{\partial y^2}, \tag{4.122}$$

$$-\frac{1}{\rho}\frac{\partial p}{\partial x} = \frac{\partial U_{\text{in}}}{\partial t} + U_{\text{in}}\frac{\partial U_{\text{in}}}{\partial x}. \tag{4.123}$$

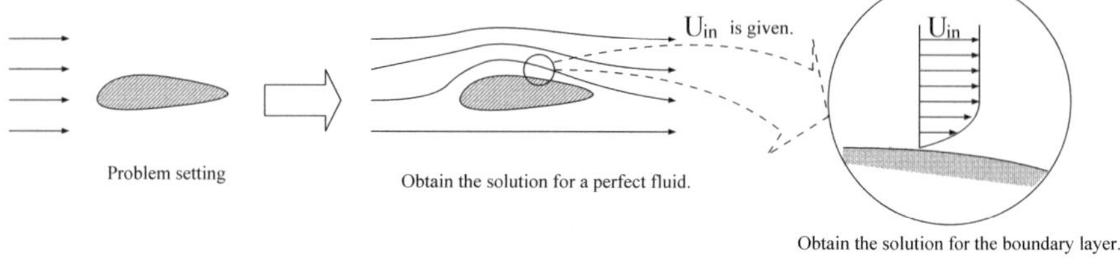

Fig. 4.19 Analysis of a flow using the boundary layer theory

This equation is known as **Prandtl's boundary layer equation**. As it connects to the internal regions over the boundary layer, the following boundary conditions apply:

$$\left.\begin{aligned} &u = 0,\ v = 0 \text{ at } y = 0, \\ &u \to U_{\text{in}}(x, t) \text{ as } y \to \infty . \end{aligned}\right\} \tag{4.124}$$

While pressure p is generally an unknown quantity in the Navier–Stokes solutions, p in the boundary layer equation is known and equals the pressure in the internal region. When problems for flows with a high Reynolds number are practically solved using the boundary layer theory, the following procedures are employed (Fig. 4.19):

1. In the internal region, viscosity can be ignored. Thus, at this stage, if Euler's equation is treated as a perfect fluid ignoring viscosity, $\boldsymbol{u}_{\text{in}}$ and p_{in} for the internal region can be solved.
2. Set the boundary layer p equal to the solution p_{in} within the internal region. Alternatively, set the boundary layer U_{in} equal to the solution u_{in} within the internal region at the upper border of the boundary layer, and obtain p from (4.123).
3. Assuming that p is known within the boundary layer equation, simultaneously solve (4.121) and (4.122), and determine $\boldsymbol{u}$ within the boundary layer.

4.4.3 Thickness of the Boundary Layer

In the previous section, the estimated thickness of the boundary layer was of the order δ. In reality, the boundary layer is not clearly separated from the internal region. Therefore, deciding the thickness δ of the boundary layer is not straightforward. Many researchers use the displacement thickness δ_1 or momentum thickness δ_2 (defined below) as a representative length scale of boundary layer thickness. If there is no fixed wall, the uniform flow U_0 is free; however, if a boundary layer is established by a fixed wall, a portion of the flow (near the fixed wall) is slowed, with corresponding decrease in the mass flow. We now focus on the reduction in mass flow. As a yardstick for deciding the boundary layer thickness, we use "the thickness

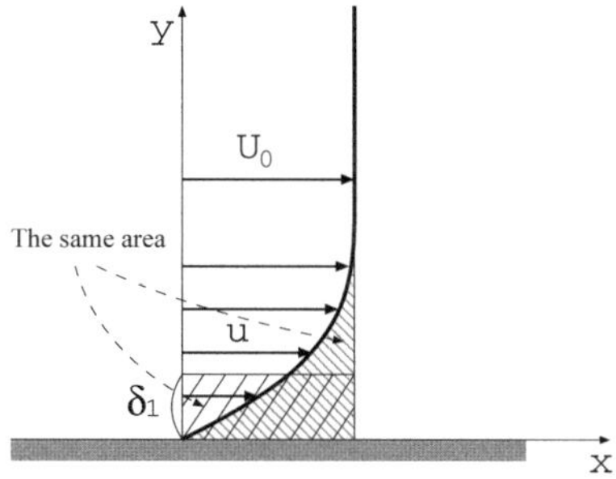

Fig. 4.20 Displacement thickness

δ_1 at which $\rho U_0 \delta_1$ equals the loss of mass flow." This thickness is called the **displacement thickness** (Fig. 4.20). Similarly, we can consider momentum reduction, and specify "the thickness δ_2 at which $\rho U_0^2 \delta_2$ equals the amount of momentum lost" as our yardstick. This thickness is called the **momentum thickness**. The displacement thickness δ_1 and the momentum thickness δ_2 are given by

$$\rho \int_0^\infty (U_0 - u) dy = \rho U_0 \delta_1 ,$$

$$\rho \int_0^\infty u \left(U_0 - u\right) dy = \rho U_0^2 \delta_2$$

i.e.,

$$\delta_1 = \int_0^\infty \left(1 - \frac{u}{U_0}\right) dy , \tag{4.125}$$

$$\delta_2 = \int_0^\infty \frac{u}{U_0} \left(1 - \frac{u}{U_0}\right) dy . \tag{4.126}$$

4.4.4 Analytical Solution of Boundary Layer Equation

In 1908, Blasius analyzed the flow along flat plates placed in the direction of uniform flow on the basis of Prandtl's boundary layer theory. The boundary layer equations on a flat plate are given by (4.121)–(4.123). For a plate in a steady uniform flow (Fig. 4.21), these are altered to

$$\frac{\partial u}{\partial x} + \frac{\partial v}{\partial y} = 0, \tag{4.127}$$

$$u \frac{\partial u}{\partial x} + v \frac{\partial u}{\partial y} = \nu \frac{\partial^2 u}{\partial y^2} \tag{4.128}$$

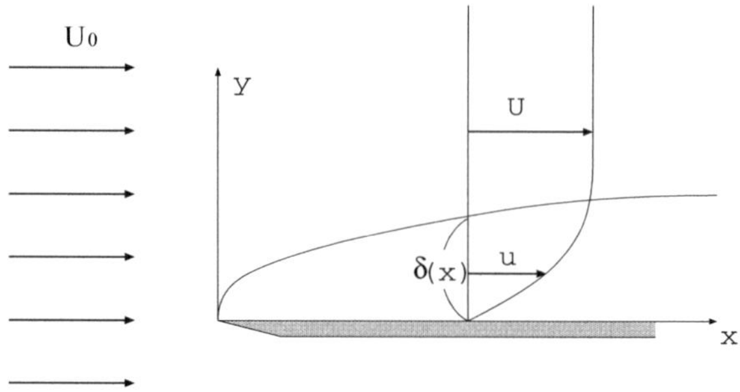

Fig. 4.21 Boundary layer formed on a flat plate

and the boundary conditions are given by

$$\left.\begin{aligned} u = v = 0 \text{ at } y = 0\,, \\ u \to U_0 \quad \text{as } y \to \infty\,. \end{aligned}\right\} \tag{4.129}$$

Introducing the following stream function Ψ

$$u = \frac{\partial \Psi}{\partial y}, \quad v = -\frac{\partial \Psi}{\partial x} \tag{4.130}$$

the equation of continuity (4.127) is automatically satisfied and (4.128) becomes

$$\frac{\partial \Psi}{\partial y}\frac{\partial^2 \Psi}{\partial x \partial y} - \frac{\partial \Psi}{\partial x}\frac{\partial^2 \Psi}{\partial y^2} = \nu \frac{\partial^3 \Psi}{\partial y^3}\,. \tag{4.131}$$

We obtain the solution by analyzing (4.131) with the boundary conditions:

$$\left.\begin{aligned} &\frac{\partial \Psi}{\partial x} = \frac{\partial \Psi}{\partial y} = 0 \text{ at } y = 0\,, \\ &\frac{\partial \Psi}{\partial y} \to U_0 \qquad \text{as } y \to \infty\,. \end{aligned}\right\} \tag{4.132}$$

Let us revisit Rayleigh's problem, introduced in Sect. 4.2. When an infinitely long flat plate instantaneously begins to move at some fixed velocity U_0, the boundary layer thickness is proportional to $\sqrt{\nu t}$, i.e.,

$$\delta \sim \sqrt{\nu t} \tag{4.133}$$

and a similar solution

$$u = U_0\{1 - \text{erf}(\eta)\} \tag{4.134}$$

is obtained as a velocity distribution, where $\eta \equiv y/(2\sqrt{\nu t})$. In the flow considered here, the boundary layer develops from the tip of the flat plate and its thickness increases with increasing distance from the tip. The fluid particles at a horizontal distance x from the plate tip receive impact from the flat plate in a period x/U_0. Therefore, the time t in Rayleigh's problem is replaced by

$$t = \frac{x}{U_0}\,. \tag{4.135}$$

From (4.133), the thickness of the boundary layer is

$$\delta \sim \sqrt{\frac{\nu x}{U_0}}\,. \tag{4.136}$$

Earlier in this chapter, (4.45), a function of $\eta \equiv y/(2\sqrt{\nu t})$, was obtained as a similar solution to Rayleigh's problem. Based on that, if we introduce a nondimensional variable,

$$\eta \equiv \frac{y}{\sqrt{\dfrac{\nu x}{U_0}}} \tag{4.137}$$

a similar solution can be obtained. Then, defining

$$\Psi \equiv \sqrt{\nu x U_0}\, f(\eta)\,, \tag{4.138}$$

(4.131) becomes

$$ff'' + 2f''' = 0\,. \tag{4.139}$$

Equation (4.139) is an ordinary differential equation known as Blasius' equation. Here, $'$ denotes the derivative with respect to η, and the boundary conditions become

$$\left.\begin{aligned} f = 0,\ f' = 0 \text{ at } \eta = 0\,, \\ f' \to 1 \qquad \text{as } \eta \to \infty\,. \end{aligned}\right\} \tag{4.140}$$

Exercise 4.5 Derive Blasius' equation (4.139) using (4.131) and (4.138).

Once function f is obtained, the velocity u and v within the boundary layer can be calculated as follows:

$$u = \frac{\partial \Psi}{\partial y} = \frac{\partial \Psi}{\partial \eta}\frac{\partial \eta}{\partial y} = U_0 f'(\eta), \tag{4.141}$$

$$v = -\frac{\partial \Psi}{\partial x} = \frac{1}{2}\sqrt{\frac{\nu U_0}{x}}\left\{\eta f'(\eta) - f(\eta)\right\}. \tag{4.142}$$

However, the ordinary differential equation (4.139) is also nonlinear and its solution cannot be expressed as an analytically simple function. Below, (4.139) and (4.140) are solved in regions of small and large y, and the results are combined to yield a solution.

Solution in the Vicinity of the Flat Plate

First, we expand $f(\eta)$ as a power series about a point on the wall surface:

$$f(\eta) = A_0 + A_1\eta + \frac{A_2}{2!}\eta^2 + \frac{A_3}{3!}\eta^3 + \cdots. \tag{4.143}$$

From the boundary conditions at $\eta = 0$, we immediately obtain

$$A_0 = A_1 = 0. \tag{4.144}$$

Substituting (4.143) into (4.139) and collecting terms in powers of η, we obtain

$$2A_3 + 2A_4\eta + \frac{2A_5 + A_2^2}{2!}\eta^2 + \frac{2A_6 + 4A_2A_3}{3!}\eta^3 + \cdots = 0. \tag{4.145}$$

For (4.145) to hold for arbitrary η^n, we require that

$$2A_3 = 0, \quad 2A_4 = 0, \quad 2A_5 + A_2^2 = 0, \quad 2A_6 + 4A_2A_3 = 0, \quad \cdots \tag{4.146}$$

Solving these expressions, we obtain

$$f(\eta) = \sum_{n=0}^{\infty}\left(-\frac{1}{2}\right)^n \frac{\alpha^{n+1}C_n}{(3n+2)!}\eta^{3n+2} \tag{4.147}$$

where

$$C_0 = 1, \ C_1 = 1, \ C_2 = 11, \ C_3 = 375, \ C_4 = 27897, \ \ldots.$$

Exercise 4.6 Use mathematical induction to prove that the coefficient A_m takes a value only when $m = 3n + 2$ $(n = 0, 1, 2, \ldots)$, and is 0 otherwise. Enumerate C_n $(n = 1\text{–}3)$ in (4.147).

Asymptotic Solution at Infinity

Since $f' \to 1$ as $\eta \to \infty$, the solution in the region of large η can be approximated by

$$f \sim f_0 = \eta - \beta. \tag{4.148}$$

Expanding f as a perturbation series, we have

$$f = f_0 + f_1 + f_2 + f_3 + \cdots. \tag{4.149}$$

When η is sufficiently large, $\varepsilon \equiv O(|f_1/f_0|)$ becomes very small. Relative to f_0, f_n is a small quantity of $O(\varepsilon^n)$, and as $\eta \to \infty$, $f_n \to 0$. Substituting (4.149) into

(4.139) and summarizing by order, we obtain

$$\left(2f_0''' + f_0 f_0''\right) + \left(2f_1''' + f_0 f_1'' + f_1 f_0''\right) + \left(2f_2''' + f_0 f_2'' + f_1 f_1'' + f_2 f_0''\right) + \cdots = 0. \tag{4.150}$$

Setting each order to 0 yields

$$\left.\begin{aligned} & 2f_0''' + f_0 f_0'' = 0\,, \\ & 2f_1''' + f_0 f_1'' + f_1 f_0'' = 0\,, \\ & 2f_2''' + f_0 f_2'' + f_1 f_1'' + f_2 f_0'' = 0\,, \\ & \cdots\cdots \end{aligned}\right\} \tag{4.151}$$

First, we note that (4.148) satisfies the zero-order equation. Next, if (4.148) is substituted to the first-order equation,

$$2f_1''' + (\eta - \beta) f_1'' = 0\,. \tag{4.152}$$

Integrating this expression, we obtain

$$f_1'' = \gamma e^{-\frac{1}{4}(\eta-\beta)^2} \tag{4.153}$$

where γ is an arbitrary constant. As $\eta \to \infty$, $f' \to 1$, but $f_0' \to 1$ also. Therefore, we can infer that $f_1' \to 0$. As $\eta \to \infty$, (4.153) can be integrated assuming $f_1' \to 0$ as a boundary condition. Then, we have

$$f_1' = -\gamma \int_\eta^\infty d\eta_\#\, e^{-\frac{1}{4}(\eta_\# - \beta)^2}\,.$$

Furthermore, integrating (4.153) as $\eta \to \infty$ and imposing the boundary condition $f_1 \to 0$, we get

$$f_1 = \gamma \int_\eta^\infty d\eta_* \int_{\eta_*}^\infty d\eta_\#\, e^{-\frac{1}{4}(\eta_\# - \beta)^2}\,. \tag{4.154}$$

From (4.148) and (4.154), the approximate solution at infinity is given by

$$f(\eta) = \eta - \beta + \gamma \int_\eta^\infty d\eta_* \int_{\eta_*}^\infty d\eta_\#\, e^{-\frac{1}{4}(\eta_\# - \beta)^2}\,. \tag{4.155}$$

Combining the Solutions

As $\eta \to \infty$, the third-order differential equation is constrained by only a single boundary condition. Consequently, the solution (4.155) includes two constants of integration, β and γ. On the other hand, when solved around the flat plate in regions of small η, the third-order differential equation is constrained by two boundary

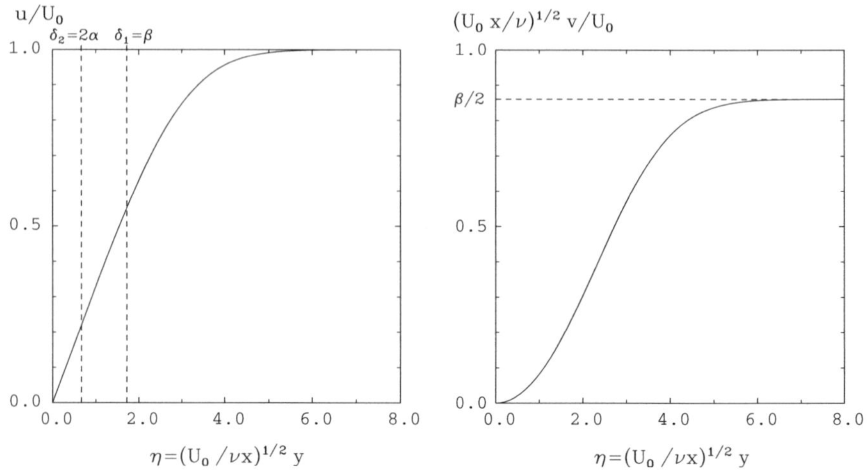

Fig. 4.22 u and v in the boundary layer on the flat plate

conditions. In this case, the solutions contain a single integration constant, α. Within the range of medium-sized η, the values f, f', and f'' of both the solutions must be continuous. Therefore, the constants α, β, and γ are obtained as follows:

$$\alpha \cong 0.332, \quad \beta \cong 1.73, \quad \gamma \cong 0.231. \tag{4.156}$$

Assuming Blasius' solution, the displacement and momentum thicknesses are, respectively, given by

$$\delta_1 \equiv \int_0^\infty \left(1 - \frac{u}{U_0}\right) dy = \sqrt{\frac{\nu x}{U_0}} \int_0^\infty \{1 - f'(\eta)\}\, d\eta$$
$$= \sqrt{\frac{\nu x}{U_0}} \lim_{\eta\to\infty} \{\eta - f(\eta)\} = \beta \sqrt{\frac{\nu x}{U_0}}, \tag{4.157}$$
$$\delta_2 \equiv \int_0^\infty \frac{u}{U_0}\left(1 - \frac{u}{U_0}\right) dy = \sqrt{\frac{\nu x}{U_0}} \int_0^\infty f'(\eta)\{1 - f'(\eta)\}\, d\eta \cong 0.664 \sqrt{\frac{\nu x}{U_0}}. \tag{4.158}$$

The coefficient 0.664 precisely equals 2α. Moreover, the flow component perpendicular to the flat plate (v) is non-zero even along the outer edge of the boundary layer (Fig. 4.22). Specifically, it is equal to

$$v_\infty = \frac{1}{2}\sqrt{\frac{\nu U_0}{x}} \lim_{\eta\to\infty} \left\{\eta f'(\eta) - f(\eta)\right\} = \frac{\beta}{2} U_0 \sqrt{\frac{\nu}{x U_0}}. \tag{4.159}$$

Exercise 4.7 Calculate $\int_0^\infty f'(\eta)\{1 - f'(\eta)\}d\eta$ by the following two methods and evaluate δ_2 in (4.158).

1. Numerically calculate the result using the f' values in Table 4.1.

2. Partially integrate, then use the relationship equation (4.139), and finally substitute $f''(0)$ obtained by (4.147) to obtain the answer.

From (4.141) and (4.147), the shear stress acting on the flat plate is given by

$$\tau_0(x) = \mu \left.\frac{\partial u}{\partial y}\right|_{y=0} = \rho\nu U_0\sqrt{\frac{U_0}{\nu x}}f''(0) = \alpha\rho U_0\sqrt{\frac{\nu U_0}{x}}\,. \tag{4.160}$$

The frictional resistance D_f acting on one side of a flat plate of length l is

$$D_f = \int_0^l \tau_0\,dx = \alpha\rho U_0\sqrt{\nu U_0}\int_0^l \frac{dx}{\sqrt{x}} = 2\alpha\rho\sqrt{\nu l}U_0^{\frac{3}{2}}\,. \tag{4.161}$$

Defining the coefficient of frictional resistance C_f by $C_f \equiv D_f\,/(\frac{1}{2}\rho U_0^2 l)$, we obtain

$$C_f = \frac{4\alpha}{\sqrt{Re_l}} \tag{4.162}$$

where $Re_l \equiv U_0 l/\nu$.

4.4.5 Separation of the Boundary Layer

As evident from (4.112), the flow perpendicular to the wall within a boundary layer is generally much smaller (approximately $O(\delta/l) \sim O(1/\sqrt{Re})$ than that along the wall. Therefore, fluid particles within the boundary layer remain close to the wall boundary and rarely venture out into the internal regions. However, in real fluids, flows within the boundary layer may escape into the internal regions. This phenomenon is called the **separation** of the boundary layer, and the point at which it occurs is called the **separation point**.

Separation requires that the flow v perpendicular to the wall is of similar magnitude to the flow u along the wall. However, in the non-dimensionalized Prandtl's boundary layer equation, while u was non-dimensionalized as $u = U_0\tilde{u}$, v was scaled by a small amount as $v = (\delta/l)U_0\tilde{v}$. Hence, for separation to occur, $\tilde{v}$ should be rendered as a large order (l/δ) in the solution to Prandtl's boundary layer equation. This equation is obtained as follows:

$$\frac{\partial\tilde{u}}{\partial\tilde{x}} + \frac{\partial\tilde{v}}{\partial\tilde{y}} = 0\,, \tag{4.163}$$

$$\frac{\partial\tilde{u}}{\partial\tilde{t}} + \tilde{u}\frac{\partial\tilde{u}}{\partial\tilde{x}} + \tilde{v}\frac{\partial\tilde{u}}{\partial\tilde{y}} = -\frac{\partial\tilde{p}}{\partial\tilde{x}} + \frac{\partial^2\tilde{u}}{\partial\tilde{y}^2}\,, \tag{4.164}$$

$$-\frac{\partial\tilde{p}}{\partial\tilde{x}} = \frac{\partial\tilde{U}_{\text{in}}}{\partial\tilde{t}} + \tilde{U}_{\text{in}}\frac{\partial\tilde{U}_{\text{in}}}{\partial\tilde{x}}\,. \tag{4.165}$$

Table 4.1 Function f obtained from [5]

η	f	f'	f''
0.0	0.00000	0.00000	0.33206
0.2	0.00664	0.06641	0.33199
0.4	0.02656	0.13277	0.33147
0.6	0.05974	0.19894	0.33008
0.8	0.10611	0.26471	0.32739
1.0	0.16557	0.32979	0.32301
1.2	0.23795	0.39378	0.31659
1.4	0.32298	0.45627	0.30787
1.6	0.42032	0.51676	0.29667
1.8	0.52952	0.57477	0.28293
2.0	0.65003	0.62977	0.26675
2.2	0.78120	0.68132	0.24835
2.4	0.92230	0.72899	0.22809
2.6	1.07252	0.77246	0.20646
2.8	1.23099	0.81152	0.18401
3.0	1.39682	0.84605	0.16136
3.2	1.56911	0.87609	0.13913
3.4	1.74696	0.90177	0.11788
3.6	1.92954	0.92333	0.09809
3.8	2.11605	0.94112	0.08013
4.0	2.30576	0.95552	0.06424
4.2	2.49806	0.96696	0.05052
4.4	2.69238	0.97587	0.03897
4.6	2.88826	0.98269	0.02948
4.8	3.08534	0.98779	0.02187
5.0	3.28329	0.99155	0.01591
5.2	3.48189	0.99425	0.01134
5.4	3.68094	0.99616	0.00793
5.6	3.88031	0.99748	0.00543
5.8	4.07990	0.99838	0.00365
6.0	4.27964	0.99898	0.00240
6.2	4.47948	0.99937	0.00155
6.4	4.67938	0.99961	0.00098
6.6	4.87931	0.99977	0.00061
6.8	5.07928	0.99987	0.00037
7.0	5.27926	0.99992	0.00022
7.2	5.47925	0.99996	0.00013
7.4	5.67924	0.99998	0.00007

(continued)

Table 4.1 (continued)

η	f	f'	f''
7.6	5.87924	0.99999	0.00004
7.8	6.07923	1.00000	0.00002
8.0	6.27923	1.00000	0.00001
8.2	6.47923	1.00000	0.00001
8.4	6.67923	1.00000	0.00000
8.6	6.87923	1.00000	0.00000
8.8	7.07923	1.00000	0.00000

In the continuity equation (4.163), as $\partial\tilde{v}/\partial\tilde{y}$ is of $O(l/\delta)$, $\partial\tilde{u}/\partial\tilde{x}$ is also of $O(l/\delta)$. The magnitude of each term in the equation of motion is then

$$\underbrace{\frac{\partial\tilde{u}}{\partial\tilde{t}}}_{O(1)}+\underbrace{\tilde{u}\frac{\partial\tilde{u}}{\partial\tilde{x}}}_{O(l/\delta)}+\underbrace{\tilde{v}\frac{\partial\tilde{u}}{\partial\tilde{y}}}_{O(l/\delta)}=-\underbrace{\frac{\partial\tilde{p}}{\partial\tilde{x}}}_{O(1)}+\underbrace{\frac{\partial^2\tilde{u}}{\partial\tilde{y}^2}}_{O(1)} \tag{4.166}$$

and the main balance is equivalent to

$$\tilde{u}\frac{\partial\tilde{u}}{\partial\tilde{x}}+\tilde{v}\frac{\partial\tilde{u}}{\partial\tilde{y}}=0\,. \tag{4.167}$$

By the equation of continuity (4.163), and (4.167) is rendered as

$$\tilde{u}\frac{\partial\tilde{u}}{\partial\tilde{x}}+\tilde{v}\frac{\partial\tilde{u}}{\partial\tilde{y}}=-\tilde{u}\frac{\partial\tilde{v}}{\partial\tilde{y}}+\tilde{v}\frac{\partial\tilde{u}}{\partial\tilde{y}}=\tilde{v}^2\frac{\partial}{\partial\tilde{y}}\left(\frac{\tilde{u}}{\tilde{v}}\right)=0\,,$$

$$\tilde{u}=C(\tilde{x})\tilde{v}\,. \tag{4.168}$$

Therefore,

$$\frac{\partial\tilde{u}}{\partial\tilde{y}}=C(\tilde{x})\frac{\partial\tilde{v}}{\partial\tilde{y}}=-C(\tilde{x})\frac{\partial\tilde{u}}{\partial\tilde{x}}\,. \tag{4.169}$$

Setting $\tilde{y}=0$ (boundary surface) in (4.169), we obtain

$$\left.\frac{\partial\tilde{u}}{\partial\tilde{y}}\right|_{\tilde{y}=0}=-C(\tilde{x})\left.\frac{\partial\tilde{u}}{\partial\tilde{x}}\right|_{\tilde{y}=0}=0\,. \tag{4.170}$$

That is, $\partial u/\partial y=0$ at the separation point.

Finally, we investigate the direction of the pressure gradient at the separation point. The boundary layer equation of motion (4.164) is $\tilde{u}=\tilde{v}=0$ at the wall surface $\tilde{y}=0$; hence, we have

$$\left.\frac{\partial^2 \tilde{u}}{\partial \tilde{y}^2}\right|_{\tilde{y}=0} = \frac{\partial \tilde{p}}{\partial \tilde{x}} . \tag{4.171}$$

At the separation point, both $\partial \tilde{u}/\partial \tilde{y}|_{\tilde{y}=0}$ and $\tilde{u}(0)$ become 0. Therefore, if $\partial^2 \tilde{u}/\partial \tilde{y}^2|_{\tilde{y}=0}$ is positive (negative), then $\tilde{u}$ can also be regarded as positive (negative) in the region $\tilde{y} > 0$ (if this is not the case, we need to consider a significantly distorted velocity profile). If the main flow travels in the positive x-direction ($\tilde{U}_{\text{in}} > 0$), $\tilde{u}$ is positive wherever $\tilde{y}$ is large. Consequently, we may consider that $\partial^2 \tilde{u}/\partial \tilde{y}^2|_{\tilde{y}=0} > 0$, and from (4.171), we have

$$\frac{\partial \tilde{p}}{\partial \tilde{x}} > 0 . \tag{4.172}$$

Furthermore, in steady flows, (4.165) gives

$$\frac{\partial \tilde{U}_{\text{in}}}{\partial \tilde{x}} < 0 . \tag{4.173}$$

In other words, in the neighborhood of the separation point, pressure increases in the direction of main flow and the magnitude of the main flow is reduced (Fig. 4.23).

According to Lagrange's vortex theorem (see Sect. 2.1), unless vortices originally existed, it does not develop into a perfect fluid. The same thing is approximately true for the internal region of viscous fluids. If it is originally a potential flow (e.g., sufficiently far), a vortex does not form easily. However, when the separation of the boundary layer occurs, a large vorticity flows into the internal region (Fig. 4.24). When the supply of vorticity that was created by separation is nonstationary, the flow inside the interior region will present a complex appearance.

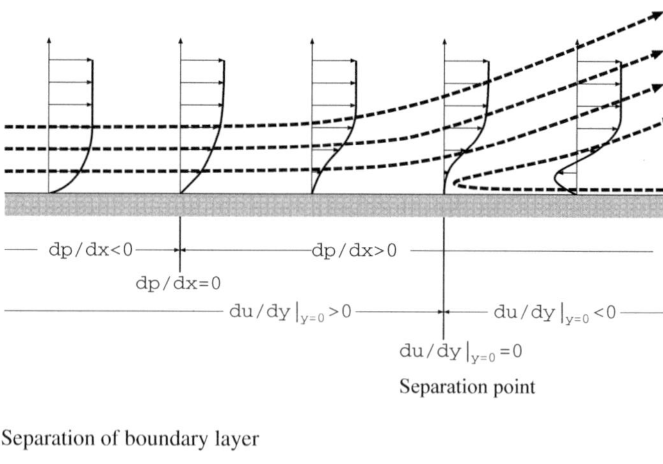

Fig. 4.23 Separation of boundary layer

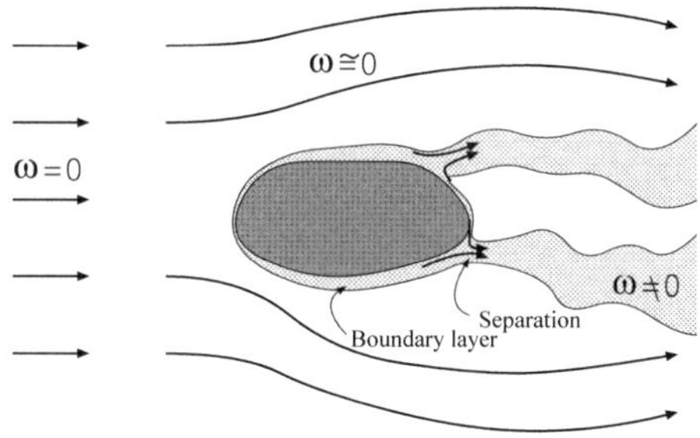

Fig. 4.24 Penetration of vorticity into inner regions

Note 4.3 (*Literature for further study*)
For further studies, textbooks [1, 2, 7, 10, 11, 14, 15, 22] may be referred. The advanced readers may also refer some other textbooks [3, 4, 6, 8, 9, 12, 13, 16–21], although they are written in Japanese.

References

1. Batchelor, G.K.: An Introduction to Fluid Dynamics. Cambridge University Press, Cambridge (1967)
2. Gill, A.E.: Atmosphere-Ocean Dynamics. Academic Press, New York (1982)
3. Gotoh, T.: Fundamentals of Turbulence Theory. Asakura Publishing, Tokyo (1998)
4. Hino, M.: Fluid Dynamics. Asakura Publishing, Tokyo (1992)
5. Howarth, L.: Proc. Roy. Soc. London A **164**, 547–579 (1938)
6. Imai, I.: Fluid Dynamics, vol. I. Shokabo, Tokyo (1973)
7. Jeffreys, H.: Cartesian Tensors. Cambridge University Press, Cambridge (1969)
8. Kambe, T. (ed.): Fluid Dynamics. Shokabo, Tokyo (1995)
9. Kimura, R.: Introduction to Geophysical Fluid Dynamics. Tokyodo Shuppan, Tokyo (1983)
10. Landau, L.D., Lifshitz, E.M.: Fluid Mechanics. Butterworth-Heinemann, Oxford (1987)
11. Newman, J.N.: Marine Hydrodynamics. The MIT Press, Cambridge (1977)
12. Ogura, Y.: Principles of Atmospheric Dynamics. University of Tokyo Press (1978)
13. Ogura, Y.: Basic Theory of Meso-Scale Meteorology. University of Tokyo Press (1997)
14. Pedlosky, J.: Geophysical Fluid Dynamics, 2nd edn. Springer, Heidelberg (1986)
15. Prandtl, L., Tietjens, O.G.: Fundamentals of Hydro- and Aeromechanics (translated by Rosenhead, L.). Dover, New York (1957)
16. Tatsumi, T.: Turbulence. Makishoten (1962)
17. Tatsumi, T.: Fluid Dynamics. Baifukan, Tokyo (1982)
18. Tatsumi, T., Goto, K.: Stability Theory of Flow. Sangyo-Tosho, Tokyo (1976)
19. Toba, Y. (ed.): Atmosphere-Ocean Interactions. University of Tokyo Press (1996)
20. Tomita, Y.: Introduction to Fluid Dynamics. Yokendo, Tokyo (1975)
21. Unoki, S.: Coastal Physical Oceanography. Tokai University Press, Tokyo (1993)
22. Whitham, G.B.: The Navier–Stokes equations of motion. In: Rosenhead, L. (ed.) Laminar Boundary Layers. Part I. Oxford University Press, Oxford (1963)

Chapter 5
Turbulent Flow

Low Reynolds number flows are generally regular and ordered, and are known as **laminar flows**. As the Reynolds number increases, the flow becomes more spatially and temporally disordered. Such a disorderly flow is called a **turbulent flow**. Unlike laminar flows, turbulent flows are unstable to small disturbances, irregular in time and space, and highly nonlinear. Furthermore, turbulent flows are characterized by nonsteady three-dimensional motion, strong transportability, large energy dissipation ability, and many degrees of freedom established from movements of various scales.

The reader may recognize that describing such disorderly flows is extremely difficult. Actually, a precise description of turbulent flows is impossible. However, the characteristics of irregular phenomena exhibit some degree of statistical regularity because of their randomness. Statistical mechanics makes no attempt to describe the irregular motions of individual molecules, but instead describes the statistical thermal characteristics of the particle ensemble. Similarly, we discuss turbulence in terms of its statistical characteristics such as average velocity, dispersion, and correlation. This chapter explores the application of these concepts to turbulent flows.

5.1 Transition from Laminar Flow to Turbulent Flow

5.1.1 Flow in a Pipe

Flow in a pipe discovered by Hagen and Poiseuille is an exact solution to the Navier–Stokes equation under the assumption of steady, homogenous flow along the axis of a pipe. This solution can be established regardless of the Reynolds number. Based on this solution, as given in (4.33), flow volume increases proportionally to the pressure gradient. However, in a real flow, beyond a certain Reynolds number, it is no longer proportional to the pressure gradient, and flow volume is less than that

D.E.S.S.T. Kyushu Univ., *Fluid Dynamics for Global Environmental Studies*,
DOI 10.1007/978-4-431-56499-7_5

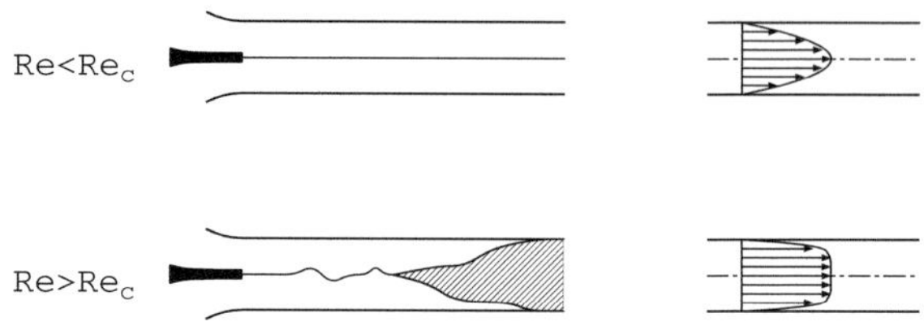

Fig. 5.1 Laminar (*top*) and turbulent (*bottom*) flows

expected from the pressure gradient. This is a very serious problem in waterworks supply.

Osborne Reynolds observed the flow patterns when colored water was injected near the entrance of a cylindrical tube. By varying the water velocity and tube diameter, he noticed that water flowed slowly as a single thin line parallel to the pipe's axis. In this state, flow was laminar and its quantity obeyed the Hagen-Poiseuille law. As the flow velocity increased beyond a certain limit, the colored water line suddenly became disordered and the color spread across the entire cross section of the pipe. Under these conditions, the flow in the pipe transitioned to a turbulent state, and the flow velocity developed a large gradient near the pipe walls and became homogeneous from the center to the walls, in contrast with the parabolic form given in (4.31). This homogeneous velocity profile was attributable to high momentum transfer. Consequently, the water experienced higher resistance from the pipe walls and the flow was less than that in laminar flows under the same pressure gradient (Fig. 5.1). Reynolds showed that the tubular flow transitions from laminar to turbulent at a critical Reynolds number depended on the pipe diameter d, the cross-sectional mean flow velocity U, and the coefficient of kinematic viscosity ν:

$$Re \equiv \frac{Ud}{\nu}. \tag{5.1}$$

The critical Reynolds number at which the flow changes from laminar to turbulent is denoted as Re_c. If the turbulence in the fluid injected into the pipe is made sufficiently small, Re_c increases; however, it is empirically known that Re_c is limited at approximately 2000.

5.1.2 Transition in Flow Behind a Cylinder

The pattern of a homogeneous flow after passing through a cylinder also depends on the Reynolds number ($Re \equiv Ud/\nu$). When $Re \leq 1$, the flow passes steadily and symmetrically along the cylinder surface. In the range $Re = 1$–10, the flow is similarly steady; however, a pair of vortices (twin vortices) forms downstream of the cylinder. The vortex size increases with increasing Reynolds number, becoming large at $Re = 10$–10^2. Oscillations now appear downstream of the cylinder, and two lines of vortices appear alternatively. This formation is known as a **Kármán's vortex**

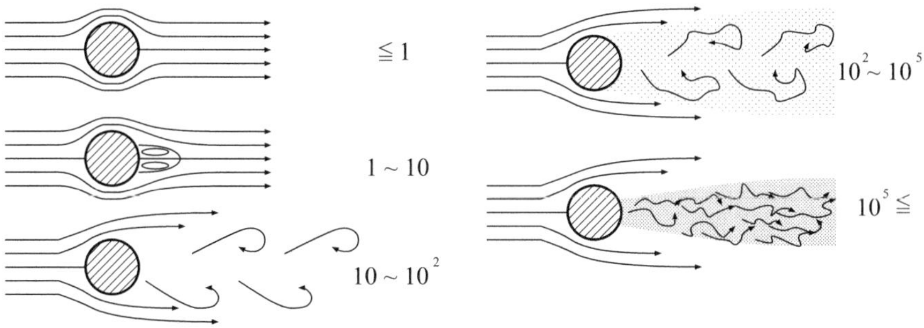

Fig. 5.2 Reynolds number and state of flows passing a cylinder

street. Such flows are nonsteady but periodic. When $Re = 10^2$–10^5, the vortex street becomes turbulent and establishes a nonsteady, nonperiodic wake. As Re increases further ($Re > 10^5$), the **wake** behind the cylinder becomes perfectly turbulent. The boundary layer formed on the surface of the cylinder also becomes turbulent and the width of the wake reduces (see Fig. 5.2).

5.2 Stability Theory of Flow

Several solutions of viscous laminar flows were developed in Chap. 4. However, as indicated in a previous section, the flows developing in many real situations are nonlaminar. This inconsistency does not invalidate the approximation of the solution sought in Chap. 4. Sufficiently accurate solutions can often be obtained by increasing the accuracy of the laminar approximation (even exact solutions were presented in Sect. 4.2). If a solvable problem exactly describes the real situation, the flows will be accurately described by the analytical solution. This is rarely the case in the natural world, when irregularity exists in natural flows however small. Such small irregularities are called **disturbances**. At a high Reynolds number, small disturbances can dramatically alter the flow behavior. This state characterizes **unstable** flows. Conversely, **stable** flows retain their original state when perturbed by external disturbances. Indeed, in order for viscous flow to exist, the flow must be stable against external disturbances. This section investigates the stability of flows subjected to very small (relative to the main flow) external disturbances, a condition known as **linear stability**.

The equation of continuity and the Navier–Stokes equation for an incompressible viscous fluid are, respectively, given by

$$\nabla \cdot \boldsymbol{u} = 0, \tag{5.2}$$

$$\frac{\partial \boldsymbol{u}}{\partial t} + (\boldsymbol{u} \cdot \nabla)\boldsymbol{u} = \boldsymbol{K} - \frac{1}{\rho}\nabla p + \nu \nabla^2 \boldsymbol{u}\,. \tag{5.3}$$

Denoting the velocity and pressure of a main laminar flow (for which stability is being sought) by $\boldsymbol{U}(\boldsymbol{r})$ and $P(\boldsymbol{r})$, respectively, $\boldsymbol{U}$ and P are steady-state solutions to (5.3) and (5.2), respectively, and satisfy

$$\nabla \cdot \boldsymbol{U} = 0, \tag{5.4}$$

$$(\boldsymbol{U} \cdot \nabla)\boldsymbol{U} = \boldsymbol{K} - \frac{1}{\rho}\nabla P + \nu\nabla^2\boldsymbol{U}\,. \tag{5.5}$$

Now suppose that disturbance with velocity $\boldsymbol{u}'(\boldsymbol{r}, 0)$ and pressure $p'(\boldsymbol{r}, 0)$ is applied to this steady flow. Then,

$$\left.\begin{aligned} \boldsymbol{u}(\boldsymbol{r}, t) &= \boldsymbol{U}(\boldsymbol{r}) + \boldsymbol{u}'(\boldsymbol{r}, t), \\ p(\boldsymbol{r}, t) &= P(\boldsymbol{r}) + p'(\boldsymbol{r}, t) \end{aligned}\right\} \tag{5.6}$$

satisfies both the original Navier–Stokes equation and the equation of continuity. Substituting (5.6) into (5.3) and (5.2), and subtracting (5.5) and (5.4) from each equation, the equations in terms of $\boldsymbol{u}'$ and p' become

$$\nabla \cdot \boldsymbol{u}' = 0, \tag{5.7}$$

$$\frac{\partial \boldsymbol{u}'}{\partial t} + (\boldsymbol{U} \cdot \nabla)\boldsymbol{u}' + (\boldsymbol{u}' \cdot \nabla)\boldsymbol{U} + (\boldsymbol{u}' \cdot \nabla)\boldsymbol{u}' = -\frac{1}{\rho}\nabla p' + \nu\nabla^2\boldsymbol{u}'\,. \tag{5.8}$$

Ignoring the small second-order terms introduced by the minor disturbance, we obtain

$$\nabla \cdot \boldsymbol{u}' = 0, \tag{5.9}$$

$$\frac{\partial \boldsymbol{u}'}{\partial t} + (\boldsymbol{U} \cdot \nabla)\boldsymbol{u}' + (\boldsymbol{u}' \cdot \nabla)\boldsymbol{U} = -\frac{1}{\rho}\nabla p' + \nu\nabla^2\boldsymbol{u}'\,. \tag{5.10}$$

Equations (5.9) and (5.10) are linear in $\boldsymbol{u}'$ and p', implying that their solutions may be superposed. Therefore, if even a single solution of $\boldsymbol{u}'$ or p' grows with time, these terms in the initial disturbance (however small) increase with time; thus, overall disturbance will increase over time. In this case, the main flow $\boldsymbol{U}$ is determined to be unstable. Conversely, in stable flows, all solutions $\boldsymbol{u}'$ and p' decrease with time or converge to a limit. If the disturbance ceases to grow at a certain finite limit, the main flow is called neutral.

Here we discuss the stability of two-dimensional parallel flows as an example of such a case (Fig. 5.3). The flow direction is the x-axis in a Cartesian coordinate system $\boldsymbol{r} = (x, y, z)$. Taking the y-axis as the direction of the flow variation, the velocity of the main flow is given by

$$\boldsymbol{U}(\boldsymbol{r}) = (U(y), 0, 0)\,. \tag{5.11}$$

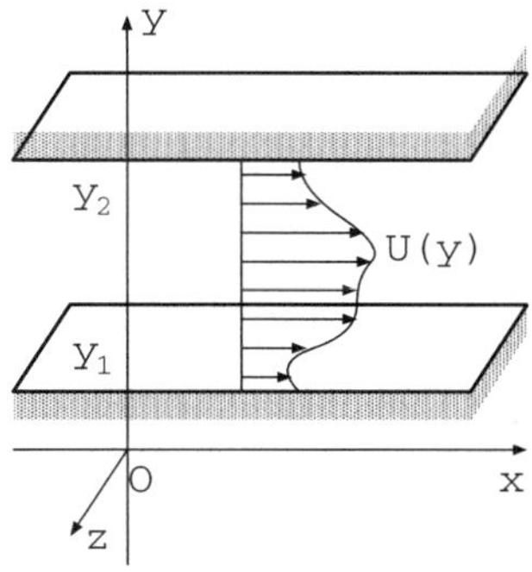

Fig. 5.3 Basic field of two-dimensional parallel flow

The velocity and pressure of the disturbance can be considered as a wave function of space (in the x- and z-directions) and time, and can be decomposed into Fourier components:

$$\begin{pmatrix} \boldsymbol{u}'(\boldsymbol{r},t) \\ p'(\boldsymbol{r},t) \end{pmatrix} = \begin{pmatrix} \hat{\boldsymbol{u}}(y) \\ \hat{p}(y) \end{pmatrix} \exp\left[i(k_x x + k_z z - \omega t)\right]. \tag{5.12}$$

Equation (5.12) expresses any arbitrarily shaped disturbance $\boldsymbol{u}'$ and p' as the sum of $\boldsymbol{u}'$ and p' components with different wavenumbers (k_x and k_z). The frequency ω need not be a real number. Separating the real and imaginary components of ω, the exp[ ] part in (5.12) can be written as

$$\begin{aligned} \exp\left[i(k_x x + k_z z - \omega t)\right] &= \exp\left[i(k_x x + k_z z - \mathrm{Re}(\omega)t) + \mathrm{Im}(\omega)t\right] \\ &= \underbrace{\exp\left[i(k_x x + k_z z - \mathrm{Re}(\omega)t)\right]}_{\text{Oscillates with amplitude 1}} \cdot \underbrace{\exp\left[\mathrm{Im}(\omega)t\right]}_{\text{Increases or decreases depending on sign}} \end{aligned} \tag{5.13}$$

Therefore, if $\mathrm{Im}(\omega) > 0$, the disturbance introduces instability; if $\mathrm{Im}(\omega) = 0$, the flow is neutral to the disturbance,[1] and if $\mathrm{Im}(\omega) < 0$, the flow is stable against the disturbance.

Substituting (5.11) and (5.12) into (5.9) and (5.10), respectively, and eliminating the respective variables $\hat{p}$, $\hat{u}$ and $\hat{w}$, we obtain the following equation in $\hat{v}$:

$$\left(U - \frac{\omega}{k_x}\right)\left[\frac{d^2}{dy^2} - \left(k_x^2 + k_z^2\right)\right]\hat{v} - \frac{d^2U}{dy^2}\hat{v} = \frac{\nu}{ik_x}\left[\frac{d^2}{dy^2} - \left(k_x^2 + k_z^2\right)\right]^2 \hat{v} \tag{5.14}$$

Equation (5.14) is called the **Orr-Sommerfeld's equation**. If the variables are nondimensionalized by the representative velocity U_0 and length L of the main flow, we obtain

[1]Strictly speaking, the flow can become unstable when $\mathrm{Im}(\omega) = 0$. If $\mathrm{Im}(\omega) = 0$ and multiple solutions exist, the disturbance develops in a linear fashion.

$$\left(\tilde{U} - \frac{\tilde{\omega}}{\tilde{k}_x}\right)\left[\frac{d^2}{d\tilde{y}^2} - \left(\tilde{k}_x^2 + \tilde{k}_z^2\right)\right]\tilde{\hat{v}} - \frac{d^2\tilde{U}}{d\tilde{y}^2}\tilde{\hat{v}}$$
$$= \frac{1}{i\tilde{k}_x Re}\left[\frac{d^2}{d\tilde{y}^2} - \left(\tilde{k}_x^2 + \tilde{k}_z^2\right)\right]^2 \tilde{\hat{v}} \tag{5.15}$$

(where $\tilde{}$ denotes dimensionless variables). Abbreviating the $\tilde{}$ sign in the above equation and rewriting

$$\left.\begin{aligned} c\ (= \mathrm{Re}(c) + i\ \mathrm{Im}(c)) &\equiv \frac{\omega}{k_x} \\ k^2 &\equiv k_x^2 + k_z^2 \end{aligned}\right\} \tag{5.16}$$

(5.15) becomes

$$(U - c)\left(\hat{v}'' - k^2\hat{v}\right) - U''\hat{v} = \frac{1}{ik_x Re}\left(\hat{v}'''' - 2k^2\hat{v}'' + k^4\hat{v}\right) \tag{5.17}$$

(here $'$ indicates d/dy). In this form, it is easily observed that $\mathrm{Im}(c) > 0$ is unstable, $\mathrm{Im}(c) = 0$ is neutral, and $\mathrm{Im}(c) < 0$ is stable.

Exercise 5.1 Derive the Orr-Sommerfeld Equation (5.14).

To obtain ω or c, we require the system parameter Re and the wavenumbers k_x and k_z. Equation (5.17) should be solved as an eigenvalue problem with appropriate boundary conditions; for example, viscous boundary conditions at $y = y_1$ and y_2.

$$\hat{v} = \frac{d\hat{v}}{dy} = 0 \quad \text{at } y = y_1, y_2\,. \tag{5.18}$$

The complex eigenvalue c is a function of k_x, k_z, and Re, and is denoted as $c(k_x, k_z, Re)$. Whether the flow remains stable when disturbed by waves of wavenumbers (k_x, k_z) depends on the sign of the imaginary component, $\mathrm{Im}(c)(k_x, k_z, Re)$; specifically,

$$\left.\begin{aligned} &\text{Stable} \quad\ \ \text{if } \mathrm{Im}(c)(k_x, k_z, Re) < 0\,, \\ &\text{Unstable if } \mathrm{Im}(c)(k_x, k_z, Re) > 0\,. \end{aligned}\right\} \tag{5.19}$$

These conditions are delineated by the neutral condition

$$\mathrm{Im}(c)(k_x, k_z, Re) = 0\,. \tag{5.20}$$

The curved surface in (k_x, k_z, Re) space traced by the neutral condition is called the **neutral surface**. (This example considers two-dimensional wavenumbers (k_x, k_z); therefore, the neutral condition defines a curved surface. The neutral surface where the wavenumber is one-dimensional is called the **neutral curve**). Disturbances in the natural world, no matter how small, are composites of many components. Thus,

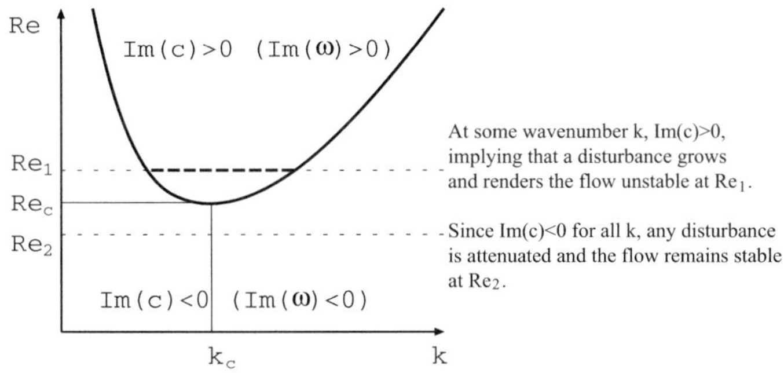

Fig. 5.4 Critial Reynolds number

disturbance of a single wavenumber is sufficient to induce an unstable flow. In other words, the linear stability of a flow of Reynolds number Re is determined by

$$\left.\begin{array}{ll}\text{Stable} & \text{if } \max\limits_{k_x,k_z} \mathrm{Im}(c)(k_x, k_z, Re) < 0\,, \\ \text{Unstable} & \text{if } \max\limits_{k_x,k_z} \mathrm{Im}(c)(k_x, k_z, Re) > 0\,. \end{array}\right\} \tag{5.21}$$

The Re value that can satisfy

$$\max_{k_x,k_z} \mathrm{Im}(c)(k_x, k_z, Re_c) = 0 \tag{5.22}$$

so that the main flow becomes neutral is called the **critical Reynolds number** Re_c of the main flow $U(y)$ (see Fig. 5.4).

5.3 Basic Equation of Turbulent Flow

When laminar flow transitions to turbulent flow as minute disturbances increase, the assumption $\boldsymbol{u}' \ll \boldsymbol{U}$ discussed in the previous section is not established, and we must construct a set of equations where $\boldsymbol{u}' \sim O(\boldsymbol{U})$. In other words, the nonlinear term $(\boldsymbol{u} \cdot \nabla)\boldsymbol{u}$ in (5.8) cannot be simply discarded as an infinitesimal quantity. Indeed, a system of equations denoting a flow field subjected to such nonlinearity exactly describes a turbulent system. Here we formulate the equations governing the turbulent motion and the temporal change in temperature or concentration of the mixture passive to the flow of the incompressible viscous fluid.

5.3.1 Average/Mean

A random variable $a(\boldsymbol{r}, t)$ can be expressed as the sum of its mean $\bar{a}(\boldsymbol{r}, t)$ and variability $a'(\boldsymbol{r}, t)$. The mean can be defined in one of the following three ways.

Time Average/Mean

This mean is obtained by measuring the random variable $a(\boldsymbol{r}_0, t)$ at some point in space $\boldsymbol{r} = \boldsymbol{r}_0$ over a sufficiently long time:

$$\bar{a}^t(\boldsymbol{r}_0, t_0) \equiv \frac{1}{T}\int_{t_0}^{t_0+T} a(\boldsymbol{r}_0, t)\, dt\,. \tag{5.23}$$

Space Average/Mean

This mean is obtained by sampling the random variable $a(\boldsymbol{r}, t_0)$ at different locations and averaging over the space $\boldsymbol{s}$ at some time $t = t_0$:

$$\bar{a}^r(\boldsymbol{r}_0, t_0) \equiv \frac{1}{V}\int_{\boldsymbol{r}_0}^{\boldsymbol{r}_0+\boldsymbol{s}} a(\boldsymbol{r}, t_0)\, d\boldsymbol{r}\,. \tag{5.24}$$

Ensemble Average/Mean

This mean is the average of repeated sampling under the same conditions:

$$\bar{a}^e(\boldsymbol{r}_0, t_0) \equiv \lim_{N\to\infty}\frac{1}{N}\sum_{n=1}^{N} a_n(\boldsymbol{r}_0, t_0) \tag{5.25}$$

where $a_n(\boldsymbol{r}_0, t_0)$ is the random variable at $\boldsymbol{r} = \boldsymbol{r}_0$ and $t = t_0$ obtained by n trials.

In most cases, when the statistical amounts are steady, we have

$$\bar{a}^e(\boldsymbol{r}_0) = \bar{a}^t(\boldsymbol{r}_0)\,. \tag{5.26}$$

Similarly, for homogeneous statistics, the following is generally true:

$$\bar{a}^e(t_0) = \bar{a}^r(t_0)\,. \tag{5.27}$$

If the statistical quantities are both steady and homogeneous, the three means are generally equal.

The mean fluctuations in the ensemble average are 0, as seen below:

$$\bar{a'}^e = \lim_{N\to\infty}\frac{1}{N}\sum_{n=1}^{N} a'_n = \lim_{N\to\infty}\frac{1}{N}\sum_{n=1}^{N}[\, a_n - \bar{a}\,] = \left[\lim_{N\to\infty}\frac{1}{N}\sum_{n=1}^{N} a_n\right] - \bar{a} = 0.$$

Additionally, for a physical quantity, $a = \bar{a}^e + a'$, $b = \bar{b}^e + b'$

$$\overline{a+b}^e = \lim_{N\to\infty}\frac{1}{N}\sum_{n=1}^{N}(a+b) = \lim_{N\to\infty}\frac{1}{N}\sum_{n=1}^{N}a + \lim_{N\to\infty}\frac{1}{N}\sum_{n=1}^{N}b = \overline{a}^e + \overline{b}^e,$$

$$\overline{a\overline{b}^e}^e = \lim_{N\to\infty}\frac{1}{N}\sum_{n=1}^{N}a\overline{b}^e = \overline{b}^e\lim_{N\to\infty}\frac{1}{N}\sum_{n=1}^{N}a = \overline{a}^e\,\overline{b}^e,$$

$$\overline{ab'}^e = \overline{(\overline{a}^e + a')b'}^e = \overline{\overline{a}^e b' + a'b'}^e = \overline{\overline{a}^e b'}^e + \overline{a'b'}^e = \overline{a}^e\underbrace{\overline{b'}^e}_{0} + \overline{a'b'}^e = \overline{a'b'}^e,$$

$$\overline{ab}^e = \overline{a(\overline{b}^e + b')}^e = \overline{a\overline{b}^e}^e + \overline{ab'}^e = \overline{a}^e\overline{b}^e + \overline{a'b'}^e.$$

Thus,

$$\overline{a'} = 0,\ \ \overline{a+b} = \overline{a} + \overline{b},\ \ \overline{a\overline{b}} = \overline{a}\,\overline{b},\ \ \overline{ab'} = \overline{a'b'},\ \ \overline{ab} = \overline{a}\,\overline{b} + \overline{a'b'}. \tag{5.28}$$

Typically, these relationships only hold for approximation in the spatial and temporal means. Moreover, when using the ensemble or time mean, the spatial derivatives or integrals (or vice versa) can be substituted by the mean operation in the sequence. When using the ensemble or space mean, time derivatives or integrals can be substituted by the mean operation. Therefore, among the three means, the ensemble mean is mathematically the easiest to manipulate. In the following discussion, we indicate the ensemble mean by $\overline{}$.

5.3.2 Equation of Fluid Motion

The basic equations for an incompressible fluid are the continuity equation and the Navier–Stokes equation:

$$\nabla\cdot\boldsymbol{u} = 0, \tag{5.29}$$

$$\frac{\partial\boldsymbol{u}}{\partial t} + (\boldsymbol{u}\cdot\nabla)\boldsymbol{u} = -\frac{1}{\rho}\nabla p + \nu\nabla^2\boldsymbol{u}. \tag{5.30}$$

Heat or substance transport is governed by the heat transportation equation

$$\frac{\partial\theta}{\partial t} + (\boldsymbol{u}\cdot\nabla)\theta = \nu_\theta\nabla^2\theta \tag{5.31}$$

where ν_θ is **diffusivity**. In tensor form, (5.29)–(5.31) are expressed as

$$\frac{\partial u_i}{\partial x_i} = 0\,, \tag{5.32}$$

$$\frac{\partial u_i}{\partial t} + u_j \frac{\partial u_i}{\partial x_j} = -\frac{1}{\rho}\frac{\partial p}{\partial x_i} + \nu \frac{\partial^2 u_i}{\partial x_j^2}\,, \tag{5.33}$$

$$\frac{\partial \theta}{\partial t} + u_j \frac{\partial \theta}{\partial x_j} = \nu_\theta \frac{\partial^2 \theta}{\partial x_j^2}\,. \tag{5.34}$$

5.3.3 Equation of Mean Flow in Turbulence

First, we express flow velocity, pressure, and temperature as the sum of their statistical means $\overline{u_i}$, $\overline{p}$, and $\overline{\theta}$ and fluctuations u'_i, p', and θ', respectively:

$$u_i = \overline{u_i} + u'_i, \quad p = \overline{p} + p', \quad \theta = \overline{\theta} + \theta'\,. \tag{5.35}$$

Substituting (5.35) into the equation of continuity (5.32), the Navier–Stokes (momentum) equation (5.33), and the heat transport equation (5.34), and taking the means, the equations governing the mean flow velocity, pressure, and temperature are derived as follows:

Equation of Continuity

Taking the mean of both sides of (5.32), we obtain

$$\frac{\partial \overline{u_i}}{\partial x_i} = 0\,. \tag{5.36}$$

Subtracting (5.36) from (5.32), we obtain the continuity equation for the fluctuation velocity:

$$\frac{\partial u'_i}{\partial x_i} = 0\,. \tag{5.37}$$

This equation will be frequently used in a later transformation.

Equation of Momentum

Substituting (5.35) into (5.33) and taking the average gives

$$\overline{\frac{\partial}{\partial t}\left(\overline{u_i} + u'_i\right) + \left(\overline{u_j} + u'_j\right)\frac{\partial}{\partial x_j}\left(\overline{u_i} + u'_i\right)} = \overline{-\frac{1}{\rho}\frac{\partial}{\partial x_i}\left(\overline{p} + p'\right) + \nu \frac{\partial^2 \overline{u_i}}{\partial x_j^2} + \nu \frac{\partial^2 u'_i}{\partial x_j^2}}\,.$$

Therefore,

$$\frac{\partial \overline{u_i}}{\partial t} + \overline{u_j}\frac{\partial \overline{u_i}}{\partial x_j} + \overline{u'_j \frac{\partial u'_i}{\partial x_j}} = -\frac{1}{\rho}\frac{\partial \overline{p}}{\partial x_i} + \nu \frac{\partial^2 \overline{u_i}}{\partial x_j^2}\,.$$

Using the continuity equation of the fluctuation velocity (5.37), we get

$$\overline{u'_j \frac{\partial u'_i}{\partial x_j}} = \overline{\frac{\partial (u'_j u'_i)}{\partial x_j}} - \overline{u'_i \underbrace{\frac{\partial u'_j}{\partial x_j}}_{0}} = \frac{\partial}{\partial x_j} \overline{u'_j u'_i}\,.$$

This yields the following momentum equation for the mean flow:

$$\frac{\partial \overline{u_i}}{\partial t} + \overline{u_j} \frac{\partial \overline{u_i}}{\partial x_j} = -\frac{1}{\rho} \frac{\partial \overline{p}}{\partial x_i} + \frac{1}{\rho} \frac{\partial}{\partial x_j} \left(\mu \frac{\partial \overline{u_i}}{\partial x_j} - \rho \overline{u'_j u'_i} \right). \tag{5.38}$$

Comparing (5.38) with the Navier–Stokes equation (5.33), we find that the momentum equation of the mean flow is the same as the Navier–Stokes equation whose shear stress tensor of $\mu \partial u_i / \partial x_j$ $(=\tau_{ij})$ is replaced with

$$\mu \frac{\partial \overline{u_i}}{\partial x_j} - \rho \overline{u'_j u'_i} \left(= \overline{\tau_{ij}} - \rho \overline{u'_j u'_i} \right). \tag{5.39}$$

In other words, the turbulent mean flow receives stress $-\rho \overline{u'_j u'_i}$ from the disturbances, in addition to the mean stress $\overline{\tau_{ij}}$ caused by molecular viscosity. This stress introduced by turbulence is called the **Reynolds stress**, in honor of Reynolds, its discoverer.

Heat Transport Equation

Substituting (5.35) in (5.34) and taking the mean gives

$$\overline{\frac{\partial}{\partial t}\left(\overline{\theta} + \theta'\right) + \left(\overline{u_j} + u'_j\right) \frac{\partial}{\partial x_j}\left(\overline{\theta} + \theta'\right)} = \nu_\theta \overline{\frac{\partial^2 \left(\overline{\theta} + \theta'\right)}{\partial x_j^2}}.$$

Therefore,

$$\frac{\partial \overline{\theta}}{\partial t} + \overline{u_j} \frac{\partial \overline{\theta}}{\partial x_j} = \frac{\partial}{\partial x_j} \left(\nu_\theta \frac{\partial \overline{\theta}}{\partial x_j} - \overline{u'_j \theta'} \right). \tag{5.40}$$

5.3.1 Equations Relating to the Variation Components

Equation Governing Turbulent Kinetic Energy $\overline{u'_i u'_i}/2$

Multiplying (5.33) by u'_i and taking the mean, we get

$$\overline{u'_i \frac{\partial (\overline{u_i} + u'_i)}{\partial t} + u'_i (\overline{u_j} + u'_j) \frac{\partial (\overline{u_i} + u'_i)}{\partial x_j}} = \overline{-\frac{1}{\rho} u'_i \frac{\partial (\overline{p} + p')}{\partial x_i} + \nu u'_i \frac{\partial^2 (\overline{u_i} + u'_i)}{\partial x_j^2}}\,.$$

Expanding each term gives

$$\underbrace{\overline{u'_i\frac{\partial \overline{u_i}}{\partial t}}}_{0}+\underbrace{\overline{u'_i\frac{\partial u'_i}{\partial t}}}_{\frac{\partial}{\partial t}\left(\frac{1}{2}\overline{u'_iu'_i}\right)}+\underbrace{\overline{u'_i\overline{u_j}\frac{\partial \overline{u_i}}{\partial x_j}}}_{0}+\underbrace{\overline{u'_i\overline{u_j}\frac{\partial u'_i}{\partial x_j}}}_{\overline{u_j}\frac{\partial}{\partial x_j}\left(\frac{1}{2}\overline{u'_iu'_i}\right)}+\underbrace{\overline{u'_iu'_j\frac{\partial \overline{u_i}}{\partial x_j}}}_{\overline{u'_iu'_j}\frac{\partial \overline{u_i}}{\partial x_j}}+\overline{u'_iu'_j\frac{\partial u'_i}{\partial x_j}}$$

$$=-\frac{1}{\rho}\underbrace{\overline{u'_i\frac{\partial \overline{p}}{\partial x_i}}}_{0}-\frac{1}{\rho}\overline{u'_i\frac{\partial p'}{\partial x_i}}+\underbrace{\overline{\nu u'_i\frac{\partial^2 \overline{u_i}}{\partial x_j^2}}}_{0}+\overline{\nu u'_i\frac{\partial^2 u'_i}{\partial x_j^2}}.$$

Using the relations

$$\overline{u'_iu'_j\frac{\partial u'_i}{\partial x_j}}=\overline{u'_j\frac{\partial}{\partial x_j}\left(\frac{1}{2}u'_iu'_i\right)}=\overline{\frac{\partial}{\partial x_j}\left(\frac{1}{2}u'_iu'_iu'_j\right)-\frac{1}{2}u'_iu'_i\underbrace{\frac{\partial u'_j}{\partial x_j}}_{0}}=\frac{\partial}{\partial x_j}\left(\frac{1}{2}\overline{u'_iu'_iu'_j}\right),$$

$$-\frac{1}{\rho}\overline{u'_i\frac{\partial p'}{\partial x_i}}=-\frac{1}{\rho}\overline{\left[\frac{\partial}{\partial x_i}(u'_ip')-p'\underbrace{\frac{\partial u'_i}{\partial x_i}}_{0}\right]}=-\frac{1}{\rho}\frac{\partial}{\partial x_i}\left(\overline{u'_ip'}\right)$$

$$\overline{\nu u'_i\frac{\partial^2 u'_i}{\partial x_j^2}}=\nu\left[\overline{\frac{\partial}{\partial x_j}\left(u'_i\frac{\partial u'_i}{\partial x_j}\right)-\left(\frac{\partial u'_i}{\partial x_j}\right)^2}\right]=\nu\frac{\partial}{\partial x_j}\left(\overline{u'_i\frac{\partial u'_i}{\partial x_j}}\right)-\nu\overline{\left(\frac{\partial u'_i}{\partial x_j}\right)^2},$$

this becomes

$$\frac{\partial}{\partial t}\left(\frac{1}{2}\overline{u'_iu'_i}\right)+\overline{u_j}\frac{\partial}{\partial x_j}\left(\frac{1}{2}\overline{u'_iu'_i}\right)$$
$$=-\frac{\partial}{\partial x_j}\left(\frac{1}{\rho}\overline{u'_jp'}+\frac{1}{2}\overline{u'_iu'_iu'_j}-\nu\overline{u'_i\frac{\partial u'_i}{\partial x_j}}\right)-\overline{u'_iu'_j}\frac{\partial \overline{u_i}}{\partial x_j}-\nu\overline{\left(\frac{\partial u'_i}{\partial x_j}\right)^2}. \tag{5.41}$$

Alternatively, using the relations

$$\frac{\partial}{\partial x_j}\left(\overline{u'_i\frac{\partial u'_i}{\partial x_j}}\right)=\frac{\partial}{\partial x_j}\left(\overline{u'_i\frac{\partial u'_i}{\partial x_j}+u'_i\frac{\partial u'_j}{\partial x_i}}\right)-\frac{\partial}{\partial x_j}\left(\overline{u'_i\frac{\partial u'_j}{\partial x_i}}\right)$$
$$=\frac{\partial}{\partial x_j}\left[\overline{u'_i\left(\frac{\partial u'_i}{\partial x_j}+\frac{\partial u'_j}{\partial x_i}\right)}\right]-\overline{\frac{\partial u'_i}{\partial x_j}\frac{\partial u'_j}{\partial x_i}}+\overline{u'_i\frac{\partial}{\partial x_i}\underbrace{\left(\frac{\partial u'_j}{\partial x_j}\right)}_{0}},$$

$$\overline{u'_i u'_j}\frac{\partial \overline{u_i}}{\partial x_j} = \overline{u'_j u'_i}\frac{\partial \overline{u_j}}{\partial x_i} = \frac{1}{2}\overline{u'_i u'_j}\left(\frac{\partial \overline{u_i}}{\partial x_j} + \frac{\partial \overline{u_j}}{\partial x_i}\right)$$

$$-\overline{\left(\frac{\partial u'_i}{\partial x_j}\right)^2} - \overline{\frac{\partial u'_i}{\partial x_j}\frac{\partial u'_j}{\partial x_i}} = -\frac{1}{2}\overline{\left(\frac{\partial u'_i}{\partial x_j}\right)^2} - \frac{1}{2}\overline{\left(\frac{\partial u'_j}{\partial x_i}\right)^2} - \overline{\frac{\partial u'_i}{\partial x_j}\frac{\partial u'_j}{\partial x_i}} = -\frac{1}{2}\overline{\left(\frac{\partial u'_i}{\partial x_j} + \frac{\partial u'_j}{\partial x_i}\right)^2},$$

(5.41) is expressed as

$$\frac{\partial}{\partial t}\left(\frac{1}{2}\overline{u'_i u'_i}\right) + \overline{u_j}\frac{\partial}{\partial x_j}\left(\frac{1}{2}\overline{u'_i u'_i}\right)$$
$$= -\frac{\partial}{\partial x_j}\left(\frac{1}{\rho}\overline{u'_j p'} + \frac{1}{2}\overline{u'_i u'_i u'_j} - 2\nu\overline{u'_i s'_{ij}}\right) - \overline{u'_i u'_j}\,\overline{s_{ij}} - 2\nu\overline{s'_{ij} s'_{ij}} \tag{5.42}$$

where $\overline{s_{ij}}$ and $\overline{s'_{ij}}$ are defined as follows:

$$\overline{s_{ij}} \equiv \frac{1}{2}\left(\frac{\partial \overline{u_i}}{\partial x_j} + \frac{\partial \overline{u_j}}{\partial x_i}\right), \quad s'_{ij} \equiv \frac{1}{2}\left(\frac{\partial u'_i}{\partial x_j} + \frac{\partial u'_j}{\partial x_i}\right).$$

Equation Governing Heat Dispersion $\overline{\theta'^2}$

Multiplying (5.34) by $2\theta'$ and taking the mean gives the heat dispersion equation

$$\frac{\partial \overline{\theta'^2}}{\partial t} + \overline{u_j}\frac{\partial \overline{\theta'^2}}{\partial x_j} = -\frac{\partial}{\partial x_j}\left(\overline{\theta'^2 u'_j} - \nu_\theta\frac{\partial \overline{\theta'^2}}{\partial x_j}\right) - 2\overline{\theta' u'_j}\frac{\partial \overline{\theta}}{\partial x_j} - 2\nu_\theta\overline{\frac{\partial \theta'}{\partial x_j}\frac{\partial \theta'}{\partial x_j}}. \tag{5.43}$$

Equation Governing Turbulent Momentum Flux

First, (5.33) is multiplied by u'_k and the equation with i and k switched is added. Then, the equation is obtained by taking the average ((5.41) with the contraction $k = i$).

$$\frac{\partial \overline{u'_i u'_k}}{\partial t} + \overline{u_j}\frac{\partial \overline{u'_i u'_k}}{\partial x_j} = -\overline{u'_i u'_j}\frac{\partial \overline{u_k}}{\partial x_j} - \overline{u'_k u'_j}\frac{\partial \overline{u_i}}{\partial x_j} - \frac{\partial \overline{u'_i u'_k u'_j}}{\partial x_j}$$
$$-\frac{1}{\rho}\left[\frac{\partial \overline{p' u'_k}}{\partial x_i} + \frac{\partial \overline{p' u'_i}}{\partial x_k} - \overline{p'\left(\frac{\partial u'_i}{\partial x_k} + \frac{\partial u'_k}{\partial x_i}\right)}\right] + \nu\frac{\partial^2 \overline{u'_i u'_k}}{\partial x_j^2} - 2\nu\overline{\frac{\partial u'_i}{\partial x_j}\frac{\partial u'_k}{\partial x_j}}. \tag{5.44}$$

Equation Governing Heat Flux Caused by Turbulence

This is obtained as the mean of (5.33) $\times \theta' +$ (5.34) $\times u'_i$.

$$\frac{\partial \overline{\theta' u'_i}}{\partial t} + \overline{u_j} \frac{\partial \overline{\theta' u'_i}}{\partial x_j} = -\overline{u'_i u'_j} \frac{\partial \overline{\theta}}{\partial x_j} - \overline{\theta' u'_j} \frac{\partial \overline{u_i}}{\partial x_j} - \frac{\partial \overline{\theta' u'_i u'_j}}{\partial x_j}$$
$$-\frac{1}{\rho} \left[\frac{\partial \overline{p'\theta'}}{\partial x_i} - \overline{p' \frac{\partial \theta'}{\partial x_i}} \right] + \nu \overline{\theta' \frac{\partial^2 u'_i}{\partial x_j^2}} + \nu_\theta \overline{u'_i \frac{\partial^2 \theta'}{\partial x_j^2}} . \tag{5.45}$$

Exercise 5.2 Derive (5.43), (5.44), and (5.45).

5.4 Closure

At the beginning of this chapter, we mentioned that we shoulds manifest changes in a statistical quantity, for example, the mean velocity or temperature of turbulence. From this perspective, we revisit (5.38) and (5.40). To evolve the mean flow velocity $\overline{u_i}$ or temperature $\overline{\theta}$ over time, the computation of $\overline{u_i}$ or $\overline{\theta}$ is insufficient and we require the secondary correlations of variations, such as $\overline{u'_i u'_j}$ or $\overline{\theta' u'_j}$. That is, the mean quantities $\overline{u_i}$ and $\overline{\theta}$ do not provide a complete solution. Then, evolving the second order correlations $\overline{u'_i u'_j}$ and $\overline{\theta' u'_j}$ over time by (5.44) and (5.45), we find that third-order correlations such as $\overline{u'_i u'_j u'_k}$ and $\overline{\theta' u'_i u'_j}$ emerge. Thus, when seeking the higher-order properties of disturbance, we require ever higher-order moments, and the system of equations expands unbounded (see Fig. 5.5).

To halt this expansion, we impose limiting assumptions. Deciding when to terminate the link to higher-order moments (Fig. 5.6) is called the **closure problem** of turbulence. Below, we introduce representative methods for completing (closing) the equations governing the mean statistics of turbulence.

Coefficient of Eddy Viscosity

The amount of momentum carried or heat (or substance) transported by molecular motion is given by $\nu \partial u_i / \partial x_j$ and $\nu_\theta \partial \theta / \partial x_j$, where ν and ν_θ are the coefficients of kinematic viscosity and molecular diffusivity, respectively. Analogous to molecular viscosity or molecular diffusion, the amount of momentum carried or heat transported by turbulence is proportional to the gradients of their respective means, providing a closing equation for solving turbulent flows. In other words, considering the **eddy viscosity** coefficient ν_t or the **eddy diffusivity** $\nu_{\theta t}$, assume that

$$-\overline{u'_i u'_j} = \nu_t \frac{d\overline{u}_i}{dx_j}, \quad -\overline{\theta' u'_j} = \nu_{\theta t} \frac{d\overline{\theta}}{dx_j} . \tag{5.46}$$

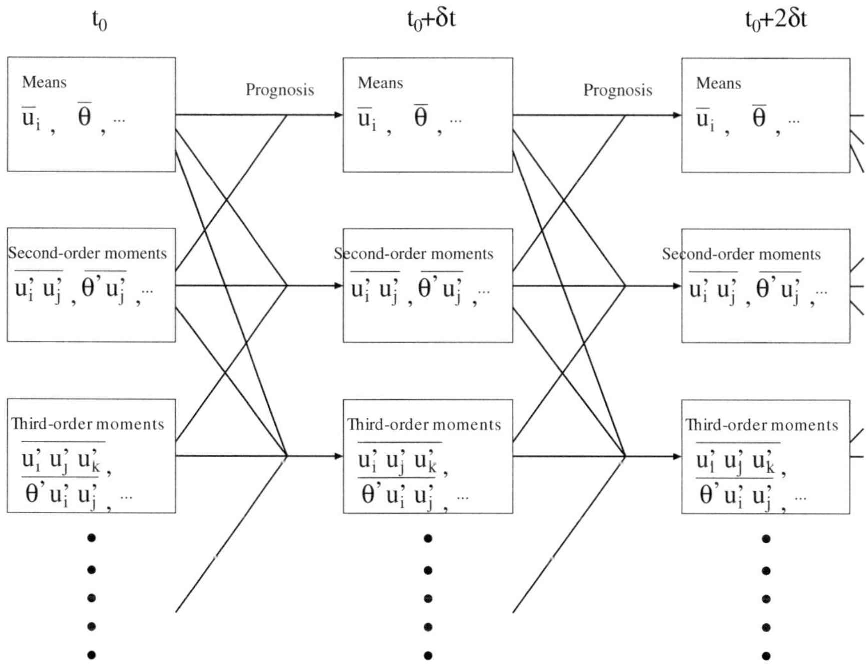

Fig. 5.5 Closure problem of turbulence (the equation set never closes)

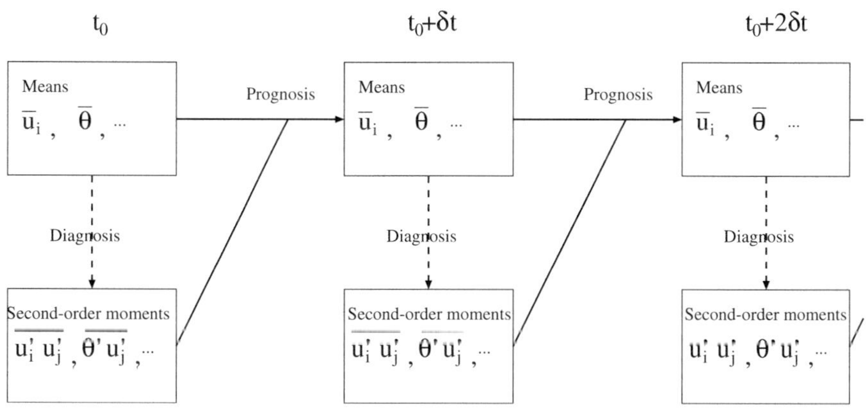

Fig. 5.6 Closure using a turbulent model

Substituting (5.46) into (5.38) and (5.40), the respective equations governing the mean flow turbulence and heat transfer are

$$\frac{\partial \overline{u}_i}{\partial t} + \overline{u}_j \frac{\partial \overline{u}_i}{\partial x_j} = -\frac{1}{\rho}\frac{\partial \overline{p}}{\partial x_i} + \frac{\partial}{\partial x_j}\left(\nu_t \frac{\partial \overline{u}_i}{\partial x_j}\right) \tag{5.47}$$

$$\frac{\partial \overline{\theta}}{\partial t} + \overline{u}_j \frac{\partial \overline{\theta}}{\partial x_j} = \frac{\partial}{\partial x_j}\left(\nu_{\theta t} \frac{\partial \overline{\theta}}{\partial x_j}\right). \tag{5.48}$$

If turbulence is well developed, the diffusion due to molecular motion is small relative to the turbulent transport, and $\nu \ll \nu_t, \nu_\theta \ll \nu_{\theta t}$. Under these conditions, we can ignore the molecular diffusion terms in (5.47) and (5.48). In closure methods using the eddy viscosity coefficient, the molecular viscosity coefficient needs merely to be replaced by the eddy viscosity coefficient.

Mixing Length Theory

Just as heat conduction or molecular diffusion occurs by physical processes such as motion of gas molecules, the irregular motion of turbulence transports physical quantities from one layer to another. In turbulent flows, the velocity of the turbulence is equivalent to the molecular velocity. Moreover, a mass of fluid carried by turbulent motion initially retains the physical characteristics of its original layer, but beyond a certain distance, its characteristics begin to merge into the new environment. The distance at which this merging begins—the so-called **mixing length**—is equivalent to the mean free path of molecular motion.

The mean temperature distribution $\overline{\theta}$ is regarded as a function of z alone (see Fig. 5.7). Denoting the turbulent velocity in the z-direction by w', the heat flux transported in the z-direction per unit time through a unit area perpendicular to the z-axis at some point z is given by

$$\overline{\left(\overline{\theta} + \theta'\right) w'} = \overline{\theta' w'}\,. \tag{5.49}$$

Flows enter this unit area from bottom to top and from top to bottom. The magnitude of the velocity, denoted as $|w'|$, is statistically the same in both flow directions. Moreover, if the mixing length is denoted as l', the mass of air passing through this surface from bottom to top retains the characteristics of the fluid layer approximately $l'/2$ below. The temperature of the fluid at its original location is approximately $\overline{\theta} - (l'/2)d\overline{\theta}/dz$. Therefore, the heat flux passing through this surface from bottom to top is given by

$$\left[\overline{\theta}(z) - \frac{l'}{2}\frac{d\overline{\theta}}{dz}\right]|w'|\,. \tag{5.50}$$

Similarly, the heat flux passing this surface from above is

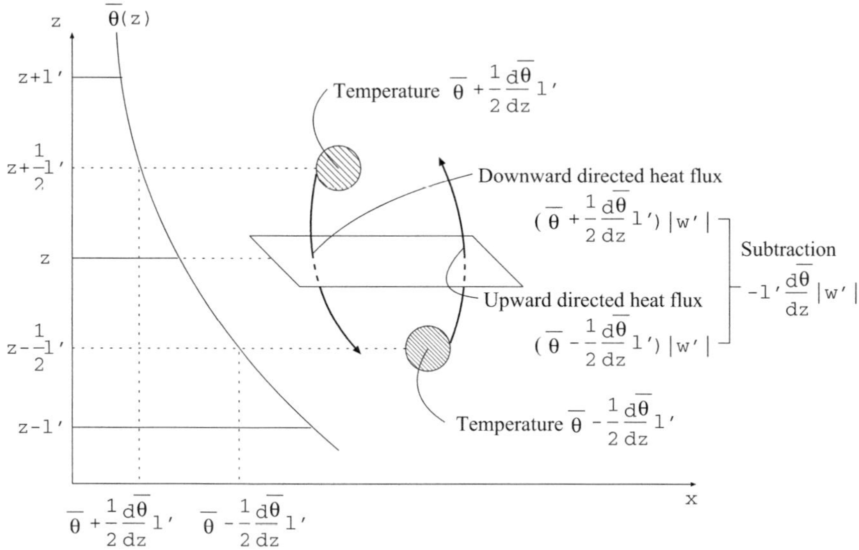

Fig. 5.7 Mixing length theory

$$\left[\overline{\theta}(z) + \frac{l'}{2}\frac{d\overline{\theta}}{dz}\right]|w'|\,. \tag{5.51}$$

The heat transported in the z-axis direction through this unit area per unit time is the difference between the heat fluxes given by (5.50) and (5.51):

$$\overline{\theta' w'} = -l'\frac{d\overline{\theta}}{dz}|w'| \sim -l'\sqrt{\overline{w'^2}}\,\frac{d\overline{\theta}}{dz}\,. \tag{5.52}$$

The heat flux $\overline{\theta' w'}$ carried by turbulence is proportional to the mean temperature gradient $d\overline{\theta}/dz$. The proportionality coefficient $l'\sqrt{\overline{w'^2}}$ merely replaces the eddy diffusivity coefficient $\nu_{\theta t}$.

A similar argument applies to the z-axis transport of momentum ρu in the x-axis direction $-\rho\overline{u'w'}$:

$$\rho\overline{u'w'} \sim -\rho l'\sqrt{\overline{w'^2}}\,\frac{d\overline{u}}{dz} \tag{5.53}$$

in which the eddy viscosity coefficient is again replaced by $l'\sqrt{\overline{w'^2}}$. Furthermore, if u' is considered to arise from the average velocity $\overline{u} + l'd\overline{u}/dz$, originally reserved by the fluid mass shifting away from the average velocity $\overline{u}$ at its current position, the turbulence strength $\sqrt{\overline{u'^2}}$ or $\sqrt{\overline{w'^2}}$ is evaluated as

$$\sqrt{\overline{u'^2}} \sim \sqrt{\overline{w'^2}} \sim l' \left| \frac{d\overline{u}}{dz} \right| . \tag{5.54}$$

Substituting (5.54) into (5.53) gives

$$- \rho \overline{u'w'} = \rho l'^2 \left| \frac{d\overline{u}}{dz} \right| \frac{d\overline{u}}{dz} . \tag{5.55}$$

Expressing Reynolds stress in the form (5.55) provides another means of closing the equation of motion. This approach, conceptualized by Prandtl, is called the **mixing length theory**.

5.5 Shear-Flow Turbulence Near a Wall (Ground)

Turbulence in flows with spatially inhomogeneous mean velocity is called **shear-flow turbulence**. It frequently develops near the walls or ground, where velocity gradients are large. This section investigates shear turbulence around walls and the ground, and the resulting velocity distribution of the mean flow.

Consider flows along a flat wall (or ground). The x-axis specifies the direction of mean flow and the z-axis is perpendicular to the wall (ground). The mean flow is assumed to be steady and homogeneous in the x-axis direction, and thus develops no pressure gradients. As

$$\frac{\partial}{\partial t} = 0, \quad \frac{\partial}{\partial x} = 0, \quad \overline{w} = 0,$$

the x-component of (5.38) becomes

$$\tau = \mu \frac{d\overline{u}}{dz} - \rho \overline{u'w'} = \text{const} = \tau_0 , \tag{5.56}$$

where τ_0 is the shear stress acting on the wall (ground surface). Instead of τ_0, the quantity with the units of velocity

$$u_* \equiv \sqrt{\frac{\tau_0}{\rho}} \tag{5.57}$$

is defined. This value, u_*, is often used for discussions, and is called **friction velocity**.

Viscous Sublayer

First, we consider a region extremely close to the wall (ground). In this region, viscid fluid flows predominate and the Reynolds stress term in (5.56) can be discarded, yielding

$$\nu\frac{d\overline{u}}{dz}=\frac{\tau_0}{\rho}=u_*^2\,. \tag{5.58}$$

Solving (5.58), we get

$$\overline{u}=\frac{u_*^2}{\nu}z\,. \tag{5.59}$$

In other words, the mean flow very close to the wall surface (ground) is linearly distributed. This layer is called a **viscous sublayer**.

Logarithmic Velocity Distribution

At a small distance from the wall surface, the Reynolds stress $-\rho\overline{u'w'}$ predominates, and the following approximation is valid:

$$-\overline{u'w'}=u_*^2\,. \tag{5.60}$$

Applying Prandtl's mixing length theory (5.55) to (5.60), we obtain

$$l'^2\left(\frac{d\overline{u}}{dz}\right)^2=u_*^2\,. \tag{5.61}$$

Furthermore, the size of the turbulent vortex is limited by the presence of the wall (ground). Therefore, assuming that the mixing length l' is proportional to the distance z from the wall surface (Fig. 5.8), we have $l'=\kappa z$ (the proportionality constant κ is called the Kármán's universal constant).

$$\kappa z\left(\frac{d\overline{u}}{dz}\right)=u_*\,. \tag{5.62}$$

Solving (5.62) gives

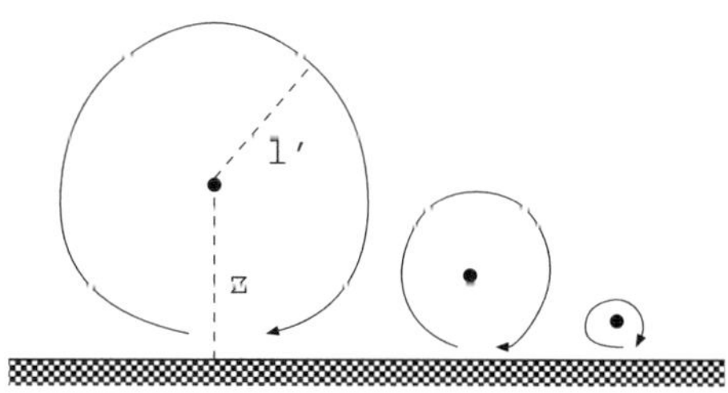

Fig. 5.8 Turbulent flow vortex near a wall (ground)

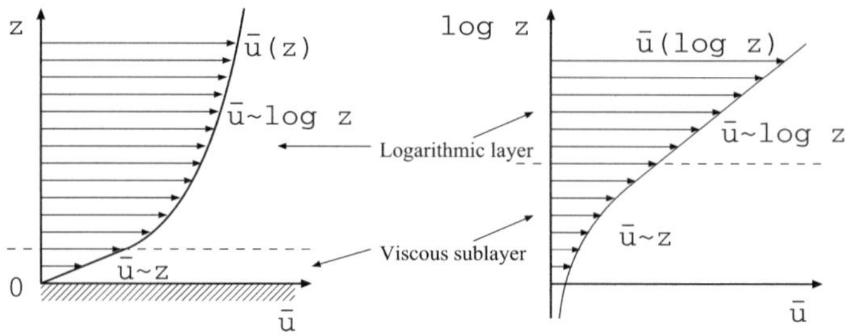

Fig. 5.9 Mean velocity distribution near a wall (ground)

$$\overline{u} = \frac{u_*}{\kappa} \log z + C\,. \tag{5.63}$$

Therefore, the velocity is logarithmically distributed in this range (Fig. 5.9).

5.6 Law of Similarity in Homogeneous Isotropic Turbulence

Thus far, we have seen that the closure problem of turbulence cannot be solved without imposing assumptions. Even the simplest turbulences (those with homogeneous and isotropic statistics) are not exempt from this problem. However, homogeneous and isotropic turbulences are characterized by several remarkable natural world phenomena, which are captured in an assumption proposed by Kolmogorov. Kolmogorov's argument, known as the law of similarity, is discussed in this section.[2]

5.6.1 *Homogeneous Isotropic Turbulence*

Turbulence theory essentially ignores the details of the flow, and instead focuses on its statistical properties. When statistical quantities are independent of location $\boldsymbol{r}$, the turbulence is called **homogeneous**. If the same statistical quantities are independent of direction, it is called isotropic. Turbulence that is both homogeneous and isotropic is referred to as **homogeneous isotropic turbulence**[3] (see Fig. 5.10).

---

[2]Most textbooks provide more systematic arguments using velocity correlation tensors. Here we limit our argument to the minimum requirement for the energy Eq. (5.84) required by the law of similarity. Readers who are unfamiliar with this topic may nonetheless find the calculations overly complicated. Such readers are recommended to skip Sects. 5.6.2 and 5.6.3 (but should read the conclusions to this section (5.84) and the sections beyond Sect. 5.6.4.

[3]Isotropic turbulence is homogeneous by definition, and is therefore sometimes simply called isotropic turbulence.

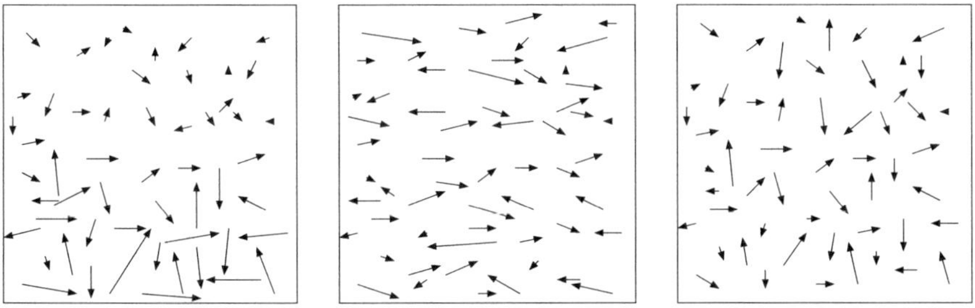

Non-homogeneous Homogeneous, non-isotropic Homogeneous, isotropic

Fig. 5.10 "Homogeneity" and "Isotropy" of turbulence

5.6.2 *Fourier Component Form of the Equation*

The equation satisfied by $\tilde{\boldsymbol{u}}(\boldsymbol{k}, t)$ is derived by Fourier transforming the continuity equation (5.32) and Navier–Stokes equation (5.33). Here $\tilde{\boldsymbol{u}}(\boldsymbol{k}, t)$ and its Fourier transform $\boldsymbol{u}(\boldsymbol{r}, t)$ are related as follows:

$$\boldsymbol{u}(\boldsymbol{r}, t) = \int \tilde{\boldsymbol{u}}(\boldsymbol{k}, t) \exp(i\boldsymbol{k} \cdot \boldsymbol{r})\, d\boldsymbol{k}, \tag{5.64}$$

$$\tilde{\boldsymbol{u}}(\boldsymbol{k}, t) = \frac{1}{(2\pi)^3} \int \boldsymbol{u}(\boldsymbol{r}, t) \exp(-i\boldsymbol{k} \cdot \boldsymbol{r})\, d\boldsymbol{r}\,. \tag{5.65}$$

First, we Fourier transform both sides of the continuity equation (5.32)

$$\widetilde{\nabla \cdot \boldsymbol{u}} = 0\,. \tag{5.66}$$

Using the Fourier transform identity $\widetilde{\partial A/\partial x_i} = ik_i\tilde{A}$, (5.66) becomes

$$i\boldsymbol{k} \cdot \tilde{\boldsymbol{u}} = 0\,. \tag{5.67}$$

Equation (5.67) implies that $\boldsymbol{k}$ and $\tilde{\boldsymbol{u}}$ are orthogonal.

Fourier transforming both sides of the Navier–Stokes equation, we obtain

$$\widetilde{\frac{\partial \boldsymbol{u}}{\partial t}} + \widetilde{(\boldsymbol{u} \cdot \nabla)\boldsymbol{u}} = -\frac{1}{\rho}\widetilde{\nabla p} + \nu\widetilde{\nabla^2 \boldsymbol{u}}\,. \tag{5.68}$$

Under Fourier transform, the terms $\widetilde{\partial \boldsymbol{u}/\partial t}$, $\widetilde{\nabla p}$, and $\widetilde{\nabla^2 \boldsymbol{u}}$, respectively, become $\partial\tilde{\boldsymbol{u}}/\partial t$, $i\boldsymbol{k}\tilde{p}$, and $-k^2\tilde{\boldsymbol{u}}$ (where $k^2 = |\boldsymbol{k}|^2 = k_j^2$). The second term on the left-hand side (advection term) is nonlinear and hence slightly complicated. However, using the relationship

$$
\begin{aligned}
\widetilde{(\boldsymbol{u}\cdot\nabla)\boldsymbol{u}} &\equiv \frac{1}{(2\pi)^3}\int d\boldsymbol{r}\,(\boldsymbol{u}\cdot\nabla)\boldsymbol{u}\exp(-i\boldsymbol{k}\cdot\boldsymbol{r}) \\
&= \frac{1}{(2\pi)^3}\int d\boldsymbol{r}\,\boldsymbol{u}\cdot\nabla\left[\int d\boldsymbol{k}'\,\tilde{\boldsymbol{u}}(\boldsymbol{k}')\exp\left(i\boldsymbol{k}'\cdot\boldsymbol{r}\right)\right]\exp(-i\boldsymbol{k}\cdot\boldsymbol{r}) \\
&= \frac{1}{(2\pi)^3}\int d\boldsymbol{r}\left[\int d\boldsymbol{k}'(\boldsymbol{u}\cdot i\boldsymbol{k}')\tilde{\boldsymbol{u}}(\boldsymbol{k}')\exp\left(i\boldsymbol{k}'\cdot\boldsymbol{r}\right)\right]\exp(-i\boldsymbol{k}\cdot\boldsymbol{r}) \\
&= \int d\boldsymbol{k}'\,\tilde{\boldsymbol{u}}(\boldsymbol{k}')\;\;i\boldsymbol{k}'\cdot\underbrace{\left[\frac{1}{(2\pi)^3}\int d\boldsymbol{r}\,\boldsymbol{u}\exp\left[i(\boldsymbol{k}'-\boldsymbol{k})\cdot\boldsymbol{r}\right]\right]}_{\tilde{\boldsymbol{u}}(\boldsymbol{k}-\boldsymbol{k}')} \\
&= i\int d\boldsymbol{k}'\,\tilde{\boldsymbol{u}}(\boldsymbol{k}')\left(\boldsymbol{k}'\cdot\tilde{\boldsymbol{u}}(\boldsymbol{k}-\boldsymbol{k}')\right)
\end{aligned}
$$

the fully Fourier-transformed Navier–Stokes equation becomes

$$
\frac{\partial\tilde{\boldsymbol{u}}(\boldsymbol{k})}{\partial t}+i\int d\boldsymbol{k}'\,\tilde{\boldsymbol{u}}(\boldsymbol{k}')\left(\boldsymbol{k}'\cdot\tilde{\boldsymbol{u}}(\boldsymbol{k}-\boldsymbol{k}')\right)=-\frac{i}{\rho}\boldsymbol{k}\tilde{p}-\nu k^2\tilde{\boldsymbol{u}}(\boldsymbol{k})\,. \tag{5.69}
$$

Using (5.67), the inner product of (5.69) and $\boldsymbol{k}$, given by $\boldsymbol{k}\cdot\tilde{\boldsymbol{u}}$, becomes zero. Thereby, $\tilde{p}$ can be eliminated from (5.69) as follows.

$$
\frac{\partial}{\partial t}\underbrace{\boldsymbol{k}\cdot\tilde{\boldsymbol{u}}(\boldsymbol{k})}_{0}+i\int d\boldsymbol{k}'\;\left(\boldsymbol{k}\cdot\tilde{\boldsymbol{u}}(\boldsymbol{k}')\right)\left(\boldsymbol{k}'\cdot\tilde{\boldsymbol{u}}(\boldsymbol{k}-\boldsymbol{k}')\right)=-\frac{i}{\rho}k^2\tilde{p}-\nu k^2\underbrace{\boldsymbol{k}\cdot\tilde{\boldsymbol{u}}(\boldsymbol{k})}_{0}
$$

$$
-\frac{i}{\rho}\,\tilde{p}=\frac{i}{k^2}\int d\boldsymbol{k}'\;\left(\boldsymbol{k}\cdot\tilde{\boldsymbol{u}}(\boldsymbol{k}')\right)\left(\boldsymbol{k}'\cdot\tilde{\boldsymbol{u}}(\boldsymbol{k}-\boldsymbol{k}')\right)\,.
$$

Inserting $\tilde{p}$ into (5.69)

$$
\begin{aligned}
&\frac{\partial\tilde{\boldsymbol{u}}(\boldsymbol{k})}{\partial t}+i\int d\boldsymbol{k}'\,\tilde{\boldsymbol{u}}(\boldsymbol{k}')\left(\boldsymbol{k}'\cdot\tilde{\boldsymbol{u}}(\boldsymbol{k}-\boldsymbol{k}')\right) \\
&\qquad = i\int d\boldsymbol{k}'\;\left(\boldsymbol{k}\cdot\tilde{\boldsymbol{u}}(\boldsymbol{k}')\right)\left(\boldsymbol{k}'\cdot\tilde{\boldsymbol{u}}(\boldsymbol{k}-\boldsymbol{k}')\right)\frac{\boldsymbol{k}}{k^2}-\nu k^2\tilde{\boldsymbol{u}}(\boldsymbol{k}),
\end{aligned}
$$

i.e.,

$$
\left(\frac{\partial}{\partial t}+\nu k^2\right)\tilde{\boldsymbol{u}}(\boldsymbol{k})=-i\int d\boldsymbol{k}'\;\left(\boldsymbol{k}'\cdot\tilde{\boldsymbol{u}}(\boldsymbol{k}-\boldsymbol{k}')\right)\left[\tilde{\boldsymbol{u}}(\boldsymbol{k}')-\frac{\boldsymbol{k}}{k^2}\left(\boldsymbol{k}\cdot\tilde{\boldsymbol{u}}(\boldsymbol{k}')\right)\right].
$$

Furthermore, setting the wavenumber to $\boldsymbol{k}-\boldsymbol{k}'$ in (5.67), we get

$$
\boldsymbol{k}'\cdot\tilde{\boldsymbol{u}}(\boldsymbol{k}-\boldsymbol{k}')=\boldsymbol{k}\cdot\tilde{\boldsymbol{u}}(\boldsymbol{k}-\boldsymbol{k}')
$$

which leads to

$$\left(\frac{\partial}{\partial t}+\nu k^2\right)\tilde{\boldsymbol{u}}(\boldsymbol{k})=-i\int d\boldsymbol{k}'\ \left(\boldsymbol{k}\cdot\tilde{\boldsymbol{u}}(\boldsymbol{k}-\boldsymbol{k}')\right)\left[\tilde{\boldsymbol{u}}(\boldsymbol{k}')-\frac{\boldsymbol{k}}{k^2}\left(\boldsymbol{k}\cdot\tilde{\boldsymbol{u}}(\boldsymbol{k}')\right)\right]. \quad (5.70)$$

This equation indicates that a wave with wavenumber $\boldsymbol{k}$ is attenuated proportionally to νk^2 by viscosity, and is further generated by the interaction between waves with wavenumbers $\boldsymbol{k}-\boldsymbol{k}'$ and $\boldsymbol{k}'$.

5.6.3 Energy Equation

First, we derive the equation describing the (kinetic) energy per unit mass. The energy E is given by $(1/2V)\int u^2\,d\boldsymbol{r}$.[4] In terms of $\tilde{\boldsymbol{u}}$, the Fourier component of $\boldsymbol{u}$, this becomes

$$\begin{aligned}E\equiv\frac{1}{2V}\int u^2\,d\boldsymbol{r}&=\frac{1}{2V}\int d\boldsymbol{r}\,\boldsymbol{u}(\boldsymbol{r})\left[\int d\boldsymbol{k}\,\tilde{\boldsymbol{u}}(\boldsymbol{k})\exp(i\boldsymbol{k}\cdot\boldsymbol{r})\right]\\&=\frac{1}{2V}\int d\boldsymbol{k}\,\tilde{\boldsymbol{u}}(\boldsymbol{k})\underbrace{\int d\boldsymbol{r}\,\boldsymbol{u}(\boldsymbol{r})\exp(i\boldsymbol{k}\cdot\boldsymbol{r})}_{(2\pi)^3\tilde{\boldsymbol{u}}(-\boldsymbol{k})}\\&=\frac{(2\pi)^3}{2V}\int d\boldsymbol{k}\,\tilde{\boldsymbol{u}}(\boldsymbol{k})\tilde{\boldsymbol{u}}(-\boldsymbol{k})\,.\end{aligned}$$

Setting

$$E(\boldsymbol{k})\equiv\frac{(2\pi)^3}{2V}\tilde{\boldsymbol{u}}(\boldsymbol{k})\tilde{\boldsymbol{u}}(-\boldsymbol{k})\,, \quad (5.71)$$

we have

$$E=\int d\boldsymbol{k}\,E(\boldsymbol{k}) \quad (5.72)$$

where $E(\boldsymbol{k})$ is the **energy spectrum** of a wave with wavenumber $\boldsymbol{k}$.

Now, we derive the equation governing the temporal change in kinetic energy density. Expressing (5.70) in tensor form, we have

$$\left(\frac{\partial}{\partial t}+\nu k^2\right)\tilde{u}_i(\boldsymbol{k})=-i\int d\boldsymbol{k}'\,k_m\tilde{u}_m(\boldsymbol{k}-\boldsymbol{k}')\left(\tilde{u}_i(\boldsymbol{k}')-\frac{k_ik_l}{k^2}\tilde{u}_l(\boldsymbol{k}')\right). \quad (5.73)$$

Multiplying both sides by $\tilde{u}_i(-\boldsymbol{k})$, we obtain

[4] As we here consider an infinite region, V and $\int u^2 dr$ are also infinite. Strictly speaking, this situation should be treated by limiting the range of $(1/2V)\int u^2 dr$ and taking the limit as $V\to\infty$.

$$
\begin{aligned}
&\tilde{u}_i(-\boldsymbol{k})\frac{\partial \tilde{u}_i(\boldsymbol{k})}{\partial t} + \nu k^2 \tilde{u}_i(\boldsymbol{k})\tilde{u}_i(-\boldsymbol{k}) \\
&\quad = -ik_m \int d\boldsymbol{k}'\, \tilde{u}_m(\boldsymbol{k}-\boldsymbol{k}')\Big(\tilde{u}_i(\boldsymbol{k}')\tilde{u}_i(-\boldsymbol{k}) - \frac{k_l k_l}{k^2} \underbrace{k_i \tilde{u}_i(-\boldsymbol{k})}_{0}\, \tilde{u}_l(\boldsymbol{k}')\Big) \\
&\quad = -ik_m \int d\boldsymbol{k}'\, \tilde{u}_i(-\boldsymbol{k})\tilde{u}_i(\boldsymbol{k}')\tilde{u}_m(\boldsymbol{k}-\boldsymbol{k}')\,. \qquad (5.74) \\
&\Big(\quad = -ik_m \int d\boldsymbol{k}'\, \tilde{u}_i(-\boldsymbol{k})\tilde{u}_i(-\boldsymbol{k}')\tilde{u}_m(\boldsymbol{k}+\boldsymbol{k}')\Big)
\end{aligned}
$$

Reversing the sign of the wavenumber $\boldsymbol{k}$ in (5.74) (i.e., replacing $\boldsymbol{k}$ with $-\boldsymbol{k}$ gives

$$
\tilde{u}_i(\boldsymbol{k})\frac{\partial \tilde{u}_i(-\boldsymbol{k})}{\partial t} + \nu k^2 \tilde{u}_i(\boldsymbol{k})\tilde{u}_i(-\boldsymbol{k}) = ik_m \int d\boldsymbol{k}'\, \tilde{u}_i(\boldsymbol{k})\tilde{u}_i(\boldsymbol{k}')\tilde{u}_m(-\boldsymbol{k}-\boldsymbol{k}')\,. \qquad (5.75)
$$

Adding (5.74) and (5.75), we get

$$
\begin{aligned}
&\left(\frac{\partial}{\partial t} + 2\nu k^2\right)\tilde{u}_i(\boldsymbol{k})\tilde{u}_i(-\boldsymbol{k}) \\
&\quad = ik_m \int d\boldsymbol{k}'\, \left\{\tilde{u}_i(\boldsymbol{k})\tilde{u}_i(\boldsymbol{k}')\tilde{u}_m(-\boldsymbol{k}-\boldsymbol{k}') - \tilde{u}_i(-\boldsymbol{k})\tilde{u}_i(-\boldsymbol{k}')\tilde{u}_m(\boldsymbol{k}+\boldsymbol{k}')\right\}.
\end{aligned}
$$

The above equation can be re-expressed as

$$
\begin{aligned}
&\left(\frac{\partial}{\partial t} + 2\nu k^2\right)\frac{(2\pi)^3}{2V}\tilde{u}_i(\boldsymbol{k})\tilde{u}_i(-\boldsymbol{k}) \\
&\quad = -k_m \frac{(2\pi)^3}{V}\int d\boldsymbol{k}'\, \mathrm{Im}\left[\tilde{u}_i(\boldsymbol{k})\tilde{u}_i(\boldsymbol{k}')\tilde{u}_m(-\boldsymbol{k}-\boldsymbol{k}')\right]. \qquad (5.76)
\end{aligned}
$$

Here we have used the Fourier transform identity $\tilde{A}(-\boldsymbol{k}) = \tilde{A}^*(\boldsymbol{k})$ for real function $A(r)$ (directly obtained as the complex conjugate of (5.65)). From (5.71), the left-hand side of (5.76) can be written as $(\partial/\partial t + 2\nu k^2)E(\boldsymbol{k})$, and the right-hand side becomes

$$
T(\boldsymbol{k}) \equiv -k_m \frac{(2\pi)^3}{V}\int d\boldsymbol{k}'\, \mathrm{Im}\big[\tilde{u}_i(\boldsymbol{k})\tilde{u}_i(\boldsymbol{k}')\tilde{u}_m(-\boldsymbol{k}-\boldsymbol{k}')\big]\,. \qquad (5.77)
$$

Therefore, (5.76) can be written as

$$
\left(\frac{\partial}{\partial t} + 2\nu k^2\right)E(\boldsymbol{k}) = T(\boldsymbol{k})\,. \qquad (5.78)
$$

In (5.78), the wave energy with a certain wavenumber is attenuated by viscosity proportional to $2\nu k^2$, and additionally, it is created by a term $T(\boldsymbol{k})$. Integrating $T(\boldsymbol{k})$ with respect to $\boldsymbol{k}$, we obtain

$$\int d\boldsymbol{k}\, T(\boldsymbol{k}) = -\frac{(2\pi)^3}{V} \int d\boldsymbol{k} \int d\boldsymbol{k}'\, k_m \mathrm{Im}\big[\tilde{u}_i(\boldsymbol{k})\tilde{u}_i(\boldsymbol{k}')\tilde{u}_m(-\boldsymbol{k}-\boldsymbol{k}')\big]\,.$$

Swapping $\boldsymbol{k}$ and $\boldsymbol{k}'$, we obtain

$$\int d\boldsymbol{k}\, T(\boldsymbol{k}) = -\frac{(2\pi)^3}{V} \int d\boldsymbol{k} \int d\boldsymbol{k}'\, k'_m \mathrm{Im}\big[\tilde{u}_i(\boldsymbol{k})\tilde{u}_i(\boldsymbol{k}')\tilde{u}_m(-\boldsymbol{k}-\boldsymbol{k}')\big]\,.$$

By adding these two equations, we obtain

$$2\int d\boldsymbol{k}\, T(\boldsymbol{k}) = -\frac{(2\pi)^3}{V} \int d\boldsymbol{k} \int d\boldsymbol{k}'\, \mathrm{Im}\Big[\tilde{u}_i(\boldsymbol{k})\tilde{u}_i(\boldsymbol{k}')\underbrace{(k_m + k'_m)\tilde{u}_m(-\boldsymbol{k}-\boldsymbol{k}')}_{0}\Big] = 0\,. \tag{5.79}$$

Thus, the generation term $T(\boldsymbol{k})$ disappears following integration over all wavenumbers $\boldsymbol{k}$, and does not directly appear in the equation that expresses the change in the whole energy obtained by integrating (5.78) with the wavenumbers:

$$\frac{d}{dt}E = -\int d\boldsymbol{k}\, 2\nu k^2 E(\boldsymbol{k}). \tag{5.80}$$

$T(k)$ expresses an energy transfer among the wavenumbers, and is called the **energy transfer function**.

Isotropic turbulence is independent of the angular direction of the statistical means. In this case, we should use the one-dimensional energy spectrum/transfer function, which is an integration of (5.71) and (5.77) in the angular direction:

$$E(k) \equiv \int d\theta_k \int d\phi_k\, E(\boldsymbol{k}), \quad T(k) \equiv \int d\theta_k \int d\phi_k\, T(\boldsymbol{k}). \tag{5.81}$$

Integrating $E(k)$ from 0 to infinity with respect to k, the total energy is given by

$$\int_0^\infty E(k)\, dk = E\,. \tag{5.82}$$

Thus, through integration, $T(k)$ becomes 0;

$$\int_0^\infty T(k)\, dk = 0\,. \tag{5.83}$$

The energy equation takes the form

$$\left(\frac{\partial}{\partial t} + 2\nu k^2\right) E(k) = T(k)\,. \tag{5.84}$$

The quantity $\Pi(k)$, denoting the flow of energy from the low-wavenumber side to the high-wavenumber side, is defined as

$$\Pi(k) \equiv -\int_0^k T(k')\,dk' \tag{5.85}$$

from which we obtain the following:

$$\left(\frac{\partial}{\partial t} + 2\nu k^2\right) E(k) = -\frac{d\Pi(k)}{dk}. \tag{5.86}$$

Exercise 5.3 Consider the energy equation (5.84). The initial condition is a unit energy for each of $k = k_1$ and $k = k_2$ (suppose here that $k_2 = 2k_1$). All other wavenumbers initially have zero energy.

1. Assuming no $T(k)$, how does the total energy change over time? Obtain E as a function of t.
2. Suppose that $T(k_1) = -T(k_2) = -\alpha(E(k_1) - E(k_2))$ (also suppose that $\alpha = 4\nu k_1^2$). That is, the transferred energy moves from the higher energy side to the lower energy side, and is proportional to the difference in the energies of k_1 and k_2. What happens to the total energy?
3. After a very long time $t \to \infty$, which of the total energies 1 and 2 is attenuated more quickly?

5.6.4 *Energy Cascade*

Let us return to (5.84), which shows the energy change in isotropic turbulence. The term $2\nu k^2 E(k)$ describes the energy dissipated by viscosity. The dispersion is more effective at larger wavenumbers (smaller scales). On the other hand, the $T(k)$ term in (5.83) merely transfers energy from one wavenumber to another, and makes no contribution to the total energy change. Nevertheless, the $T(k)$ term is related to the energy dissipation process. If the $T(k)$ term transfers energy in waves with small wavenumbers (large scale) to those with large wavenumbers (small scale), the overall energy remains unchanged; however, the energy is transported from a region of little viscosity effect to that of predominant viscosity effect. In this way, the $T(k)$ term indirectly increases the efficiency of energy dissipation (Fig. 5.11).

The following scenario can be imagined as a process that generates such effective energy attenuation. "The energy injected into small wavenumbers (large scale) by external forces gradually flows into large wavenumbers (small scale) under the effect of $T(k)$. Once energy has been transferred into wavenumbers within a certain size range, it is dissipated by viscous effects." This energy flow from small to large wavenumbers is called an **energy cascade** (Fig. 5.12). An important parameter in the

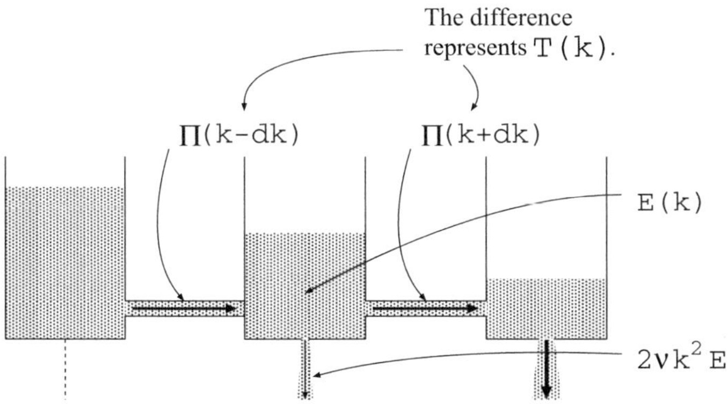

Fig. 5.11 Role of energy transfer function $T(k)$

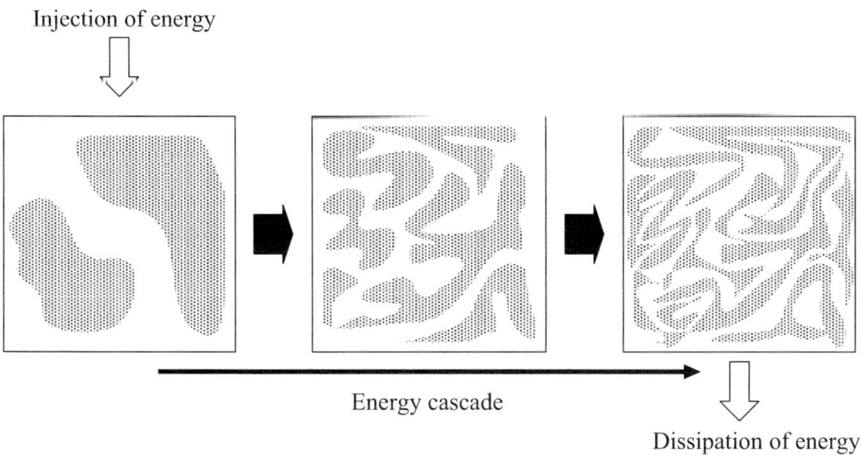

Fig. 5.12 Energy flow in real spaces

law of similarity is the **energy dissipation rate** ε, expressing the energy dissipation per unit time per unit mass.

5.6.5 Law of Similarity in Three-Dimensional Isotropic Turbulent Flows

Andrey Kolmogorov proposed the following hypothesis regarding homogeneous isotropic turbulence under such conditions.

Kolmogorov's First Hypothesis: Although an equilibrium state may not be established in the small wavenumber (large scale) region, the large wavenumber (small scale) region admits a range of wavenumbers at which homogeneous, isotropic, and steady equilibrium is maintained. This equilibrium state is determined by the energy dissipation rate ε and the coefficient of viscosity ν.

This range of scale is somewhat smaller than the external scale L that supplies the turbulent energy (in terms of wavenumbers, larger than $1/L$). This range of wavenumbers is called the **universal equilibrium range**. The energy dissipation rate ε has dimensions $[\mathrm{L}^2\mathrm{T}^{-3}]$ and the coefficient of viscosity has dimensions $[\mathrm{L}^2\mathrm{T}^{-1}]$. Hence, the characteristic time scale $t_d[\mathrm{T}]$ and spatial scale $l_d[\mathrm{L}]$ are, respectively, given by

$$t_d = \varepsilon^{-\frac{1}{2}}\nu^{\frac{1}{2}} \tag{5.87}$$

$$l_d = \varepsilon^{-\frac{1}{4}}\nu^{\frac{3}{4}} \tag{5.88}$$

l_d and its inverse $k_d \equiv l_d^{-1}$ are called the **Kolmogorov length** and the **Kolmogorov wavenumber**, respectively. From Kolmogorov's first hypothesis, if the physical quantities that are nondimensionalized by these scales are expressed in terms of independent variables that are nondimensionalized by the same scales, they should be described by universal functions. For instance, the dimensions of the energy spectrum $E(k)$ $[\mathrm{L}^3\mathrm{T}^{-2}]$ can be constructed as $\varepsilon^{\frac{1}{4}}\nu^{\frac{5}{4}}$. Hence, we have

$$E(k)\,/\,(\varepsilon^{\frac{1}{4}}\nu^{\frac{5}{4}}) = F(k/k_d)$$

from which the energy spectrum $E(k)$ can be expressed as

$$E(k) = \varepsilon^{\frac{1}{4}}\nu^{\frac{5}{4}}F(k/k_d) = \varepsilon^{\frac{2}{3}}k^{-\frac{5}{3}}G(k/k_d)\,. \tag{5.89}$$

In (5.89), $F(x)$ is a universal nondimensional function and $G(x) \equiv x^{\frac{5}{3}}F(x)$.

Kolmogorov's Second Hypothesis: At a sufficiently high Reynolds number, a wavenumber region exists in which viscosity exerts a negligible effect within the universal equilibrium range.

At wavenumbers similar to or larger than the Kolmogorov number k_d, viscosity-induced dissipation becomes appreciable. In this case, regions with wavenumbers smaller than k_d can be within the range of Kolmogorov's second hypothesis. Such a wavenumber region is called an **inertial range** or **inertial subrange**. Kolmogorov's second hypothesis implies that $E(k)$ in (5.89) must be a function independent of ν

within the inertial range. However, the $E(k)$ term in the second equation of (5.89), $E(k) = \varepsilon^{\frac{2}{3}} k^{-\frac{3}{5}} G(k/k_d)$, depends on ν through the $G(k/k_d)$ term. Thus, $G(k/k_d) =$ constant $\equiv C$. Consequently, (5.89) becomes

$$E(k) = C\varepsilon^{\frac{2}{3}} k^{-\frac{5}{3}}. \tag{5.90}$$

Thus, we have derived an important characteristic of three-dimensional homogeneous isotropic turbulence; within the inertial range, the energy spectrum is proportional to the $(-5/3)$rd power of the wavenumber.

Energy injected on a large scale by external forces is carried to the high-wavenumber side via the energy cascade. Energy transported through the inertial range is unaffected by viscosity. Thus, the injected energy is transported with negligible dissipation. Once it reaches regions of higher wavenumbers, it becomes dissipated by viscosity. Under equilibrium conditions, the injected energy ε_L, the energy transported in the inertial range $\Pi(k)$, and the dissipated energy ε are all equal (see Figs. 5.13 and 5.14).

Exercise 5.4 Given that the energy spectrum $E(k)$ in the inertial range depends only on ε and k, the relation (5.90) can be directly obtained from dimensional analysis. Derive (5.90) in this way.

The energy spectrum $E(1/L)$ corresponding to the spatial scale of energy injection, i.e., the global flow scale L (in terms of wavenumber, $1/L$), becomes $\varepsilon^{\frac{2}{3}} L^{\frac{5}{3}}$ by extending the relation (5.90). Denoting the velocity normalized by this length scale as U, the energy spectrum is approximated by $U^2/(1/L)$ (the energy per unit wavenumber); thus, $U^2 L \sim \varepsilon^{\frac{2}{3}} L^{\frac{5}{3}}$, i.e., $U \sim \varepsilon^{\frac{1}{3}} L^{\frac{1}{3}}$. Consequently, the Reynolds number can be expressed as

$$Re \equiv \frac{UL}{\nu} \sim \varepsilon^{\frac{1}{3}} L^{\frac{4}{3}} \nu^{-1} \tag{5.91}$$

where the ratio of external scale L and the Kolmogorov length (5.88) is given as

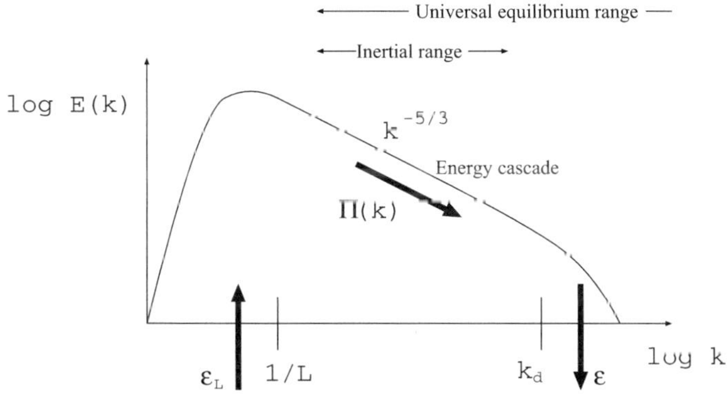

Fig. 5.13 Energy flow in three-dimensional turbulence

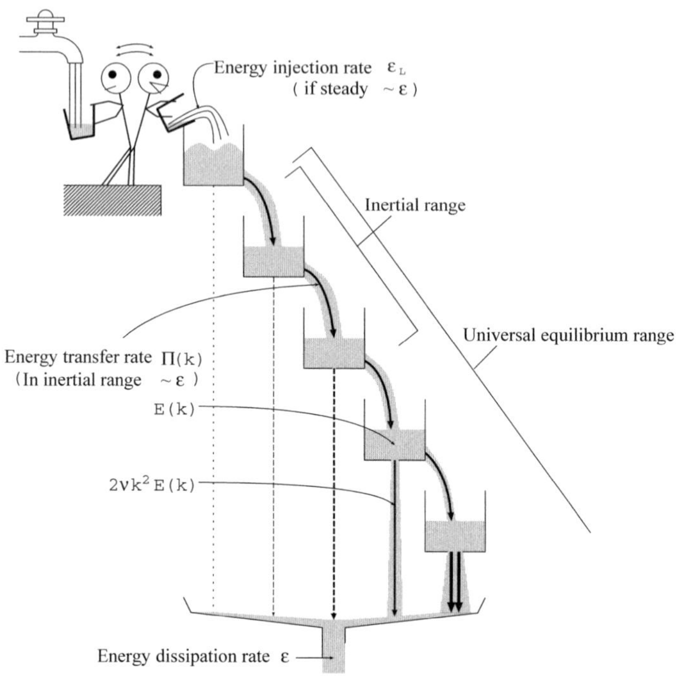

Fig. 5.14 Schematics of energy flow

$$\frac{L}{l_d} \sim Re^{\frac{3}{4}} . \tag{5.92}$$

Clearly, this ratio is proportional to the Reynolds number raised to the power 3/4.

5.6.6 Law of Similarity in Two-Dimensional Isotropic Turbulent Flows

When considering a two-dimensional turbulent flow, (5.84), which describes energy change, is the same as in the three-dimensional case. However, in a two-dimensional turbulent flow, the situation is completely different from the energy cascade in a three-dimensional turbulent flow. This is because in a two-dimensional flow, because of strong shape-related restrictions, conserved quantity called enstrophy exists in addition to energy.

In the absence of viscosity, we have confirmed energy conservation by Fourier transforming the energy equation. Alternatively, we can integrate the Navier–Stokes equation (5.33) to derive the energy conservation directly. Multiplying both sides of (5.33) by u_i and utilizing the relationship

$$u_i \frac{\partial u_i}{\partial t} = \frac{\partial}{\partial t}\left(\frac{1}{2}u^2\right)$$

$$u_i u_j \frac{\partial u_i}{\partial x_j} = u_j \frac{\partial}{\partial x_j}\left(\frac{1}{2}u^2\right) = \frac{\partial}{\partial x_j}\left(\frac{1}{2}u^2 u_j\right) - \frac{1}{2}u^2 \underbrace{\frac{\partial u_j}{\partial x_j}}_{0} = \frac{\partial}{\partial x_j}\left(\frac{1}{2}u^2 u_j\right)$$

$$u_i \frac{\partial p}{\partial x_i} = \frac{\partial}{\partial x_i}(pu_i) - p\underbrace{\frac{\partial u_i}{\partial x_i}}_{0} = \frac{\partial}{\partial x_i}(pu_i)$$

we obtain

$$\frac{\partial}{\partial t}\left(\frac{1}{2}u^2\right) + \frac{\partial}{\partial x_j}\left(\frac{1}{2}u^2 u_j\right) = -\frac{1}{\rho}\frac{\partial}{\partial x_i}(pu_i) + \nu u_i \frac{\partial^2 u_i}{\partial x_j^2}. \tag{5.93}$$

Integrating this result over the region gives

$$\frac{d}{dt}\int \frac{1}{2}u^2\, d\boldsymbol{r} + \int \frac{\partial}{\partial x_j}\left(\frac{1}{2}u^2 u_j\right) d\boldsymbol{r}$$
$$= -\frac{1}{\rho}\int \frac{\partial}{\partial x_i}(pu_i)\, d\boldsymbol{r} + \nu \int \frac{\partial}{\partial x_j}\left(u_i \frac{\partial u_i}{\partial x_j}\right) d\boldsymbol{r} - \nu \int \left(\frac{\partial u_i}{\partial x_j}\right)^2 d\boldsymbol{r}. \tag{5.94}$$

The second term on the left-hand side and the first and second terms on the right-hand side can be eliminated by applying Gauss' theorem with appropriate boundary conditions. Thus, we obtain

$$\frac{d}{dt}\int \frac{1}{2}u^2\, d\boldsymbol{r} = -\nu \int \left(\frac{\partial u_i}{\partial x_j}\right)^2 d\boldsymbol{r}. \tag{5.95}$$

Equation (5.95) shows that when viscosity is minimal, the mean energy density

$$E = \frac{1}{2S}\int u^2\, d\boldsymbol{r} \tag{5.96}$$

is maintained.[5]

Recall the equation of vorticity (1.131):

$$\frac{D\omega}{Dt} \equiv \frac{\partial \omega}{\partial t} + u_j \frac{\partial \omega}{\partial x_j} = \nu \frac{\partial^2 \omega}{\partial x_j^2}. \tag{5.97}$$

---

[5]Note that here we divide by S, not V, because we are dealing with two-dimensional volumes. Of course, the integration over $d\boldsymbol{r}$ is similarly performed in two dimensions.

By a similar transformation, we obtain

$$\frac{d}{dt}\int \frac{1}{2}\omega^2\, d\boldsymbol{r} = -\nu \int \left(\frac{\partial \omega}{\partial x_i}\right)^2 d\boldsymbol{r}\,. \tag{5.98}$$

In situations of minimal viscosity, the following is conserved:

$$Q \equiv \frac{1}{2S}\int \omega^2\, d\boldsymbol{r}\,. \tag{5.99}$$

This Q is called **enstrophy**.

Exercise 5.5 Derive (5.98) from (5.97).

Similar to (5.71) and (5.72), we define the enstrophy for each wavenumber:

$$Q(\boldsymbol{k}) \equiv \frac{(2\pi)^2}{2S}\tilde{\omega}(\boldsymbol{k})\tilde{\omega}(-\boldsymbol{k}) \tag{5.100}$$

which yields

$$Q = \int d\boldsymbol{k}\; Q(\boldsymbol{k})\,. \tag{5.101}$$

$E(k)$ and $Q(k)$ are simply related as follows.

Since the vorticity ω was originally defined as $\omega \equiv \partial v/\partial x - \partial u/\partial y$, its Fourier components are given by

$$\tilde{\omega}(\boldsymbol{k}) = i\left(k_x\tilde{v}(\boldsymbol{k}) - k_y\tilde{u}(\boldsymbol{k})\right)\,. \tag{5.102}$$

Therefore, using the Fourier-transformed continuity equation (5.67), we obtain the relation

$$\begin{aligned}
\tilde{\omega}(\boldsymbol{k})\tilde{\omega}(-\boldsymbol{k}) &= \left[k_x\tilde{v}(\boldsymbol{k}) - k_y\tilde{u}(\boldsymbol{k})\right]\left[k_x\tilde{v}(-\boldsymbol{k}) - k_y\tilde{u}(-\boldsymbol{k})\right] \\
&= \left(k_x^2 + k_y^2\right)\left[\tilde{u}(\boldsymbol{k})\tilde{u}(-\boldsymbol{k}) + \tilde{v}(\boldsymbol{k})\tilde{v}(-\boldsymbol{k})\right] \\
&\qquad - \underbrace{\left[k_x\tilde{u}(\boldsymbol{k}) + k_y\tilde{v}(\boldsymbol{k})\right]}_{0}\underbrace{\left[k_x\tilde{u}(-\boldsymbol{k}) + k_y\tilde{v}(-\boldsymbol{k})\right]}_{0} \\
&= \left(k_x^2 + k_y^2\right)\left[\tilde{u}(\boldsymbol{k})\tilde{u}(-\boldsymbol{k}) + \tilde{v}(\boldsymbol{k})\tilde{v}(-\boldsymbol{k})\right] \\
&= k^2\tilde{\boldsymbol{u}}(\boldsymbol{k})\tilde{\boldsymbol{u}}(-\boldsymbol{k})
\end{aligned} \tag{5.103}$$

which indicates

$$Q(\boldsymbol{k}) = k^2 E(\boldsymbol{k})\,. \tag{5.104}$$

In two-dimensional turbulence, the energy cascade seen in three-dimensional flows does not occur because it is strongly restricted by the condition where enstrophy and energy related by (5.104) for each wavenumber are both preserved (in the absence of viscosity). Let us suppose that such an energy cascade has occurred. In the energy cascade process, the movement of energy among different wavenumbers is essential, whereas viscosity dissipation is unimportant. Hence, we can assume that enstrophy and energy are both preserved. However, if we suppose that energy injected into low wavenumbers is transferred to higher wavenumbers, even if the magnitude of $E(k)$ is unchanged, $Q(k)$ $(=k^2E(k))$ increases with increasing k. In other words, if energy is transferred from the low-wavenumber side to the high-wavenumber side without loss, the enstrophy inevitably increases, violating the property of two-dimensional fluids, according to which both enstrophy and energy are preserved. To maintain both energy and enstrophy, one part of the energy must be transferred to the low-wavenumber side, while the remainder passes to the high-wavenumber side (see Fig. 5.15).

If this constraint is quantitatively investigated in detail, we find that in two-dimensional turbulent flows with a large Reynolds number which dissipates at appreciably high wavenumbers, most of the energy is transferred to low wavenumbers. On the other hand, if energy is transferred to sufficiently low wavenumbers with respect to the energy-injection wavenumber, much of the enstrophy is transferred to the high wavenumber side. Therefore, in complete contrast to three-dimensional turbulence, energy flow in two-dimensional turbulent fluids is transferred to low wavenumbers (large scale). This phenomenon is called **inverse cascade** or **upward cascade**.

In two-dimensional turbulence, energy and enstrophy are transferred within two inertial ranges (Fig. 5.16). The energy inertial range exists at wavenumbers lower than

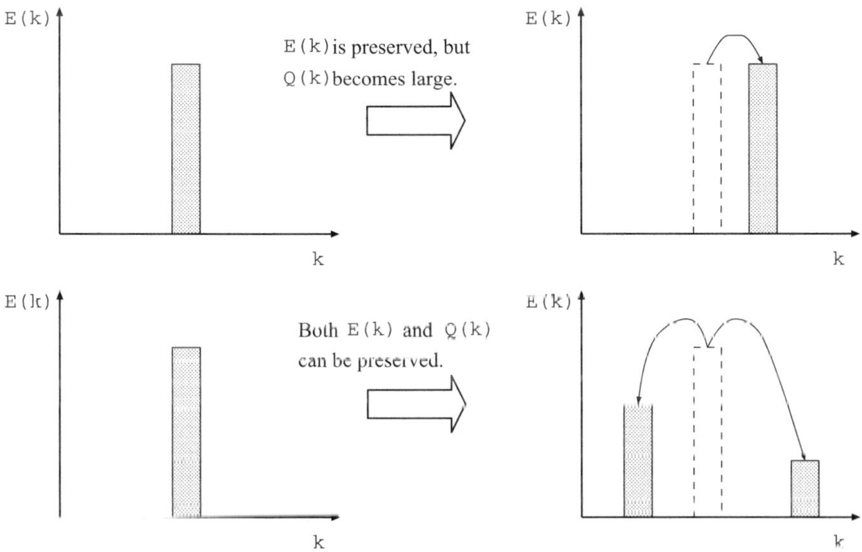

Fig. 5.15 Energy transfer in regions of preserved energy and enstrophy

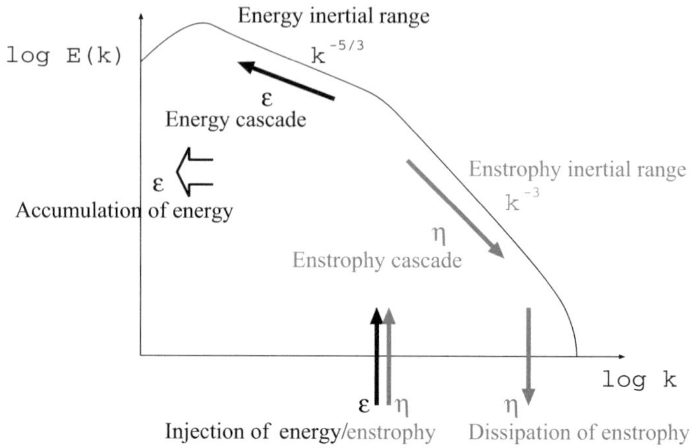

Fig. 5.16 Energy flow in two-dimensional turbulent flow

the wavenumber injected by the external force. The enstrophy inertial range exists at the high-wavenumber side. Within the energy inertial range, flow is characterized by the energy transfer rate ε, and the energy spectrum is proportional to the $-(5/3)$rd power of the wavenumber, identical to three-dimensional turbulence (5.90). On the other hand, within the enstrophy inertial range, flow is characterized by the **enstrophy dissipation rate** η and becomes proportional to the -3rd power of the wavenumber:

$$E(k) = C\eta^{\frac{2}{3}}k^{-3}\,. \tag{5.105}$$

Exercise 5.6 Within the enstrophy inertial range, the energy spectrum $E(k)$ is determined solely by parameters η and k. Use this fact to derive (5.105) by dimensional analysis.

We have been discussing two-dimensional turbulent flows as the last topic of homogeneous isotropic turbulent flows. At this stage, the reader might rightfully wonder whether such things actually exist. As we live in a three-dimensional world, strictly two-dimensional flows are difficult to imagine. However, as elaborated in Chaps. 6 and 7, the atmosphere and oceans are largely affected by stratification and rotation, and the flows in both media are largely homogeneous in the vertical direction (an example is the Taylor-Proudman theorem, to be presented in Sect. 7.3.2. Among the large-scale motions, atmospheric and oceanic flows exhibit particularly strong two-dimensional characteristics. Therefore, the flows in these systems may be frequently approximated by two-dimensional flow theories. Indeed, the atmospheric kinetic energy spectrum is proportional to the -3rd power of the wavenumber on scales around 1,000 km, as predicted by the two-dimensional turbulence theory.

5.7 Turbulent Flow Analysis

To conduct research on actual turbulence, or turbulence-related research, numerical calculations are required. Numerical turbulence analysis is conducted using one of three broad methods: direct numerical simulation (DNS), Reynolds-averaged numerical simulation (RANS), and large eddy simulation (LES).

Direct Numerical Simulation (DNS)

DNS is a direct numerical simulation method (i.e., the Navier–Stokes equations are directly solved without averaging) that does not use the turbulence model. Of course, there is no problem if simulation can be conducted with this method; however, this method requires huge computational resources. Furthermore, DNS requires a thin grid size (a scale similar to the Kolmogorov length) so that the energy entering the system can be properly dissipated. As seen in (5.92), the ratio of the flow scale to the Kolmogorov length increases proportionally to the Reynolds number raised to the power of 3/4. Hence, the number of grid points N that resolves the three-dimensional turbulence increases proportionally to the Reynolds number raised to the power of 9/4:

$$N = O(Re^{9/4}) . \tag{5.106}$$

Therefore, the use of DNS in real calculations is limited to turbulence with a low Reynolds number. Turbulence with a high Reynolds number requires some sort of turbulence model.

Reynolds-Averaged Numerical Simulation/Reynolds-Averaged Navier–Stokes Equation Model; Ensemble Average (RANS)

RANS completes turbulence equations by evaluating the ensemble-averaged Reynolds stress, as shown in Sects. 5.3 and 5.4. It can be broadly subdivided into eddy viscosity models (which evaluate eddy viscosity) and Reynolds stress models (which evaluate eddy viscosity by regressing against Reynolds stress). Eddy viscosity models are further subdivided into 0-equation models, 1-equation models, 2-equation models, or higher, on the basis of the number of input variables (calculated by integrating over time) to the eddy viscosity. 0-Equation models include methods using eddy viscosity coefficients and those based on the mixing length hypothesis. An example of a 2-equation model is the k-ε model.

Large Eddy Simulation (LES); Subgrid Closure

LES is a subgrid closure model. It adopts a spatial averaging process called filtering or volume averaging, and divides the flow field into a grid scale and a subgrid scale. Direct calculation is conducted on the grid scale, whereas the subgrid scale is modeled.

Note 5.1 (*Literature for further study*)
For further studies, textbooks [1, 2, 8, 9, 12, 13] may be referred. The advanced readers may also refer some other textbooks [3–7, 10, 11, 14–17], although they are written in Japanese.

References

1. Batchelor, G.K.: An Introduction to Fluid Dynamics. Cambridge University Press, Cambridge (1967)
2. Gill, A.E.: Atmosphere-Ocean Dynamics. Academic Press, Amsterdam (1982)
3. Gotoh, T.: Fundamentals of Turbulence Theory. Asakura Publishing, Tokyo (1998)
4. Hino, M.: Fluid Dynamics. Asakura Publishing, Tokyo (1992)
5. Imai, I.: Fluid Dynamics, vol. I. Shokabo, Tokyo (1973)
6. Kanbe, T. (ed.): Fluid Dynamics. Shokabo, Tokyo (1995)
7. Kimura, R.: Introduction to Geophysical Fluid Dynamics. Tokyodo Shuppan, Tokyo (1983)
8. Landau, L.D., Lifshitz, E.M.: Fluid Mechanics. Butterworth-Heinemann, Oxford (1987)
9. Newman, J.N.: Marine Hydrodynamics. The MIT Press, Cambridge (1977)
10. Ogura, Y.: Principles of Atmospheric Dynamics. University of Tokyo Press, Tokyo (1978)
11. Ogura, Y.: Basic Theory of Meso-scale Meteorology. University of Tokyo Press, Tokyo (1997)
12. Pedlosky, J.: Geophysical Fluid Dynamics, 2nd edn. Springer, New York (1986)
13. Prandtl, L., Tietjens, O.G.: Fundamentals of Hydro- and Aeromechanics (translated by Rosenhead, L.). Dover, New York (1957)
14. Tatsumi, T.: Turbulence. Makishoten, Tokyo (1962)
15. Tatsumi, T.: Fluid Dynamics. Baifukan, Tokyo (1982)
16. Tatsumi, T., Goto, K.: Stability Theory of Flow. Sangyo-Tosho, Tokyo (1976)
17. Tomita, Y.: Introduction to Fluid Dynamics. Yokendo, New York (1975)

Chapter 6
Dynamics of Stratified Fluids

In previous chapters, we have essentially ignored the effect of external forces on fluid motion. As shown in (4.3), potential forces such as gravity do not require explicit treatment, as they can be incorporated into pressure terms. However, our previous derivations assumed fixed density ρ. When the density is inhomogeneous, even potential forces cannot simply be absorbed into pressure terms, and external forces such as gravity exert impact on the fluid motion.[1]

Consider an inhomogeneous fluid density within a gravitational field. Denser parts are more affected by gravity than lower-density parts, and tend to displace downward. Thus, fluids in a gravity field tend to have higher density in the lower strata. For example, when heating water to prepare a bath, the hottest (lightest) water rises to the top, while the cold (heavy) water remains at the bottom. This state, in which the fluid density mainly changes in the direction of gravity (z-direction), is known as **stratification**. In this case, stratification is highly stable, and its reversal requires excessive energy input. Anybody who has stepped into a manually heated bath has probably experienced the shock of cold water beneath a warm surface. Although the lightest fluids may reach the bottom in nature, this phenomenon is local or transient. As the fluid motions of the atmosphere and oceans are strongly affected by density stratification (the focus of this chapter), it plays an important role in understanding the motions of the atmosphere and oceans along with the effects of the Earth's rotation (to be considered in the next chapter).

[1]Free surfaces can be considered to be an extreme case of density changes, and fluids with free surfaces are also affected by gravity. As the wave treatment in Chap. 3 relied to a large extent on gravitational restoring forces, the effect of gravity was touched upon in that chapter. However, homogenous fluid density was assumed in Chap. 3, and the gravitational effects were absorbed into the boundary conditions at the free surface, and were thus not a part of the equation of motion.

D.E.S.S.T. Kyushu Univ., *Fluid Dynamics for Global Environmental Studies*,
DOI 10.1007/978-4-431-56499-7_6

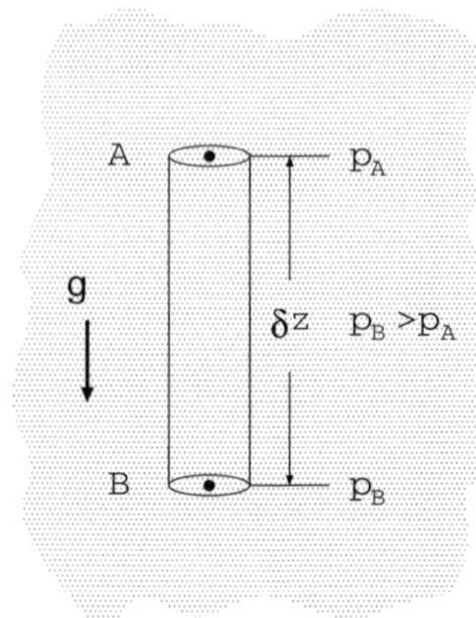

Fig. 6.1 Hydrostatic pressure

6.1 Hydrostatic Equilibrium

The effects of a gravitational field on fluids appear clearly on the pressure distribution. Let us compare the pressure at two points A and B vertically separated by δz within a stationary fluid (Fig. 6.1). The pressure p_B at the lower point B is the same as the pressure p_A at point A plus gravity on the fluid column existing between A and B ($p_B = p_A + \rho g \delta z$). Thus, the pressure increases with depth. Setting $\boldsymbol{u} = 0$ in the vertical component of the Navier–Stokes equation, this relationship is mathematically described by

$$\frac{\partial p}{\partial z} = -\rho g \,. \tag{6.1}$$

The equilibrium state (6.1) is called **hydrostatic balance**. The pressure that is determined from density following this relationship is called the **hydrostatic pressure**.

The hydrostatic balance is strictly limited to stationary fluids. However, fluids in slight motion approximately satisfy (6.1) if the inertial and viscous terms are sufficiently smaller than the pressure and gravitational acceleration terms. In the large-scale phenomena of the atmosphere and oceans, a hydrostatic relationship is established with extremely high precision.

Exercise 6.1 The vertical distribution of p and ρ is not immediately apparent from (6.1) but can be obtained from the state equation, expressing the relation between p and ρ.

1. Obtain $p(z)$ when ρ = constant. At what depth does the water pressure become 100 atm higher than that at the sea surface? (Let 1 atm $\cong$ 1,000 hPa, and express your answer in meters)
2. Show that the pressure of an isothermal ideal gas obeying the equation of state $p = \rho R T$ is $p(z) = p(0)\exp(-z/H)$ and express H in terms of T, R, and g. This H is called the scale height. Given that $R = 2.9 \times 10^2\,\mathrm{J\,kg^{-1}\,K^{-1}}$ for air and assuming the temperature T as 270 K, what is the scale height of the atmosphere?

6.2 Hydrostatic Stability

As the horizontal component of $\boldsymbol{u}$ in a stationary fluid subject to a gravitational field is 0, the Navier–Stokes equation becomes

$$\frac{\partial p}{\partial x} = \frac{\partial p}{\partial y} = 0 . \tag{6.2}$$

Therefore, the pressure of a stationary fluid in a gravitational field is constant in the horizontal direction and depends only on the vertical coordinate z. Such fluids also obey the hydrostatic relation (6.1), and ρ and other thermodynamic quantities are similarly functions of z alone. Real atmosphere or oceans are not stationary; however, their thermodynamic quantities such as temperature and density are primarily dependent on the vertical axis z, and these systems behave much like stratified fluids.

Stratified fluids whose thermodynamic quantities change mainly in the vertical direction exist in stable states that are robust to slight disturbances (for example, a layer of hot water on top of cool water) or unstable states, in which the entire fluid layer vertically flips when subjected to small movement. In this section, let us investigate the state of a fluid layer with the following method. We suppose that the distributions of thermodynamic quantities such as density, pressure, and entropy are functions of the vertical coordinate z alone; that is, $\rho(z), p(z)$, and $S(z)$. The state equation of the fluid is $\rho = f(p, S)$. Consider raising a small fluid parcel from height $z = z_0$ to an incrementally higher level $z_0 + \delta z$ (Fig. 6.2). This vertical change is adiabatic, and the pressure is rapidly adjusted by the fluid pressure surrounding the elevated fluid parcel. Intrinsically, movement of the fluid parcel will alter the pressure of the surrounding fluid, but we regard this effect as negligible. This method is called the **parcel method**.

Under these assumptions, the pressure in the fluid parcel elevated by δz is $p(z_0 + \delta z)$ but the entropy $S(z_0)$ remains unchanged. Hence, the density becomes

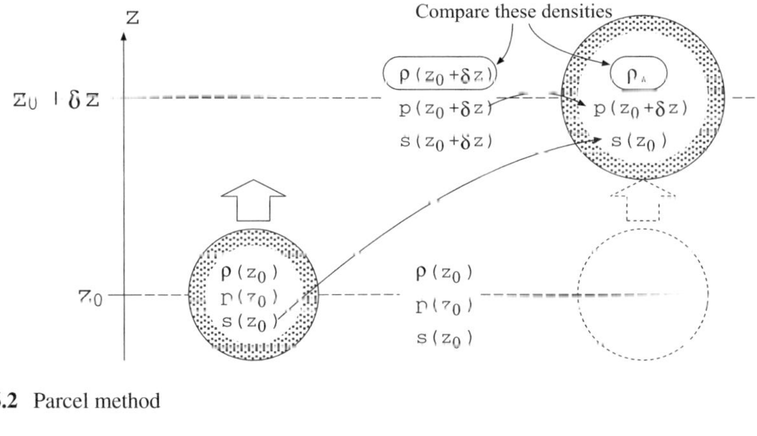

Fig. 6.2 Parcel method

$\rho_* = f(p(z_0 + \delta z), S(z_0))$. If δz is assumed to be sufficiently small, the following approximation holds:

$$\begin{aligned}\rho_* &= f\big(p(z_0 + \delta z), S(z_0)\big) \\ &\sim f\big(p(z_0), S(z_0)\big) + \frac{\partial f}{\partial p}\frac{dp}{dz}\,\delta z \\ &= \rho(z_0) + \left(\frac{\partial \rho}{\partial p}\right)_S \frac{dp}{dz}\,\delta z\,.\end{aligned} \tag{6.3}$$

Meantime, the density of the surrounding fluid is

$$\begin{aligned}\rho(z_0 + \delta z) &= f\big(p(z_0 + \delta z), S(z_0 + \delta z)\big) \\ &\sim f\big(p(z_0), S(z_0)\big) + \frac{\partial f}{\partial p}\frac{dp}{dz}\,\delta z + \frac{\partial f}{\partial S}\frac{dS}{dz}\,\delta z \\ &= \rho(z_0) + \left(\frac{\partial \rho}{\partial p}\right)_S \frac{dp}{dz}\,\delta z + \left(\frac{\partial \rho}{\partial S}\right)_p \frac{dS}{dz}\,\delta z\,.\end{aligned} \tag{6.4}$$

Hence, the density of the elevated fluid parcel exceeds that of the surrounding fluid by

$$\begin{aligned}\rho_* - \rho(z_0 + \delta z) &= -\left(\frac{\partial \rho}{\partial S}\right)_p \frac{dS}{dz}\,\delta z \\ &= -\left(\frac{\partial T}{\partial S}\right)_p \left(\frac{\partial \rho}{\partial T}\right)_p \frac{dS}{dz}\,\delta z = \frac{T\alpha\rho}{c_p}\frac{dS}{dz}\,\delta z\,.\end{aligned} \tag{6.5}$$

Here c_p and α are specific heat at constant pressure and the coefficient of thermal expansion, respectively, given by $c_p \equiv T(\partial S/\partial T)_p$ and $\alpha \equiv -(1/\rho)(\partial \rho/\partial T)_p$. As c_p and α of a substance are generally positive, (6.5) indicates that the raised fluid parcel is denser than the surrounding fluid if $dS/dz > 0$. In this case, the raised parcel is accelerated downward by gravity. If the parcel falls below its original position, it enters the fluid of higher density and is lifted upward by buoyancy. These upward-downward movements are known as **Brunt–Väisälä oscillations**. The period of the Brunt–Väisälä oscillations can be determined from the equation of motion of the fluid parcel:

$$\rho\,\frac{d^2}{dt^2}\,\delta z = -\big[\,\rho_* - \rho(z_0 + \delta z)\,\big]g = -\frac{T\alpha\rho g}{c_p}\frac{dS}{dz}\,\delta z\,. \tag{6.6}$$

The angular frequency of this oscillation is

$$N^2 = \frac{T\alpha g}{c_p}\frac{dS}{dz} \tag{6.7}$$

and the period is given by $2\pi/N$, where N is called the **Brunt–Väisälä frequency**. From the speed of sound c_s, defined as $c_s^2 = (\partial p/\partial \rho)_S$, we get

$$d\rho = \left(\frac{\partial \rho}{\partial S}\right)_p dS + \left(\frac{\partial \rho}{\partial p}\right)_S dp = -\frac{T\alpha\rho}{c_p} dS + \frac{1}{c_s^2}\, dp\,. \tag{6.8}$$

From (6.8) and the hydrostatic equation (6.1), we obtain the following relation:

$$\frac{d\rho}{dz} = -\frac{T\alpha\rho}{c_p}\frac{dS}{dz} + \frac{1}{c_s^2}\frac{dp}{dz} = -\frac{T\alpha\rho}{c_p}\frac{dS}{dz} - \frac{\rho g}{c_s^2} \tag{6.9}$$

from which the Brunt–Väisälä frequency is given by

$$N^2 = -\frac{g}{\rho}\frac{d\rho}{dz} - \frac{g^2}{c_s^2}\,. \tag{6.10}$$

Thus far, we have assumed that $dS/dz > 0$, and therefore, $N^2 > 0$. In this case, a slightly elevated or lowered fluid parcel will descend or rise, respectively, seeking its original position. Such a fluid is referred to as **statically stable**. In contrast, a fluid in the state $dS/dz < 0$ implies that $N^2 < 0$. If a small fluid parcel is raised, it becomes less dense than the surrounding fluid and continues to rise under buoyancy forces. Fluids in this state are referred to as **statically unstable**. An archetypal example of an unstable fluid phenomenon is convection.

Note 6.1 In meteorology, the term *entropy* is often replaced with the term ***potential temperature***, defined as "the temperature at which the air pressure is adiabatically altered to a reference pressure p_0 (usually 1,000 hPa)". Atmospheric gases frequently obey the ideal gas state equation

$$p = \rho RT \tag{6.11}$$

(where R is the gas constant per unit mass of air) while simultaneously satisfying the thermodynamic relationship:

$$dS = \left(\frac{\partial S}{\partial T}\right)_p dT + \underbrace{\left(\frac{\partial S}{\partial p}\right)_T}_{\text{From Maxwell's equation } -(\partial(1/\rho)/\partial T)_p} dp = \frac{c_p}{T}\, dT + \frac{1}{\rho^2}\underbrace{\left(\frac{\partial \rho}{\partial T}\right)_p}_{\text{From the state equation } -p/RT^2} dp = \frac{c_p}{T}\, dT - \frac{R}{p}\, dp\,. \tag{6.12}$$

Therefore, during an adiabatic change where S remains unchanged, $(c_p/T)dT - (R/p)dp = 0$, i.e.,

$$Tp^{-R/c_p} = \text{const} \tag{6.13}$$

is maintained and the potential temperature θ, by definition, becomes

$$\theta = T\left(\frac{p_0}{p}\right)^{R/c_p}. \tag{6.14}$$

Differentiating the logarithm of (6.14)

$$\frac{d\theta}{\theta} = \frac{dT}{T} - \frac{R}{c_p}\frac{dp}{p} \tag{6.15}$$

and comparing with (6.12), we find that $dS = c_p d\theta/\theta$. Defined in this way, the potential temperature θ depends only on the thermodynamic function, entropy S; the temperature and entropy are related as follows:

$$S = c_p \log\theta\,. \tag{6.16}$$

Furthermore, in an ideal gas, $p = \rho RT$. Hence, the coefficient of thermal expansion α becomes

$$\alpha \equiv -\frac{1}{\rho}\left(\frac{\partial\rho}{\partial T}\right)_p = \frac{1}{\rho}\frac{p}{RT^2} = \frac{1}{T} \tag{6.17}$$

and the Brunt–Väisälä frequency N is expressed as

$$N^2 = \frac{T\alpha g}{c_p}\frac{dS}{dz} = \frac{g}{c_p}\cdot\frac{c_p}{\theta}\frac{d\theta}{dz} = \frac{g}{\theta}\frac{d\theta}{dz}. \tag{6.18}$$

Alternatively, from (6.15), we obtain

$$N^2 = \frac{g}{\theta}\frac{d\theta}{dz} = \frac{g}{T}\frac{dT}{dz} - \frac{gR}{pc_p}\underbrace{\frac{dp}{dz}}_{\text{From the equation of hydrostatic pressure } -\rho g} = \frac{g}{T}\left(\frac{dT}{dz} + \frac{g}{c_p}\right). \tag{6.19}$$

Exercise 6.2 For every 100 m increase in altitude, find the minimum decrease in temperature at which the atmosphere becomes statically unstable. Express your answer in K. Suppose that the specific heat at constant pressure is $c_p = 1.0 \times 10^3\,\mathrm{J\,K^{-1}\,kg^{-1}}$. This temperature gradient specifies the temperature decrease in adiabatically lifted air, and is hence called the **adiabatic lapse rate**.

At the beginning of this chapter, the higher layers of stable fluids were assumed to be occupied by hotter or lighter fluids. Such qualitative discussions cannot be directly applied to the general stratified fluids, which are considered in this subsection. As density changes with pressure, a fluid moving in the vertical direction may not maintain its original density or temperature. Although the atmosphere is fundamentally statically stable, its temperature is usually a decreasing function of height ((6.19) shows that, even when $dT/dz < 0$, N^2 may be positive). Under strong insolation in summer, cumulonimbus clouds often form as the atmosphere becomes unstable, but even for such an unstable atmosphere, the density usually does not increase with elevation ((6.10) shows that, even when $d\rho/dz < 0$, N^2 may be negative). More accurately, we must say that "the higher it is, stability depends on a larger value of entropy (or potential temperature)".

6.3 Boussinesq Approximation

The current chapter considers fluids of inhomogeneous density, whose motions cannot be described by incompressible fluids with fixed ρ. Our analysis begins with a system of equations describing compressible fluid motion:

$$\frac{\partial \rho}{\partial t} + (\boldsymbol{u} \cdot \nabla)\rho + \rho(\nabla \cdot \boldsymbol{u}) = 0, \tag{6.20}$$

$$\frac{\partial \boldsymbol{u}}{\partial t} + (\boldsymbol{u} \cdot \nabla)\boldsymbol{u} = -\frac{1}{\rho}\nabla p + \nu\left[\nabla^2 \boldsymbol{u} + \frac{1}{3}\nabla(\nabla \cdot \boldsymbol{u})\right] + \boldsymbol{g}, \tag{6.21}$$

$$\frac{\partial S}{\partial t} + (\boldsymbol{u} \cdot \nabla)S = \frac{k}{\rho T}\nabla^2 T, \tag{6.22}$$

$$f(p, \rho, T) = 0 \tag{6.23}$$

(see (1.114)–(1.117)). However, (6.21) includes gravitational acceleration $\boldsymbol{g} = (0, 0, -g)$ as an external force and (6.22) omits the heat Φ induced by viscosity dissipation, which exerts little effect in this case.

Equations (6.20)–(6.23) certainly describe the motions of fluids with varying density. However, these equations account for numerous phenomena, and are too complicated to investigate fluid motions in practice. Instead, we consider internal gravity waves and convective motion as specific examples of stratified fluid motion. In these motions, the density changes caused by pressure changes are unimportant, although temperature-induced density changes may be significant. Therefore, we seek a system of equations that describes these phenomena as simply as possible. We begin by deriving an approximate governing equation satisfying the following conditions.

1. The thermodynamic variables do not vary significantly.
2. Pressure alters independently of the other thermodynamic variables. For example, density and other thermodynamic variables do not respond to pressure changes.

If density variations are small, ρ can be regarded as the sum of a constant component ρ_0 and its deviation ρ'; that is, $\rho = \rho_0 + \rho'(\boldsymbol{r}, t)$ (where $\rho_0 \gg \rho'$). As the density differentials in the first and second terms of the continuity Eq. (6.20) are small, the third term of (6.20) dominates, and can be approximated as

$$\nabla \cdot \boldsymbol{u} = 0\,. \tag{6.24}$$

Next, the pressure is decomposed into a sum of the hydrostatic pressure $p_h(z)$ of a fluid with density ρ_0 and its translation $p_d(\boldsymbol{r}, t)$; that is, $p = p_h(z) + p_d(\boldsymbol{r}, t)$. p and ρ

are then substituted into (6.21). Considering the pressure gradient and gravitational acceleration terms

$$-\frac{1}{\rho}\nabla p+\boldsymbol{g}=\underbrace{\frac{-1}{\rho_0+\rho'}}_{}\nabla(p_h+p_d)+\boldsymbol{g}\ \sim -\frac{1}{\rho_0}\left(1-\frac{\rho'}{\rho_0}\right)(\nabla p_h+\nabla p_d)+\boldsymbol{g}$$

After expanding by ρ'/ρ_0, terms of higher than second order are ignored.

$$\sim -\underbrace{\frac{1}{\rho_0}\nabla p_h}_{g}+\frac{\rho'}{\rho_0}\underbrace{\frac{1}{\rho_0}\nabla p_h}_{g}-\frac{1}{\rho_0}\nabla p_d+\frac{\rho'}{\rho_0}\frac{1}{\rho_0}\nabla p_d+\boldsymbol{g}$$

$$=\frac{\rho'}{\rho_0}\boldsymbol{g}-\frac{1}{\rho_0}\nabla p_d+\underbrace{\frac{\rho'}{\rho_0}\frac{1}{\rho_0}\nabla p_d}_{}\ \sim \frac{\rho'}{\rho_0}\boldsymbol{g}-\frac{1}{\rho_0}\nabla p_d \tag{6.25}$$

Since this term is much smaller than the second term, it can be ignored.

as well as the approximated continuity Eq. (6.24), the equation of motion (6.21) can be approximated as

$$\frac{\partial \boldsymbol{u}}{\partial t}+(\boldsymbol{u}\cdot\nabla)\boldsymbol{u}=-\frac{1}{\rho_0}\nabla p_d+\nu\nabla^2\boldsymbol{u}+\frac{\rho'}{\rho_0}\boldsymbol{g}\,. \tag{6.26}$$

The state Eq. (6.23) can be expressed in several forms. Expressing ρ in terms of p and T and adopting the relationship $(\partial\rho/\partial T)_p=-\alpha\rho$, we get

$$d\rho=\left(\frac{\partial\rho}{\partial p}\right)_T dp-\alpha\rho\, dT\,. \tag{6.27}$$

Assuming that pressure alters independently of the other variables, we can write $d\rho=-\alpha\rho dT$. When density deviation is smaller, T, α, ρ, and c_p can be regarded as constant and the perturbed density approximately satisfies the following linear equation:

$$\rho'=-\alpha_0\rho_0 T'\,. \tag{6.28}$$

Similarly, T can be expressed in terms of variables p and S, and rearranged through the relation $(\partial T/\partial S)_p=T/c_p$ to give

$$dT=\left(\frac{\partial T}{\partial p}\right)_S dp+\frac{T}{c_p}\,dS\,. \tag{6.29}$$

Therefore, in the same approximation,

$$T'=\frac{T_0}{c_{p0}}S'\,. \tag{6.30}$$

Finally, we consider the energy equation (6.22). Expressing the entropy and temperature as sums of the standard values and their deviation, that is, $S=S_0+S'$ and

$T = T_0 + T'$, the term with the highest order is

$$\frac{\partial S'}{\partial t} + (\boldsymbol{u} \cdot \nabla) S' = \frac{k_0}{\rho_0 T_0} \nabla^2 T' . \tag{6.31}$$

Using the relation (6.30), (6.31) can be rearranged to yield

$$\frac{\partial T'}{\partial t} + (\boldsymbol{u} \cdot \nabla) T' = \frac{k_0}{\rho_0 c_{p0}} \nabla^2 T' . \tag{6.32}$$

Rewriting (6.24), (6.26), (6.28) and (6.32) by removing the prime and subscript d from the notations of the deviations, we derive the following system of approximation equations:

$$\nabla \cdot \boldsymbol{u} = 0 , \tag{6.33}$$

$$\frac{\partial \boldsymbol{u}}{\partial t} + (\boldsymbol{u} \cdot \nabla) \boldsymbol{u} = -\frac{1}{\rho_0} \nabla p + \nu \nabla^2 \boldsymbol{u} + \frac{\rho}{\rho_0} \boldsymbol{g} , \tag{6.34}$$

$$\rho - \rho_0 = -\alpha_0 \rho_0 (T - T_0) , \tag{6.35}$$

$$\frac{\partial T}{\partial t} + (\boldsymbol{u} \cdot \nabla) T = \kappa_0 \nabla^2 T . \tag{6.36}$$

Here $\kappa \equiv k/(\rho c_p)$ is the thermal diffusion coefficient. The expressions (6.33)–(6.36) are collectively called the **Boussinesq approximation**, and fluids obeying this approximation are called **Boussinesq fluids**.

The system of equations describing compressible fluids, (6.20)–(6.23), admits sound waves, which are characterized by pressure-dependent density changes. However, sound waves play no important role in convection or internal gravity waves. Their existence merely complicates theoretical analysis and numerical calculations. The Boussinesq approximation assumes that pressure changes independently of the other thermodynamic variables. Equivalently, it regards the speed of sound as infinite; that is, $1/c_s^2 \equiv (\partial \rho / \partial p)_S = 0$, thereby eliminating any phenomenon related to sound waves. By ignoring pressure-induced density changes and retaining density changes caused by other thermodynamic quantities (such as temperature), the Boussinesq approximation partially captures the characteristics of incompressible fluids, and provides a concise description of stratified fluids.

Since the Boussinesq approximation assumes infinitely fast sound, the second term in the Brunt–Väisälä frequency (6.10) can be ignored and the undifferentiated ρ can be approximated by ρ_0; therefore,

$$N^2 = -\frac{g}{\rho_0} \frac{d\rho}{dz} . \tag{6.37}$$

Furthermore, the variable T can be eliminated from (6.35) and (6.36) to get

$$\frac{\partial \rho}{\partial t} + (\boldsymbol{u} \cdot \nabla)\rho = \kappa \nabla^2 \rho\,. \tag{6.38}$$

Equations (6.33), (6.34) and (6.38) constitute the most basic equations. These equations are frequently applied in theoretical treatments of oceanic flows.

Note 6.2 Pressure increases induce remarkable rises in air density. Assumption 2 that "pressure changes independently of other thermodynamic variables," is clearly violated in atmospheric systems. However, the following method allows us to derive an equation equivalent to the Boussinesq approximation. If Assumption 1, "the thermodynamic variables do not vary significantly" holds, the equations of continuity and entropy can be transformed into (6.24) and (6.31), respectively, as in the Boussinesq approximation.

We now introduce a function of pressure Π called the **Exner function**.

$$\Pi \equiv \left(\frac{p}{p_0}\right)^{R/c_p} \left(= \frac{T}{\theta}\right)\,. \tag{6.39}$$

Differentiating (6.39), we obtain

$$d\Pi = \frac{R}{c_p}\frac{p^{R/c_p-1}}{p_0^{R/c_p}}\,dp = \frac{R}{c_p}\frac{\Pi}{p}\,dp = \frac{RT}{c_p p \theta}\,dp = \frac{1}{c_p \rho \theta}\,dp\,. \tag{6.40}$$

Hence, $-(1/\rho)\nabla p = -c_p\theta\nabla\Pi$. Using this and (6.24), (6.21) can be rewritten as

$$\frac{\partial \boldsymbol{u}}{\partial t} + (\boldsymbol{u} \cdot \nabla)\boldsymbol{u} = -c_p\theta\nabla\Pi + \nu\nabla^2\boldsymbol{u} + \boldsymbol{g}\,. \tag{6.41}$$

Similar to the previous analysis, when the pressure was decomposed into the hydrostatic pressure and its deviation, we can split Π as $\Pi = \Pi_h(z) + \Pi_d(\boldsymbol{r}, t)$. In this form, Π is the sum of a component corresponding to the hydrostatic pressure, which satisfies $0 = -c_p\theta_0 d\Pi_h/dz - g$ and its deviation $\Pi_d(\boldsymbol{r}, t)$. Π and $\theta = \theta_0 + \theta'$ are then substituted into (6.41). The pressure gradient and gravitational acceleration terms are then approximated as

$$-c_p\theta\nabla\Pi + \boldsymbol{g} \sim -\frac{\theta'}{\theta_0}\boldsymbol{g} - c_p\theta_0\nabla\Pi_d \tag{6.42}$$

and (6.41) further becomes

$$\frac{\partial \boldsymbol{u}}{\partial t} + (\boldsymbol{u} \cdot \nabla)\boldsymbol{u} = -c_p\theta_0\nabla\Pi_d + \nu\nabla^2\boldsymbol{u} - \frac{\theta'}{\theta_0}\boldsymbol{g}\,. \tag{6.43}$$

The equation of entropy (6.31) indicates that the flux carried by diffusion is proportional to the temperature gradient. When considering eddy diffusions in atmospheric motions, it is frequently more appropriate to regard diffusion flux as proportional to the gradient of entropy or potential temperature. Thus, we replace the right-hand side of (6.31) by $\kappa\nabla^2 S'$, yielding the alternative expression

$$\frac{\partial S'}{\partial t} + (\boldsymbol{u}\cdot\nabla)S' = \kappa\nabla^2 S' \,. \tag{6.44}$$

Multiplying both sides by θ_0/c_p and using the relationship $S' = (c_p/\theta_0)\theta'$ derived from (6.16), we obtain

$$\frac{\partial \theta'}{\partial t} + (\boldsymbol{u}\cdot\nabla)\theta' = \kappa\nabla^2 \theta' \,. \tag{6.45}$$

The equation set (6.24), (6.43), and (6.45) can be rewritten in terms of their original quantities without the prime or subscript $_d$:

$$\nabla\cdot\boldsymbol{u} = 0, \tag{6.46}$$

$$\frac{\partial \boldsymbol{u}}{\partial t} + (\boldsymbol{u}\cdot\nabla)\boldsymbol{u} = -c_p\theta_0\nabla\Pi + \nu\nabla^2\boldsymbol{u} - \frac{\theta}{\theta_0}\boldsymbol{g}, \tag{6.47}$$

$$\frac{\partial \theta}{\partial t} + (\boldsymbol{u}\cdot\nabla)\theta = \kappa\nabla^2 \theta \,. \tag{6.48}$$

Comparing (6.24), (6.43) and (6.48) or their equivalents (6.46)–(6.48) with the system of Boussinesq equations (6.33), (6.34) and (6.38), and making the following replacements: $p \to \Pi$, $1/\rho_0 \to c_p\theta_0$, and $\rho/\rho_0 \to -\theta/\theta_0$, we find that the above system of equations is identical to the Boussinesq approximation. In air, whose density can never be determined independently of pressure (for example, from temperature alone), the Boussinesq approximation may initially appear as an invalid approximation that ignores the prerequisite conditions. However, since the pressure has been replaced by the Exner function, and the buoyancy induced by pressure changes (causing density change) has been absorbed into the pressure gradient (the Exner function gradient), this treatment is applicable to atmospheric motions.

The above derivation excludes Assumption 2, that "the pressure changes independently of the other thermodynamic variables." However, Assumption 1, that "the thermodynamic variables do not vary significantly," was adopted in (6.42) for θ and in the equation of continuity for ρ. These assumptions suitably describe shallow atmospheric phenomena, but may not adequately describe phenomena more than several kilometers above the ground. For example, in cumulonimbus clouds, the density gradient ρ' is no longer sufficiently small compared with the absolute value, and the equation of continuity cannot be approximated by (6.24). Hence, when considering phenomena at these scales, we commonly replace (6.46) by an equation that accounts for the z-dependency of the density ρ_0 as follows:

$$\nabla\cdot(\rho_0(z)\,\boldsymbol{u}) = 0 \,. \tag{6.49}$$

Equations (6.49), (6.47) and (6.48) are collectively called the **anelastic approximation**. This approximation maintains the essential characteristics of the Boussinesq approximation, such as removing sound waves.

Exercise 6.3 By substituting $\Pi = \Pi_h(z) + \Pi_d(\boldsymbol{r}, t)$ and $\theta = \theta_0 + \theta'$, derive the approximate equation of (6.42).

Large-scale motions of the atmosphere and oceans are commonly approximated by assuming a vertical hydrostatic equilibrium. Meteorologists and oceanographers refer to the corresponding equation as the **primitive equation**. Alternatively, if the relevant phenomena are limited to quasi-geostrophic flows (see Sect. 7.4.4), the **quasi-geostrophic equation** may be adopted. These systems of approximation equations are validated only when the horizontal scale is considerably greater than the vertical scale. Within the atmosphere and oceans, where the horizontal and vertical scales are similar at several kilometers, phenomena are modeled by the anelastic (atmosphere) and Boussinesq (ocean) approximations. Of course, none of these approximations are suitable for describing sound wave phenomena. If sound waves are important, the compressible fluid equations must be applied in their original form.

6.4 Internal Gravity Waves

In stratified fluids, waves stabilized by gravity also propagate in the vertical direction. This section investigates the properties of such waves in stratified fluids. The basic equation assumes a Boussinesq fluid influenced by gravity, whose diffusion and viscosity can be ignored. Excluding the viscosity terms from (6.33), (6.34) and (6.38) and writing out each component, we obtain

$$\frac{\partial u}{\partial x} + \frac{\partial v}{\partial y} + \frac{\partial w}{\partial z} = 0, \tag{6.50}$$

$$\frac{\partial u}{\partial t} + u\frac{\partial u}{\partial x} + v\frac{\partial u}{\partial y} + w\frac{\partial u}{\partial z} = -\frac{1}{\rho_0}\frac{\partial p}{\partial x}, \tag{6.51}$$

$$\frac{\partial v}{\partial t} + u\frac{\partial v}{\partial x} + v\frac{\partial v}{\partial y} + w\frac{\partial v}{\partial z} = -\frac{1}{\rho_0}\frac{\partial p}{\partial y}, \tag{6.52}$$

$$\frac{\partial w}{\partial t} + u\frac{\partial w}{\partial x} + v\frac{\partial w}{\partial y} + w\frac{\partial w}{\partial z} = -\frac{1}{\rho_0}\frac{\partial p}{\partial z} - \frac{\rho}{\rho_0}g, \tag{6.53}$$

$$\frac{\partial \rho}{\partial t} + u\frac{\partial \rho}{\partial x} + v\frac{\partial \rho}{\partial y} + w\frac{\partial \rho}{\partial z} = 0. \tag{6.54}$$

In a static field, the pressure and density must satisfy the hydrostatic pressure relation:

$$\frac{1}{\rho_0}\frac{d\overline{p}}{dz} = -\frac{\overline{\rho}}{\rho_0}g. \tag{6.55}$$

Substituting each variable into the basic equation with the following replacements: $u = u'$, $v = v'$, $w = w'$, $p = \overline{p} + p'$, $\rho = \overline{\rho} + \rho'$, and ignoring the infinitesimally small second-order terms, we get

$$\frac{\partial u'}{\partial x} + \frac{\partial v'}{\partial y} + \frac{\partial w'}{\partial z} = 0\,, \tag{6.56}$$

$$\frac{\partial u'}{\partial t} = -\frac{1}{\rho_0}\frac{\partial p'}{\partial x}\,, \tag{6.57}$$

$$\frac{\partial v'}{\partial t} = -\frac{1}{\rho_0}\frac{\partial p'}{\partial y}\,, \tag{6.58}$$

$$\frac{\partial w'}{\partial t} = -\frac{1}{\rho_0}\frac{\partial p'}{\partial z} - \frac{\rho'}{\rho_0}g, \tag{6.59}$$

$$\frac{\partial \rho'}{\partial t} - \frac{\rho_0 N^2}{g}w' = 0\,. \tag{6.60}$$

Here $N^2 = -(g/\rho_0)d\overline{\rho}/dz$ is obtained from the Brunt–Väisälä frequency of a Boussinesq fluid (6.37). For simplicity, we assume that N is constant. Further assuming a solution of the form $u'(x, y, z, t) = \hat{u}\exp[i(k_x x + k_y y + k_z z - \omega t)]$, we have

$$ik_x\hat{u} + ik_y\hat{v} + ik_z\hat{w} = 0\,, \tag{6.61}$$

$$-i\omega\hat{u} = -\frac{1}{\rho_0}ik_x\hat{p}\,, \tag{6.62}$$

$$-i\omega\hat{v} = -\frac{1}{\rho_0}ik_y\hat{p}\,, \tag{6.63}$$

$$-i\omega\hat{w} = -\frac{1}{\rho_0}ik_z\hat{p} - \frac{\hat{\rho}}{\rho_0}g\,, \tag{6.64}$$

$$-i\omega\hat{\rho} - \frac{\rho_0 N^2}{g}\hat{w} = 0\,. \tag{6.65}$$

Eliminating $\hat{u}$ and $\hat{v}$ from $ik_x \times (6.62) + ik_y \times (6.63) + i\omega \times (6.61)$, we obtain

$$-\,k_z\omega\hat{w} = \frac{1}{\rho_0}(k_x^2 + k_y^2)\hat{p}\,. \tag{6.66}$$

On the other hand, eliminating $\hat{\rho}$ from $i\omega \times (6.64) + (g/\rho_0) \times (6.65)$ gives

$$(\omega^2 - N^2)\hat{w} - \frac{k_z\omega}{\rho_0}\hat{p}\,. \tag{6.67}$$

Hence, from $-k_z\omega \times (6.66) + (k_x^2 + k_y^2) \times (6.67)$, we obtain an equation involving a single variable, $\hat{w}$:

$$\omega^2(k_x^2 + k_y^2 + k_z^2)\hat{w} - N^2(k_x^2 + k_y^2)\hat{w} = 0 \tag{6.68}$$

from which the dispersion relation is obtained as

$$\omega^2 = \frac{N^2 k_h^2}{k_h^2 + k_z^2}$$

i.e.,

$$\omega = \pm \frac{N k_h}{(k_h^2 + k_z^2)^{\frac{1}{2}}} \,. \tag{6.69}$$

Clearly, the frequency of the internal gravity wave is always lower than the Brunt–Väisälä frequency (note that $k_h^2 \equiv k_x^2 + k_y^2$ represents the total horizontal wavenumber).

The phase velocities along the axial directions, c_x, c_y, and c_z, are given by

$$\left.\begin{aligned} c_x &\equiv \frac{\omega}{k_x} = \pm \frac{N k_h}{k_x (k_h^2 + k_z^2)^{\frac{1}{2}}} \,, \\ c_y &\equiv \frac{\omega}{k_y} = \pm \frac{N k_h}{k_y (k_h^2 + k_z^2)^{\frac{1}{2}}} \,, \\ c_z &\equiv \frac{\omega}{k_z} = \pm \frac{N k_h}{k_z (k_h^2 + k_z^2)^{\frac{1}{2}}} \,. \end{aligned}\right\} \tag{6.70}$$

The group velocity is

$$\begin{aligned} (c_{gx}, c_{gy}, c_{gz}) &\equiv \left(\frac{\partial \omega}{\partial k_x}, \frac{\partial \omega}{\partial k_y}, \frac{\partial \omega}{\partial k_z} \right) \\ &= \left(\pm \frac{N k_x k_z^2}{k_h (k_h^2 + k_z^2)^{\frac{3}{2}}}, \pm \frac{N k_y k_z^2}{k_h (k_h^2 + k_z^2)^{\frac{3}{2}}}, \mp \frac{N k_z k_h}{(k_h^2 + k_z^2)^{\frac{3}{2}}} \right) . \end{aligned} \tag{6.71}$$

Note that c_z and c_{gz} have opposite signs. Therefore, when the internal gravity wave propagates energy upward, its phase descends from greater heights. Moreover,

$$\boldsymbol{c}_g \cdot \boldsymbol{k} \equiv (c_{gx}, c_{gy}, c_{gz}) \cdot (k_x, k_y, k_z) = 0 \,. \tag{6.72}$$

Equation (6.72) indicates that the energy travels perpendicular to the wavenumber vector and parallel to the lines of equal phase.

Figure 6.3 shows the construction of this wave when $k_x > 0$, $k_y = 0$, and $k_z > 0$. The disturbance pressure is high along phase line A and has velocity toward the bottom-right corner. Conversely, where the disturbance pressure is low at phase line C, the velocity vector points toward the top-left corner. Along phase lines B and

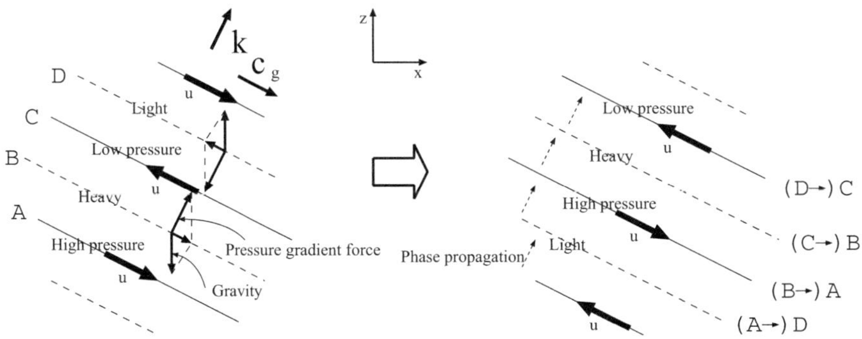

Fig. 6.3 Structure and propagation of internal gravity waves

D, the disturbance density is large and small, respectively, and both phase lines are established at zero fluid velocity. We now investigate the propagation of this wave in detail. Along phase line C, the fluid particles move from bottom-right to top-left; however, the density of the basic field is stably stratified, and therefore bottom heavy. Consequently, advection increases the density along phase line C, which means that the current phase line C gradually changes to phase line B characterized by large density. On the other hand, the forces acting on the fluid along phase line B are the pressure gradient (from the bottom-left region of high pressure to the top-right region of low pressure) and gravity (acting downward, since the disturbance density is large). Therefore, the combined force points toward the right-bottom corner. In the left panel of Fig. 6.3, the stationary fluid particle at B experiences a combined pressure gradient and gravitational force in the bottom-right direction, which impels it to move in that direction. Furthermore, as mentioned above, the density of phase line C increases, imposing additional pressure along phase line B immediately below, which in turn means that parcels at phase line B changes to hold characteristics of higher pressure and a velocity pointing toward bottom-right, i.e., characteristics for phase line A. In this manner, the phase of a wave structured as shown in the figure is propagated from the bottom-left to the top-right. Moreover, if the wave structure is carefully examined, fluid particles in high-pressure region point toward the bottom-right, while those in low-pressure region toward the top-left. This indicates that the top-left fluid does net work on bottom-right fluid and that energy is transported from top-left to bottom-right. In this situation, the group velocity described by (6.71) should point toward the bottom-right corner, exactly matching the direction of energy flow (It is an evident conclusion from the idea that "the group velocity describes the propagation of energy," as learned in Sects. 3.4–3.6).

The restoring force against gravity (buoyancy) serves an essential role in propagating these waves. Since the density inside the fluid continuously changes, it is disturbed by fluid displacement, turning gravity (or buoyancy) to restoring force. The wave has an oscillatory structure vertically as well as horizontally. Thus, the waves considered in this section are called **internal gravity waves**. Incidentally, the waves described in Chap. 3 are also restored by gravity. Recall that gravity exerts a restoring force on the displacement of the free surface, and the wave is exponentially attenuated in the vertical direction. Such waves are sometimes referred to as **external gravity waves**, in contrast to internal gravity waves.

Although we have thus far considered fluids in the basic field, a similar argument holds in the presence of a basic field. Here we consider the simplest case, in which U is homogeneous in the x-direction (constant and positively valued). In this case, the linearized equation replaces $\partial/\partial t$ in (6.56)–(6.60) by $\partial/\partial t + U\partial/\partial x$. Therefore, we need to only replace ω in (6.61)–(6.65) with $\omega - Uk_z$. By transformations similar to those used to obtain (6.69), we obtain the following dispersion relation:

$$\omega = Uk_x \pm \frac{Nk_h}{(k_h^2 + k_z^2)^{\frac{1}{2}}} . \tag{6.73}$$

The group velocity is (6.71) with an additional term $(U, 0, 0)$:

$$(c_{gx}, c_{gy}, c_{gz}) = \left(U \pm \frac{Nk_x k_z^2}{k_h (k_h^2 + k_z^2)^{\frac{3}{2}}}, \pm \frac{Nk_y k_z^2}{k_h (k_h^2 + k_z^2)^{\frac{3}{2}}}, \mp \frac{Nk_z k_h}{(k_h^2 + k_z^2)^{\frac{3}{2}}} \right) . \tag{6.74}$$

Waves created by wind striking a mountain are called **mountain waves**. Such waves can be regarded as internal gravity waves overlapped with a basic flow velocity. If the wind is steady (of course, force from immobile mountains is steady), the generated mountain waves will also be steady. In this case, $\omega = 0$, indicating that, in contrast to the positive x-component of the group velocity c_{gx}, the x-component of the group velocity opposing the wind $c_{gx} - U$ is negative. Furthermore, since the disturbance energy is generated from the mountain below, the wave energy should be propagated upward, and the z-component of the group velocity c_{gz} can be regarded as positive. The condition $c_{gz} > 0$ implies that k_x and k_z have the same sign. Given the direction of the group velocity and the wavenumber vector obtained above, a stable mountain wave should be shaped as illustrated in Fig. 6.4.

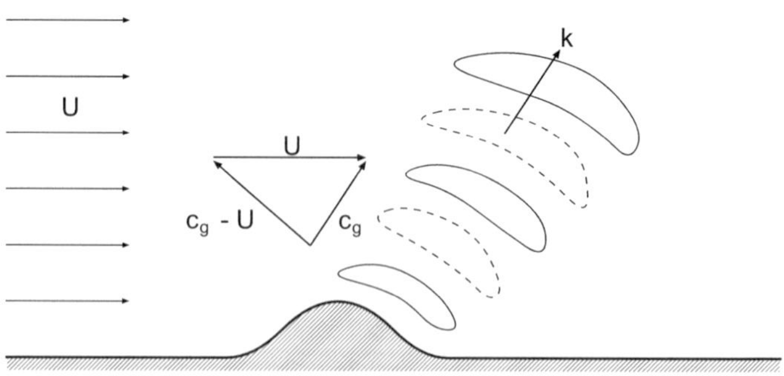

Fig. 6.4 Mountain wave

Exercise 6.4 As a simple case, consider that wave is homogeneous in the y-direction (i.e., $k_y = 0$). In this case, (6.73) and (6.74) reduce to

$$\omega = Uk_x \pm \frac{Nk_x}{(k_x^2 + k_z^2)^{\frac{1}{2}}},$$

$$(c_{gx}, c_{gy}, c_{gz}) = \left(U \pm \frac{Nk_z^2}{(k_x^2 + k_z^2)^{\frac{3}{2}}}, \quad 0, \quad \mp \frac{Nk_z k_x}{(k_x^2 + k_z^2)^{\frac{3}{2}}} \right).$$

Show that when $\omega = 0$, $c_{gx} > 0$ and $c_{gx} - U < 0$. Furthermore, show that k_x and k_z have the same sign when $c_{gz} > 0$.

6.5 Benard Convection

The previous section considered internal gravity waves as a state of motion occurring within a stably stratified fluid. We now consider the motions of unstably stratified fluids. Broadly speaking, unstable stratification occurs when the fluid becomes heavy (cold) at the top and lightweight (warm) at the bottom. Such stratification is unnatural and is expected to naturally reverse as the heavy fluid descends while the light fluid ascends. In particular, fluids that are steadily heated from below and cooled at the surface develop (vertical) convection motions. The simplest convection is the **Benard convection** or **Rayleigh–Benard convection**, which is adequately described by linear theory. This type of convection is introduced next.

6.5.1 Fundamental Equation and Boundary Conditions

The fundamental equations of a Boussinesq fluid are

$$\nabla \cdot \boldsymbol{u} = 0, \tag{6.75}$$

$$\frac{\partial \boldsymbol{u}}{\partial t} + (\boldsymbol{u} \cdot \nabla)\boldsymbol{u} = -\frac{1}{\rho_0}\nabla p + \nu \nabla^2 \boldsymbol{u} + \frac{\rho}{\rho_0}\boldsymbol{g}, \tag{6.76}$$

$$\frac{\partial T}{\partial t} + (\boldsymbol{u} \cdot \nabla)T = \kappa \nabla^2 T, \tag{6.77}$$

$$\rho - \rho_0 = -\alpha_0 \rho_0 (T - T_0)\,. \tag{6.78}$$

Assuming that the upper and lower surfaces are solid walls maintained at constant temperature (Fig. 6.5), the boundary conditions are

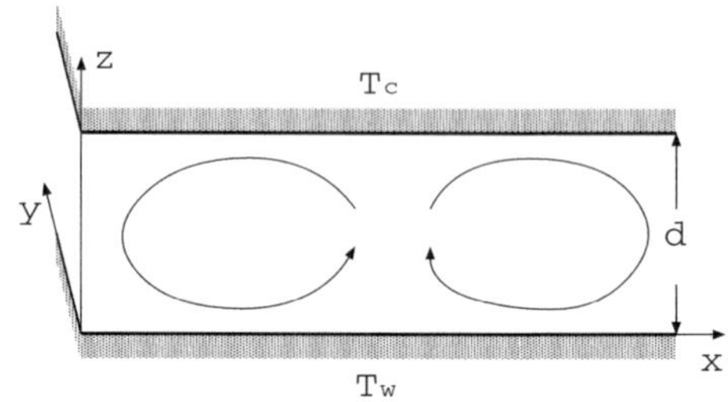

Fig. 6.5 Problem setting in Benard convection

$$\left.\begin{aligned} T &= T_w \ \text{ at } z = 0, \\ T &= T_c \ \text{ at } z = d\,. \end{aligned}\right\} \tag{6.79}$$

Moreover, since the boundary is impermeable to the fluid, we establish additional boundary conditions,

$$w = 0 \quad \text{at } z = 0,\ d\,. \tag{6.80}$$

Furthermore, in the simplest scenario, we impose a zero-stress condition.

Therefore, we have

$$\frac{\partial u}{\partial z} + \frac{\partial w}{\partial x} = \frac{\partial v}{\partial z} + \frac{\partial w}{\partial y} = 0 \quad \text{at } z = 0,\ d\,. \tag{6.81}$$

As $w = 0$ is satisfied anywhere on the boundary in the x, y-direction, $\partial w/\partial x$ and $\partial w/\partial y$ also vanish, and the boundary conditions become

$$\frac{\partial u}{\partial z} = \frac{\partial v}{\partial z} = 0 \quad \text{at } z = 0,\ d. \tag{6.82}$$

6.5.2 *Nondimensionalization*

Each variable is nondimensionalized as follows:

$$\left.\begin{aligned} &\boldsymbol{x} = d\tilde{\boldsymbol{x}},\ \ t = \frac{d^2}{\nu}\tilde{t},\ \ \boldsymbol{u} = \frac{\nu}{d}\tilde{\boldsymbol{u}},\ \ T - T_w = (T_w - T_c)\,\tilde{T}, \\ &p + \rho_0 g z\left[\,1 - \alpha_0(T_w - T_0)\,\right] = \rho_0\frac{\nu^2}{d^2}\tilde{p}\,. \end{aligned}\right\} \tag{6.83}$$

The nondimensionalized equations obtained by substituting (6.83) into the fundamental equation, where ρ was eliminated by substituting (6.78) into (6.76), are given as

$$\tilde{\nabla} \cdot \tilde{\boldsymbol{u}} = 0, \tag{6.84}$$

$$\frac{\partial \tilde{\boldsymbol{u}}}{\partial \tilde{t}} + (\tilde{\boldsymbol{u}} \cdot \tilde{\nabla})\tilde{\boldsymbol{u}} = -\tilde{\nabla}\tilde{p} + \tilde{\nabla}^2\tilde{\boldsymbol{u}} + \frac{Ra}{Pr}\tilde{T}\boldsymbol{e}_z, \tag{6.85}$$

$$\frac{\partial \tilde{T}}{\partial \tilde{t}} + (\tilde{\boldsymbol{u}} \cdot \tilde{\nabla})\tilde{T} = \frac{1}{Pr}\tilde{\nabla}^2\tilde{T}. \tag{6.86}$$

The dimensionless quantities

$$Pr \equiv \frac{\nu}{\kappa}, \quad Ra \equiv \frac{\alpha_0 g(T_w - T_c)d^3}{\nu\kappa} \tag{6.87}$$

are called the **Prandtl** and **Rayleigh numbers**, respectively. These numbers characterize the system.

The boundary conditions are given by

$$\tilde{T} = 0 \quad \text{at } \tilde{z} = 0, \quad \tilde{T} = -1 \quad \text{at } \tilde{z} = 1, \tag{6.88}$$

$$\tilde{w} = \frac{\partial \tilde{u}}{\partial \tilde{z}} = \frac{\partial \tilde{v}}{\partial \tilde{z}} = 0 \quad \text{at } \tilde{z} = 0,\ 1. \tag{6.89}$$

6.5.3 Basic Field

In this problem setting, regardless of whether the system is stable or unstable, a stationary solution exists at hydrostatic equilibrium. We begin by stating this stationary solution.

Because the solution we seek is at hydrostatic equilibrium, $\tilde{\boldsymbol{u}}$ vanishes and the other variables depend on $\tilde{z}$ alone. Given this assumption, among the dimensionless Eqs. (6.84)–(6.86), the $\tilde{x}$ and $\tilde{y}$ components of (6.85) and (6.84) are always established and the remaining two equations become

$$0 = -\frac{d\tilde{p}}{d\tilde{z}} + \frac{Ra}{Pr}\tilde{T}, \tag{6.90}$$

$$0 = \frac{1}{Pr}\frac{d^2\tilde{T}}{d\tilde{z}^2}. \tag{6.91}$$

Equation (6.91) is solved using the boundary conditions (6.88). Substituting the resulting $\tilde{T}$ into (6.90), we obtain $\tilde{p}$:

$$\tilde{T} = -\tilde{z}, \quad \tilde{p} = \tilde{p}_0 - \frac{1}{2}\frac{Ra}{Pr}\tilde{z}^2 \tag{6.92}$$

where $\tilde{p}_0$ is the pressure at $\tilde{z} = 0$. This solution describes the heat conduction in a stationary fluid, whereby heat ascends from the bottom to the top through conduction alone.

6.5.4 Equation for Small Disturbances

Above, we obtained the heat conduction solutions $\overline{T}$ and $\overline{p}$. We now investigate the linear stability of these solutions using the method described in Sect. 5.2. We first separate $\tilde{T}$, $\tilde{p}$, and $\tilde{u}$ into their basic solutions plus disturbances as follows:

$$\tilde{T} = \overline{T} + T', \quad \tilde{p} = \overline{p} + p', \quad \tilde{\boldsymbol{u}} = \boldsymbol{u}' . \tag{6.93}$$

These equations are substituted into the system of Eqs. (6.84)–(6.86) and the boundary conditions (6.88) and (6.89). Ignoring the infinitesimal second-order terms, we obtain a linear equation in terms of the minute disturbance

$$\tilde{\nabla} \cdot \boldsymbol{u}' = 0, \tag{6.94}$$

$$\frac{\partial \boldsymbol{u}'}{\partial \tilde{t}} = -\tilde{\nabla} p' + \tilde{\nabla}^2 \boldsymbol{u}' + \frac{Ra}{Pr} T' \boldsymbol{e}_z, \tag{6.95}$$

$$\frac{\partial T'}{\partial \tilde{t}} - w' = \frac{1}{Pr} \tilde{\nabla}^2 T', \tag{6.96}$$

and boundary conditions.

$$T' = w' = \frac{\partial u'}{\partial \tilde{z}} = \frac{\partial v'}{\partial \tilde{z}} = 0 \quad \text{at } \tilde{z} = 0,\ 1 . \tag{6.97}$$

6.5.5 Elimination of Variables

To express our equation in terms of w' alone, we eliminate variables from the continuity equations of T', p', and $\boldsymbol{u}'$, given by (6.94)–(6.96), as follows:

First, we apply the $\tilde{\nabla}\cdot$ operator to both sides of (6.95). Using the continuity equation (6.94) and $\tilde{\nabla} \cdot (T' \boldsymbol{e}_z) = \partial T' / \partial \tilde{z}$

$$0 = -\tilde{\nabla}^2 p' + \frac{Ra}{Pr} \frac{\partial T'}{\partial \tilde{z}} \tag{6.98}$$

and differentiating with respect to z, we get

$$0 = -\tilde{\nabla}^2 \frac{\partial p'}{\partial \tilde{z}} + \frac{Ra}{Pr} \frac{\partial^2 T'}{\partial \tilde{z}^2} . \tag{6.99}$$

Applying the $\tilde{\nabla}^2$ operator to the $\tilde{z}$-component of (6.95)

$$\frac{\partial}{\partial \tilde{t}}\tilde{\nabla}^2 w' = -\tilde{\nabla}^2\frac{\partial p'}{\partial \tilde{z}} + \tilde{\nabla}^2\tilde{\nabla}^2 w' + \frac{Ra}{Pr}\tilde{\nabla}^2 T' \tag{6.100}$$

and substituting (6.99) into (6.100), we obtain an equation in T' and w' alone:

$$\left(\frac{\partial}{\partial \tilde{t}} - \tilde{\nabla}^2\right)\tilde{\nabla}^2 w' = \frac{Ra}{Pr}\left(\frac{\partial^2}{\partial \tilde{x}^2} + \frac{\partial^2}{\partial \tilde{y}^2}\right)T' \,. \tag{6.101}$$

Next, we apply $[\partial/\partial\tilde{t} - (1/Pr)\tilde{\nabla}^2]$ to (6.101), and substitute (6.96) to eliminate T'. The resulting equation is a function of w' only:

$$\left(\frac{\partial}{\partial \tilde{t}} - \tilde{\nabla}^2\right)\left(\frac{\partial}{\partial \tilde{t}} - \frac{1}{Pr}\tilde{\nabla}^2\right)\tilde{\nabla}^2 w' = \frac{Ra}{Pr}\left(\frac{\partial^2}{\partial \tilde{x}^2} + \frac{\partial^2}{\partial \tilde{y}^2}\right)w' \,. \tag{6.102}$$

Furthermore, we must express the boundary condition (6.97) in terms of w'. Since $\partial u'/\partial\tilde{z} = \partial v'/\partial\tilde{z} = 0$ is satisfied everywhere on the boundary in the $\tilde{x}$ and $\tilde{y}$ directions, we have

$$\frac{\partial^2 u'}{\partial\tilde{z}\partial\tilde{x}} = \frac{\partial^2 v'}{\partial\tilde{z}\partial\tilde{y}} = 0 \,. \tag{6.103}$$

Differentiating the equation of continuity (6.94) with respect to $\tilde{z}$, we obtain

$$\frac{\partial^2 u'}{\partial\tilde{z}\partial\tilde{x}} + \frac{\partial^2 v'}{\partial\tilde{z}\partial\tilde{y}} + \frac{\partial^2 w'}{\partial\tilde{z}^2} = 0 \,.$$

Thus, the boundary condition becomes

$$\frac{\partial^2 w'}{\partial\tilde{z}^2} = 0 \,. \tag{6.104}$$

Furthermore, since $T' = w' = \partial^2 w'/\partial\tilde{z}^2 = 0$ everywhere along the boundary, each of the following terms in (6.101)

$$\left(\frac{\partial}{\partial \tilde{t}} - \frac{\partial^2}{\partial \tilde{x}^2} - \frac{\partial^2}{\partial \tilde{y}^2}\right)\left(\frac{\partial^2}{\partial \tilde{x}^2} + \frac{\partial^2}{\partial \tilde{y}^2}\right)w', \quad \left(\frac{\partial^2}{\partial \tilde{x}^2} + \frac{\partial^2}{\partial \tilde{y}^2}\right)\frac{\partial^2 w'}{\partial\tilde{z}^2},$$
$$\left(\frac{\partial}{\partial \tilde{t}} - \frac{\partial^2}{\partial \tilde{x}^2} - \frac{\partial^2}{\partial \tilde{y}^2}\right)\frac{\partial^2 w'}{\partial\tilde{z}^2}, \quad \left(\frac{\partial^2}{\partial \tilde{x}^2} + \frac{\partial^2}{\partial \tilde{y}^2}\right)T'$$

vanishes, and we are left with the fourth-order differential term in z

$$\frac{\partial^4 w'}{\partial\tilde{z}^4} = 0 \,. \tag{6.105}$$

Summarizing the above, the boundary condition in terms of w' is

$$w' = \frac{\partial^2 w'}{\partial \tilde{z}^2} = \frac{\partial^4 w'}{\partial \tilde{z}^4} = 0 \quad \text{at } \tilde{z} = 0,\ 1\,. \tag{6.106}$$

6.5.6 Fourier Components of the Equation

The Fourier components of the vertical velocity w' are

$$w'(\tilde{\boldsymbol{r}}, \tilde{t}) = \hat{w}(\tilde{z}) \exp\big[\, i(\tilde{k}_x \tilde{x} + \tilde{k}_y \tilde{y} - \tilde{\omega}\tilde{t}\,)\,\big]\,. \tag{6.107}$$

Substituting (6.107) into (6.102) yields an ordinary differential equation in terms of $\hat{w}$:

$$\left(\frac{d^2}{d\tilde{z}^2} - \tilde{k}_h^2 + i\tilde{\omega}\right)\left(\frac{d^2}{d\tilde{z}^2} - \tilde{k}_h^2 + iPr\tilde{\omega}\right)\left(\frac{d^2}{d\tilde{z}^2} - \tilde{k}_h^2\right)\hat{w} = -Ra\tilde{k}_h^2\hat{w} \tag{6.108}$$

where $\tilde{k}_h^2 \equiv \tilde{k}_x^2 + \tilde{k}_y^2$ is the (nondimensional) total horizontal wavenumber. The corresponding boundary condition is

$$\hat{w} = \frac{d^2\hat{w}}{d\tilde{z}^2} = \frac{d^4\hat{w}}{d\tilde{z}^4} = 0 \quad \text{at } \tilde{z} = 0,\ 1\,. \tag{6.109}$$

An eigenfunction that satisfies the boundary condition (6.109) is

$$\hat{w} = W \sin(n\pi\tilde{z}) \tag{6.110}$$

where n is a natural number. Substituting (6.110) into (6.108), we obtain an algebraic expression for the eigen value $\tilde{\omega}$:

$$(n^2\pi^2 + \tilde{k}_h^2 - i\tilde{\omega})(n^2\pi^2 + \tilde{k}_h^2 - iPr\tilde{\omega})(n^2\pi^2 + \tilde{k}_h^2) = Ra\tilde{k}_h^2\,. \tag{6.111}$$

Note 6.3 Expressing the eigenfunction in the form (6.110) can be justified as follows. Equation (6.108) is a homogeneous linear differential equation with constant coefficients, whose solutions can be expressed in the form $\exp(m\tilde{z})$. Substituting this form into (6.108) gives

$$(m^2 - \tilde{k}_h^2 + i\tilde{\omega})(m^2 - \tilde{k}_h^2 + iPr\tilde{\omega})(m^2 - \tilde{k}_h^2) = -Ra\tilde{k}_h^2\,. \tag{6.112}$$

Equation (6.112) is a cubic algebraic equation in m^2 and yields six solutions of m that $\pm m_1$, $\pm m_2$, and $\pm m_3$. The general solution of (6.108) can be expressed as a linear combination of $\exp(\pm m_i\tilde{z})$. Alternatively, the solution is a linear combination of $\sinh(m_i\tilde{z})$ and $\cosh(m_i\tilde{z})$:

$$\hat{w} = \sum_{i=1}^{3} S_i \sinh(m_i \tilde{z}) + \sum_{i=1}^{3} C_i \cosh(m_i \tilde{z}) \,. \tag{6.113}$$

Suppose that (6.113) satisfies the boundary condition (6.106) under the following conditions:

$$\mathsf{A} \equiv \begin{pmatrix} 0 & 0 & 0 & 1 & 1 & 1 \\ 0 & 0 & 0 & m_1^2 & m_2^2 & m_3^2 \\ 0 & 0 & 0 & m_1^4 & m_2^4 & m_3^4 \\ \sinh m_1 & \sinh m_2 & \sinh m_3 & \cosh m_1 & \cosh m_2 & \cosh m_3 \\ m_1^2 \sinh m_1 & m_2^2 \sinh m_2 & m_3^2 \sinh m_3 & m_1^2 \cosh m_1 & m_2^2 \cosh m_2 & m_3^2 \cosh m_3 \\ m_1^4 \sinh m_1 & m_2^4 \sinh m_2 & m_3^4 \sinh m_3 & m_1^4 \cosh m_1 & m_2^4 \cosh m_2 & m_3^4 \cosh m_3 \end{pmatrix} \tag{6.114}$$

$$\boldsymbol{x} \equiv {}^t(S_1\ S_2\ S_3\ C_1\ C_2\ C_3) \,. \tag{6.115}$$

Then our system of equations can be summarized as a matrix:

$$\mathsf{A}\boldsymbol{x} = \boldsymbol{0} \,. \tag{6.116}$$

To obtain a nontrivial solution, the determinant of matrix A is set to 0; then,

$$(m_1^2 - m_2^2)^2 (m_2^2 - m_3^2)^2 (m_3^2 - m_1^2)^2 \sinh m_1 \ \sinh m_2 \ \sinh m_3 = 0 \,. \tag{6.117}$$

The function $\sinh x$ is 0 only for $in\pi$ (where n is an integer)[2]; thus, any of m_i $(i = 1, 2, 3)$ equals $in\pi$. Without loss of generality, we arbitrarily take $m_1 = in\pi$. Substituting m_1 in (6.116), we obtain $S_2 = S_3 = C_1 = C_2 = C_3 = 0$. The solution becomes $S_1 \sinh(in\pi\tilde{z})$ (or $iS_1 \sin(n\pi\tilde{z})$). As sin 0 is a trivial solution, and $\sin(n\pi\tilde{z})$ and $\sin(-n\pi\tilde{z})$ are identical but opposite in sign, we obtain the independent solutions (6.110).

6.5.7 *Onset of Convective Motion*

As discussed in Sect. 5.2, if the imaginary part of the eigenvalues $\tilde{\omega}$ satisfy $\mathrm{Im}(\tilde{\omega}) < 0$, a minute disturbance is attenuated (stable), whereas if $\mathrm{Im}(\tilde{\omega}) > 0$, it is amplified (unstable). The two states are separated by the neutral state $\mathrm{Im}(\tilde{\omega}) = 0$. Let us now obtain the conditions under which disturbance is neutral.

---

[2] $\sinh x = [\exp x - \exp(-x)]/2$, however as $|\exp x| = \exp(\mathrm{Re}(x))$, in order to have $\sinh x = 0$, at least, it is required that $|\exp x| = |\exp(-x)|$. In other words, the zero of $\sinh x$ can only reside on the imaginary axis. Because $\sinh(ix) = i \sin x$, the only permissible zero points are $in\pi$ (where n is an integer).

The nondimensional oscillation frequency $\tilde{\omega}$ is a complex number. In fact, the $\tilde{\omega}$ obtained from (6.111) is purely imaginary, as demonstrated below. Multiplying both sides of (6.111) by $(n^2\pi^2 + \tilde{k}_h^2 - iPr\tilde{\omega}^*)$, we get

$$(n^2\pi^2 + \tilde{k}_h^2 - i\tilde{\omega})(n^2\pi^2 + \tilde{k}_h^2)\left[(n^2\pi^2 + \tilde{k}_h^2 + Pr\mathrm{Im}(\tilde{\omega}))^2 + Pr^2\mathrm{Re}(\tilde{\omega})^2\right]$$
$$= Ra\tilde{k}_h^2(n^2\pi^2 + \tilde{k}_h^2 - iPr\tilde{\omega}^*) \qquad (6.118)$$

from which imaginary parts were extracted to obtain

$$\mathrm{Re}(\tilde{\omega})\Big\{(n^2\pi^2 + \tilde{k}_h^2)\big[\,(n^2\pi^2 + \tilde{k}_h^2 + Pr\,\mathrm{Im}(\tilde{\omega}))^2$$
$$+ Pr^2\,\mathrm{Re}(\tilde{\omega})^2\,\big] + Ra\,Pr\,\tilde{k}_h^2\Big\} = 0\,. \qquad (6.119)$$

As the term inside { } is positive, $\mathrm{Re}(\tilde{\omega}) = 0$. Therefore, while the usual condition for neutrality is $\mathrm{Im}(\tilde{\omega}) = 0$, here it suffices to solve $\tilde{\omega} = 0$. Setting $\tilde{\omega} = 0$ in (6.111) gives

$$(n^2\pi^2 + \tilde{k}_h^2)^3 = Ra\tilde{k}_h^2\,. \qquad (6.120)$$

Expressing Ra satisfying this condition as a function of $\tilde{k}_h$, we get

$$Ra = \frac{(n^2\pi^2 + \tilde{k}_h^2)^3}{\tilde{k}_h^2}\,. \qquad (6.121)$$

In many natural numbers n, Ra is relative to $n = 1$,

$$Ra = \frac{(\pi^2 + \tilde{k}_h^2)^3}{\tilde{k}_h^2} \qquad (6.122)$$

becomes the smallest. Additionally, if $\tilde{k}_h$ of $Ra = (\pi^2 + \tilde{k}_h^2)^3/\tilde{k}_h^2$ is changed, its minimum value is

$$Ra_c = \frac{27\pi^4}{4} \cong 657.5 \;\; (\text{for } \tilde{k}_h = \tilde{k}_c \equiv \frac{\pi}{\sqrt{2}} \cong 2.221)\,. \qquad (6.123)$$

Exercise 6.5 Confirm that (6.123) gives the minimum Ra of (6.122).

6.5.8 Disturbance Developing Near Critical Rayleigh Number

Equation (6.122) draws a single curve on the $\tilde{k}_h - Ra$-plane. However, in the area where Rayleigh number Ra is smaller than this curve, no eigenmode satisfies $\mathrm{Im}(\tilde{\omega}) > 0$; thus, it is stable. In contrast, in the area with a large Rayleigh number, there is at least one eigenmode ($n = 1$) that satisfies $\mathrm{Im}(\tilde{\omega}) > 0$; thus, it is unstable. The curve (6.122) is a neutral curve (see Sect. 5.2) delineating the neutral stability condition at that boundary.

Now consider a gentle increase in the Rayleigh number. While the Rayleigh number remains small, the system is stable at all wavenumbers $\tilde{k}_h$. Under these conditions, any disturbance will be rapidly attenuated and the system will continue to conduct heat. This fluid layer is statically unstable from the perspective from Sect. 6.1. Such unstable layers are bounded above and below, and viscosity and heat dissipation exists. Thus, despite its unstable nature, a stable condition where disturbance does not develop is maintained. However, once the Rayleigh number reaches its critical value Ra_c indicated from (6.123), the situation changes. At this point, most disturbance is attenuated; however, the disturbance of wavenumber $\tilde{k}_c$ remains in the system. As the Rayleigh number increases slightly beyond Ra_c, disturbances around the wavenumber $\tilde{k}_c$ will persist and increase. In other words, as the Rayleigh number is slowly increased, the disturbance of wavenumber k_c increases and becomes visible as the system reaches Ra_c. This transition marks the beginning of Benard convection. Similar to the critical Reynolds number (see Sect. 5.2), this Ra_c is called the **critical Rayleigh number** and $\tilde{k}_c$ is called the **critical wavenumber**.

In this discussion, we have considered a simple situation where there was no stress along with the top and bottom boundary conditions. However, if we attempt to conduct an experiment like this, it is more suitable to consider adhesive conditions in the bottom layer rather than bottom slip boundary conditions. Problems in which one or both boundaries have adhesive conditions are slightly more complicated, but can be solved in a manner similar to the stress-free case to obtain $Ra_c \cong 1100.7$, $\tilde{k}_c \cong 2.682$ (one adhesive boundary) and $Ra_c \cong 1707.8$, $\tilde{k}_c \cong 3.117$ (two adhesive boundaries). (see Fig. 6.6.)

Exercise 6.6 Show that for an adhesive boundary, the boundary condition (6.106) becomes

$$w' = \frac{\partial w'}{\partial \tilde{z}} = \left(\frac{\partial}{\partial \tilde{t}} - \tilde{\nabla}^2\right)\tilde{\nabla}^2 w' = 0\,.$$

Note that in (6.120), we set $\tilde{\omega} = 0$ to investigate the beginning of the convection. In this case, the flow is characterized only by the Rayleigh number Ra (the Prandtl number Pr disappears). Thus, the conditions at the onset of a flow current are determined solely by the Rayleigh number. Of course, if the critical Rayleigh number is exceeded and the convective motion begins to amplify, the Prandtl number begins to exert an effect. Moreover, (6.108) is expressed in terms of the horizontal wavenumber $\tilde{k}_h$, and is independent of the $\tilde{k}_x$ and $\tilde{k}_y$ directions. The critical wavenumber $\tilde{k}_c$ can

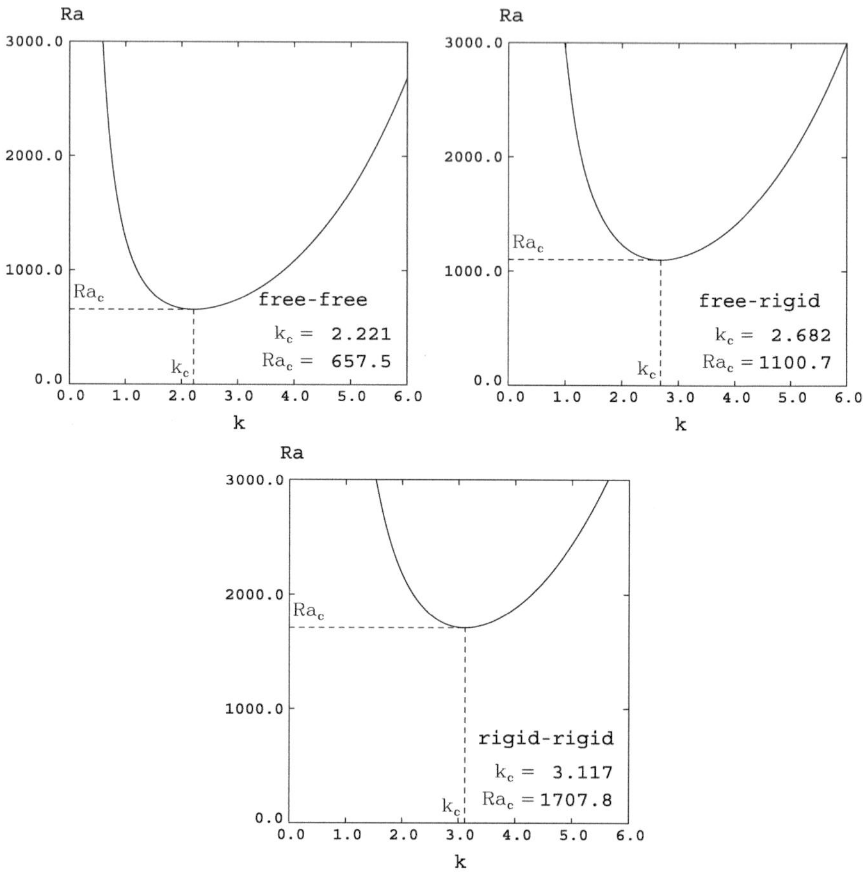

Fig. 6.6 Neutral curves in Benard convection *Left top* Both *top* and *bottom* boundaries are stress-free. *Right top* One boundary is stress-free, the other is adhesive. *Left bottom* Both *top* and *bottom* boundaries are adhesive

Fig. 6.7 Shapes of various convection cells with the same total wavenumber

also be expressed in the form of $\tilde{k}_h$. Thus, provided that the total wavenumber is $\tilde{k}_h$, regardless of the combinations of $\tilde{k}_x$ and $\tilde{k}_y$ or even their superposition, it describes the disturbance at the onset of convection. For example, in

$$\left.\begin{array}{ll}\sin\frac{\pi}{\sqrt{2}}\tilde{x}, & \frac{2}{3}\sin\frac{\pi}{2\sqrt{2}}\tilde{x}\sin\frac{\sqrt{3}\pi}{2\sqrt{2}}\tilde{y}+\frac{1}{3}\sin\frac{\pi}{\sqrt{2}}\tilde{x},\\ \sin\frac{\pi}{2}\tilde{x}\sin\frac{\pi}{2}\tilde{y}, & \frac{2}{3}\sin\frac{\pi}{2\sqrt{2}}\tilde{x}\sin\frac{\sqrt{3}\pi}{2\sqrt{2}}\tilde{y}+\frac{1}{3}\cos\frac{\pi}{\sqrt{2}}\tilde{x}\end{array}\right\} \tag{6.124}$$

each unit of convection (called a **convection cell**), respectively, takes the form of a roll, an equilateral triangle, a square, and a hexagon, depending on the $(\tilde{k}_x, \tilde{k}_y)$ components in modes $(\pi/\sqrt{2}, 0)$, $(\pi/2, \pi/2)$, $(\pi/(2\sqrt{2}), \sqrt{3}\pi/(2\sqrt{2}))$ or their superposition (Fig. 6.7). In all cases, the total wavenumber $\tilde{k}_h$ is $\tilde{k}_h = \pi/\sqrt{2}$. From the above arguments, we cannot discern the shape of the convection cell that develops and realizes. In fact, the shape of the convection is governed by secondary effects such as nonlinear effects or slight basic field shears, which are not considered here. Benard convection in a free-surface fluid tank in the laboratory frequently develops honeycombed hexagonal cells. Such cells are considered to be created by the surface tension of the fluid or the temperature dependence of the coefficient of viscosity. On the other hand, in winter monsoon outbreak over a warm sea, streaks of clouds are often seen on weather satellite images (these are visual manifestations of the uprising regions of roll-shaped convection cells). At the onset of the outbreak, the basic flow velocity is subjected to vertical shearing. If this shear is included in the linear problem describing the onset of convection, the fastest-growing disturbance is a wavenumber vector oriented perpendicular to the shear (leading to a roll-type disturbance whose axis is the shear direction).

Note 6.4 (*Literature for further study*)
For further studies textbooks [1–3, 5, 9–12] may be referred. The advanced readers may also refer some other books [4, 6–8, 13–17], although they are all written in Japanese.

References

1. Abarbanel, H.D., Young, W.R. (eds.): General Circulation of the Ocean. Springer, New York (1986)
2. Batchelor, G.K.: An Introduction to Fluid Dynamics. Cambridge University Press, Cambridge (1967)
3. Gill, A.E.: Atmosphere-Ocean Dynamics. Academic Press, Cambridge (1982)
4. Kimura, R.: Introduction to Geophysical Fluid Dynamics. Tokyodo Shuppan, Tokyo (1983)
5. Newman, J.N.: Marine Hydrodynamics. The MIT Press, Cambridge (1977)
6. Ogura, Y.: Principles of Atmospheric Dynamics. University of Tokyo Press, Tokyo (1978)
7. Ogura, Y.: General Meteorology. University of Tokyo Press, Tokyo (1984)
8. Ogura, Y.: Basic Theory of Meso-scale Meteorology. University of Tokyo Press, Tokyo (1997)

9. Open University. Oceanography Course Team: Ocean Circulation. Pergamon Press (1989)
10. Pedlosky, J.: Geophysical Fluid Dynamics, 2nd edn. Springer, New York (1986)
11. Pond, S., Pickard, G.L.: Introductory Dynamical Oceanography, 2nd edn. Pergamon Press, Oxford (1983)
12. Prandtl, L., Tietjens, O.G.: Fundamentals of Hydro- and Aeromechanics (translated by Rosenhead, L.). Dover, New York (1957)
13. Sekine, Y.: Introduction to Physical Oceanography. Seizando Shoten, Tokyo (1995)
14. Tatsumi, T., Goto, K.: Stability Theory of Flow. Sangyo-Tosho, Tokyo (1976)
15. Tomita, Y.: Introduction to Fluid Dynamics. Yokendo, Tokyo (1975)
16. Unoki, S.: Coastal Physical Oceanography. Tokai University Press, Japan (1993)
17. Unoki, S., Kubota, M.: Waves and Currents in the Ocean. Tokai University Press, Japan (1996)

Chapter 7
Dynamics of Rotating Fluids

In the previous chapters, We have learned a wide range of basic fluid dynamics, including fundamental equations, wave motion, turbulence, and stratified fluids. The final target of this textbook is "geophysical fluids," under the influence of both general fluid dynamic concepts and rotational effects. "Rotational effects" are introduced in the first instance when describing the fluid in a rotating system such as the Earth. This effects can not be found in non-rotating system. This term can be derived by rearranging the fundamental equations so that "a fluid in a rotating system is described by an observer in the same rotating system."

In Sect. 7.1 of this chapter, we derive the fundamental equation of rotating fluids by rearranging the fundamental equation introduced in Chap. 1. Section 7.2 will simplify the equation by introducing approximations commonly used in geophysical fluid dynamics, and will derive steady-state and wave solutions to the fundamental equations. Sections 7.3–7.6 are devoted to the physical phenomena underlying these solutions. Finally, Sect. 7.7 introduces the additional effect of stratification, in preparation for Chap. 8, which deals with real climate change and general circulations.

7.1 Fundamental Equation in a Rotating System

In previous chapters, we studied the fundamental equation for an observer in a stationary system describing fluids in that system. However, the description of rotating fluids such as those on the Earth become more complicated when viewed from a fixed coordinate frame in a stationary system. To simplify these problems, we define a coordinate system that rotates with the fluid (rotating coordinate system), and describe motions of the rotating fluid that is relatively not rotating with respect to this new coordinate. In this section, we will examine how fundamental equations are modified when we use this new rotating coordinate system.

D.E.S.S.T. Kyushu Univ., *Fluid Dynamics for Global Environmental Studies*,
DOI 10.1007/978-4-431-56499-7_7

7.1.1 Transformation from Fixed Coordinate Frame to Rotating Coordinate System

Consider a vector $\boldsymbol{A}$ rotating at angular velocity $\boldsymbol{\Omega}$ in a stationary coordinate system (Fig. 7.1). As time advances from t to $t + \delta t$, the vector $\boldsymbol{A}(t)$ apparently alters to $\boldsymbol{A}(t + \delta t)$, with no change in magnitude but a change in direction. Consider an arc $\tilde{A}$ joining $\boldsymbol{A}(t)$ and $\boldsymbol{A}(t + \delta t)$ on a plane orthogonal to $\boldsymbol{\Omega}$ and joining the endpoints of $\boldsymbol{A}(t)$ and $\boldsymbol{A}(t + \delta t)$. The internal angle of the arc, $\delta\theta$ is $|^{\Omega}|\delta t$, and the radius of the arc can be expressed as $|\boldsymbol{A}| \sin\gamma$, where the angle γ is formed by $\boldsymbol{\Omega}$ and $\boldsymbol{A}$. Hence, the length of this arc $\tilde{A}$ is $\delta\theta|\boldsymbol{A}| \sin\gamma$.

If δt is sufficiently small, the arc $\tilde{A}$ approximates the vector $\delta\boldsymbol{A} \equiv \boldsymbol{A}(t+\delta t)-\boldsymbol{A}(t)$. In other words, $\delta\boldsymbol{A}$ is orthogonal to both $\boldsymbol{A}$ and $\boldsymbol{\Omega}$, and its magnitude is $|\delta\boldsymbol{A}| \simeq \delta t|\boldsymbol{\Omega}||\boldsymbol{A}| \sin\gamma$. The differential as $\delta t \to 0$ is given by[1]

$$\left(\frac{d\boldsymbol{A}}{dt}\right)_{\mathrm{F}} = \lim_{\delta t \to 0} \frac{\delta\boldsymbol{A}}{\delta t} = \boldsymbol{\Omega} \times \boldsymbol{A}\,. \tag{7.1}$$

On the other hand, an observer rotating with the same angular velocity $\boldsymbol{\Omega}$ will not recognize that $\boldsymbol{A}$ is rotating, and rather will view it as stationary. In this rotating reference system, the time differential becomes[2]

$$\left(\frac{d\boldsymbol{A}}{dt}\right)_{\mathrm{R}} = 0 \tag{7.2}$$

which is simpler than that observed by a stationary observer in the fixed coordinate frame.

For completeness, let us also consider a more general case in which the change in $\boldsymbol{A}$ can be observed even from a rotating system. In this scenario, as the vector is linearly separable, it can be considered that a fluctuating component observed from the rotating system $(d\boldsymbol{A}/dt)_{\mathrm{R}}$ added to the component $\boldsymbol{A} + \delta\boldsymbol{A}$ appeared stationary in the rotating system (see Fig. 7.1). Therefore, for an arbitrary vector $\boldsymbol{A}$, we have

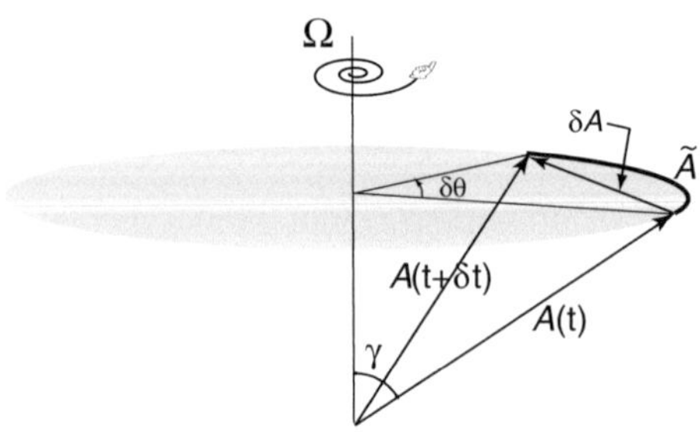

Fig. 7.1 Vector $\boldsymbol{A}$ is rotating at angular velocity $\boldsymbol{\Omega}$ when viewed in a fixed coordinate frame. During a short time δt, $\boldsymbol{A}$ changes by $\delta\boldsymbol{A}$. If δt is sufficiently small, $\delta\boldsymbol{A}$ is orthogonal to $\boldsymbol{\Omega}$ as well as $\boldsymbol{A}$, and can be written as $\boldsymbol{\Omega} \times \boldsymbol{A}$

[1] $(d/dt)_{\mathrm{F}}$ refers to "derivative in Lagrangian notation in a fixed reference frame."

[2] $(d/dt)_{\mathrm{R}}$ refers to "derivative in Lagrangian notation in a rotating reference system."

$$\left(\frac{d\boldsymbol{A}}{dt}\right)_{\mathrm{F}} = \left(\frac{d\boldsymbol{A}}{dt}\right)_{\mathrm{R}} + \boldsymbol{\Omega} \times \boldsymbol{A}\,. \tag{7.3}$$

From (7.3), we observe that the difference in time derivative is only $\boldsymbol{\Omega}$ between a fixed coordinate frame and a coordinate system rotating at $\boldsymbol{\Omega}$.

7.1.2 Velocity and Acceleration in a Rotating Coordinate System

We can use (7.3) to compare the velocity and acceleration in a rotating coordinate system with those in a fixed coordinate frame.

First, the time derivative of the position vector $\boldsymbol{r}$ is expressed as

$$\left(\frac{d\boldsymbol{r}}{dt}\right)_{\mathrm{F}} = \left(\frac{d\boldsymbol{r}}{dt}\right)_{\mathrm{R}} + \boldsymbol{\Omega} \times \boldsymbol{r}\,. \tag{7.4}$$

Therefore, the observed velocities $\boldsymbol{u}_{\mathrm{F}}$ and $\boldsymbol{u}_{\mathrm{R}}$, observed in the fixed and rotating coordinate systems, respectively, are related as

$$\boldsymbol{u}_{\mathrm{F}} = \boldsymbol{u}_{\mathrm{R}} + \boldsymbol{\Omega} \times \boldsymbol{r}\,. \tag{7.5}$$

The acceleration, i.e., the time derivative of velocity, is given by

$$\left(\frac{d\boldsymbol{u}_{\mathrm{F}}}{dt}\right)_{\mathrm{F}} = \left(\frac{d\boldsymbol{u}_{\mathrm{F}}}{dt}\right)_{\mathrm{R}} + \boldsymbol{\Omega} \times \boldsymbol{u}_{\mathrm{F}}\,. \tag{7.6}$$

However, this expression is inconvenient because the subscripts R and F remain mixed in the right-hand side. Namely, the flow velocity $\boldsymbol{u}_{\mathrm{F}}$ observed in the stationary system cannot be inferred from the rotating system. To express the right-hand side in terms of $\boldsymbol{u}_{\mathrm{R}}$ alone, we substitute (7.5) into this expression to get

$$\left(\frac{d\boldsymbol{u}_{\mathrm{F}}}{dt}\right)_{\mathrm{F}} = \left(\frac{d\boldsymbol{u}_{\mathrm{R}}}{dt}\right)_{\mathrm{R}} + 2\boldsymbol{\Omega} \times \boldsymbol{u}_{\mathrm{R}} + \boldsymbol{\Omega} \times (\boldsymbol{\Omega} \times \boldsymbol{r}) + \left(\frac{d\boldsymbol{\Omega}}{dt}\right)_{\mathrm{R}} \times \boldsymbol{r}\,. \tag{7.7}$$

The second and third terms on the right-hand side of (7.7) are called **Coriolis acceleration** and centripetal acceleration terms, respectively. Within the equation of motion, these accelerations are transposed to their respective external forces, known as the **Coriolis force** and centrifugal force, respectively.[3]

In ordinary geophysical fluid dynamics, the Earth's angular velocity $\boldsymbol{\Omega}$ is assumed to be sufficiently stable; hence, we can ignore the final term on the right-hand side of (7.7). Moreover, the third term on the right-hand side is equivalent to the centrifugal

[3]In this chapter, "Coriolis acceleration" and "Coriolis force" are often used interchangeably.

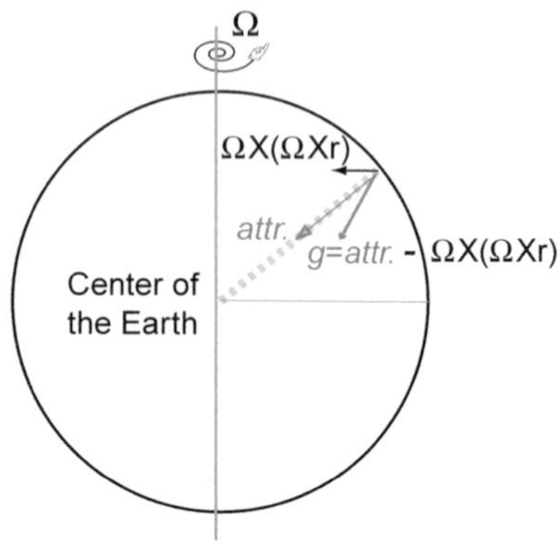

Fig. 7.2 Gravity (g) is contributed by the Earth's gravitational attractive force (*attr.*) and the centrifugal force $-\boldsymbol{\Omega} \times (\boldsymbol{\Omega} \times r)$ due to rotation. In this diagram, the centrifugal force has been exaggerated for illustration purposes. In reality, the centrifugal force is considerably smaller than the gravitational pull so that the direction of $\vec{\mathbf{g}}$ (*vertically downwards*) is almost toward the center of the Earth

force imposed by the Earth's rotation, which is usually treated as the Earth's gravitational force together with the Earth's gravitational attractive force (see Fig. 7.2). Basically, the transformation in the rotating and fixed coordinate systems are dominated by the Coriolis acceleration (the second term on the right-hand side of (7.7)).

7.1.3 *Fundamental Navier-Stokes Equation in a Rotating System*

Based on the above analysis, we can now express the Navier-Stokes equation in a rotating system. That is, we reform the Navier-Stokes equation (1.89) from Chap. 1 in terms of $\boldsymbol{u}_{\mathrm{R}}$ of a rotating system.

If time is fixed in both rotating and stationary coordinate systems, the coordinate frame is merely rotated, and the spatial derivative at any moment takes the same vector expression in both systems. Therefore, to transform (1.89) into a rotating system, we add the Coriolis acceleration (7.7) and exchange $\boldsymbol{u}_{\mathrm{F}}$ and $\boldsymbol{u}_{\mathrm{R}}$ using (7.5). However, the vector $\boldsymbol{\Omega} \times \boldsymbol{r}$ in (7.5) disappears if $\nabla \cdot$ or ∇^2 is applied. Thus, the Navier-Stokes equation in a rotating system can be considered as the Navier-Stokes equation in the stationary frame with an additional Coriolis acceleration term:

$$\frac{D\boldsymbol{u}_{\mathrm{R}}}{Dt} + 2\boldsymbol{\Omega} \times \boldsymbol{u}_{\mathrm{R}} = \boldsymbol{K} - \frac{1}{\rho}\nabla p + \frac{\mu}{\rho}\nabla^2 \boldsymbol{u}_{\mathrm{R}} + \frac{\mu}{3\rho}\nabla(\nabla \cdot \boldsymbol{u}_{\mathrm{R}})\,. \tag{7.8}$$

In (7.8), the material derivative is defined as

$$\frac{D}{Dt} = \left(\frac{\partial}{\partial t}\right)_{\mathrm{R}} + \boldsymbol{u}_{\mathrm{R}} \cdot \nabla \tag{7.9}$$

which uses the Eulerian field in the rotating coordinate system, namely, with the subscript R alone.

Note 7.1 Although we stated that the spatial derivative at a given instant is the same in vector notation in both fixed and rotating coordinate systems, the vector components will differ between the frames since the axial directions are rotated. Furthermore, these directions change in time, which complicate the description of rotating fluids from a fixed coordinate frame.

Moreover, although the vector expressions of the external forces $\boldsymbol{K}$ are identical in fixed and rotating coordinate systems, their functional forms may differ. For example, if an external force $\boldsymbol{K}$ is a function of $\boldsymbol{u}_{\mathrm{F}}$ in the fixed coordinate frame, it is often expressed as a different function if expressed with $\boldsymbol{u}_{\mathrm{R}}$.

Equation of Continuity

We now consider the other fundamental equation of fluid dynamics: the equation of continuity. The scalar time derivative takes the same form in both rotating and fixed coordinate systems. Therefore, if we consider the equation of continuity from Chap. 1 (1.38) in the rotating system, we get

$$\frac{D\rho}{Dt} + \rho \nabla \cdot \boldsymbol{u}_{\mathrm{R}} = 0\,. \tag{7.10}$$

Thus, ultimately, the form is exactly the same in a rotating system. For incompressible fluids, of course this equation holds, hence, regardless of whether the system is fixed or rotating, we have

$$\nabla \cdot \boldsymbol{u}_{\mathrm{R}} = 0\,. \tag{7.11}$$

Vorticity Equation

The vorticity equation in a rotating system can be derived using the arguments developed in Sect. 1.10, which simply involve rearranging the Navier-Stokes equation.

The Navier-Stokes equation in a rotating system (7.8) is essentially the same as the fixed-frame equation (1.89) with an additional Coriolis acceleration term $2\boldsymbol{\Omega} \times \boldsymbol{u}_{\mathrm{R}}$. Therefore, the transformed version of (1.89) in Sect. 1.10, namely (1.121), becomes

$$\begin{aligned}\frac{\partial \boldsymbol{u}_{\mathrm{R}}}{\partial t} &+ \nabla\left(\frac{1}{2}|\boldsymbol{u}_{\mathrm{R}}|^2 + \Pi - \frac{\nu}{3}\nabla \cdot \boldsymbol{u}_{\mathrm{R}}\right) \\ &= \boldsymbol{u}_{\mathrm{R}} \times (\boldsymbol{\omega}_{\mathrm{R}} + 2\boldsymbol{\Omega}) - \frac{1}{\rho}\nabla p + \nu \nabla^2 \boldsymbol{u}_{\mathrm{R}}\end{aligned} \tag{7.12}$$

where the term $-2\boldsymbol{\Omega} \times \boldsymbol{u}_{\mathrm{R}} = \boldsymbol{u}_{\mathrm{R}} \times 2\boldsymbol{\Omega}$ has been added to the first term on the right-hand side. This transformation merely replaces ω in (1.121) with $\boldsymbol{\omega}_{\mathrm{R}} + 2\boldsymbol{\Omega}$ ($\equiv \boldsymbol{\omega}_{\mathrm{a}}$). Here $\boldsymbol{\omega}_{\mathrm{a}}$ is the sum of vorticity $\boldsymbol{\omega}_{\mathrm{R}} = \nabla \times \boldsymbol{u}_{\mathrm{R}}$ and the **planetary vorticity** imposed by the self-rotation of the coordinate system ($2\boldsymbol{\Omega}$), and is called **absolute vorticity**. For distinguishing purposes, $\boldsymbol{\omega}_{\mathrm{R}}$ is often called **relative vorticity**.

Again following the arguments in Sect. 1.10, we can derive the vorticity equation (1.127) in the rotating coordinate system. The derivation differs from that in Sect. 1.10 only by the inclusion of the planetary vorticity $2\boldsymbol{\Omega}$. However, given that $\nabla \cdot \boldsymbol{\Omega} = 0$

and $\nabla^2 \boldsymbol{\Omega} = 0$, and assuming that rotation angular velocity is time invariant, we can set $\partial \boldsymbol{\Omega}/\partial t = 0$; then, the vorticity equation becomes

$$\frac{D\boldsymbol{\omega}_{\mathrm{a}}}{Dt} = (\boldsymbol{\omega}_{\mathrm{a}} \cdot \nabla)\boldsymbol{u}_{\mathrm{R}} - (\nabla \cdot \boldsymbol{u}_{\mathrm{R}})\boldsymbol{\omega}_{\mathrm{a}} + \frac{1}{\rho^2}\nabla\rho \times \nabla p + \nu\nabla^2\boldsymbol{\omega}_{\mathrm{a}}\,. \tag{7.13}$$

Here to maintain consistency with Chap. 1, it was assumed that the external force $\boldsymbol{K}$ is the conservative force expressed with potential Π. However, a more general form of Navier-Stokes equation, in terms of an arbitrary external force $\boldsymbol{K}$, is written as

$$\frac{D\boldsymbol{\omega}_{\mathrm{a}}}{Dt} = (\boldsymbol{\omega}_{\mathrm{a}} \cdot \nabla)\boldsymbol{u}_{\mathrm{R}} - (\nabla \cdot \boldsymbol{u}_{\mathrm{R}})\boldsymbol{\omega}_{\mathrm{a}} + \frac{1}{\rho^2}\nabla\rho \times \nabla p + \nu\nabla^2\boldsymbol{\omega}_{\mathrm{a}} + \nabla \times \boldsymbol{K}\,. \tag{7.14}$$

Note that (7.14) reduces to (7.13) when $\boldsymbol{K}$ is a conservative force. The fundamental fluid equation includes other equations such as the energy equation; however, the derivation of these equations is omitted in this chapter.

Exercise 7.1 Obtain the work performed by the Coriolis force as part of the energy calculation.

7.2 Simplification of Equations

In this section, the fundamental equation of rotating fluids derived in the previous section is simplified to a form that is commonly employed in geophysical fluids. In subsequent discussions, a rotating coordinate system is mainly assumed; therefore, we omit the postscript R. To simplify the analysis, we assume gravity as the sole external force $\boldsymbol{K}$ and an incompressible fluid. Furthermore, to isolate the effect of rotation in the absence of other restricting phenomena, we assume that density is a constant (ρ_0) fixed in time and space. The case of density stratification will be considered in Sect. 7.7.

7.2.1 Local Cartesian Coordinates

The vector components of the Navier-Stokes equation (7.8), the equation of continuity (7.11), and the vorticity equation (7.14) can be expressed in an arbitrary coordinate system. Meanwhile, spherical coordinates would be most suitable for geophysical fluids, which are generally regarded as a "thin fluid layer adjacent to the Earth's surface."

Note 7.2 The Earth's radius R is approximately 6.4×10^6 m, whereas the vertical scale H of the atmosphere and oceans is at most 10–10^4 m. Therefore, geophysical fluids can be approximated as a thin layer of a fluid limited to the area close to the Earth's surface.

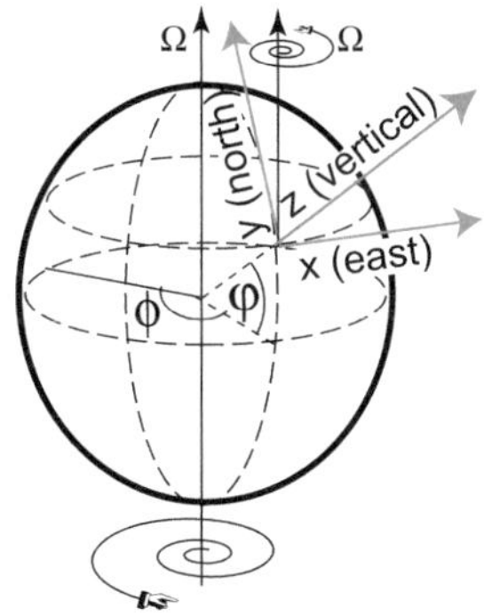

Fig. 7.3 Local Cartesian coordinate system with the origin places at a specific point on a spherical surface. In the local orthogonal coordinate system, the x-, y-, and z-axes point *east*, *north*, and *upwards*, respectively. Far from the origin, the x-y surface no longer approximates the spherical surface; hence, this simplification is not suitable at extremely large scales

In a "thin fluid layer," the vertical scale H can be regarded as extremely small compared with the horizontal scale L of geophysical fluids. The ratio of vertical scale to horizontal scale is called the aspect ratio. Geophysical fluids are characterized by a large aspect ratio ($L/H \simeq 10^2$–10^5).

Such large aspect ratio implies that, in geophysical fluids, the horizontal motion dominates over the vertical motion. Therefore, we often ignore the vertical flow components relative to the horizontal flow components in geophysical fluids, except for some processes where vertical flows are important, such as convection. Details are presented in Sect. 7.2.2.

However, as shown in the Appendix, the spherical coordinate representation is so complicated that the simple **local Cartesian coordinates** are often applied in many situations. These coordinates are obtained by simplifying spherical curvilinear coordinates to rectilinear coordinates.

In local Cartesian coordinates, the origin of the spherical coordinates is commonly shifted to an adequate convenient position on the spherical surface; the easterly, northerly, and vertical directions are designated as the x-, y-, and z-axes, respectively (see Fig. 7.3). That is, assuming the radius of the sphere as R, the radial coordinates in the spherical coordinate transform to $r = R + z$ and we can set $dr = dz$. Moreover, as $|z| \ll R$, $r \simeq R$; thus, we can write $dx = R \cos\varphi \, d\phi$ and $dy = R \, d\varphi$. Therefore, the Cartesian and spherical coordinates are related as $\boldsymbol{u} = (u, \ v, \ w) = (dx/dt, \ dy/dt, \ dz/dt) = (R \cos\varphi \, d\phi/dt, \ R \, d\varphi/dt, \ dr/dt) = (u_\phi, \ u_\varphi, \ u_r)$.

It should be reminded that the local Cartesian coordinates are linear coordinates but the spherical coordinates are not. In other words, the x-y surface is the tangential plane of the spherical surface. Close to the origin, the spherical coordinates are well approximated by the local Cartesian coordinates. Apart from the origin, the spherical and tangential surfaces deviate; therefore, the spherical surface can no longer be adequately described by the local Cartesian coordinates. In this sense, the replacement of a spherical surface with a planar surface is strictly "local."

The components of the vector equation are easily expressed in Cartesian coordinates, which are linear and orthogonal. The angular velocity components at latitude φ are given by $\boldsymbol{\Omega} = (0, \ \Omega \cos\varphi, \ \Omega \sin\varphi)$; therefore, the Coriolis acceleration introduced at the beginning of this chapter can be expressed as

$$2\boldsymbol{\Omega} \times \boldsymbol{u} = 2\Omega(w\cos\varphi - v\sin\varphi,\ u\sin\varphi,\ -u\cos\varphi)\,. \tag{7.15}$$

Furthermore, the individual components x, y, and z of the Navier-Stokes equation (7.8) are given by

$$\frac{Du}{Dt} + 2\Omega w\cos\varphi - 2\Omega v\sin\varphi = -\frac{1}{\rho_0}\frac{\partial p}{\partial x} + \nu_{\mathrm{h}}\nabla_{\mathrm{h}}^2 u + \nu_z\frac{\partial^2 u}{\partial z^2}\,, \tag{7.16}$$

$$\frac{Dv}{Dt} + 2\Omega u\sin\varphi = -\frac{1}{\rho_0}\frac{\partial p}{\partial y} + \nu_{\mathrm{h}}\nabla_{\mathrm{h}}^2 v + \nu_z\frac{\partial^2 v}{\partial z^2}\,, \tag{7.17}$$

$$\frac{Dw}{Dt} - 2\Omega u\cos\varphi = -\frac{1}{\rho_0}\frac{\partial p}{\partial z} - g + \nu_{\mathrm{h}}\nabla_{\mathrm{h}}^2 w + \nu_z\frac{\partial^2 w}{\partial z^2}\,. \tag{7.18}$$

In these equations, the viscosity term is divided into horizontal viscosity $\nabla_{\mathrm{h}}^2 = \partial^2/\partial x^2 + \partial^2/\partial y^2$ and vertical viscosity $\partial^2/\partial z^2$ components. Actually, the "viscosity" appearing in the Navier-Stokes equation of geophysical fluids is the "eddy viscosity" introduced in Sect. 5.4, rather than the "molecular viscosity" discussed in Sect. 1.7. As will be discussed in Sect. 7.3.2, vortexes in barotropic geophysical fluids behave as vertically homogeneous columns. Thus, flow behavior is fundamentally different in the horizontal and vertical directions, and must be appropriately distinguished.

Comparing the spherical coordinate representation from the Appendix with (7.16)–(7.18), we observe an additional Coriolis acceleration term and the omission of several terms. As the axes of both coordinate systems intercept at the origin, these terms account for the curvature of the coordinates. In particular, missing terms on the left side are called metric acceleration, in which the products of the flow velocities are divided by R. In geophysical fluids, metric acceleration is frequently small enough to be ignored.

Exercise 7.2 On the Earth, what magnitude of the flow velocity will ensure that metric acceleration becomes significantly larger than the Coriolis acceleration?

7.2.2 *Elimination of Vertical Flow*

As already mentioned, the vertical flow component w is considerably smaller than the horizontal components u and v. Therefore, the system of equations can be reduced to a two-dimensional expression of horizontal flows as follows.

Navier-Stokes Equation

The horizontal components (7.16) and (7.17) of the Navier-Stokes equation include vertical flow component w within their $w\partial/\partial z$ terms of material derivatives and the Coriolis acceleration term of (7.16). The latter is significantly smaller than the Coriolis acceleration induced by the horizontal flow component u, and can be ignored. Moreover, in Sect. 7.3.2, we will show that the vertical gradients of the horizontal flow components u and v are small, and thus, together with additional influence by multiplying small w, the vertical advection term becomes particularly smaller than the other terms of material derivatives. Thus, material derivatives without the vertical advection term, $D_h/Dt \equiv \partial/\partial t + u\partial/\partial x + v\partial/\partial y$, are used to express the horizontal components of the Navier-Stokes equation as

$$\frac{D_h u}{Dt} - fv = -\frac{1}{\rho_0}\frac{\partial p}{\partial x} + \nu_h \nabla_h^2 u + \nu_z \frac{\partial^2 u}{\partial z^2}, \tag{7.19}$$

$$\frac{D_h v}{Dt} + fu = -\frac{1}{\rho_0}\frac{\partial p}{\partial y} + \nu_h \nabla_h^2 v + \nu_z \frac{\partial^2 v}{\partial z^2}. \tag{7.20}$$

Here a new latitude-dependent quantity called the **Coriolis parameter** $f \equiv 2\Omega \sin\varphi$ is introduced. This is the most important quantity for the rotational effect, and is further discussed in Sect. 7.2.4.

In the remaining z-component equation (7.18), the acceleration and viscosity terms of vertical flows w and the Coriolis acceleration term $2\Omega u \cos\varphi$ are all negligible compared with the gravitational acceleration g. Thus, the hydrostatic equilibrium equation

$$0 = -\frac{1}{\rho_0}\frac{\partial p}{\partial z} - g \tag{7.21}$$

can be established. As shown in Sect. 6.1, this relation holds even in stratified fluids.

Exercise 7.3 If the Coriolis acceleration is not negligible relative to g in (7.18), the velocity $|u|$ must be extremely large. On the Earth, what magnitude of $|u|$ will be required to balance the z-component of the Coriolis acceleration with the gravitational acceleration?

Equation of Continuity

In the equation of continuity (7.11) for an incompressible fluid in local Cartesian coordinates, the vertical flow is in the form of vertical derivatives, which may not be negligible:

$$\frac{\partial u}{\partial x} + \frac{\partial v}{\partial y} + \frac{\partial w}{\partial z} = 0. \tag{7.22}$$

To eliminate the vertical flow, we vertically integrate (7.22) from the bottom of the fluid $z = -H$ to its upper surface $z = \eta$ to obtain

$$\int_{-H}^{\eta} \left(\frac{\partial u}{\partial x} + \frac{\partial v}{\partial y}\right) dz + w|_\eta - w|_{-H} = 0. \tag{7.23}$$

Therefore, to exclude the vertical flow from the equation of continuity, the vertical flows on the floor and surface of the fluid, $w|_{-H}$ and $w|_\eta$, respectively, must be expressed in terms of the horizontal flow only.

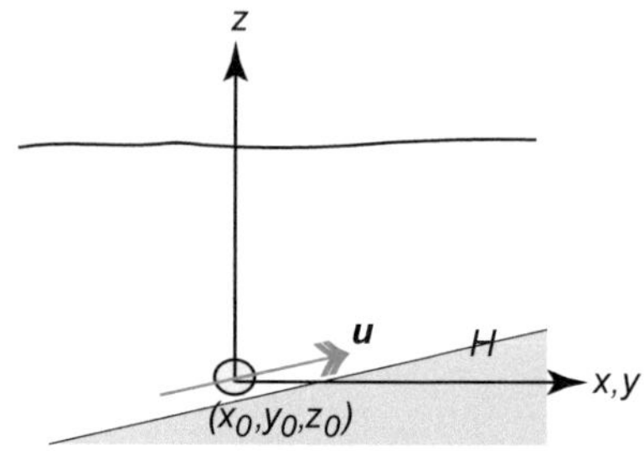

Fig. 7.4 Movement of fluid particles near the sea floor (x_0, y_0, z_0). Flow is driven upward along the floor surface by horizontal flow climbing over the inclined surface

Moreover, denoting an arbitrary horizontal position (x_0, y_0) of fluids on the bottom surface by (x_0, y_0, z_0), we have $z_0 = -H(x_0, y_0, t)$. If a flowing fluid particle does not detach from the bottom surface near position (x_0, y_0, z_0), its material derivative is given by

$$\frac{D(z_0 + H)}{Dt} = 0\,. \tag{7.24}$$

As the vertical velocity component at the base surface $w|_{-H}$ is Dz_0/Dt, the temporal change Dz_0/Dt in the vertical position of fluid particles z_0 is given by rearranging (7.24):

$$w|_{-H} + \left.\frac{D_\mathrm{h} H}{Dt}\right|_{-H} = w|_{-H} + \frac{\partial H}{\partial t} + u|_{-H}\frac{\partial H}{\partial x} + v|_{-H}\frac{\partial H}{\partial y} = 0\,. \tag{7.25}$$

As an example, let us consider a situation where there is no vertical movement of the bottom surface over time, such as at the sea floor without earthquakes. From $\partial H/\partial t = 0$,

$$w|_{-H} = u|_{-H}\frac{\partial(-H)}{\partial x} + v|_{-H}\frac{\partial(-H)}{\partial y}\,. \tag{7.26}$$

Vertical flow component on the bottom surface is caused by horizontal flows along the upward and downward slopes of the bottom (see Fig. 7.4). Similarly, at the fluid surface, $z_0 = \eta(x_0, y_0, t)$ holds; therefore,

$$\frac{D(z_0 - \eta)}{Dt} = w|_\eta - \left.\frac{D_\mathrm{h}\eta}{Dt}\right|_\eta = 0\,. \tag{7.27}$$

Substituting these into $w|_\eta$ or $w|_{-H}$ of (7.23), the continuity equation is expressed solely in terms of horizontal two-dimensional flows:

$$\int_{-H}^{\eta}\left(\frac{\partial u}{\partial x} + \frac{\partial v}{\partial y}\right)dz + \left.\frac{D_\mathrm{h}\eta}{Dt}\right|_\eta + \left.\frac{D_\mathrm{h} H}{Dt}\right|_{-H} = 0\,. \tag{7.28}$$

Now consider the x derivative of the x-directional volume transport Q_x, or the vertical integration of u. Accounting that the boundaries of integration, η and H, themselves are functions of x, we obtain

$$\frac{\partial Q_x}{\partial x} = \frac{\partial}{\partial x}\left(\int_{-H}^{\eta} u\, dz\right) = \frac{\partial \eta}{\partial x}\, u|_{\eta} + \frac{\partial H}{\partial x}\, u|_{-H} + \int_{-H}^{\eta} \frac{\partial u}{\partial x}\, dz\,. \tag{7.29}$$

A similar expression can be derived for $\partial Q_y/\partial y$. By substituting both forms into (7.28), we obtain the simplification

$$\frac{\partial Q_x}{\partial x} + \frac{\partial Q_y}{\partial y} + \frac{\partial h}{\partial t} = 0 \tag{7.30}$$

where $h \equiv \eta + H$ is the height of the fluid column. Equation (7.30) states that the fluid column height h changes as the divergence and convergence of the horizontal transport. Moreover, using the vertical mean flows,

$$\bar{u} = \frac{1}{\eta + H}\int_{-H}^{\eta} u\, dz = \frac{Q_x}{h}\,, \tag{7.31}$$

$$\bar{v} = \frac{1}{\eta + H}\int_{-H}^{\eta} v\, dz = \frac{Q_y}{h}\,, \tag{7.32}$$

(7.30) can be rearranged to

$$h\left(\frac{\partial \bar{u}}{\partial x} + \frac{\partial \bar{v}}{\partial y}\right) + \frac{\partial h}{\partial t} + \bar{u}\frac{\partial h}{\partial x} + \bar{v}\frac{\partial h}{\partial y} = 0\,. \tag{7.33}$$

Alternatively,

$$h\left(\frac{\partial \bar{u}}{\partial x} + \frac{\partial \bar{v}}{\partial y}\right) + \frac{D_{\bar{h}} h}{Dt} = 0 \tag{7.34}$$

where $D_{\bar{h}}/Dt \equiv \partial/\partial t + \bar{u}\partial/\partial x + \bar{v}\partial/\partial y$ is the material derivative of the vertical mean flow.

Vorticity Equation

The vorticity vector has three components. Among them, the vertical component $\zeta \equiv \partial v/\partial x - \partial u/\partial y$ is expressed in terms of horizontal flows alone, and hence is appropriate for this section. Partially differentiating (7.20) and (7.19) with respect to x and y, respectively, and subtracting (i.e., calculating $\partial(7.20)/\partial x - \partial(7.19)/\partial y$), we obtain

$$\frac{D_{\mathrm{h}}\zeta}{Dt} + (f + \zeta)\left(\frac{\partial u}{\partial x} + \frac{\partial v}{\partial y}\right) + \beta v = \nu_{\mathrm{h}}\nabla_{\mathrm{h}}^2\zeta + \nu_z\frac{\partial^2 \zeta}{\partial z^2}\,. \tag{7.35}$$

As f is a function of latitude φ only, all derivatives vanish except the y-derivative, which becomes $\beta \equiv df/dy = df/(Rd\varphi) = (2\Omega/R)\cos\varphi$.

Note 7.3 The remaining two components of the vorticity vector are the vertical derivatives of the horizontal flow. As will be discussed later in Sect. 7.3.2, these derivatives become 0 in a steady flow of homogenous density. Conversely, if its horizontal density gradient is established, these vorticity components become meaningful.

The above discussion relates to the $\nabla\rho \times \nabla p$ term in the general vector format of the vorticity equation (7.14). As the pressure gradient ∇p is nearly vertical, this vector product basically expresses the horizontal vorticity component. This term is non negligible if the isobaric and isopycnic surfaces are not parallel (such a fluid is called a baroclinic fluid). By contrast, in barotropic fluids (see Sect. 1.5.2), the density ρ is a function of the pressure p alone, implying that $\nabla\rho$ and ∇p are parallel.

Potential Vorticity Equation

We now consider the vertical integration of (7.35) from the bottom to the surface of the fluid and state the mean as

$$\frac{D_{\bar{\mathrm{h}}}\bar{\zeta}}{Dt} + (f+\bar{\zeta})\left(\frac{\partial \bar{u}}{\partial x} + \frac{\partial \bar{v}}{\partial y}\right) + \beta\bar{v} = \nu_{\mathrm{h}}\nabla_{\mathrm{h}}^2\bar{\zeta} + \frac{\nu_z}{h}\left(\left.\frac{\partial\zeta}{\partial z}\right|_{\eta} - \left.\frac{\partial\zeta}{\partial z}\right|_{-H}\right). \tag{7.36}$$

The stress at a horizontal surface is defined as

$$\frac{\boldsymbol{\tau}^z}{\rho_0} = \left(\frac{\tau_x^z}{\rho_0}, \frac{\tau_y^z}{\rho_0}\right) \equiv \left(\nu_z\frac{\partial u}{\partial z}, \nu_z\frac{\partial v}{\partial z}\right) \tag{7.37}$$

so that we can express the last two terms on the right-hand side of (7.36) as $(\nabla_{\mathrm{h}} \times \boldsymbol{\tau}^z)_z = \partial\tau_y^z/\partial x - \partial\tau_x^z/\partial y$, using the relation $\partial\zeta/\partial z = \partial(\partial v/\partial z)/\partial x - \partial(\partial u/\partial z)/\partial y$. Moreover, replacing the second term on the left-hand side with (7.34) obtained from the equation of continuity, (7.36) can be rearranged as

$$\frac{D_{\bar{\mathrm{h}}}\bar{\zeta}}{Dt} - \frac{f+\bar{\zeta}}{h}\frac{D_{\bar{\mathrm{h}}}h}{Dt} + \beta\bar{v} = \nu_{\mathrm{h}}\nabla_{\mathrm{h}}^2\bar{\zeta} + \frac{1}{h\rho_0}\left(\nabla_{\mathrm{h}} \times (\boldsymbol{\tau}^{\mathrm{t}} - \boldsymbol{\tau}^{\mathrm{b}})\right)_z, \tag{7.38}$$

where $\boldsymbol{\tau}^{\mathrm{t}}$ and $\boldsymbol{\tau}^{\mathrm{b}}$ are the stresses at the top and bottom fluid surfaces, respectively. Writing

$$\beta\bar{v} = \frac{df}{dy}\bar{v} = \frac{\partial f}{\partial t} + \bar{u}\frac{\partial f}{\partial x} + \bar{v}\frac{\partial f}{\partial y} = \frac{D_{\bar{\mathrm{h}}}f}{Dt}, \tag{7.39}$$

(7.38) can be summarized as

$$h\frac{D_{\bar{\mathrm{h}}}}{Dt}\left(\frac{f+\bar{\zeta}}{h}\right) = \nu_{\mathrm{h}}\nabla_{\mathrm{h}}^2\bar{\zeta} + \frac{1}{h\rho_0}\left(\nabla_{\mathrm{h}} \times (\boldsymbol{\tau}^{\mathrm{t}} - \boldsymbol{\tau}^{\mathrm{b}})\right)_z. \tag{7.40}$$

The term $(f+\bar{\zeta})/h$ is the absolute vorticity $f+\bar{\zeta}$ normalized by the fluid height h, and is called the **absolute potential vorticity**. From (7.40), we find that the vertically averaged relative vorticity $\bar{\zeta}$ is governed by four factors: change in the Earth's planetary vorticity f induced by movement in the south-north direction, change in the height h of the fluid column, dispersion by the horizontal viscosity

$\nu_h \nabla_h^2 \zeta$, and torque introduced by stress at the top and bottom surfaces $\left(\nabla_h \times \boldsymbol{\tau}^t\right)_z - \left(\nabla_h \times \boldsymbol{\tau}^b\right)_z$. If the dispersion and torque are 0, the vorticity does not interact with any external factor, and the absolute potential vorticity is conserved.

Note 7.4 In an interior region where the viscosity can be neglected, the absolute potential vorticity would be almost conserved. In real oceans, the absolute potential vorticity within the internal mid layers remains essentially constant over a wide range (as discussed in the Prandtl-Batchelor theorem; see Note 4.2).

Exercise 7.4 Show that the direction of flow in the mid- to high-latitude oceanic region is almost parallel to the f/H isoline, if the absolute potential vorticity is conserved and the horizontal scale of the flow is large enough.

The equation (7.40) has been derived from the x- and y-components of the Navier-Stokes equation and the equation of continuity. Together with the remaining z-component of the Navier-Stokes equation, or the hydrostatic equilibrium equation (7.21), the absolute potential vorticity essentially enables the description of barotropic geophysical fluids.

7.2.3 Linearization of Equation

As demonstrated above, the effect of rotation is summarized in the Coriolis acceleration term. If this term is negligible, the system of equations becomes identical to that described in earlier chapters, and the fluid can be treated as a non-rotating fluid.

Whether the fluid should be treated as a rotating fluid depends on the magnitude ratios of the fluid acceleration and Coriolis acceleration terms on the left-hand side of the Navier-Stokes equation. The acceleration term in the Navier-Stokes equations (7.19) and (7.20) can be separated into the temporal acceleration term and the advection term. Assuming characteristic time scale T, horizontal special scale L, and horizontal velocity scale U, the ratio of the temporal variation term and the Coriolis acceleration term becomes $|\frac{U}{T}/(fU)| = 1/(|f|T)$, and the corresponding ratio of the advection term to the Coriolis term becomes $|U\frac{U}{L}/(fU)| = U/(|f|L)$. These nondimensional quantities, particularly the latter, are called Rossby numbers. If the Rossby number is large, the effects of rotation are relatively small. Conversely, if the Rossby number is small, that is, the time scale T exceeds $1/|f|$ and the flow velocity U is smaller than $|f|L$, or if the horizontal scale L exceeds $U/|f|$, rotation exerts a noticeable effect.

In particular, the advection term is diminished at small $U/(|f|L)$. Since the advection term is a product of the horizontal velocities, it introduces nonlinearity into the differential equations (7.19) and (7.20), which makes analytically solving the equations difficult. However, in geophysical fluids, the horizontal scale L is sufficiently large and the fluid velocity U is sufficiently small such that the advection term can be ignored, allowing a linearized equation with analytic solutions.

7.2.4 The f-plane and β-plane Approximations

The Coriolis parameter f introduced in Sect. 7.2.2 is a function of latitude ϕ, and therefore changes sinusoidally in the y-direction. Such curvatures are inconvenient to represent in the local Cartesian coordinate system, which is more suited to phenomena occurring on smaller scales than the spherical coordinate system. Therefore, we frequently assume a fixed $f_0 \equiv 2\Omega \sin \varphi_0$ at some latitude φ_0. This approximation is called the **f-plane approximation**.

When the sign of f reverses, as crossing the Equator, or when the south-north scale is fairly large, f cannot be regarded as fixed. In this case, f is frequently approximated by a linear function of β_0, denoting the rate of change in f with respect to y at latitude φ_0:

$$f(\varphi) \simeq f(\varphi_0) + \frac{1}{R} \left.\frac{df}{d\varphi}\right|_{\varphi_0} R(\varphi - \varphi_0) = f_0 + \beta_0(y - y_0)\,. \tag{7.41}$$

This approximation is called the **β-plane approximation**.

Exercise 7.5

1. Calculate f_0 and β_0 at the Equator and at the North Pole.
2. Determine the required range of the latitude φ_0 such that $f_0 > \beta_0 L$ holds over a horizontal scale L of approximately 1,000 km.

The β-plane approximation is more precise than the f-plane approximation. Therefore, all assumptions of the f-plane are locally applicable to the β-plane, but the converse is not true. For example, the absolute vorticity (7.38) expresses the rate of change in planetary vorticity as $\beta\bar{v}$; however, this assumption is not applicable in the f-plane approximation. The effects induced by changes in the latitude of f are generally known as the **β-effect**.

7.2.5 Special Solution Used in This Chapter

The next and following sections in this chapter are devoted to solving the Navier-Stokes equations (7.19) and (7.20) and the absolute vorticity potential equation (7.40) under special conditions such as steady flow fields. These particular solutions will be related to real phenomena of geophysical fluids.

We first consider phenomena in the f-plane approximation, and derive their steady-state and wave solutions. We then extend the problem to the β-plane approximation and again derive the steady-state and wave solutions (see Table 7.1). Note that descriptions with the coordinate systems other than the local Cartesian coordinates, such as the spherical coordinates, are out of our scope; however, they are essentially similar to the descriptions with the β-plane approximation.

Table 7.1 Particular solutions and corresponding phenomena treated in this chapter

	Steady solution	Wave solution
f-plane	Sect. 7.3 Geostrophic flow, Ekman's drift current	Sect. 7.4 Inertial gravity waves, Kelvin waves
β-plane	Sect. 7.6 Oceanic general circulation	Sect. 7.5 Rossby waves (temporal development of quasi-geostrophic component)

7.3 Steady Field in the f-plane Approximation

First, we consider a steady-state solution in the f-plane approximation. The flow is a generalized viscous flow, but for simplicity, we assume that the stress is applied only at the upper surface of the fluid, and that viscosity along the side and bottom boundaries can be ignored. This condition is equivalent to offshore oceanic regions with steady winds. The horizontal viscosity is weak in the interior region, which is far from the coastal boundaries. Because the ocean current is also weak, there is little friction on the seafloor, and the only significant stress is introduced by wind at the sea surface.

7.3.1 Geostrophic Flow

Following Sect. 4.4.2, in which we discussed flows along the boundary layer, we now obtain the solution (u_g, v_g) in the interior region where the viscosity (u_g, v_g) is zero.

Ignoring the viscosity terms in the Navier-Stokes equations (7.19) and (7.20) and assuming a small Rossby number to linearize the equations, we have, under the f-plane approximation

$$-f_0 v_g = -\frac{1}{\rho_0}\frac{\partial p}{\partial x}, \tag{7.42}$$

$$f_0 u_g = -\frac{1}{\rho_0}\frac{\partial p}{\partial y}. \tag{7.43}$$

This flow, in which the Coriolis force and pressure gradient are balanced, is called a **geostrophic current** or **geostrophic wind**.

Note 7.5 If the ocean surface is disturbed by high-velocity winds such as typhoons, the advection term cannot be ignored. For example, when modeling a vortex in plane polar coordinates, the radial advection term u_θ^2/r (namely, the centripetal force) must be accounted to balance with the radial pressure gradient $(1/\rho_0)\partial p/\partial r$ and the Coriolis force $f u_\theta$. In a high-pressure vortex, the Coriolis force and centripetal force act in

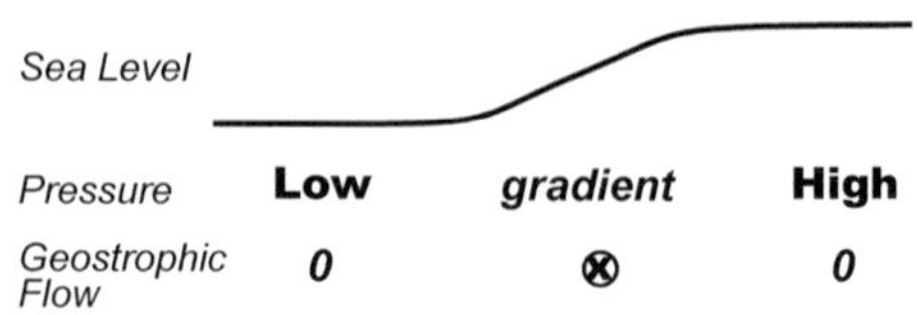

Fig. 7.5 Elevation of the fluid surface (pressure magnitude) and direction of geostrophic flow in the Northern Hemisphere

the same direction, whereas, in a low-pressure vortex, they oppose each other. In the latter scenario, the pressure gradient matches the Coriolis force at subgeostropic flow velocities (velocities below geostrophic flow). In the former scenario, the velocities can become supergeostropic (velocities above geostrophic flow). Such geostrophic wind with the effect of centripetal force included is called **gradient wind**.

Moreover, since the pressure p is in the hydrostatic equilibrium, (7.42) and (7.43) can be written as,

$$-f_0 v_g = -g\frac{\partial \eta}{\partial x}, \tag{7.44}$$

$$f_0 u_g = -g\frac{\partial \eta}{\partial y}, \tag{7.45}$$

where η denotes the elevation of the surface. These equations describe the presence of a steady elevation of the fluid surface η whose gradient induces a geostrophic flow. For example, in the Northern Hemisphere ($f_0 > 0$), if η is lower at the northern than at the southern latitudes, then $-g\partial\eta/\partial y > 0$, implying that $u_g > 0$ and produces an easterly flow. In this way, the direction of the geostrophic flow is orthogonal to the gradient of the fluid's surface height η depending on the sign of f_0. In the Northern (Southern) Hemisphere, the higher pressure lies to the right (left) of the geostrophic flow, depending on the sign of f_0 (the pressure gradient and geostrophic flow in the Northern Hemisphere are illustrated in Fig. 7.5).

Exercise 7.6 Consider a 1.0 m/s geostrophic flow of width 100 km at northern latitudes of 30°. What is the corresponding change in η (in m)?

In the absence of rotational effects, as seen in Chap. 3, the elevation of the fluid surface will propagate as a wave and will eventually smooth out. However, if the elevation is forcibly maintained for a "long time," it becomes surrounded by geostrophic flow. If this elevation is in **geostrophic balance**, the Coriolis force is balanced by the pressure gradient force and the elevation is maintained by the surrounding geostrophic flow in the absence of any net force. The "long time" required to reach geostrophic equilibrium is roughly estimated under the condition that the Coriolis acceleration term is larger than the temporal variation term, i.e., $T \gg 1/|f_0|$.

On the other hand, the spatial scale required for geostrophic balance is quantified by the **Rossby radius of deformation**, a_R:

$$a_R = \frac{\sqrt{gH}}{|f_0|}. \tag{7.46}$$

In (7.46), a_R is the distance propagated by the shallow water waves investigated in Sect. 3.2 traveling with phase velocity $\sqrt{gH}$ in a period $1/|f_0|$. If the elevation of the fluid surface is not in geostrophic balance and propagates as shallow water waves, it takes a period $1/|f_0|$ to travel over the radius of deformation a_R. Consequently, for any surface elevation with large spatial scales $L \gg a_R$, rotational effects precede the propagation of the shallow water waves, and thus the geostrophic balance is reached to maintain the surface elevation.

Note 7.6 In reality, the situation is more complicated. For example, if the surface elevation is initially provided without the velocity of the fluid, the above argument can still hold. Meanwhile, if the velocity distribution of the fluid is initially provided with no surface elevation, geostrophic balance is more easily reached over smaller spatial scales (we will obtain an inertial oscillation, rather than a geostrophic current, if a large-scale velocity is initially provided; see Sect. 7.4.2).

Now if $L \gg a_R$, the Rossby number becomes $U/(|f_0|L) \ll U/(|f_0|a_R) = U/\sqrt{gH}$ (in terms of the Rossby number). The right-hand side of this relation is the **Froude number**. We are now dealing with small U; therefore, the Froude number is smaller than 1 and the Rossby number becomes extremely small. Thus, the two assumptions of $L \gg a_R$ and small Rossby number both hold without contradiction.

7.3.2 *Taylor-Proudman Theorem*

Let us differentiate (7.42) with respect to the vertical component z. Using the hydrostatic pressure equation (7.21), we obtain

$$-f_0 \frac{\partial v_g}{\partial z} = -\frac{1}{\rho_0} \frac{\partial^2 p}{\partial x \partial z} = -\frac{\partial g}{\partial x} = 0 \,. \tag{7.47}$$

Similarly, from (7.43), we have

$$\frac{\partial u_g}{\partial z} = 0 \,. \tag{7.48}$$

Partially differentiating (7.42) and (7.43) with respect to y and x, respectively, we get

$$f_0 \left(\frac{\partial u_g}{\partial x} + \frac{\partial v_g}{\partial y} \right) = 0 \,. \tag{7.49}$$

From (7.49) and the equation of continuity (7.22), the vertical flow satisfies

$$\frac{\partial w_g}{\partial z} = 0 \,. \tag{7.50}$$

Collectively, (7.47), (7.48), and (7.50) state that none of the three flow components change in the vertical direction. This is called the **Taylor-Proudman theorem**.

Note 7.7 Suppose that we have a stratified fluid of varying density ρ. In this case, the z-derivative of (7.42) is

$$-f_0\frac{\partial v_g}{\partial z} = -\frac{\partial}{\partial z}\left(\frac{1}{\rho}\frac{\partial p}{\partial x}\right) = -\frac{1}{\rho}\frac{\partial^2 p}{\partial x \partial z} - \frac{\partial p}{\partial x}\frac{\partial}{\partial z}\left(\frac{1}{\rho}\right). \tag{7.51}$$

The second term on the right-hand side can be ignored in most cases. Equation (7.51) then becomes a Boussinesq approximation, as discussed in Sect. 6.3. Although variation in the density ρ is accounted for in the variation in the pressure p (first term on the right-hand side of (7.51)), $1/\rho$ itself (the second term on the right-hand side) can be regarded as a constant.

Under the hydrostatic equilibrium (7.21), the Eq. (7.51), which is equivalent to (7.47), becomes

$$-f_0\frac{\partial v_g}{\partial z} = \frac{g}{\rho}\frac{\partial \rho}{\partial x}. \tag{7.52}$$

Similarly, the equivalent of (7.48) is

$$f_0\frac{\partial u_g}{\partial z} = \frac{g}{\rho}\frac{\partial \rho}{\partial y}. \tag{7.53}$$

Equations (7.52) and (7.53) describe the relation between the horizontal density distribution and the vertical shear of the horizontal flow velocity, and are called the **thermal wind equations**.

If a geostrophic flow passes along an inclined bottom surface, a vertical flow $w_g\big|_{-H}$ is generated (see (7.26)). By the Taylor-Proudman theorem (7.50), the same vertical flow must be present at the fluid's upper surface η. Thus, from (7.27), we require that

$$w_g\big|_{-H} = w_g\big|_{\eta} = \left.\frac{D_h\eta}{Dt}\right|_{\eta} = \frac{\partial \eta}{\partial t} + u_g\frac{\partial \eta}{\partial x} + v_g\frac{\partial \eta}{\partial y}. \tag{7.54}$$

However, from the geostrophic flow equations (7.44) and (7.45), we have

$$u_g\frac{\partial \eta}{\partial x} + v_g\frac{\partial \eta}{\partial y} = -\frac{g}{f_0}\frac{\partial \eta}{\partial y}\frac{\partial \eta}{\partial x} + \frac{g}{f_0}\frac{\partial \eta}{\partial x}\frac{\partial \eta}{\partial y} = 0. \tag{7.55}$$

Moreover, the flow is steady-state, and hence $\partial\eta/\partial t = 0$; thus,

$$w_g\big|_{\eta} = w_g\big|_{-H} = 0. \tag{7.56}$$

Equation (7.56) states that the vertical flow is 0 everywhere at all times. Therefore, the geostrophic flow cannot penetrate the bottom inclined surface. Clearly, in this situation, the barotropic geostrophic flow moves along the H isocline in the f-plane. Equivalently, the potential vorticity is maintained on the f-plane (see Exercise 7.4).

7.3.3 Ekman's Drift Current

We now consider the flow inside a boundary layer (u_{E}, v_{E}). As the boundary layer is subject to vertical viscosity, a vertical shear should develop. This shear decreases u_{E} and v_{E} with depth, and u_{E} and v_{E} should exert no effect on the interior regions below a certain depth from the surface. When the viscosity is balanced by the Coriolis acceleration, u_{E} and v_{E} are called **Ekman's drift currents**, expressed by the following equations:

$$- f_0 v_{\mathrm{E}} = \nu_z \frac{\partial^2 u_{\mathrm{E}}}{\partial z^2}, \tag{7.57}$$

$$f_0 u_{\mathrm{E}} = \nu_z \frac{\partial^2 v_{\mathrm{E}}}{\partial z^2}. \tag{7.58}$$

To solve (7.57) and (7.58), we introduce a complex number $\boldsymbol{U} \equiv u_{\mathrm{E}} + i v_{\mathrm{E}}$ and calculate (7.57) – i(7.58) as follows:

$$f_0(u_{\mathrm{E}} + i v_{\mathrm{E}}) = \nu_z \frac{\partial^2}{\partial z^2}(v_{\mathrm{E}} - i u_{\mathrm{E}}) = -i\nu_z \frac{\partial^2}{\partial z^2}(u_{\mathrm{E}} + i v_{\mathrm{E}}) \tag{7.59}$$

which yields the simple differential equation

$$f_0 \boldsymbol{U} = -i\nu_z \frac{\partial^2 \boldsymbol{U}}{\partial z^2}. \tag{7.60}$$

Using the complex of (7.37), which specifies the stress $\boldsymbol{\tau}$ imposed at the fluid surface $z = \eta$, the boundary condition of (7.60) is given by

$$\frac{1}{\rho_0}\boldsymbol{T} \equiv \frac{1}{\rho_0}\left(\tau_x + i\tau_y\right) = \nu_z \left.\frac{\partial \boldsymbol{U}}{\partial z}\right|_{\eta}. \tag{7.61}$$

Suppose that solutions to (7.60) take the following form:

$$\boldsymbol{U} = \boldsymbol{A} \exp\{\boldsymbol{B}(z - \eta)\} \tag{7.62}$$

(where $\boldsymbol{A}$ and $\boldsymbol{B}$ are complex numbers), and the case in the Northern Hemisphere ($f_0 > 0$). Substituting $\boldsymbol{U}$, i.e., (7.62), into (7.60) gives

$$f_0 = -i\nu_z \boldsymbol{B}^2. \tag{7.63}$$

Using the relationship $\sqrt{i} = (1 + i)/\sqrt{2}$, $\boldsymbol{B}$ is obtained as

$$\boldsymbol{B} = \pm\sqrt{\frac{f_0}{2\nu_z}}(1 + i). \tag{7.64}$$

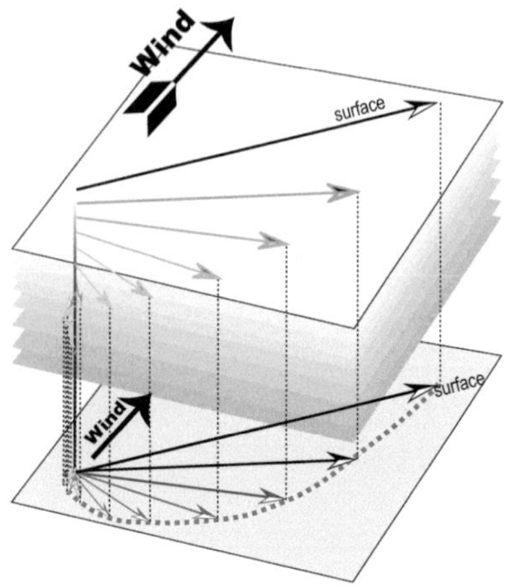

Fig. 7.6 Schematic of Ekman spiral driven by wind blowing over the sea surface

The boundary flow must satisfy $|U| \to 0$ as $z \to -\infty$; therefore, by selecting Re $\boldsymbol{B} > 0$, we have

$$\boldsymbol{B} = \sqrt{\frac{f_0}{2\nu_z}}(1+i) \equiv \frac{1+i}{D_{\mathrm{E}}}. \tag{7.65}$$

As both the real and imaginary parts of (7.65) are positive, the flow reduces its speed as the depth increases (as z decreases), and rotates clockwise. This distinct structure is called an **Ekman spiral** (see Fig. 7.6). This structure gives an impression as if the drift current were rotated by the Coriolis force while the drag stress at the surface proceeds downward by viscosity.

At a depth $D_{\mathrm{E}} = \sqrt{2\nu_z/f_0}$ from the fluid upper surface, the flow velocity reduces to $1/e$ of its magnitude at the upper surface. At this depth, the flow rotates clockwise by 1 rad ($\simeq 57.3°$). With respect to the surface, D_{E} is a measure of the thickness of the Ekman boundary layer and is called the Ekman depth. Furthermore, substituting (7.62) into the boundary condition (7.61) at $z = \eta$, we obtain

$$\frac{1}{\rho_0}\boldsymbol{T} = \nu_z \boldsymbol{A}\boldsymbol{B}. \tag{7.66}$$

Substituting (7.65) into this expression, we get

$$\boldsymbol{A} = \frac{\boldsymbol{T}}{\rho_0\nu_z}\sqrt{\frac{2\nu_z}{f_0}}\frac{1}{1+i} = \frac{1-i}{\rho_0\sqrt{2f_0\nu_z}}\boldsymbol{T} = \frac{1}{\rho_0\sqrt{f_0\nu_z}}\boldsymbol{T}\,e^{-\frac{\pi}{4}i}. \tag{7.67}$$

The final exponential in (7.67) clarifies that Ekman's drift current at the sea surface is shifted clockwise by 45° relative to the wind stress T (see Fig. 7.6).

Exercise 7.7 The above arguments were developed for the Northern Hemisphere. Find out which part of these arguments must be changed for the Southern Hemisphere, where $f_0 < 0$.

7.3.4 Vertical Mean Flow and Ekman Transport

From (7.42), (7.57), and (7.48), we find that

$$-f_0(v_g + v_E) = -\frac{1}{\rho_0}\frac{\partial p}{\partial x} + \nu_z \frac{\partial^2(u_g + u_E)}{\partial z^2}. \tag{7.68}$$

Similarly, from (7.43), (7.58), and (7.47), we obtain

$$f_0(u_g + u_E) = -\frac{1}{\rho_0}\frac{\partial p}{\partial y} + \nu_z \frac{\partial^2(v_g + v_E)}{\partial z^2}. \tag{7.69}$$

In these expressions, $u = u_g + u_E$ and $v = v_g + v_E$ form the general solution of the Navier-Stokes equation linearized under the f-plane approximation.

As the analysis of Sect. 7.2.2 was based on the vertical mean flow $\bar{u}$, $\bar{v}$, the vertical mean flow here is also

$$\bar{u} = \frac{1}{h}\left(\int_{-H}^{\eta} u_g\, dz + \int_{-H}^{\eta} u_E\, dz\right) = u_g + \frac{Q_x^E}{h} \tag{7.70}$$

$$\bar{v} = \frac{1}{h}\left(\int_{-H}^{\eta} v_g\, dz + \int_{-H}^{\eta} v_E\, dz\right) = v_g + \frac{Q_y^E}{h} \tag{7.71}$$

u_g, v_g are constant in the vertical direction and are thus unchanged by averaging. In this case, we require only the volume transport Q^E due to Ekman's drift current. Similar to the previous section, we introduce a complex number $\boldsymbol{Q}^E = Q_x^E + iQ_y^E$ to get,

$$\boldsymbol{Q}^E = \int_{-H}^{\eta} \boldsymbol{U}\, dz = \frac{\boldsymbol{A}}{\boldsymbol{B}}\left[\exp\{\boldsymbol{B}(z-\eta)\}\right]_{-H}^{\eta} \simeq \frac{\boldsymbol{A}}{\boldsymbol{B}} \tag{7.72}$$

where it is assumed that the Ekman depth D_E is significantly thinner than H, and $\boldsymbol{U}|_{-H} \simeq 0$ is assumed. Substituting previously obtained (7.65) and (7.67) into (7.72), we obtain

$$\boldsymbol{Q}^E = \frac{-i\boldsymbol{T}}{f_0\rho_0} = \frac{\boldsymbol{T}}{f_0\rho_0}e^{-\frac{\pi}{2}i} = \frac{1}{f_0\rho_0}(\tau_y - i\tau_x) = Q_x^E + iQ_y^E. \tag{7.73}$$

Clearly from (7.73), the volume transport Q^{E} is proportional to the magnitude of the applied stress $|\boldsymbol{T}|$; however, its direction is shifted clockwise (in the Northern Hemisphere) by 90°.

The **Ekman transport** is the volume transport Q^{E} multiplied by the density:

$$M_x^{\mathrm{E}} = \rho_0 Q_x^{\mathrm{E}} = \frac{\tau_y}{f_0}, \tag{7.74}$$

$$M_y^{\mathrm{E}} = \rho_0 Q_y^{\mathrm{E}} = -\frac{\tau_x}{f_0}. \tag{7.75}$$

The density-independent transport makes the comparison easier between fluids of different densities, such as the atmosphere and oceans. For example, in winds above the sea, the Ekman transports in the atmosphere and oceans are of equal magnitude but in opposite directions, and the same momentum is exchanged between the two fluids.

7.3.5 *Ekman Upwelling and Downwelling*

In the previous section, we derived the solutions for the vertical mean flow. Now we consider the equation of continuity in Sect. 7.2.2. Since $Q_x = hu_{\mathrm{g}} + Q_x^{\mathrm{E}}$ and $Q_y = hv_{\mathrm{g}} + Q_y^{\mathrm{E}}$ are defined, substituting them into the flow Eq. (7.30), we obtain

$$\frac{\partial h}{\partial t} + \frac{\partial Q_x^{\mathrm{E}}}{\partial x} + \frac{\partial Q_y^{\mathrm{E}}}{\partial y} + h\left(\frac{\partial u_{\mathrm{g}}}{\partial x} + \frac{\partial v_{\mathrm{g}}}{\partial y}\right) + u_{\mathrm{g}}\frac{\partial h}{\partial x} + v_{\mathrm{g}}\frac{\partial h}{\partial y} = 0. \tag{7.76}$$

Following the argument of Sect. 7.3.2, the last two terms of this expression are 0; therefore, we have

$$\frac{\partial h}{\partial t} + \frac{\partial Q_x^{\mathrm{E}}}{\partial x} + \frac{\partial Q_y^{\mathrm{E}}}{\partial y} + h\left(\frac{\partial u_{\mathrm{g}}}{\partial x} + \frac{\partial v_{\mathrm{g}}}{\partial y}\right) = 0. \tag{7.77}$$

Moreover, as seen in (7.49), the geostrophic flow does not converge or diverge in the f-plane approximation; therefore, we can write

$$-\frac{\partial h}{\partial t} = \frac{\partial Q_x^{\mathrm{E}}}{\partial x} + \frac{\partial Q_y^{\mathrm{E}}}{\partial y}. \tag{7.78}$$

Thus, the change in the fluid column height $h = \eta + H$ is generated by the convergence/divergence of Ekman volume transport. As the cause for convergence/divergence in Ekman volume transport, (7.73) implies that,

$$\frac{\partial Q_x^{\mathrm{E}}}{\partial x} + \frac{\partial Q_y^{\mathrm{E}}}{\partial y} = \frac{1}{f_0\rho_0}\left(\frac{\partial \tau_y}{\partial x} - \frac{\partial \tau_x}{\partial y}\right) \tag{7.79}$$

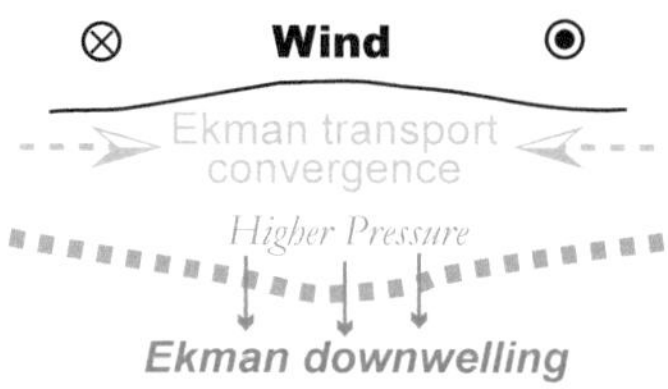

Fig. 7.7 Scenario in which torque develops in the stress imposed by wing above the sea. In this case, Ekman transport converges in the central region, generating Ekman downwelling

suggesting a torque in the stress $\boldsymbol{\tau}$. Furthermore, we can consider the other causes, such as interference of transport by a shore, or in equatorial regions where the direction of Ekman transport becomes symmetric with respect to the Equator due to reversing of the sign of f.[4]

In this section, we consider only the steady case, for which $\partial h/\partial t = 0$. Therefore, when convergence/divergence occurs in the Ekman transport, no steady-state solutions are allowed in the f-plane approximation.

As an example, consider that a negative torque develops in the stress, as shown in Fig. 7.7, and that the Ekman transport converges. As the Ekman transport is mainly constrained within the Ekman boundary layer, convergence implies that descending flow is generated in the lower parts of the Ekman boundary layer. This phenomenon, called **Ekman downwelling** or **sinking**,[5] can be derived from the equation of continuity. In the event of Ekman downwelling, a downward flow exists at the upper surface of the interior region (see Fig. 7.7); however, this solution fails because it violates the Taylor-Proudman theorem, which prohibits shrinkage of the interior flow column in a steady field. Moreover, in the hydrostatic pressure approximation, the pressure in the interior region just below the bottom of the Ekman layer can be considered to increase with time. These temporal variations will be reflected in the geostrophic flow u_g, v_g, again violating the assumption of a steady field in the f-plane. This argument will be discussed in more detail in Sects. 7.4.4 and 7.6.1.

If the direction of the wind is aligned along the shore, the direction of the Ekman transport is orthogonal to the shore. In this case, the Ekman transport is suppressed by the shore, causing convergence/divergence and consequent Ekman downwelling/upwelling. This phenomenon is called **coastal downwelling/upwelling**, and is accompanied by a change in the pressure distribution in the offshore direction, inducing a "geostrophic" flow called a **coastal jet stream** (see Sect. 7.4.4).

Note 7.8 During upwelling along the coast, cold water of the lower layer rich in nutrients is supplied to the layer near the sea surface where the sunlight penetrates (the **euphotic layer**), which becomes rich in planktons and ideal for fishing. For example, if a south-easterly trade wind blows in the Peruvian ocean in the Southern Hemisphere, the Ekman transport occurs in the southwestward direction, inducing

[4]However, equatorial regions are difficult to treat in the f-plane approximation. The reversing the sign of f at the Equator, or non-constant f, is already an issue in the β-plane approximation.

[5]Conversely, the upward flow generated during divergence of the Ekman transport is called Ekman upwelling or pumping.

upwelling along the coast. The water temperature will decrease and the region will become an excellent fishing zone. If the trade wind is weakened, for example, during an **El Niño** event, the upwelling will be weakened and thus the fishing catch will decline.

Moreover, if easterly winds blow at the Equator, the Ekman transport in the Northern (Southern) Hemisphere is directed toward the north (south), and hence diverges, causing an Ekman upwelling. Such an event is called equatorial upwelling.

7.4 Time Variation Field in the f-plane Approximation

We now consider the temporal variations in the f-plane approximation. We would account for the time variation term in the fundamental equations, but instead ignore the viscosity for simplicity. Here the discussion is restricted to the dynamics of barotropic flow in the interior region, and we assume that the fluid bottom surface $z = -H$ is constant for all times. In this scenario, there is no reason for the horizontal flow to change in the vertical direction; hence, we set $\bar{u} = u$ and $bar v = v$.

7.4.1 *Inertial Gravity Waves*

Under the above conditions, the linear equations of motion and continuity for the barotropic flow are given by

$$\frac{\partial u}{\partial t} - f_0 v = -g\frac{\partial \eta}{\partial x}, \tag{7.80}$$

$$\frac{\partial v}{\partial t} + f_0 u = -g\frac{\partial \eta}{\partial y}, \tag{7.81}$$

$$\frac{\partial u}{\partial x} + \frac{\partial v}{\partial y} = -\frac{1}{H}\frac{\partial \eta}{\partial t}, \tag{7.82}$$

where η is assumed to be sufficiently smaller than H, that is, $h = H + \eta$ is approximated by H. The variation in η is included on the right-hand side of (7.81) and (7.80) as the pressure gradient, but is ignored in our treatment of $1/h$, or the height of the fluid column itself. This approximation of h is similar to the Boussinesq approximation of the density ρ, and is called rigid lid approximation.

The wave motion solutions to (7.80)–(7.82) are given by

$$u = \mathrm{Re}\left[\boldsymbol{U} \exp i(kx + ly - \omega t)\right], \tag{7.83}$$

$$v = \mathrm{Re}\left[\boldsymbol{V} \exp i(kx + ly - \omega t)\right], \tag{7.84}$$

$$\eta = \mathrm{Re}\left[\boldsymbol{Z} \exp i(kx + ly - \omega t)\right], \tag{7.85}$$

where $(k,\ l)$ are the wavenumber vectors and ω is the frequency. We now determine the dispersion relations among k, l, and ω. In (7.83)–(7.85), $\boldsymbol{U}$, $\boldsymbol{V}$, $\boldsymbol{Z}$ are all complex numbers and a phase shift may occur among u, v, and η.

The dispersion relation of wave motion solution (for $\omega \neq 0$) is

$$\omega^2 = f_0^2 + gH(k^2 + l^2)\,. \tag{7.86}$$

(The derivation of (7.86) is left to the readers as Exercise 7.8.) The form of (7.86) is identical to the shallow water wave dispersion relation $\omega^2 = gH(k^2 + l^2)$, but contains an additional term to account for the Earth's rotation. Such waves are called **inertial gravity waves**, Poincaré waves, or Sverdrup waves. At short wavelengths, $\omega^2 \simeq gH(k^2+l^2)$ because $gH(k^2+l^2) \gg f_0^2$, and the wave behaves as an "ordinary" shallow water wave. As the wavelength increases, $k^2 + l^2$ decreases, and the effect of rotation becomes relatively important. To maintain the condition $gHk^2 \leq f_0^2$, the wavelength must exceed $2\pi\sqrt{gH}/|f_0| = 2\pi a_\mathrm{R}$. Therefore, we can apply the argument, developed in Sect. 7.3.1, that rotational effects become significant once the spatial scale exceeds the radius of deformation a_R.

Exercise 7.8 Derive the dispersion relation (7.86) for inertial gravitational waves from the fundamental equations (7.80)–(7.82).

7.4.2 *Inertial Oscillations*

As an extreme example of the long-wavelength inertial gravity waves, we consider infinite wavelength; that is, $k = l = 0$. From the dispersion relation (7.86), one can see that $\omega = |f_0|$ in this case. Moreover, given that $k = l = 0$, all of u, v, and η are spatially homogeneous; hence, the right-hand sides of (7.80) and (7.81), as well as the left-hand side of (7.82), become 0. Therefore, the right-hand side of (7.82) is also 0, and η remains fixed in time and space. On the other hand, (7.80) and (7.81) are reduced to

$$\frac{\partial u}{\partial t} = f_0 v\,, \tag{7.87}$$

$$\frac{\partial v}{\partial t} = \quad f_0 u\,, \tag{7.88}$$

whose solutions are sinusoidal functions with the frequency $|f_0|$. This phenomenon, in which only the flow velocity fluctuates at a frequency $|f_0|$ with no change in the height of the fluid upper surface η, is called **inertial oscillation**. For example, if an ocean is disturbed by wind and develops an initial flow that is spatially homogeneous, the flow velocity continues inertial oscillations without reaching geostrophic balance.

Inertial oscillation alters only the flow direction; therefore, a material advected by a flow of velocity U traces a circular orbit with radius $U/|f_0|$ (see Fig. 7.8). The flow induced by the inertial oscillation is spatially homogeneous and has no relative

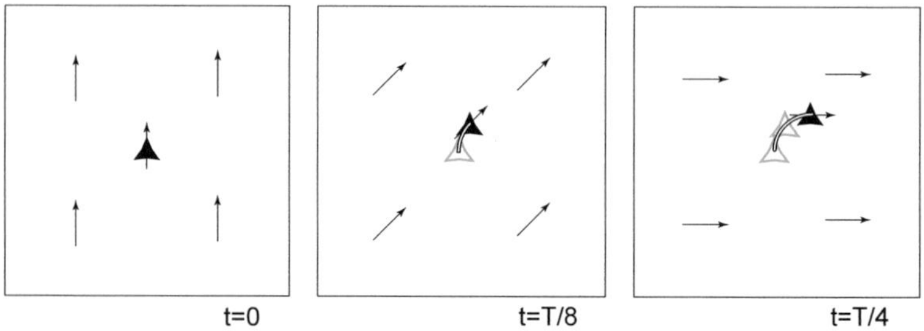

Fig. 7.8 Schematic of material advection and flow distribution throughout the first 1/4 period of an inertial oscillation

vorticity. Hence, as shown in Fig. 7.8, the material floating with a circular path does not rotate around itself, heading in the same direction.

Exercise 7.9 Determine the inertial oscillation and the trajectory of the advected material when $u = U_0$ and $v = \eta = 0$.

7.4.3 Kelvin Waves

Inertial gravity waves are ordinary shallow water waves that are subjected to rotation. In this section, as a special case, we consider a situation where Coriolis force in one direction no longer works. For example, if orthogonal flows are prevented by the presence of a seashore in the x-direction ($v = 0$), the Coriolis acceleration in the x-direction (along the shore) vanishes and (7.80) is reduced to a simple shallow water wave equation. Moreover, the y-component governed by (7.81) has no acceleration term in v, and thus becomes a geostrophic flow equation. Under these conditions, the wave equations simplify to

$$\frac{\partial u}{\partial t} = -g\frac{\partial \eta}{\partial x}, \tag{7.89}$$

$$f_0 u = -g\frac{\partial \eta}{\partial y}, \tag{7.90}$$

$$\frac{\partial u}{\partial x} = -\frac{1}{H}\frac{\partial \eta}{\partial t}. \tag{7.91}$$

Solving these equations similar to the previous section, we can write

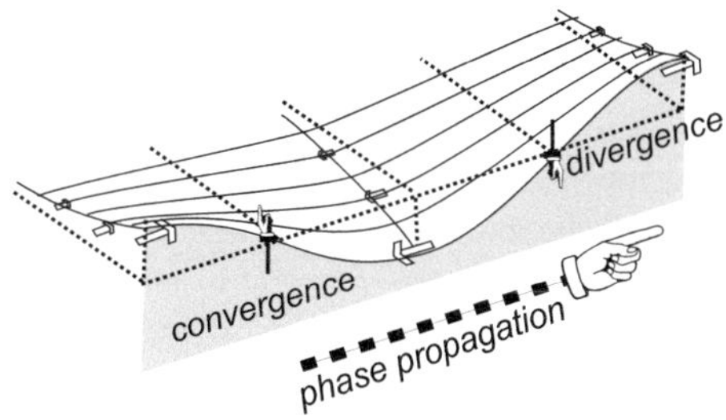

Fig. 7.9 Kelvin waves progressing along seashores in the Northern Hemisphere. Peaks (troughs) will propagate into the convergence (divergence) regions where the sea surface rises (falls)

$$\eta = \eta_0 \exp\left(-|f_0 \frac{k}{\omega} y|\right) \exp i\,(kx - \omega t)\ , \tag{7.92}$$

$$u = g\frac{k}{\omega}\eta_0 \exp\left(-|f_0 \frac{k}{\omega} y|\right) \exp i\,(kx - \omega t)\ , \tag{7.93}$$

$$\omega^2 = gHk^2\,. \tag{7.94}$$

The waves described by (7.92)–(7.94) are called **Kelvin waves**, and are commonly observed in **tidal waves** around the shore.

As implied by (7.89), the structures of Kelvin waves and ordinary shallow water waves are identical along the shore direction. However, unlike ordinary shallow water waves, the amplitude of Kelvin waves exponentially decreases in the offshore direction, as shown in Fig. 7.9. This reduction generates seaward gradients of η. In ordinary shallow water waves, seaward velocity v is accelerated to diminish the gradients. However, under the conditions discussed here, this movement is prohibited by the existence of the shore. Instead, the Coriolis force imposed by the flow u along the shore is balanced with the seaward pressure gradient in Kelvin waves.

As shown in Fig. 7.9, the maximum speed of the geostrophic flow along the shore u is found at the largest seaward slope of η, namely, at the peaks and troughs. Meanwhile, at the nodes where no geostrophic flows are present, either divergences or convergences occur, depending on the signs of f_0 and the seaward gradient of η. Governed by (7.91), the sea surface height η rises (or falls) in regions of convergence (or divergence); therefore, the direction of propagation of those peaks and troughs is fixed. For example, in the Northern (Southern) Hemisphere, Kelvin waves propagate such that the shore lies to the right (left).

As discussed for Ekman upwelling in Sect. 7.3.5, flows tends to be prohibited from crossing the Equator where the sign of f_0 switches; in this sense, the Equator behaves as shores. Kelvin waves (here known as **equatorial Kelvin waves**) propagate along the Equator only to the east. Specifically, Kelvin waves move along the Equator shores in the Southern (Northern) Hemisphere and the Equator in a clockwise (anticlockwise) direction.

The coefficient of the exponential function y, $|f_0 k/\omega|$, is given by $|f_0|/\sqrt{gH} = 1/a_R$, where we have used the dispersion relation (7.94). Therefore, the amplitude of Kelvin waves decreases to $1/e$ at a location that is away from the Equator or shores by a_R (radius of deformation).

Note 7.9 In the discussion on Kelvin waves, the direction along the shore was taken as the x-axis. In Sect. 7.2.1, the x-axis of the local Cartesian coordinate system was considered as the eastern direction. In the f-plane approximation, the axis of the coordinate system is freely rotated. Meanwhile, the β-plane approximation considers the directional change in the latitude of f. Therefore, in the β-plane approximation, the meridional and zonal directions are inevitably different, but in the f-plane approximation, where f is assumed to be fixed, we only require that the Coriolis force is orthogonal to the direction of propagation. In this scenario, the latitude and longitude need not be clearly distinguished.

7.4.4 *Quasi-geostrophic Component*

From the dispersion relation (7.86) of inertial gravity waves, we observe that $\omega \geq |f_0|$. In other words, the period of the wave solution is always shorter than $2\pi/|f_0|$. Conversely, fluctuations with longer periods do not form wave solutions of the form (7.83)–(7.85), and instead approximately develop geostrophic flows (solutions of (7.44) or (7.45)) in which the temporal variation term has been neglected.

In practice, the flow does change in time, although the changes are small. Therefore, rather than becoming a completely steady geostrophic flow, there exists a minor discrepancy (called the **ageostrophic component**) that accounts for small temporal changes. This slowly changing geostrophic current in an interior region is called a **quasi-geostrophic current**.

In Sect. 7.3.5, we discussed the temporal changes in a fluid column of height h corresponding to the convergence/divergence of a steady Ekman transport. If the period of these temporal changes exceeds $2\pi/|f_0|$, the interior flow establishes a quasi-geostrophic flow that is in approximate geostrophic balance at each moment in time, as well as evolves over a long time. In the example of Fig. 7.7, where the Ekman transport converges, the pressure applied within the interior region increases over time. This small temporal change can be accounted for in a quasi-geostrophic flow, maintaining the approximate geostrophic balance between the pressure gradient and Coriolis force at every moment.

The temporal variations term in the Navier-Stokes equation is extremely small compared with the Coriolis acceleration and pressure gradient. Thus, the time development of the quasi-geostrophic component is not easily determined, and we seek innovative alternatives such as removing the geostrophic balance component from the Navier-Stokes equation. For this purpose, we adopt a vorticity equation whose pressure gradient term can be eliminated because of cross derivatives. This vorticity equation is in fact more useful than the Navier-Stokes equation itself in the treatment of geophysical fluids that frequently become quasi-geostrophic.

In the absence of stress, the vorticity equation sought from (7.80)–(7.82) is given by

$$\frac{\partial}{\partial t}\left(\zeta - \frac{f_0}{H}\eta\right) = 0\,. \tag{7.95}$$

Approximating u and v as quasi-geostrophic flows that momentarily establish geostrophic balance despite slowly varying over a long time (the quasi-geostrophic approximation), we have $\zeta \simeq g/f_0 \nabla_{\mathrm{h}}^2 \eta$. Therefore, (7.95) can be rewritten as

$$\left(\nabla_{\mathrm{h}}^2 - \frac{1}{a_{\mathrm{R}}^2} \right) \frac{\partial \eta}{\partial t} \simeq 0 . \tag{7.96}$$

From (7.96), we understand that the spatial scale L of the time variation in η would be of the same order of the radius of deformation a_{R}. However, the temporal development of the quasi-geostrophic component itself cannot be determined from (7.96). In fact, the change in vorticity requires the more sophisticated β-plane approximation, which accounts for the effect of planetary vorticity. The β-plane approximation to the quasi-geostrophic component will be discussed in Sect. 7.5.1.

7.5 Time Variation Field in β-plane Approximation

Let us extend our discussion of the time-dependent field to the β-plane. Consistent with Sect. 7.4, we ignore the viscosity term and assume that $H \gg |\eta|$. To simplify the problem, we further assume that the bottom fluid surface ($z = -H$) is flat and fixed in both time and space.

7.5.1 *Quasi-geostrophic Component in the β-plane*

If the vorticity equation (7.95) in the f-plane approximation shown in Sect. 7.4.4 is extended to the β-plane approximation, we get

$$\frac{\partial \zeta}{\partial t} - \frac{f_0}{H} \frac{\partial \eta}{\partial t} + \beta_0 v = 0 . \tag{7.97}$$

Introducing the quasi-geostrophic component approximation such that $v = (g/f_0)\, \partial \eta / \partial x$ and $\zeta = (g/f_0) \nabla_{\mathrm{h}}^2 \eta$ are satisfied at each moment, we can express (7.97) as a function of η alone:

$$\frac{\partial}{\partial t} \nabla_{\mathrm{h}}^2 \eta - \frac{f_0^2}{gH} \frac{\partial \eta}{\partial t} + \beta_0 \frac{\partial \eta}{\partial x} = 0 . \tag{7.98}$$

Note 7.10 If the generalized vorticity equation (7.38) in Sect. 7.2.2 is linearized and the right-hand side is neglected, we find that the second term $\zeta + f$ is replaced with f_0 in (7.97). Actually, the linearization assumes that the Rossby number $U/(|f_0|L) \ll 1$. Thus, $|\zeta| \lll |f_0|$ is derived naturally and $f_0 + \zeta \simeq f_0$ is satisfied.

Note here that the third term on the left-hand side of (7.97) and (7.98) contains no time derivatives. Because of these terms, we can describe the temporal development of the quasi-geostrophic flow.

Let us estimate the ratio of the first and second terms (time variation terms) to the third term. The ratio of the first to the third term is $1/(\beta_0 LT)$, whereas that of the second to the third term is $(L/a_{\mathrm{R}})^2 \times 1/(\beta_0 LT)$. The dimensionless quantity $1/(\beta_0 LT)$ corresponds to the Rossby number $1/(|f_0|T)$, which has been frequently used in our previous discussions on quasi-geostrophic flows. Here $|f_0|$ is replaced by $\beta_0 L$. Similarly, we can specify a Rossby number for the spatial scale, $U/(|f_0|L)$, as $U/(\beta_0 L^2)$. These quantities (particularly the latter) are called **local Rossby numbers**.

At middle to high latitudes where $|f_0|$ is of a certain magnitude, $\beta_0 L < f_0$ holds even over a large horizontal scale L (see Exercise 7.5). Furthermore, if the situation is limited to the case where $L > a_{\mathrm{R}}$, when geostrophic balance is established, we have $1/|f_0| < 1/(\beta_0 L) < (L/a_{\mathrm{R}})^2 \times 1/(\beta_0 L)$. Therefore, even if the time variation scale T of the quasi-geostrophic component exceeds $1/|f_0|$, the ratios of the first and second terms to the third term in (7.98) are $\simeq 1$ as long as the time scale remains shorter than $(L/a_{\mathrm{R}})^2 \times 1/(\beta_0 L)$. Thus, even for the temporal development of the quasi-geostrophic component that could not be determined in the f-plane approximation, the component can likely be denoted as a balance between the second (or first) and the third terms of (7.98) under the β-plane approximation.

7.5.2 Rossby Waves

If (7.85) is given as a wave solution to (7.98), we obtain the following dispersion relation:

$$\left(k^2 + l^2 + \frac{1}{a_{\mathrm{R}}^2}\right) = -\beta_0 \frac{k}{\omega}. \tag{7.99}$$

The waves satisfying this dispersion relation are called **Rossby waves** or **planetary waves**.

The most unique feature of these waves is the propagation direction of their phase velocity c. As the left-hand side of (7.99) is positive, the right-hand side must also be positive. On the other hand, $\beta_0 \geq 0$ in both the Northern and Southern Hemispheres; therefore, $k/\omega = 1/c$ must be negative. Furthermore, all phases of Rossby waves propagate to the west.

The group velocity in the x-direction is given by

$$\frac{\partial \omega}{\partial k} = \frac{-\beta_0\left(-k^2 + l^2 + 1/a_{\mathrm{R}}^2\right)}{\left(k^2 + l^2 + 1/a_{\mathrm{R}}^2\right)^2}. \tag{7.100}$$

This quantity is negative when k is small and positive when k is large. For this reason, waves of long wavelength (small k) are generated on the west side of the disturbance source, and energy is propagated westward. Conversely, waves of short wavelength (large k) are generated on the east side, and energy propagates eastward. As the phase always propagates westward, short Rossby waves generated on the east side of the disturbance source appear to propagate toward the source of disturbance for a long distance.

Note 7.11 This type of phenomenon can be experimentally reproduced in a rotating water tank. However, as the rotating angular velocity in a water tank is spatially invariant, it does not generally incorporate the β-effect. The β-effect can be simply induced by tilting the base surface. A sloped base causes the fluid depth H to vary. The resulting change in the absolute potential vorticity imitates the change in planetary vorticity β because both the effects satisfy the potential vorticity preservation equation (7.40). The β-effect generated by the bottom slope is called the **topographic β-effect**.

7.5.3 Long Rossby Waves

If the wavelengths become extremely long, $k^2 = l^2 \simeq 0$, (7.99) is simplified to

$$c = \frac{\omega}{k} = -\beta_0 a_R^2 \,. \tag{7.101}$$

The waves satisfying this condition are called **long Rossby waves**, and are characterized by extremely large L/a_R. Under these conditions, the first term in the original vorticity equation (7.98) is negligible and the changes in potential vorticity due to planetary vorticity (third term) and the stretching/shrinking of the fluid column (second term) balance. The vorticity equation is

$$\frac{f_0^2}{gH}\frac{\partial \eta}{\partial t} = \beta_0 \frac{\partial \eta}{\partial x} = \beta_0 \frac{f_0}{g} v \,. \tag{7.102}$$

According to (7.102), if the sea surface height is tilted in the east-west direction, the fluid column temporally stretches/shrinks to compensate for $\beta_0 v$ induced by planetary vorticity accompanying the south-north flow v. As illustrated in Fig. 7.10, in the Northern Hemisphere, a southward (northward) flow develops on the west (east) side of a trough where η is low. This flow is accompanied by a decrease (increase) in planetary vorticity on the west (east) side and consequent shrinking (stretching) of the fluid column on the west (east) side. These phenomena propagate the peaks and troughs of η toward the west.

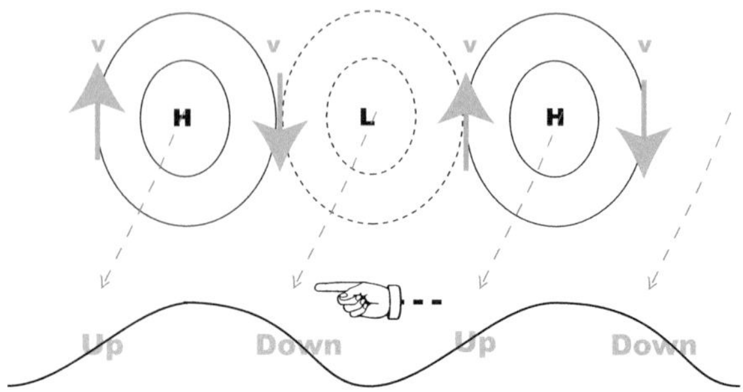

Fig. 7.10 Map view of long Rossby waves developed in the Northern Hemisphere, and horizontal view from the south. Sea surface height η decreases in the southward flow regions and rises in the northward flow regions, such that both peaks and troughs propagate westward

7.6 Steady Field in the β-plane Approximation

We now consider a steady field in the β-plane approximation. Similar to the f-plane case developed in Sect. 7.3, suppose that a steady stress $\boldsymbol{\tau}$ is applied at the upper surface of a fluid with vertical viscosity. For simplicity, we assume that depth H is constant and significantly greater than $|\eta|$.

7.6.1 Steady Vorticity Equation in the β-plane

Assuming a linear and steady-state flow condition, the vorticity equation (7.38) for a vertical mean flow is simplified to

$$\beta_0 \bar{v} = \frac{1}{\rho_0 H} \left(\nabla_\mathrm{h} \times \boldsymbol{\tau}\right)_z . \tag{7.103}$$

Under these conditions, the torque introduced by the stress $\boldsymbol{\tau}$ exactly balances the planetary vorticity induced by the vertical mean flow.

As demonstrated in Sect. 7.3.5, the torque introduced by the stress $\boldsymbol{\tau}$ equates to the convergence/divergence of the Ekman flow transport, which must equal zero to permit steady solutions in the f-plane approximation. In this section, we extend this argument to the β-plane approximation. The geostrophic flow u_g, v_g tends to follow the contour of η, in either f- or β-plane approximations; thus, in a steady state, we have

$$\frac{\partial Q_x^\mathrm{E}}{\partial x} + \frac{\partial Q_y^\mathrm{E}}{\partial y} + h\left(\frac{\partial u_\mathrm{g}}{\partial x} + \frac{\partial v_\mathrm{g}}{\partial y}\right) = 0 . \tag{7.104}$$

Equation (7.104) states that divergence/convergence also forms within interior geostrophic flows as a result of Ekman downwelling/upwelling accompanying the convergence/divergence of the Ekman transport.

In the f-plane approximation, the Taylor-Proudman theorem (7.49) only permitted the convergence/divergence of geostrophic flow for 0. However, in the β-plane approximation, if (7.42) and (7.43) are differentiated with respect to y and x, respectively, in the β-plane approximation, we get

$$f_0\left(\frac{\partial u_g}{\partial x}+\frac{\partial v_g}{\partial y}\right)+\beta_0 v_g=0\,, \qquad (7.105)$$

which indicates that the convergence/divergence of geostrophic flow (and the accompanying stretching/shrinking of the fluid column) is canceled by the planetary vorticity $\beta_0 v_g$. Incidentally, this relationship can also be derived by conserving the absolute potential vorticity in the interior, non-viscous region, namely, f/h (see Exercise 7.4).

Now, if the Ekman volume transport (7.73) is multiplied by f and the convergence/divergence is calculated under the β-plane approximation, we have

$$\begin{aligned}\left.\frac{\partial(fQ_x^{\mathrm{E}})}{\partial x}\right|_{y_0}+\left.\frac{\partial(fQ_y^{\mathrm{E}})}{\partial y}\right|_{y_0} &= f_0\left(\frac{\partial Q_x^{\mathrm{E}}}{\partial x}+\frac{\partial Q_y^{\mathrm{E}}}{\partial y}\right)+\beta_0 Q_y^{\mathrm{E}}\\ &= \frac{1}{\rho_0}(\nabla_{\mathrm{h}}\times\boldsymbol{\tau})_z \end{aligned} \qquad (7.106)$$

and (7.104) becomes

$$-\frac{\beta_0}{f_0}Q_y^{\mathrm{E}}+\frac{1}{f_0\rho_0}(\nabla_{\mathrm{h}}\times\boldsymbol{\tau})_z-h\frac{\beta_0}{f_0}v_g=0\,.$$

Rearranging gives

$$\beta_0\left(v_g+\frac{Q_y^{\mathrm{E}}}{h}\right)=\frac{1}{\rho_0 h}(\nabla_{\mathrm{h}}\times\boldsymbol{\tau})_z\,. \qquad (7.107)$$

Thus, the vertical average flow in (7.103), or the Ekman current in addition to the geostrophic flow, can be derived with the approximation $h=H+\eta\simeq H$.

This analysis differs from that of the f-plane approximation. In particular, a steady-state solution can be obtained in the β-plane approximation even if convergence/divergence is present in the Ekman transport. In this steady-state solution, the stretching/shrinking of the interior fluid column accompanying the Ekman upwelling/downwelling is canceled by the change in planetary vorticity induced by the south-north flow.

7.6.2 Western Intensification

To investigate the difference in interior flow in the f-plane and β-plane approximations, we consider a rectangular ocean with a steady wind providing negative torque blown onto the sea surface. The situation is illustrated in Fig. 7.11. For simplicity, we also assume that this rectangular sea is in the Northern Hemisphere.

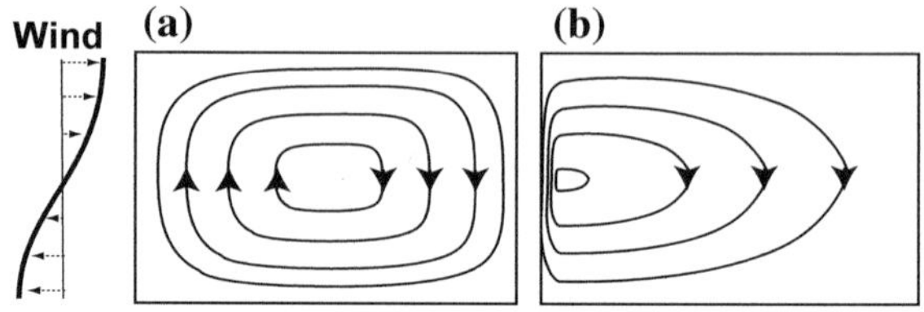

Fig. 7.11 Western intensification for a wind providing negative vorticity blown onto a rectangular sea in the Northern Hemisphere. **a** f-plane approximation and **b** β-plane approximation

As seen in Fig. 7.7, the negative torque in the wind stress causes the Ekman transport to converge in the central region. Consequently, the Ekman layer tends to stretch. However, if convergence stretches the Ekman layer according to (7.104), the water column shrinks in the interior region. Since the absolute vorticity potential is maintained in the interior region, the absolute vorticity $f + \zeta$ decreases (recall that our hypothetical sea is situated in the Northern Hemisphere). Additionally, the wind is steady, and therefore $f + \zeta$ decreases at a constant (time-independent) rate.

The f-Surface Approximation

In this approximation, the Rossby number $U/(|f_0|L)$ is assumed to be small; therefore, $|f_0| \gg |\zeta|$ is true. On the other hand, the Coriolis parameter f remains constant on the f-plane. Thus, the change in absolute vorticity must be solely contributed by fluctuations in the relative vorticity ζ. In this case, ζ decreases with time, and a clockwise circulation is created (see Fig. 7.11a).

This result is consistent with the interpretation that a clockwise circulation forms when pressure in the interior region at the center of a rectangular sea increases by convergence of the Ekman layer. However, this circulation will not become steady. The flow velocity will increase over time and will be strengthened to the extent that friction at the bottom or at the horizontal boundary reaches non-negligible levels.

The β-plane Approximation

In the β-plane, the change in absolute vorticity can be treated as a change in the planetary vorticity f, which is the dominant term. As the rate of change of f is $\beta_0 v_g$, the absolute vorticity decreases at a fixed rate over time and manifests as a steady south-downward flow with $v_g < 0$. However, the arguments for the f-plane approximation should be valid in the local β-plane; therefore, the "clockwise circulation" demanded by the pressure distribution and "southward flow" must be simultaneously satisfied. This situation requires that the circulation be centralized on the extreme west side, as shown in Fig. 7.11b.

Here southward transport of the whole sea is returned to the north at the west coast. However, a northward flow is prohibited by (7.103). At the western boundary, we need to introduce an exceptional flow which does not satisfy previous assumptions. Such flows, called **western boundary currents**, are represented by the **Kuroshio** along the south coast of Japan and the **Gulf Stream** in the Gulf of Mexico and along the east coast of the US. The western boundary currents are strong flows, and thus have non-negligible vorticity advection, bottom friction, horizontal viscosity, relative vorticity, and other parameters that have been hitherto ignored. Balance of these parameters permits the northward flow only in the western boundary. The existence of these strong western boundary currents is called westward intensification. Although the case of negative torque in the Northern Hemisphere was considered in the above discussion, **western intensification** will always be required regardless of the sign of torque from wind stress. In both the Southern and Northern Hemispheres, western intensification will always be required.

Exercise 7.10 How is the circulation different in the Southern Hemisphere? Consider a wind introducing negative torque to an oceanic system, as in Fig. 7.11, and determine whether westward intensification will occur.

Note 7.12 From the torque balance viewpoint of the entire circulation in the β-plane approximation, the torque supplied by the wind stress diverges through friction in the western boundary region, and the system becomes steady. This situation is similar to the f-plane approximation, in which friction is required to achieve a steady-state solution.

7.6.3 Sverdrup Transport

The southward transport within the interior region, where (7.103) holds, can be written as

$$M^{\mathrm{Sv}} = \rho_0 H \bar{v} = \frac{1}{\beta_0} (\nabla_{\mathrm{h}} \times \boldsymbol{\tau})_z \,. \tag{7.108}$$

Transport returned to the north by the western boundary current is equal to the sum of these southward transports in the eastern interior region. This transport is called **Sverdrup transport**. Similar to Ekman transport given by (7.74) and (7.75), Sverdrup transport is independent of the viscosity coefficient ν_z and is determined solely from the torque of the wind-induced stress.

The difference between the Ekman and Sverdrup transports is that the former is proportional to the wind stress $\boldsymbol{\tau}$ itself, whereas the latter is proportional to its torque $(\nabla_{\mathrm{h}} \times \boldsymbol{\tau})_z$. This difference arises because the Sverdrup transport indicates the strength of the circulation generated by the convergence/divergence of the Ekman transport.

7.7 Rotating Stratified Fluid and General Circulation of thc Oceans

Throughout this chapter, we have focused on the rotational effects on constant-density fluids. Here we extend our discussion to stratified rotating fluids, drawing on the knowledge of stratified fluids acquired in Chap. 6. We restrict our analysis to a stable two-layered fluid in which a bottom layer with higher density is overlaid with a fluid of lower density. A general treatment of rotating stratified fluids is complicated and beyond the scope of this book.

7.7.1 Extension to a Two-Layered Fluid

Consider the two-layered fluid shown in Fig. 7.12. To focus on stratification effects, we assume non-viscous motion in the f-plane approximation. The equations of motion and continuity in the top layer are given by

$$\frac{\partial u_1}{\partial t} - f_0 v_1 = -\frac{1}{\rho_1}\frac{\partial p_1}{\partial x} = -g\frac{\partial \eta}{\partial x}, \tag{7.109}$$

$$\frac{\partial v_1}{\partial t} + f_0 u_1 = -\frac{1}{\rho_1}\frac{\partial p_1}{\partial y} = -g\frac{\partial \eta}{\partial y}, \tag{7.110}$$

$$\frac{\partial u_1}{\partial x} + \frac{\partial v_1}{\partial y} = -\frac{1}{D_1}\frac{\partial}{\partial t}(\eta - \xi). \tag{7.111}$$

These equations are almost identical to (7.80)–(7.82). The only difference is that, on the right-hand side of (7.111), the fluctuation in the thickness of the upper layer $h_1 = D_1 + \eta - \xi$ ($\simeq D_1$) is contributed by changes in both upper elevation η and bottom displacement ξ.

Now, the pressure p_2 on the lower layer depends on the thickness of the upper layer as

$$p_2 = \rho_1 g(D_1 + \eta - \xi) - \rho_2 g(z + D_1 - \xi), \tag{7.112}$$

therefore, the equations of continuity and motion of the bottom layer become

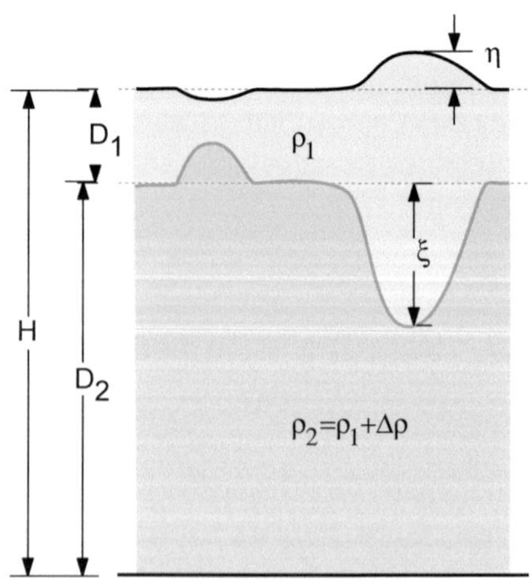

Fig. 7.12 Example of a two-layered fluid. The mean thicknesses of the *upper* and *lower* layers are denoted by D_1 and D_2 (>0), respectively, and $H = D_1 + D_2$. The elevations at the surface η and at the interface ξ are considered to be positive in the vertically upward direction

$$\frac{\partial u_2}{\partial t} - f_0 v_2 = -\frac{1}{\rho_2}\frac{\partial p_2}{\partial x} = -g^*\frac{\partial \xi}{\partial x} - g\frac{\rho_1}{\rho_2}\frac{\partial \eta}{\partial x}, \tag{7.113}$$

$$\frac{\partial v_2}{\partial t} + f_0 u_2 = -\frac{1}{\rho_2}\frac{\partial p_2}{\partial y} = -g^*\frac{\partial \xi}{\partial y} - g\frac{\rho_1}{\rho_2}\frac{\partial \eta}{\partial y}, \tag{7.114}$$

$$\frac{\partial u_2}{\partial x} + \frac{\partial v_2}{\partial y} = -\frac{1}{D_2}\frac{\partial \xi}{\partial t}, \tag{7.115}$$

where the thickness of the lower layer $h_2 = D_2 + \xi$ is approximated as D_2. The quantity $g^* \equiv g(\rho_2 - \rho_1)/\rho_2 = g\Delta\rho/\rho_2 \simeq g\Delta\rho/\rho_1$ is called **reduced gravity**. In general, because $\Delta\rho \ll \rho_1, \rho_2$, g^* is considerably smaller than g.

Now, $\Delta\rho \ll \rho_2$ implies that $\rho_1/\rho_2 \simeq 1$; therefore, the final terms on the right-hand sides of (7.113) and (7.114) are virtually the same as those of (7.109) and (7.110), respectively. Thus, we can consider that the motions in the lower and upper layers differ mainly because of the first-term ξ on the right-hand sides of (7.113) and (7.114).

In the most extreme case, if ξ and η are related by $\xi = -\eta\rho_1 g/(\rho_2 g^*) = -\eta\rho_1/\Delta\rho$, the right-hand side of (7.113) or (7.114) becomes 0. In this limit, the pressure fluctuation created by η at the top surface is canceled by the fluctuation ξ at the interface. This type of state is similar to **isostasy** in geology. In particular, during a steady state, the geostrophic flow in the lower layer disappears because no pressure gradient develops there, and the motion is strictly limited to the upper layer. Two layers that have established a state of "isostacy" are frequently referred to as "1.5 layers."

7.7.2 Internal Radius of Deformation

In the previous section, the difference between the bottom and top layers were attributed to the displacement ξ at the interface.

The difference between the lower and upper layers can be determined from (7.109)–(7.111) and (7.113)–(7.115) as follows:

$$\frac{\partial \Delta u}{\partial t} - f_0 \Delta v = -g^*\frac{\partial}{\partial x}(\xi - \eta), \tag{7.116}$$

$$\frac{\partial \Delta v}{\partial t} + f_0 \Delta u = -g^*\frac{\partial}{\partial y}(\xi - \eta), \tag{7.117}$$

$$\frac{\partial \Delta u}{\partial x} + \frac{\partial \Delta v}{\partial y} = -\frac{1}{D_1}\frac{\partial \eta}{\partial t} - \left(\frac{1}{D_1} + \frac{1}{D_2}\right)\frac{\partial \xi}{\partial t}, \tag{7.118}$$

where $\Delta u = u_2 - u_1$ and $\Delta v \equiv v_2 - v_1$. As $g^* \ll g$, we would expect larger variations in ξ, whose restoring force is the buoyancy, than those in η, whose restoring force is the gravity. Therefore, we can write $\xi - \eta \simeq \xi$, and express (7.116)–(7.118) in terms of $\xi \Delta u$, and Δv:

$$\frac{\partial \Delta u}{\partial t} - f_0 \Delta v = -g^* \frac{\partial \xi}{\partial x}, \tag{7.119}$$

$$\frac{\partial \Delta v}{\partial t} + f_0 \Delta u = -g^* \frac{\partial \xi}{\partial y}, \tag{7.120}$$

$$\frac{\partial \Delta u}{\partial x} + \frac{\partial \Delta v}{\partial y} = -\frac{1}{H^*} \frac{\partial \xi}{\partial t}, \tag{7.121}$$

where $H^* = D_1 D_2/(D_1 + D_2) = D_1 D_2/H$.

These equations are analogous to (7.80)–(7.82), and the difference is that g and H in (7.80)–(7.82) are replaced by g^* and H^*, respectively. Therefore, the arguments developed for single-layer fluids are applicable to two-layer fluids, but with some modifications. In particular, the radius of deformation in single-layer fluids, $a_\mathrm{R} = \sqrt{gH}/f_0$, becomes $a_\mathrm{R}^\mathrm{int} = \sqrt{g^* H^*}/f_0$ in two-layer fluids. To distinguish the two cases, we refer to the former (barotropic flow) and the latter (induced by internal waves in a stratified fluid) as the **external** and **internal radii of deformation**, respectively. Generally, $D_2 \gg D_1$ in oceans; therefore, $H^* = D_1 D_2/H \simeq D_1 \ll H$. Furthermore, as $g^* \ll g$, the internal radius of deformation is 1/10–1/100 of the external radius of deformation.

Note 7.13 Partially differentiating (7.111), (7.109), and (7.110) with respect to t, x, and y, respectively, and assuming no rotation, we obtain

$$\frac{\partial^2 (\eta - \xi)}{\partial t^2} = D_1 g \nabla_\mathrm{h}^2 \eta \,. \tag{7.122}$$

Similarly, in the lower layers, we get

$$\frac{\partial^2 \xi}{\partial t^2} = D_2 \nabla_\mathrm{h}^2 \left[g^* \xi + (g - g^*) \eta \right] . \tag{7.123}$$

If η and ξ are proportional and related by $\xi = \mu\eta$, both (7.122) and (7.123) become wave equations in η. By comparing their coefficients (phase velocities), we obtain a second-order equation in terms of μ:

$$c^2 = \frac{D_1 g}{1 - \mu} = \frac{D_2 (g^* \mu + g - g^*)}{\mu} \,. \tag{7.124}$$

Solving (7.124) under the approximation $(D_1 + D_2)/D_2 \simeq 1$, we obtain two values of μ: $\mu = 1$ and $\mu = -\rho_1/\Delta\rho$.

The former is equivalent to the shallow water wave in single-layer fluids. In this case, $\xi = \eta$ from $\mu = 1$, and although density is stratified, the fluid body establishes barotropic flow with isopycnal surfaces (ξ) and isobaric surfaces (η) in parallel.

The other solution is equivalent to internal gravity waves in which isostacy is established. This phase velocity is $c = \sqrt{D_1 g/(1 + \rho_1/\Delta\rho)} = \sqrt{g^* D_1} \simeq \sqrt{g^* H^*}$. Hence, the internal radius of deformation $a_\mathrm{R}^\mathrm{int}$ is equivalent to the distance propagated by an internal gravitational wave (in the absence of rotational effects) for $1/f_0$ h.

7.7.3 Baroclinic Rossby Waves

The above discussion should also be applicable to the β-plane approximation. Rossby waves, particularly long Rossby waves, conserve their potential vorticity by balancing the planetary vorticity and stretching/shrinking of the fluid column, as in (7.102). In the two-layer fluids, the variations of the interface displacement ξ are the dominant factor of the change in height of the column. A major wave solution is the **baroclinic Rossby wave**, with a_R replaced by the internal radius of deformation a_R^{int} that is related to variations in ξ. The Rossby waves shown in Sects. 7.5.2 and 7.5.3 are distinctively called **barotropic Rossby waves**, which use the external radius of deformation.

The phase velocity of long Rossby waves is proportional to the square of the radius of deformation. This implies that in oceans, where the internal radius of deformation is 10 to 100 times smaller than the external radius of deformation, the phase velocity of baroclinic long Rossby waves is two to four orders of magnitude slower than that of barotropic long Rossby waves.

7.7.4 Thermohaline Circulation and Wind-Driven Circulation

The heat received by ocean surfaces can reach to a depth of several hundred meters only. Below this warmed layer, there is a **thermocline** in which the water temperature drops drastically with depth. In layers below the thermocline, the water temperature is extremely low. Therefore, a fairly stable stratified state is maintained in real oceans (with some exceptions in polar and other special-case regions), which is well-approximated by the stable two-layered fluid considered in this section.

Besides heat, upper ocean surfaces capture momentum carried by wind stress and additional fresh water by precipitation. Furthermore, they lose fresh water by evaporation from the sea surface. Therefore, the upper layer frequently exchanges heat, fresh water, and momentum with the atmosphere. On the other hand, the aforementioned isostacy is fairly well established in the ocean; hence, motions in the upper and lower layers may be quite independent, and the interaction between the upper layer and the atmosphere is not directly conveyed to the lower layer.

Section 7.6.2 discussed on general circulations of the oceans driven by wind. Such circulations, called **wind-driven circulations**, occur only in layers above the thermocline. Below the thermocline, circulation is rather driven by the pressure gradient caused by spatial distribution of the density, or temperature and salinity. These so-called **thermohaline circulations** in the lower layer are relatively slower than the wind-driven circulations in the upper layers, and their spatial patterns are also quite different. More detailed descriptions of actual oceans are covered in Chap. 8.

Note 7.14 (*Literature for further study*)
For further studies of the rotating fluids, textbooks [1–3, 10] may be referred. The

advanced readers may prefer more practical ones [5, 9, 11]. We also recommend some other textbooks [4, 6–8, 12–16], although they are all written in Japanese.

References

1. Abarbanel, H.D., Young, W.R. (eds.): General Circulation of the Ocean. Springer, Berlin (1986)
2. Batchelor, G.K.: An Introduction to Fluid Dynamics. Cambridge University Press, Cambridge (1967)
3. Gill, A.E.: Atmosphere-Ocean Dynamics. Academic Press, USA (1982)
4. Kimura, R.: Introduction to Geophysical Fluid Dynamics. Tokyodo Shuppan, Japan (1983)
5. Newman, J.N.: Marine Hydrodynamics. The MIT Press, Cambridge (1977)
6. Ogura, Y.: Principles of Atmospheric Dynamics. University of Tokyo Press, Tokyo (1978)
7. Ogura, Y.: General Meteorology. University of Tokyo Press, Tokyo (1984)
8. Ogura, Y.: Basic Theory of Meso-scale Meteorology. University of Tokyo Press, Tokyo (1997)
9. Open University. Oceanography Course Team: Ocean Circulation. Pergamon Press, Oxford (1989)
10. Pedlosky, J.: Geophysical Fluid Dynamics (2nd ed.). Springer, Berlin (1986)
11. Pond, S., Pickard, G.L.: Introductory Dynamical Oceanography (2nd ed.). Pergamon Press, Oxford (1983)
12. Sekine, Y.: Introduction to Physical Oceanography. Seizando Shoten, Tokyo (1995)
13. Toba, Y. (ed.): Atmosphere-Ocean Interactions. University of Tokyo Press, Tokyo (1996)
14. Unoki, S.: Coastal Physical Oceanography. Tokai University Press, Tokyo (1993)
15. Unoki, S., Kubota, M.: Waves and Currents in the Ocean. Tokai University Press, Tokyo (1996)
16. Yanagi, T.: Coastal Oceanography (Revised Edition). Kouseisha-kouseikaku, Tokyo (1994)

Chapter 8
Environmental Fluid Dynamics

This chapter shows how each of the processes studied in Chaps. 1–7 actually operates on the Earth, through the circulations of energy/water/carbon over the entire Earth.

8.1 Global Energy Balance

8.1.1 Radiation Equilibrium

The Earth receives most of its energy as radiation energy from the Sun. The Sun has been actively fusing hydrogen over the past several billion years, and continues to release a fairly fixed amount of energy into space. The Earth receives approximately 2×10^{17} W of this energy, and the average solar radiation energy received by a surface perpendicular to the Sun's rays at the top of the Earth's atmosphere is 0.14 W/cm^2 (2 cal/cm^2/min) (1 W = 1 J/s = 10^7 erg/s = 1 kg $\cdot$ m^2/s^2 = 0.24 cal/s).

An additional 10^{13} W (2.0×10^{-5} W/cm^2) of the Earth's energy is sourced from geothermal fields, and approximately 10^{13} W (2.0×10^{-5} W/cm^2) is supplied by the Earth's rotation, which generates the tidal phenomenon (Fig. 8.1). Compared with solar radiation energy, however, both sources are negligible. Incidentally, the global fossil energy currently consumed by humankind is around 10^{13} W (2.0×10^{-5} W/cm^2), a mere 1/7000 of solar energy. The Sun's radiation energy covers a broad range of the electromagnetic spectrum, from X-rays with shorter wavelengths to electric waves with longer wavelengths. Incident sunlight to a substance performs one of the following three actions:

1. ignores the substance and passes through it,
2. is reflected from the surface, and
3. is absorbed by the substance.

D.E.S.S.T. Kyushu Univ., *Fluid Dynamics for Global Environmental Studies*,
DOI 10.1007/978-4-431-56499-7_8

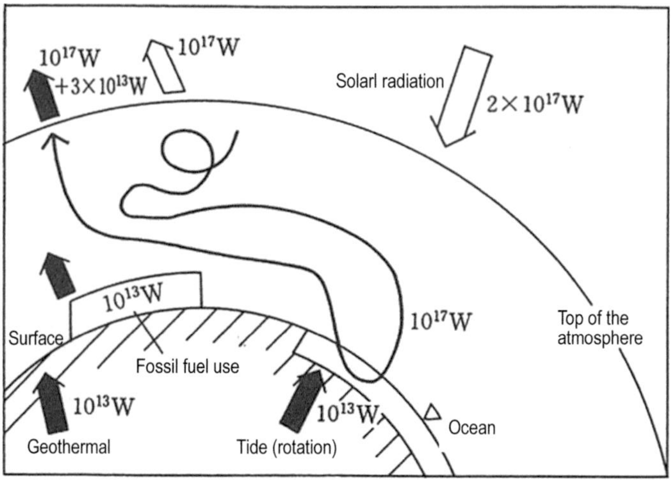

Fig. 8.1 Flow of energy on the earth

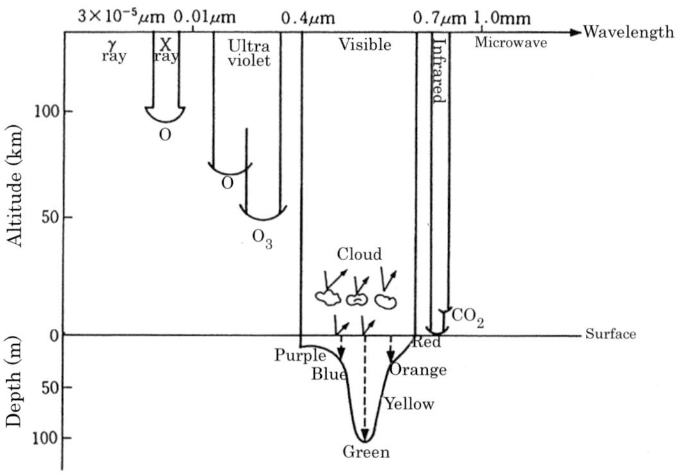

Fig. 8.2 Behavior of solar radiation energy [12]

Examples of the above processes are (Fig. 8.2)

1. visible rays on oxygen and nitrogen in the atmosphere,
2. visible rays on clouds, and
3. X-rays and ultraviolet rays on oxygen atoms and ozone molecules.

Thirty percent of the Sun's radiation energy reaching the Earth (shortwave radiation) is reflected by the atmosphere or the surface and returns to space; the remaining 70% is absorbed by the atmosphere, oceans, and surface (Fig. 8.3).

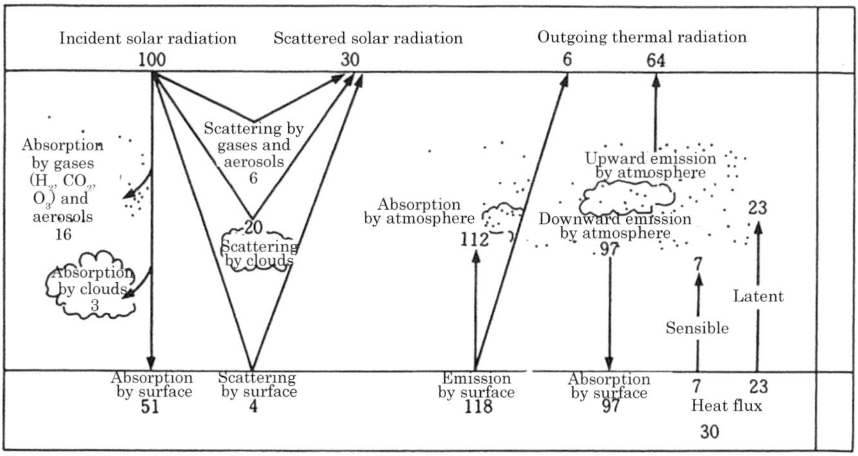

Fig. 8.3 Flux of energy between the atmosphere and surface

In contrast, the Earth also emits energy from its upper atmosphere into space. This energy, in the form of longwave infrared radiation (heat energy), is equivalent to the amount of heat that the Earth has absorbed. Thus, the equilibrium of radiation is established for the Earth as a whole. In other words, the amount of heat received by the Earth from the Sun equals that released into space. This equilibrium maintains the Earth's average temperature at a constant value. If one is higher than the other, the Earth will gradually cool down or heat up.

As shown in Fig. 8.3, the surface emits approximately twice as much heat energy as it acquires. However, most of this heat energy is absorbed by the atmosphere and returned to the surface as emitted heat energy from the lower atmosphere (downward radiation), where the temperature is relatively high.

As seen, the atmosphere is transparent to shortwave solar radiation, but is opaque to longwave radiation, and therefore obstructs heat flow from the ground surface. This property of the atmosphere is known as the **greenhouse effect**. The greenhouse effect is mainly contributed by water vapor and carbon dioxide. Both substances have strong absorption bands in various parts of the far infrared region, leading to the atmospheric greenhouse effect.

The transfer of **sensible heat** and **latent heat** from the surface to the atmosphere mainly occurs at the sea surface (Fig. 8.3). In other words, the water temperature at the sea surface is usually higher than the air temperature; thus, sensible heat is conducted from the ocean to the atmosphere. The other type of heat (latent heat) is extracted in large quantities by evaporation of water from the sea surface and transported to the atmosphere. In this way, the ocean becomes a major heat source for the atmosphere. The above-described radiation equilibrium applies only to the whole Earth. At each of the Earth's latitudes, the amount of solar heat received and lost by the Earth is unbalanced, as shown in Fig. 8.4.

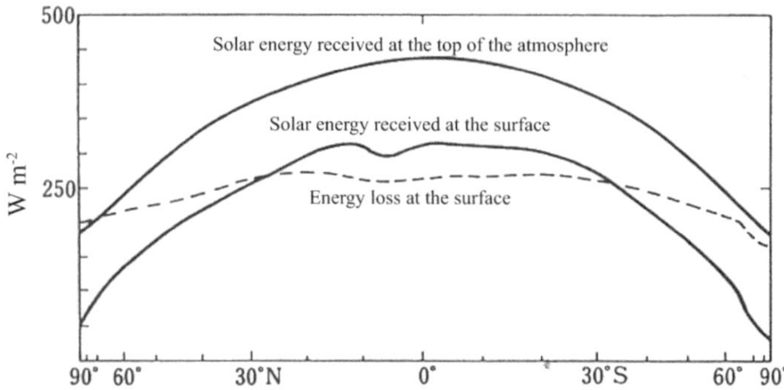

Fig. 8.4 Heat received and lost by the surface at each latitude on the earth [5]

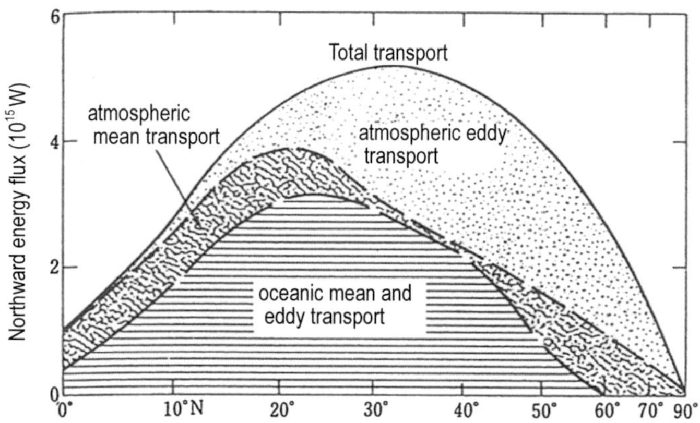

Fig. 8.5 Amount of heat transported northward [5]

In equatorial regions, the received heat is larger than the emitted heat; conversely, in polar regions, the emitted heat is larger than the received heat. Only in radiative equilibrium, the equatorial temperature must increase by 20 °C, while the polar temperature decreases by 50 °C. This unbalanced meridional heat budget is adjusted by the atmospheric and oceanic circulations, which redistribute the heat (Fig. 8.5).

Figure 8.6 demonstrates the net heat flux entering the oceans from the atmosphere through the sea surface. The net heat flux is positive around the Equator, which indicates heat flux from the atmosphere to the ocean. When negative net heat flux occurs in mid-latitude regions, heat is supplied from the oceans to the atmosphere. In particular, large negative values are observed around the Kuroshio and the Gulf Stream, and it can be understood that the western boundary current transports a large amount of heat from the equatorial regions to the mid-latitude regions.

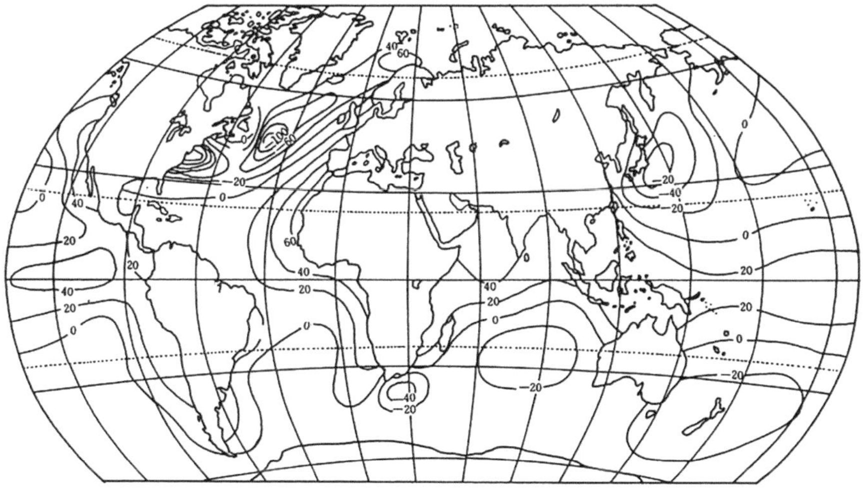

Fig. 8.6 Net heat flux entering the ocean from the atmosphere through the sea surface (kcal/cm^2/year) [6]

8.1.2 Atmospheric General Circulation

The vertical distribution of air temperature from the ground into the air is shown in Fig. 8.7. In the troposphere, the temperature decreases at a rate of approximately 0.6°C/100 m because of radiative cooling of the atmosphere at the tropopause.

The troposphere is approximately 18 km thick at low latitudes, 12 km at mid-latitudes, and 6 km at high latitudes. This latitudinal variability results from strong heating of the ground and sea surfaces around the Equator, which generates strong updraft.

On the other hand, in the upper stratosphere, temperature maxima are created by absorption of ultraviolet radiation by the ozone layer. In the lower stratosphere, there is a stratospheric jet stream at the lower stratosphere-troposphere boundary, where the wind velocity exceeds 30 m/s. The jet stream is several hundred kilometers wide and several kilometers thick.

The air mass rises by heating in the equatorial regions, and then becomes heavier by radiative cooling as it heads toward the poles, where it sinks at latitudes of around 20°. This meridional circulation is called the **Hadley cell** (Fig. 8.8). The downdraft returns to the Equator along the ground and sea surfaces from near the 20° latitude, where it is influenced by the Coriolis force and driven westward. Thus, it becomes the trade wind from northeast in the Northern Hemisphere and from southeast in the Southern Hemisphere.

On the other hand, the air mass made heavy by cooling around the poles becomes lighter as it heads toward the Equator, and rises at latitudes around 50°, creating the **Polar cell**.

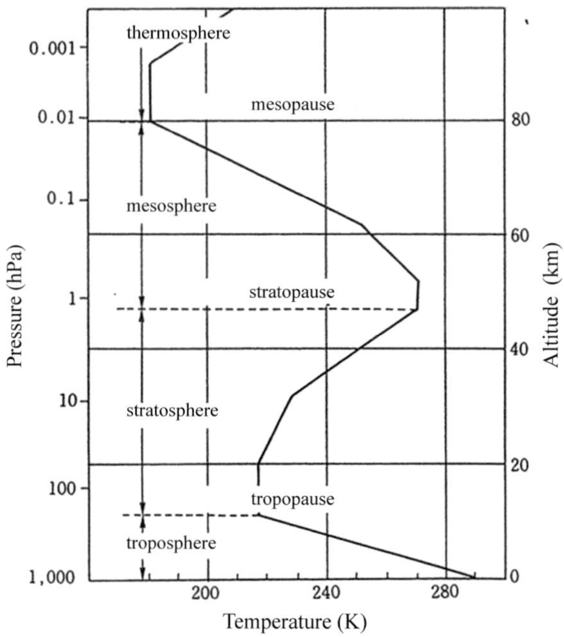

Fig. 8.7 Vertical distribution of temperature and pressure

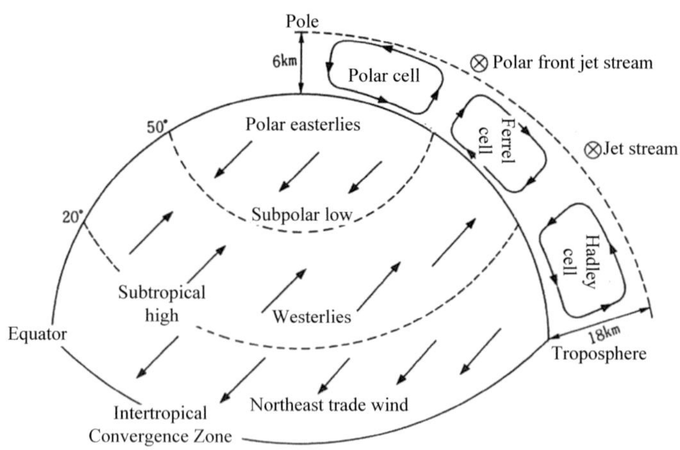

Fig. 8.8 Atmospheric general circulation

Between these meridional circulations, another meridional circulation (the **Ferrel cell**) is established in the mid-latitude regions. In this circulation, the air mass heading from latitudes 20 to 50° forms near the ground/water surface, which is also driven eastward under the Coriolis force, becoming a prevailing westerly wind. An almost identical air stream is generated in the Southern Hemisphere. Such atmospheric movements are known as the **atmospheric general circulation**. Atmospheric general circulation has a huge impact on the Earth's climate and the circulation of ocean surface layers. For example, the descending stream around latitudes of 20° forms the **Subtropical High Pressure belt** where the air pressure is constantly high and precipitation is low, encouraging the spread of deserts (e.g., northern Africa, the Middle East, Mexico, and inland Australia). Moreover, small meteorological perturbations occur in the **Intertropical Convergence Zone** (ITCZ), where the updraft develops, leading to tropical cyclones such as typhoons in the Northwest Pacific Ocean, cyclones in the Southwest Pacific and Indian Oceans, or hurricanes in the Northeast Pacific and North Atlantic Oceans. Furthermore, the north-eastern trade wind and prevailing westerlies drive the subtropical circulation on the ocean surface. This phenomenon will be discussed later in this chapter.

8.1.3 Oceanic General Circulation

The main oceanic flow distribution in the surface layers of the world's oceans (0–1,000 m) is shown in Fig. 8.9. This flow has the following characteristics:

1. The oceanic flow distribution is virtually symmetric across the Equator in the north and south.
2. A large anti-cyclonic circulation of high pressure develops in the sub-tropical zones (clockwise in the Northern Hemisphere and counterclockwise in the Southern Hemisphere).
3. Strong currents such as the Kuroshio, the Gulf Stream, the East Australian Current, and the Brazil Current are all established on the western side of the oceans (see Sect. 7.6.2; **western boundary current**).
4. Most of the ocean currents flow in virtually the same direction throughout the year; however, the Somali Current in the north-western Indian Ocean flows north east during summer and south-west during winter. This change in flow direction is caused by seasonal change in winds (monsoon).

The surface layer circulation is driven by two forces: the prevailing wind blowing across the sea and the density gradient of the seawater between the equatorial and polar regions. Intuitively, one expects that the trade wind pulls the North Equatorial Current and the prevailing westerlies pull the North Pacific Current, driving the subtropical circulation in the North Pacific Ocean. However, this is incorrect because, on the rotating Earth, seawater does not flow in the direction of the wind because of the Coriolis force.

The relationship between the prevailing wind and ocean current was explained in Chap. 7. The Ekman transport by the trade winds and prevailing westerlies accu-

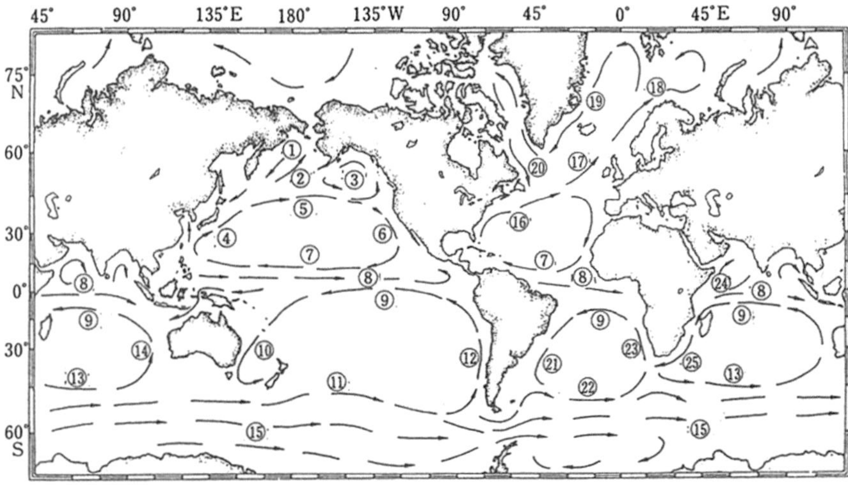

Fig. 8.9 Oceanic surface layer circulation. ① Oyashio, ② Bering Current, ③ Alaskan Stream, ④ Kuroshio, ⑤ North Pacific Current, ⑥ California Current, ⑦ North Equatorial Current, ⑧ Equatorial Countercurrent, ⑨ South Equatorial Current, ⑩ East Australian Current, ⑪ South Pacific Current, ⑫ Peru Current, ⑬ South Indian Current, ⑭ West Australian Current, ⑮ Antartic Circumpolar Current, ⑯ Gulf Stream, ⑰ North Atlantic Current, ⑱ Norwegian Current, ⑲ East Greenland Current, ⑳ Labrador Current, ㉑ Brazil Current, ㉒ South Atlantic Current, ㉓ Benguela Current, ㉔ Somali Current, ㉕ Agulhas Current

mulates seawater in the central Pacific Ocean. Subtropical circulation develops to balance the Coriolis force with the pressure gradient due to the difference in the sea surface height. The strong ocean current along the western coast of the oceans, such as the Kuroshio, are explained by the change in Coriolis force with latitude. On the other hand, a completely different circulation from that in the surface layer is established in the deep layer (~5,000 m), as shown in Fig. 8.10.

The deep waters in the world are mainly formed in the Irminger Sea, south of Greenland, and the Weddell Sea in Antarctica. In the surface layers of these seas, seawater became heavy by cooling to sink to a deeper layer and spread over the world's deep oceans. The amount of seawater sunken during the abyssal circulation is estimated to be approximately $7 \times 10^5\,\text{km}^3/\text{year}$ (22 Sv.), twenty times the total river discharges around the world. Although seawater is also cool and dense in the northern North Pacific, the shallowness and low salinity of this region prevent the formation of deep waters.

The current velocity in the **abyssal circulation** is too small for direct measurement with a current meter. However, the flow direction of deep water can be deduced by measuring the concentration ratio of carbon isotopes in the seawater; specifically, by comparing the concentration of radio isotope ^{14}C, with a half-life of 5,570 years, with that of stable isotope ^{12}C. The ratio ^{14}C/^{12}C is almost constant in the atmosphere; however, when the seawater sinks from the sea surface and loses contact with the

Fig. 8.10 Abyssal circulation in the ocean [10]

atmosphere, $^{14}C/^{12}C$ decreases exponentially with time. Consequently, we can determine, from the ratio, the time elapsed since atmospheric contact was lost; in other words, the age of the deep water. The age distribution of the seawater in the Pacific and Atlantic is shown in Fig. 8.11.

The youngest deep water is found around the Irminger Sea in the northern North Atlantic and the oldest deep water (at 2,000 years) is distributed in the mid-layers of the North Pacific. The deep water that has sunk in the Irminger Sea moves south throughout the Atlantic Ocean and to the Southern Ocean in approximately 700 years. It then takes 600 years to travel the eastern path to the south of Australia, and an additional 700 years to move up the Pacific Ocean and upwell into the mid-layers of the northern North Pacific. Seawater circulates around the globe over approximately 3,000 years, carried by the abyssal circulation in the north and south, and then gradually upwells mixing with the east/west flows of the surface circulation, before being slowly conveyed toward the high latitude regions and again sinking into the deep layers (Fig. 8.12).

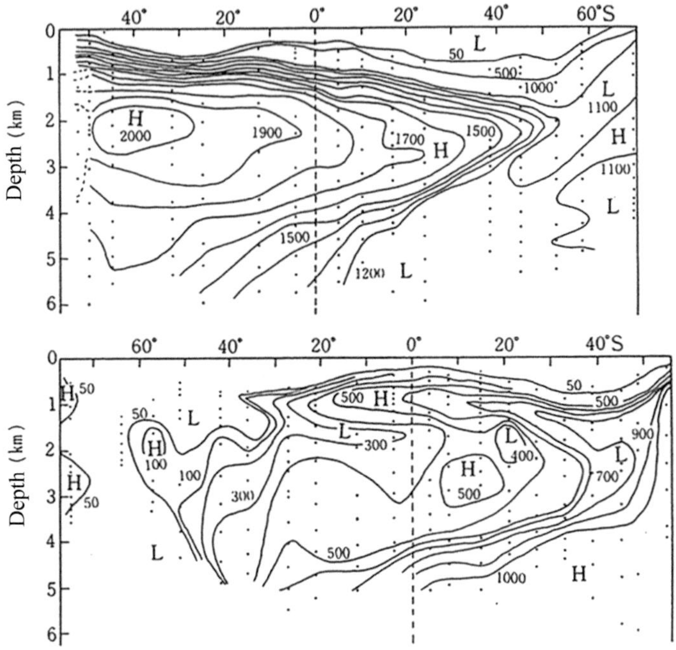

Fig. 8.11 Age of seawater (years) in the West Pacific Ocean (*above*) and West Atlantic Ocean (*bottom*) [11]

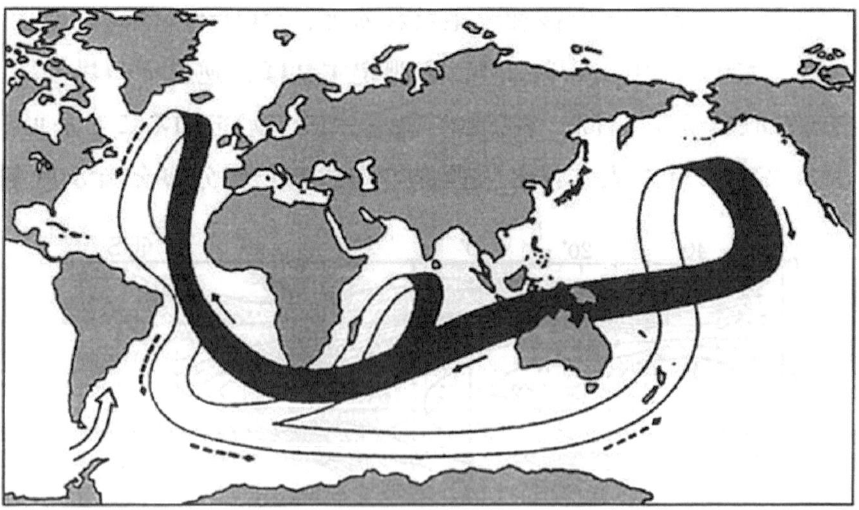

Fig. 8.12 Oceanic circulations in surface and abyssal layers [2]

8.2 Global Water Circulation

8.2.1 Freshwater Balance

The Earth contains $1{,}434 \times 10^6\,km^3$ of water. Each $1\,cm^2$ of ground surface holds 237 L. Ninety-four percent of this water is stored in the oceans as seawater (Fig. 8.13). The second largest water reserve is underground water (4.2%), followed by glaciers (1.7%). Of course, if continental glaciers melted and flowed into the sea because of global warming, the seawater reserves would increase, which would result in sea level rise.

The Earth's water is not confined to specific reservoirs, but is constantly moving. For example, moisture evaporates from the sea surface at a rate of $440 \times 10^3\,km^3$/year (equivalent to a sea level fall by 126 cm/year), and $411 \times 10^3\,km^3$/year of water falls as rain (equivalent to a sea level rise by 114 cm/year). The remaining water from the sea surface evaporation ($29 \times 10^3\,km^3$/year) is carried to the land by wind and descends as rain. A majority of the excess land precipitation ($29 \times 10^3\,km^3$/year, or 29% of the precipitation) flows into the oceans through rivers.

The time for which the water remains in the ocean (**average residence time**, ca. 3,000 years) can be obtained by dividing the seawater volume ($1{,}350 \times 10^3\,km^3$) by the evaporated seawater volume. Such a long average residence time indicates that freshwater entering the sea is less likely to become polluted. On the other hand, once polluted, it will take several thousand years for the ocean to recover from it.

Similarly, the average residence time of atmospheric water (ca., 10 days) is obtained by dividing the volume of water vapor in the atmosphere ($0.014 \times 10^6\,km^3$) by the volume precipitated on the sea and land surfaces ($512 \times 10^3\,km^3$/year). This indicates that the atmosphere is more easily polluted than the sea, but recovers much more rapidly. This can be readily understood by recalling the temporal changes in atmospheric pollution following the Chernobyl nuclear power station accident in April 1986.

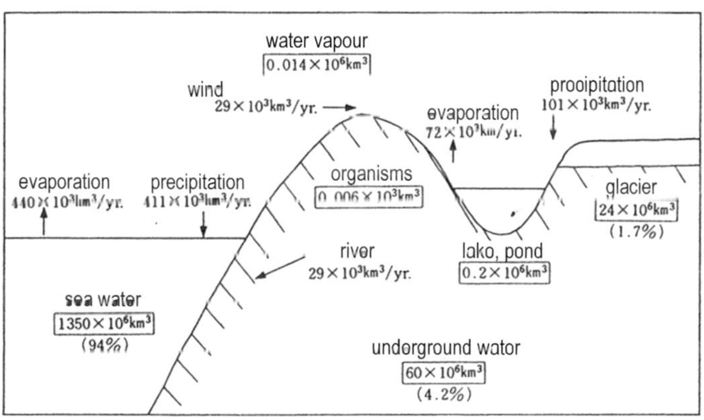

Fig. 8.13 Global water balance

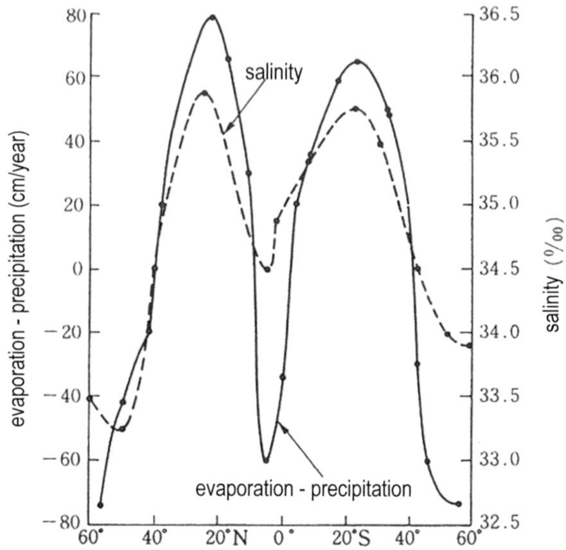

Fig. 8.14 Latitudinal distribution of the difference between evaporation and precipitation volumes, and salinity of the surface layer, distributed across earth's latitudes [3]

8.2.2 *Freshwater Transport*

Precipitation and evaporation on sea and land surfaces are heterogeneously distributed. Some regions are subject to more precipitation than evaporation; in other regions, the evaporation exceeds precipitation.

Broadly speaking, this distribution is determined by the circulation of the atmosphere described in Sect. 8.1.2. Precipitation exceeds evaporation in the equatorial convergence zone; the reverse is true in the subtropical high pressure belt (Fig. 8.14). Consequently, salinity is low in near-equatorial seawater and high in seawater around the subtropical zones. Of course, salinity is influenced not only by precipitation and evaporation above the sea but also by freshwater discharge from the land. Figure 8.15 indicates the vertically integrated annual mean water vapor content in the atmosphere obtained by subtracting the annual evaporation volume from the annual precipitation volume. Large positive values indicate where precipitation exceeds evaporation and surroundings receive an inflow of freshwater of that amount. Large positive values are focused around the Equator. Regions of large negative values (evaporation exceeding precipitation) are concentrated in bands spanning the subtropical oceans of both the hemispheres. This type of distribution is reflected in Fig. 8.14.

As part of the vapor balance in the atmosphere, precipitation leads to the release of latent heat which warms the atmosphere, thereby directly affecting the climate system. Moreover, the ocean is provided with positive buoyancy, altering its salinity distribution that drives thermohaline circulation. Therefore, more accurately, we can state that the distributions in Figs. 8.14 and 8.15 are determined by interactions between the vapor/precipitation volumes and atmospheric circulation.

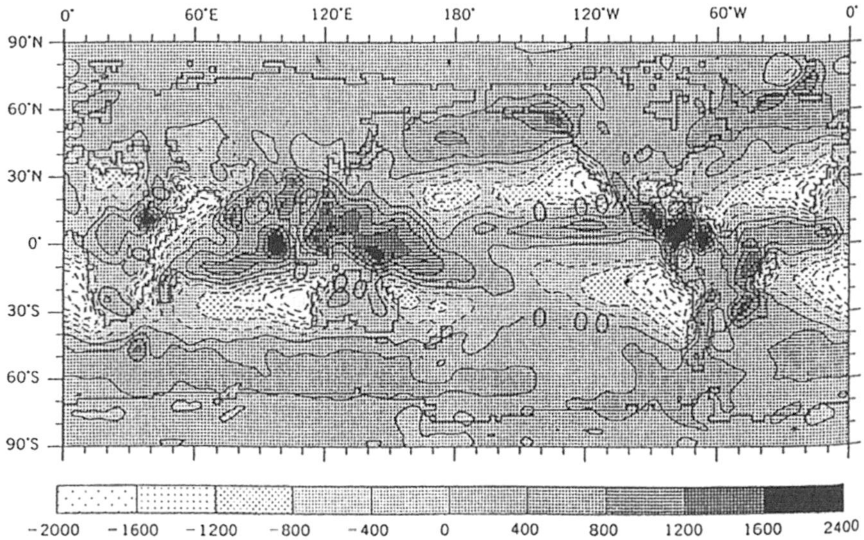

Fig. 8.15 Vertically integrated annual mean water vapor content in the atmosphere (mm/year) [9]

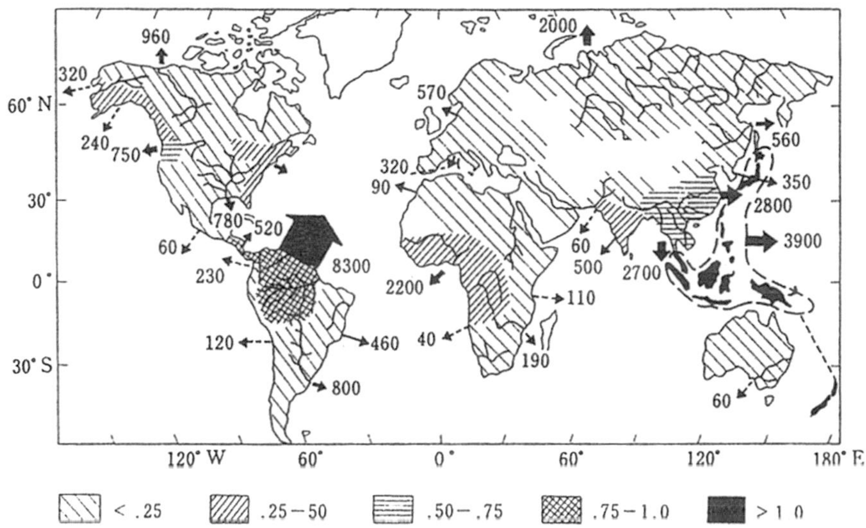

Fig. 8.16 Annual rates of river discharge from each land region (km^3/year) [7]

When precipitation on land exceeds evaporation, the excess water flows into the oceans as river discharges. The annual river discharge and the volume of water flow per unit surface area are shown in Fig. 8.16. It is noticeable that the river discharges are particularly large in eastern Asia (Asia monsoon belt) and the South American Amazon.

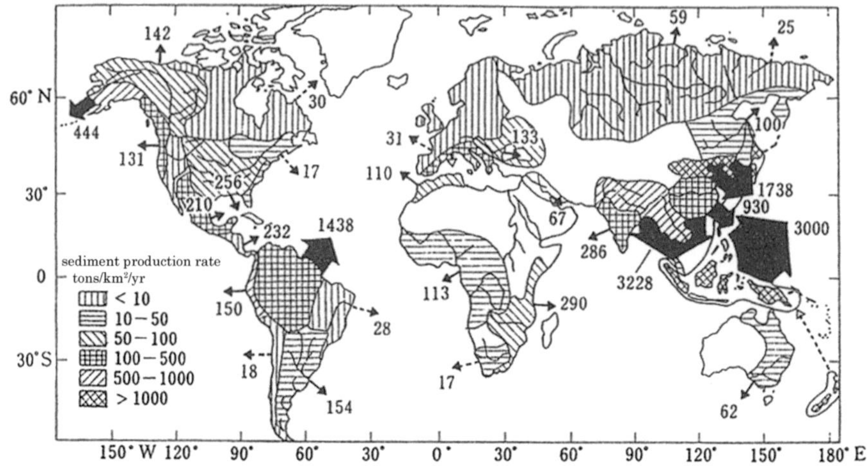

Fig. 8.17 Annual discharges of suspended matter from each land region (10^6 ton/year) [7]

Table 8.1 Annual water balance in each of the world's oceans

	Area (10^{12} m^2)	Atmospheric vapor balance	Precipitation	Evaporation	Inflow from land
			(10^{15} kg/year)		
North Pacific Ocean	76.6	7.6	129.1	121.5	8.0
South Pacific Ocean	99.1	−6.4	138.4	144.8	3.2
North Atlantic Ocean	43.5	−12.3	48.7	61.0	5.4
South Atlantic Ocean	45.8	−8.9	37.0	45.9	6.1
Indian Ocean	70.8	−8.9	84.4	93.3	0.1
Arctic Ocean	10.6	2.0	3.4	1.4	3.7
All oceans	346	−27	441	468	27
All land	163	27	139	112	

The river discharge of the Amazon, the biggest river in the world, is 8.3×10^3 km^3/year $= 0.26 \times 10^6$ m^3/s, equaling $1/150$ of the average volume transport of the Kuroshio, south of the Cape Ashizuri (approximately 40 Sv. $= 40 \times 10^6$ m^3/s). This comparison highlights the enormity of the western boundary current.

Volumes of suspended matter transported to the oceans by rivers from each land region are shown in Fig. 8.17. Here the volumes from East Asia are particularly prominent. In contrast to the Amazon River region, which is covered by tropical rainforests that protect the soil from erosion, Southeast Asia has been heavily modified by agricultural farming, which exposes the soil to erosion by heavy rainfalls.

The water balance in the major seas of the world is summarized in Table 8.1. In the North Atlantic Ocean, the evaporation exceeds the amount of water received (given by precipitation + water inflow from land). Conversely, the North Pacific Ocean receives more water (precipitation + water inflow from land) than it loses to the atmosphere. Consequently, the average salinity in the North Atlantic Ocean is approximately 2.0 psu (Practical Salinity Unit) higher than that in the North Pacific Ocean. The vapor from the North Atlantic Ocean is carried westward by trade winds through the Isthmus of Panama, and then falls to the North Pacific as precipitation. The freshwater is then transported from the tropical regions of the North Pacific Ocean to the subtropical North Pacific Ocean by the subtropical circulation and evaporates again. Meanwhile, water vapor transported to the east by the prevailing westerlies is obstructed by the Rocky Mountains and returns to the Pacific Ocean as rain. In this way, the North Pacific Ocean receives a large freshwater input from the land.

The low and high salinities in the North Pacific and North Atlantic Oceans, respectively, are partially responsible for the generation of deep water in the North Atlantic Ocean but not in the North Pacific Ocean (as discussed in Sect. 8.1.3).

8.2.3 Transport of Water and Nutrients

Elements essential to living cells, such as nitrogen and phosphorus, are generally deficient in oceans. Substances containing these elements are called nutritive substances. In coastal regions, nutritive substances are supplied from the land to the sea by rivers, and then transported to the offshore areas. However, the transport of nutritive substances is slightly different from that of water.

Inorganic dissolved states of nitrogen and phosphate, such as nitrate, nitrite, ammonium nitrogen, and phosphate-phosphorus, are called **nutrients**. Nutrients carried by river water into coastal regions are then transported offshore by estuarine circulation in the surface layer, where they are incorporated into phytoplanktons by photosynthesis in the euphotic zones (particulate organic matter). Phytoplanktons are eaten by zooplanktons, which in turn are consumed by small fish; thus, the original nutritive substances are transported to higher organisms through the food chain. As the excrement of most zooplanktons is heavier than seawater, it sinks to the seabed, and is decomposed by bacteria in the aphotic layers of the coastal region. The decomposition restores the original nutrients, which are then transported onshore-ward in the bay head by the estuarine circulation in the regions, and eventually returned to the euphotic zones via the upwelling current (Fig. 8.18).

Therefore, the mean residence time of nitrogen and phosphorus in the coastal regions is longer than that of freshwater because of the synergistic effect between the chemical-biological processes and the estuarine circulation. This mechanism is called the **nutrient-trapping mechanism** in the coastal regions.

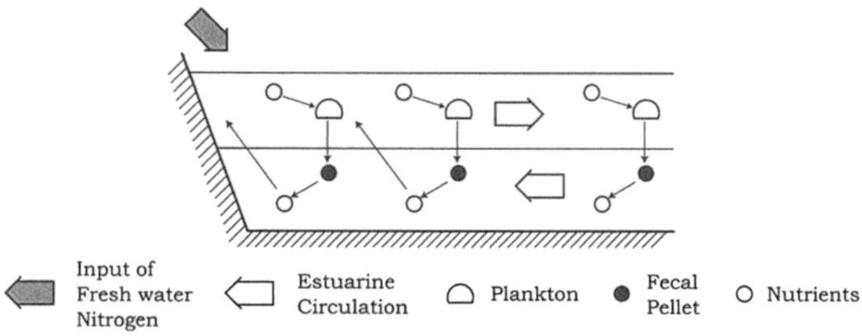

Fig. 8.18 Nutrient-trapping mechanism in coastal regions

Fig. 8.19 Catch in the major enclosed coastal regions of the world

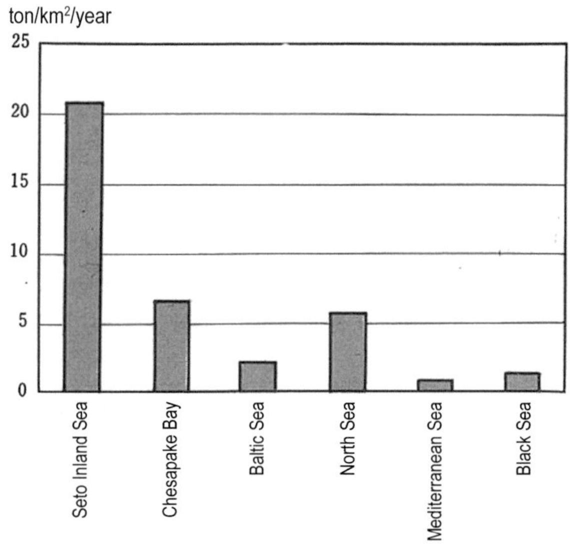

8.2.4 Water Transport and Fish Catch

The catch per unit area in the major enclosed coastal regions of the world is shown in Fig. 8.19. The Seto Inland Sea of Japan yields by far the largest catch rate (21 ton/km^2/year). The catch per unit area is chiefly determined by the water transport mechanism in each sea.

To increase the number of fish, the phytoplankton population must first be enhanced by photosynthesizing nutrients such as nitrogen and phosphorus. Most of the nitrogen and phosphorus in coastal regions are supplied from the land.

As shown in Fig. 8.20, the high evaporation in some sea regions, such as the Mediterranean, increases the salinity in the surface water, and hence the density of the seawater. The heavier saline water sinks to the aphotic bottom layers, and then flows into the Atlantic Ocean from the Strait of Gibraltar to the west. To compensate for this water flow, the low-nutrient seawater in the surface layer of the Atlantic Ocean

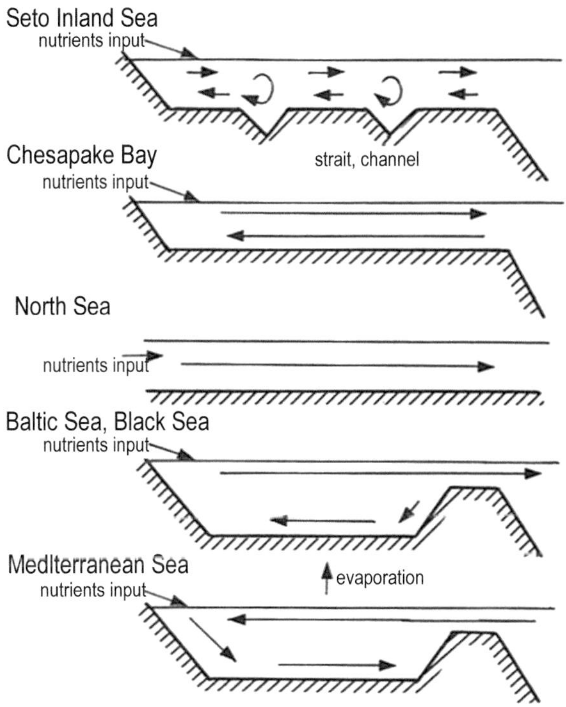

Fig. 8.20 Water transport mechanisms in different enclosed oceans

flows into the Mediterranean. Under these water movements, nutrients entering from the land are rarely consumed by photosynthesis, but are transported to the Atlantic Ocean. Therefore, the catch per unit area is extremely low in the Mediterranean. On the other hand, where precipitation on the sea surface and freshwater inflow exceed the volume evaporated from the sea surface, the conditions favor photosynthetic nutrient consumption. Examples are the Baltic Sea and the Black Sea. In these seas, freshwater from the land flows toward the oceans through the shallow surface layers of the Kattegat Strait and the Bosporus Strait to reach the Atlantic Ocean and the Mediterranean, respectively. Carried by this type of water movement, nutrients flowing in from the land are captured by photosynthesis before flowing outward.

However, in the bottom layers of the straits with a sill, there is inflow of seawater from the Atlantic Ocean and the Mediterranean to compensate for the surface outflow. In the central Baltic and Black Seas subject to this type of outflow, the bottom layers are not only below the euphotic zone but also rendered oxygen deficient by the decomposition of organic matter sunken from the surface layer, with concurrent removal of dissolved oxygen. Moreover, because of the large density difference between the surface and bottom layers, the vertical convection induced by surface cooling never reaches the seabed even in winter. Therefore, the bottom layers of both oceans are inhospitable to life and the catch per unit area is small.

By contrast, the North Sea experiences a prominent counterclockwise circulation throughout the sea from the surface to the seabed during the year. A large amount of nutrients is carried from the Atlantic Ocean, northeast of the United Kingdom, and captured by photosynthesis. Following release by decomposition, the nutrients are eventually conveyed to the Arctic Ocean from the west of Norway. Therefore, the catch per unit area is higher in the North Sea than in the Mediterranean, Baltic Sea, and Black Sea, but is lower than that in the Chesapeake Bay or the Seto Inland Sea of Japan.

As the Chesapeake Bay receives large volumes of freshwater, a vertical circulation (estuarine circulation) in which the surface layer flows to the outer sea and the bottom layer flows to the bay is dominant. The nutrients supplied from the bay head are assimilated into phytoplanktons by photosynthesis at the surface layer and transported to the mouth of the bay. Meanwhile, phytoplankton carcasses and zooplankton excrement sink to the bottom layer, and are decomposed and returned as nutrients as they flow back to the bay head because of the vertical circulation at the bottom layer. These nutrients are returned to the surface where they can be re-used in photosynthesis. Such vertical and horizontal movements between the mouth and the inner section of the bay are repeated until the nutrients eventually escape into the Atlantic Ocean. The large catch per unit area in the Chesapeake Bay is attributable to the estuarine circulation, which enhances photosynthetic nutrient processing.

In contrast, the Seto Inland Sea of Japan is characterized by numerous straits between the inner bays and the oceans. Tidal currents in the straits vertically mix the surface water with the deep water; this results in more effective nutrient recycling in the Seto Inland Sea of Japan than that in the Chesapeake Bay. The highest catch per unit area in the Seto Inland Sea is attributable to the vertical circulation and numerous straits.

8.3 Global Carbon Circulation

8.3.1 Carbon Balance

Figure 8.21 shows the distribution of carbon on the Earth. The estimated weight of carbon is 740 G ton (where G denotes giga, or 1×10^9) in the atmosphere, 550 G ton in terrestrial plants, 900 G ton in the shallow seas of the ocean, 34,000 G ton in the deep seas, and 1,200 G ton in the soil.

However, a vast majority of carbon (approximately 25,000,000 G ton) is sequestered in carbon-bearing sediments such as limestone. Excluding these sediments, approximately 38,000 G ton of carbon is available on the Earth's surface, of which 91% is stored as inorganic carbon in the deep ocean.

Of course, carbon does not remain stationary but is actively exchanged between various locations. Carbon exchange between the atmosphere and oceans is estimated to be 105 G ton/year, whereas that between the land and the atmosphere is 110 G ton/year. Thus, the average residence time of carbon in the atmosphere is $740/(105+$

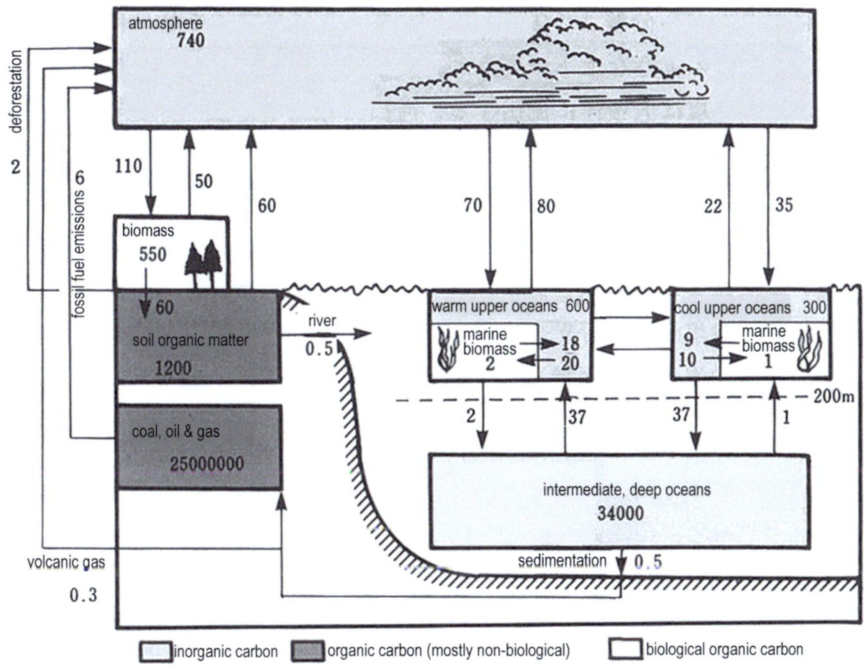

Fig. 8.21 Global-scale carbon cycle. Numbers are 10^9 tons; *arrows* denote the flux per year

110) = 3.4 years, and atmospheric carbon exchange occurs within a period just exceeding 3 years. Similarly, dividing the total weight of carbon held in seawater (34,900 G ton) by the carbon exchange velocity between the atmosphere and oceans (105 G ton/year), the average residence time of carbon in seawater is estimated as 332 years.

Global warming, which has become a worldwide concern, is caused by the burning of fossil fuels (which contain carbon in the form of coal and petrol), resulting in an increase in carbon dioxide concentration in the atmosphere. As shown in Fig. 8.22, the atmospheric carbon dioxide concentration was approximately 180 ppm in the last glacial period (20,000 years ago), but had risen to approximately 280 ppm at the end of that period (around 10,000 years ago) and remained stable thereafter.

However, as shown in Fig. 8.23, coal and petrol consumption has drastically increased since the industrial revolution in the latter half of the 19th century, and the atmospheric carbon dioxide concentration has now reached approximately 350 ppm.

The current usage of fossil fuels releases approximately 6 G tons of carbon each year (Fig. 8.21), of which 60% contributes to raising the atmospheric concentration of carbon dioxide. The destination of the remaining 40% (2.4 G tons) is undetermined and is called the **"Missing Sink" problem**. One school of academic thought posits that most of the "Missing Sink" carbon is absorbed by the oceans; others claim absorption by trees. Currently, the debate continues.

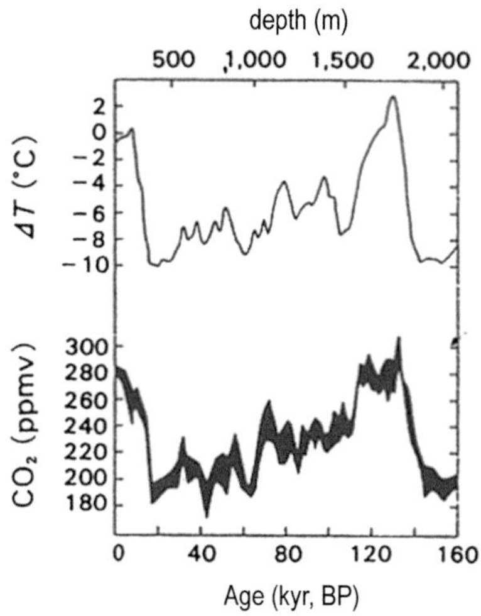

Fig. 8.22 Temperature and carbon dioxide concentration changes in the atmosphere over the past 160,000 years, recreated by the analysis of ice sheet cores collected from the Vostok base in Antarctica [1]

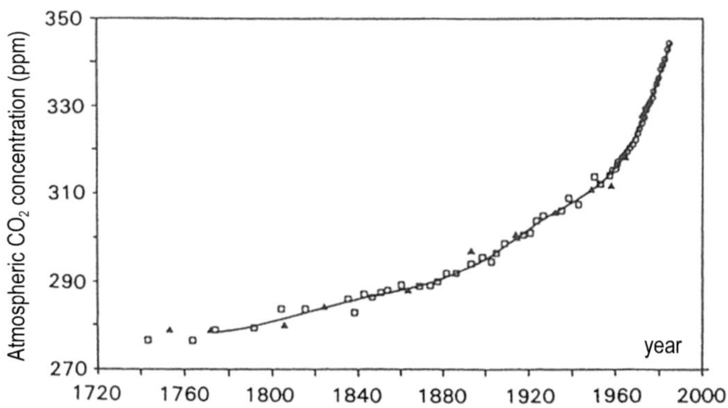

Fig. 8.23 Carbon dioxide concentration changes in the atmosphere since the 18th century, recreated by the analysis of ice sheet cores collected from the Saiple base at the South Pole [4]

8.3.2 Atmospheric Carbon

Carbon dioxide strongly absorbs light energy in the infrared region. Therefore, carbon dioxide released into the atmosphere will enhance the greenhouse effect and raise the average temperature of the Earth, as described in Sect. 8.1.1.

Atmospheric carbon dioxide reacts with water to form an acid. An identical process (known as weathering) dissolves rocks and forms carbonic acid.

$$CaSiO_3 + 3H_2O + 2CO_2 \rightarrow Ca^{2+} + 2HCO_3^{\ -} + H_4SiO_3 \tag{8.1}$$

In this equation, $CaSiO_3$ is the representative silicate within the rocks of the Earth's crust. $MgSiO_3$ undergoes a similar weathering process. The calcium, carbonic acid, and silicic acid generated by weathering are conveyed to the sea by river water.

8.3.3 Oceanic Carbon

Atmospheric carbon dioxide that reaches the ocean surface is initially transported to lower layers by mixing and homogeneously distributed at depths of approximately 300 m. Carbon dioxide in the ocean surface layer is changed to organic calcium carbonate by phytoplankton and corals, and is transported to the deeper layers through the biological pumping mechanism.

The carbon transport $(2 + 37)$ G ton/year from warm and cold shallow seas to the intermediate/deeper layers, shown in Fig. 8.21, is due to biological pumping. In the deep layers, carbon is converted to carbon dioxide by calcium carbonate dissolution and decomposition of organic matter by microorganisms. In this way, the deep waters become enriched in carbon dioxide. After circulating the oceans over 1,000–2,000 years, the deep water returns to the surface layer and releases its carbon dioxide into the atmosphere.

The photosynthetic chemical reaction undertaken by phytoplanktons is shown below:

$$106CO_2 + 16HNO_3 + H_3PO_4 + 122H_2O \rightarrow (CH_2O)_{106}(NH_3)_{16}H_3PO_4 + 138O_2 \tag{8.2}$$

This reaction indicates that the atomic ratio C : N : P in a phytoplankton cell is 106 : 16 : 1. This ratio is called the **Redfield ratio**. The synthesis of 1 mole of carbohydrate requires approximately 120 kcal of energy, which is provided by the light energy absorbed by the chlorophyll of phytoplankton cells. The opposite reaction to (8.2) is detritus decomposition.

Zooplankton eat phytoplankton and release excrements. Zooplankton frass sinks at an approximate rate of 200 m/day, which is considerably higher than the sinking velocity of phytoplankton (1 m/day). Therefore, the carbon incorporated into the organic frass is promptly transported from the ocean surface to the deep layer. Among the organic materials photosynthesized in the euphotic zone of the ocean surface layers, 20–50% are transported across the thermocline to the intermediate layers by biological pumping. The remainder is decomposed within the euphotic layers and reused in photosynthesis.

In addition to photosynthesis, inorganic substances are converted to organic materials by calcification, a process adopted by phytoplankton and corals:

$$Ca^{2+} + 2HCO_3^{-} \rightarrow CaCO_3 + H_2O + CO_2 \tag{8.3}$$

This reaction forms calcium carbonate husks such as foraminifera, coccolithophorid, and corals. The calcium carbonate product sinks and becomes marine sediments, while the carbon dioxide product escapes from the sea into the atmosphere.

As (8.3) is an equilibrium reaction that can move in either direction, it shifts to the left if carbon dioxide concentration increases in the seawater. This shift reduces carbon dioxide concentration and dissolves calcium carbonate. By contrast, if carbon dioxide concentration decreases in the seawater, the reaction shifts to the right to compensate the imbalance, and calcium carbonate is precipitated (calcification). As evident in (8.3), as calcification progresses, carbon dioxide is released, and its concentration in seawater increases.

Dissolution of carbon dioxide is enhanced with decreasing temperature and increasing pressure. This process, called the **dissolution pump**, ensures that carbon dioxide concentration is maximized in the deep seas of the polar regions and depleted in the shallow seas of the equatorial regions.

Moreover, in deep-sea regions with high pressure, (8.3) shifts to the left, and most of the remains of calcareous organisms dissolve. The depth at which the rates of carbonate supply and dissolution become the same is called the **calcium carbonate compensation depth** (CCD). The CCD in the North Pacific Ocean is ~4,000 m, implying that no carbonates exist below 4,000 m.

By contrast, in the tropical/subtropical shallow seas characterized by high water temperature and low pressure, carbon dioxide levels are low and (8.3) shifts to the right. These conditions favor calcification and the transport of carbon dioxide from the oceans to the atmosphere. For this reason, the shallow warm seas release vast quantities of carbon dioxide (80 G ton/year) into the atmosphere (Fig. 8.21).

In the short-term, calcification is an additional source of carbon dioxide; however, in the long-term, it can be regarded as a process that deposits calcium and bicarbonate ions in the form of limestone (calcium carbonate), and thereby sequesters carbon into the Earth's crust.

Since the dawn of life, calcifying cyanobacteria have created extensive structures called stromatolites. Later, sponges and bivalves evolved, and since the Mesozoic era, calcification has been dominant in corals. The calcium carbonate accumulated by these organisms became limestone, which was sealed into the Earth's crust.

In the terrestrial ecosystem, there was a time such as the Paleozoic era when coal was produced from buried wood, sealing the carbon in the Earth's crust. However, since the Mesozoic, wood-corroding organisms such as mushrooms and termites appeared; thus, wooden materials were decomposed rather than accumulated, and the sequestering of carbon into the Earth's crust was terminated.

Furthermore, since carbon dioxide is an acidic gas, excess carbon dioxide is dissolved into seawater by neutralization of weakly alkaline waters (pH = 8.3):

$$CO_2 + H_2O + {CO_3}^{2-} \rightarrow 2HCO_3{}^{-} \tag{8.4}$$

As this reaction consumes carbonate ions, the calcium carbonate undergoes a dissolution reaction:

$$CaCO_3 \rightarrow Ca^{2+} + {CO_3}^{2-} \tag{8.5}$$

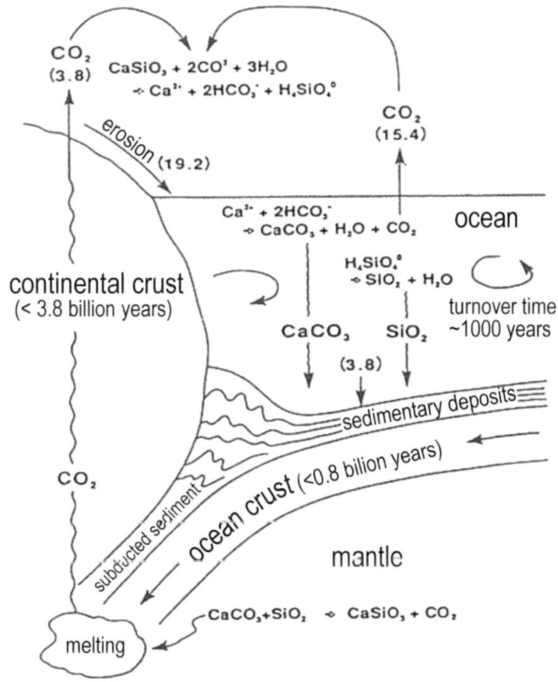

Fig. 8.24 Circulation of carbon dioxide in the Earth's surface layer. The number in parentheses is the rate of carbon movement (10^{12} gC/year) [8]

In this way, excess carbon dioxide absorbed by the seawater accumulates as dissolved inorganic carbonic acid, increasing the alkalinity of the seawater.

The above-described incorporation of atmospheric carbon dioxide into seawater by carbonate dissolution is called the **alkalinity pump**. The same process is responsible for the consumption of atmospheric carbon dioxide by carbonate weathering.

8.3.4 Carbon on the Seabed

Calcium carbonates deposited on a seabed shallower than the CCD do not remain there forever. As shown in Fig. 8.24, the Earth's oceanic crust comprises basalt that originated in the mid ocean ridge. This plateau slowly moves (by a few centimeters each year) and eventually slips beneath the continental crust. The abyssal calcium carbonate sediments behave in a similar manner to the sand placed on a belt conveyor in a plate motion. If the argillaceous component accumulates above the calcium carbonate layer during seabed motion, calcium carbonate does not dissolve even at depths below the CCD. Instead, a vast majority of calcium carbonate is dragged under the continent with the oceanic crust, and the minerals are reconstructed under the high temperatures and pressures underground. This reaction is the opposite of the weathering equation (8.3):

$$CaCO_3 + SiO_2 \rightarrow CaSiO_3 + CO_2 \tag{8.6}$$

The carbon dioxide generated by (8.6) is released into the atmosphere again in the form of volcanic gases. In this way, carbon completes a circuit on the Earth's surface. The age of the oceanic crust is at most 200 million years; therefore, carbon has repeated this cycle many times throughout the geological ages.

References

1. Baronola, J.M., et al.: Nature **329**, 408–414 (1987)
2. Broecker, W.S., et al.: Nature **315**, 21–26 (1985)
3. Defant, A.: Physical Oceanography. Pergamon Press, New York (1960)
4. Friedli, G., et al.: Geophys. Res. Lett. **11**, 1145–1148 (1984)
5. Gill, A.E.: Atmosphere-Ocean Dynamics. Academic Press, Amsterdam (1982)
6. Hsiung, J.: J. Phys. Oceanogr. **15**, 140 (1985)
7. Milliman, J.D., Meade, R.M.: J. Geol. **91**, 1–21 (1983)
8. Nozaki, Y.: Global warming and sea, University of Tokyo press (1994)
9. Oki, T.: Science **63**, 678–682 (1993)
10. Stommel, H., Arons, A.B.: Deep Sea Res. **6**, 217–233 (1960)
11. Tsunogai, S., Noriki, S.: Chemical Oceanography. Sangyo Tosho, Tokyo (1983)
12. Yanagi, T.: Coastal Oceanography (Revised Edition). Kouseisha-kouseikaku (1994)

Chapter 9
Space Plasma Environment

It is not only the atmosphere and the ocean that constitute the Earth's environment. Once you leave the Earth and travel into space, you will find that the space is not a vacuum as is often misunderstood, but is filled with a medium called plasma, an extremely high-temperature fluid (gas) that conducts electricity. Via this plasma, the Earth's environment is linked to outer space, in particular, to dynamic activities of the sun.

In this chapter, we first introduce the solar wind, a hot, tenuous plasma flow streaming from the sun at supersonic speed. Then, we discuss some basic concepts of the plasma and introduce the magnetohydrodynamic (MHD) model, which is the fluid set of equations including electromagnetic interactions. Finally, using these concepts just introduced, we overview how the Earth's environment is influenced by the outer space phenomena.

9.1 Space is Not a Vacuum

As we have learned in previous chapters, there is no doubt that the light (the electromagnetic wave) radiating from the sun is the ultimate energy source for almost every activity in the Earth environment, e.g., convection of the atmosphere and the ocean, water circulation, accumulation of fossil fuels, and the vital activity of all creatures including that of humans. However, light is not the only substance emitted by the sun. Already in the middle of the nineteenth century, it was noticed that when a solar flare, a large-scale eruption on the surface of the sun, occurs the Earth's magnetic field is disturbed after a few days of delay. This phenomena led some scientists to speculate that the solar flare generates some energetic particles, which travel all the way from the sun to the Earth, and interact with the Earth's magnetic field to trigger

---

The original version of this chapter was revised: See the "Chapter Note" section at the end of this chapter for details. The erratum to this chapter is available at 10.1007/978-4-431-56499-7_10

D.E.S.S.T. Kyushu Univ., *Fluid Dynamics for Global Environmental Studies*,
DOI 10.1007/978-4-431-56499-7_9

the observed perturbations. At the beginning of the 20th century, it was discovered that auroral activities are correlated with an increase of the solar sunspot number. Aurora is one of the few space phenomena visible by naked eyes from the Earth. When electrically-charged energetic particles in space come to the Earth and bombard the ionosphere, molecules there such as Nitrogen and Oxygen are excited and emit light. Although it is now understood that it is not the solar energetic particles that directly cause the ionization of ionospheric molecules,[1] people knew already over a century ago that the solar activity, represented by the sunspot number in this case, can influence the Earth's environment. This, in turn, can be noticed as geomagnetic field oscillations and auroral light emissions.

In the 1950s, a German astronomer Ludwig Biermann inferred that a quasi-steady flow of plasma must be streaming from the sun, based on his observations of cometary tails. Comets usually have two tails: a dust tail is the stream of gasses and dust particles sublimated from the cometary nucleus. They are left behind in the cometary orbital plane while pushed away by radiation pressure of sunlight, making a slightly curved, whitish-looking tail pointing nearly away from the sun. More blueish-looking plasma tail, or the ion tail, is the stream of gasses of cometary origin ionized by sunlight. This tail always directs away from the sun, suggesting that the tail is blown by some kind of "wind" from the sun.[2] In 1958, an American physicist Eugene Parker showed that, in the fluid set of equations including the solar gravity and the effect of radial expansion, there exists a solution that can be interpreted as the "solar wind", a continuous plasma flow streaming away from the sun. Soon after this theoretical prediction, spacecraft observation confirmed the existence of the solar wind, as predicted by Parker. Nowadays, one can find movies of the solar wind on the internet. See, for example, homepages of the SOHO mission,[3] the YOHKOH mission,[4] and the HINODE mission.[5]

9.2 The Sun

Before discussing the solar wind, let us first briefly look at the properties of the sun. The sun is an ordinary star mainly consisting of hydrogen (about 92% in terms of particle number) as the fuel for solar energy production and helium (about 8%). Its

[1]The "auroral particles" are believed to be energetic electrons and protons stored in the magnetosphere, a region created by the confinement of the Earth's magnetic field by the solar wind. Some changes of the solar wind conditions, such as the direction of the magnetic field there, triggers these particles to start precipitating into the ionosphere.

[2]The radiation pressure of sunlight is not sufficient to explain the rapid movement of the fine structure of the plasma tail such as twists and kinks.

[3]SOHO mission by ESA (European Space Agency) and NASA (National Aeronautics and Space Administration), http://sohowww.nascom.nasa.gov/.

[4]YOHKOH mission by JAXA (Japan Aerospace eXploration Agency), http://www.darts.isas.ac.jp/solar/yohkoh/data.html.

[5]HINODE mission by JAXA, http://hinode.nao.ac.jp.

Table 9.1 Typical parameters of the sun and the earth

	The sun	The earth
Radius	6.96×10^8 m	6.35×10^6 m
Mass	1.99×10^{30} kg	5.97×10^{24} kg
Average mass density	1.41×10^3 kg/m^3	5.51×10^3 kg/m^3
Escape velocity	6.17×10^5 m/s	1.12×10^4 m/s
Average surface temperature	5.78×10^3 K	3.33×10^2 K
Luminosity	3.84×10^{26} J/s	–
Rotation period (equator)	25.4 days	1.00 day

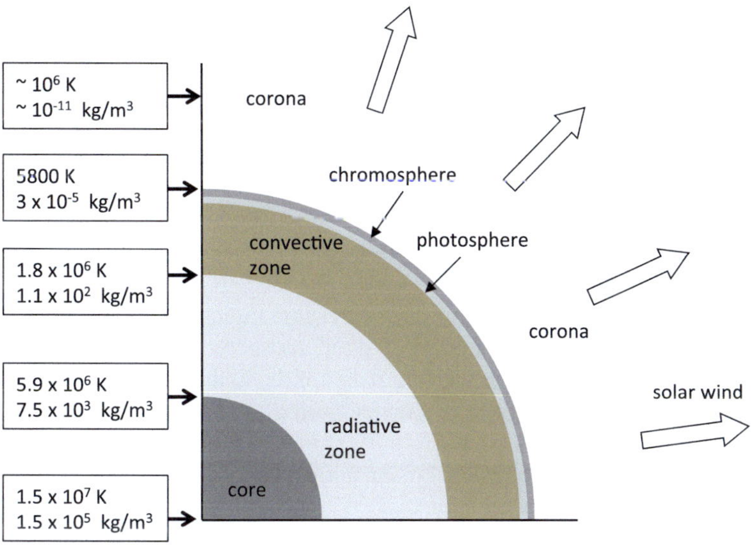

Fig. 9.1 The structure of the sun

size is about a hundred times larger and about 300 thousand times heavier than the Earth. Some basic parameters of the sun and the Earth are compared in Table 9.1. The average mass density shown in the table is evaluated simply by dividing the mass by the volume of the sun, but this is actually misleading because its interior is far from being homogeneous. Not only the mass density, but also pressure, temperature, and the energy density all strongly depend on the location, or the radial distance from the center, as different regions of the sun are dominated by different physical processes. It is safe to say that these physical processes are well understood theoretically. For an ordinary star like the sun, by giving initial conditions for its evolution such as the initial mass and the initial composition of chemical elements, one can tell fairly accurately the evolution of the star and its eventual fate. The **standard model** explained below is a result of such a theoretical computation, being fitted to the observed data of the sun, such as the solar radius, luminosity, mass, and chemical composition.

Figure 9.1 shows the structure of the sun and its surrounding region. The most internal **core** is the region extending from the center to about 20–25% of the solar radius. Since the core is at the center of the sun, it is severely compressed by the solar mass and the density there is up to 150 g/cm^3, temperature is about 15 million K (Kelvin), and the pressure is about 260 billion atm.[6] These extreme conditions enable nuclear fusion to take place at the core, in which energy is produced as hydrogen atoms convert into helium.

The region surrounding the core is the **radiative zone**, where energy is transported by radiation (energetic photons). Photons can travel only a very short distance, typically a few millimeters, before it is absorbed by the ambient gas, but they are released again in random directions at a slightly lower energy. Then they travel again a short distance and are captured by other gas. This random walk of photons eventually carries energy outward, even though a typical time scale for a single photon to traverse the radiation zone is as long as a few million years.

Adjacent to the radiation zone is the **convective zone**, where the density of the gasses is so low that convective currents can develop for more effective energy transfer than radiation. The resultant convection of the ionized gas is thought to be the source of a dynamo, by which the solar magnetic field is generated.

The visible surface of the sun, the **photosphere**, is the layer where both the gas density and the temperature are low enough so that the region is optically transparent. Below the photosphere, the sun is opaque to visible light, i.e., the rim of the sun as we observe by naked eyes corresponds to the photosphere. At the photosphere, the temperature is about 5800 K, and the main part of the light spectrum from the sun is due to the black body radiation at this temperature.

Outside the photosphere, the temperature continues to decrease as one goes further away, and at an altitude of about 500 km the temperature takes the minimum value of about 4,300K. The region up to around 2000 km from the photosphere is called the **chromosphere**, whereindexSun!chromosphere a variety of dynamic activities of the sun such as the solar flare and prominence is observed. Magnetic field plays crucial roles in the phenomena at the chromosphere.

There is a drastic increase of temperature outside the chromosphere, and in the **corona**, the region with a tenuous gas surrounding the chromosphere, the temperature is of the order of a few million K. The solar corona is the source region of the solar wind. It has been one of the central issues of solar physics for some decades to explain why the corona is so hot. The magnetic reconnection and the dissipation of Alfven waves in the corona are two plausible candidates.[7]

The solar radiation spectrum for a wide range of wavelength is shown in Fig. 9.2. The horizontal axis is the logarithm of the wave number, λ (in the unit of 1 Å $=$ 10^{-10} m $=$ 10 nm), which can be converted to the wave frequency, $f = c/\lambda$ with

[6] 1 atm $= 1.013 \times 10^5$ Pa. Pa is a SI unit equivalent to Nm^{-2}.

[7] Magnetic reconnection is a process where the magnetic energy is converted to gas thermal and kinetic energies via merging of two magnetic field lines. Alfven wave is one of the fundamental waves in the plasma, which can carry energy from the solar surface to a higher altitude by propagating into the corona, and release the energy to the gas by wave damping.

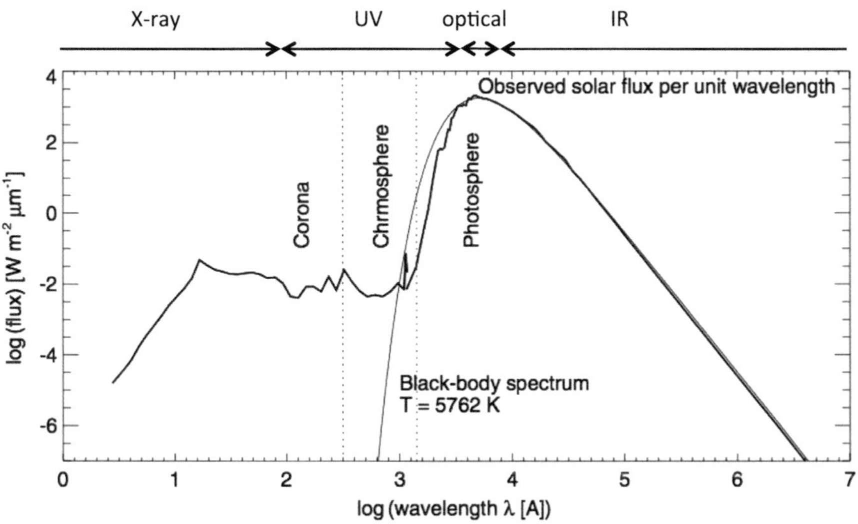

Fig. 9.2 The observed solar flux spectrum per unit wavelength superposed with a black-body spectrum of temperature $T = 5{,}762$ K [1]

$c = 3.0 \times 10^8$ m/s being the speed of light, to the photon energy, $\epsilon = hf$ with $h = 6.6 \times 10^{-34}$ Js being the Planck constant, and to equivalent temperature, $T = \epsilon/k_B$ with $k_B = 1.38 \times 10^{-23}$ JK^{-1} being the Boltzmann constant. The optical and the infra-red (IR) regimes are fairly well represented by a **black-body spectrum**,

$$B(\lambda, T) = \frac{2hc^2}{\lambda^5} \frac{1}{e^{hc/\lambda k_B T} - 1} \tag{9.1}$$

with $T \sim 6{,}000$ K, where $B(\lambda, T)$ is the amount of power per unit surface area, per unit solid angle, and per unit wavelength emitted at a wavelength λ by a black body at temperature T. In addition to this "thermal" part of the spectrum, there is substantial radiation flux detected at shorter wavelength regimes, i.e., at the ultraviolet (UV) and (soft) X-ray regimes. Since the wavelength of 100 Å corresponds to $T = 1.44 \times 10^6$ K, there must exist some physical processes that can heat the gas and maintain its temperature at a million or higher K.

9.3 Solar Wind Models

Let us consider how the solar wind can be produced. Hereafter, we describe the solar wind as a fluid with mass density ρ, speed $\boldsymbol{u}$, pressure p, and temperature T. The solar wind in reality is a plasma and it couples with the electromagnetic field, but here we assume a neutral fluid for simplicity. We also assume one-dimensionality:

variables depend only on the z-axis for cartesian coordinates or on the r-axis for spherical coordinates and allow derivatives with respect to all the other coordinates vanish.

Governing set of equations is the continuity equation (1.37),

$$\frac{\partial \rho}{\partial t} + \nabla \cdot (\rho \boldsymbol{u}) = 0 \tag{9.2}$$

the Euler equation (1.49),

$$\rho \left(\frac{\partial}{\partial t} + \boldsymbol{u} \cdot \nabla \right) \boldsymbol{u} + \nabla p = \boldsymbol{f} \tag{9.3}$$

and the equation of state (1.54),

$$p = p(\rho, T) \,. \tag{9.4}$$

In the above $\boldsymbol{f}$ is the external force (such as the gravity force) acting per unit mass. We seek a stationary ($\partial/\partial t = 0$) solution for the above set of equations.

9.3.1 Hydrostatic Equilibrium Under Uniform Gravity

When the gravity is constant,

$$\boldsymbol{f} = -\rho g \boldsymbol{z} \tag{9.5}$$

where ρ is the mass density of the gas, g is the acceleration of gravity (=constant) and $\boldsymbol{z}$ the unit vector along the z-axis, the equation set (9.2)–(9.4) yields the **hydrostatic equilibrium** solution (6.1), which can be written as

$$\rho(z) = \rho_0 \exp\left(-\frac{z}{h}\right) \tag{9.6}$$

where we have assumed an **isothermal gas**, $p = a^2 \rho$, for the equation of state (Fig. 9.3). Here, $h = a^2/g$ is the scale height, $a^2 = T/m$ the sound speed squared, m is the average mass of individual particles constituting the gas, and T is the temperature measured in the unit of energy, (i.e., 'usual' temperature times the Boltzmann constant, k_B). Although the particles try to spread into space by thermal motion they are forced to return to the ground ($z = 0$) by the gravitational force, which continues to exist no matter how high the altitude is. The eventual spread of the gas becomes of the order of the scale height. This solution, of course, is not the solar wind.

In Fig. 9.3 a group of solutions are added as dashed lines when the gas obeys the **polytropic law**, $p = p_0(\rho/\rho_0)^\gamma$, where p_0 and ρ_0 are constants and γ is the ratio of specific heats. The isothermal gas is recovered by letting $\gamma = 1$. When this relation

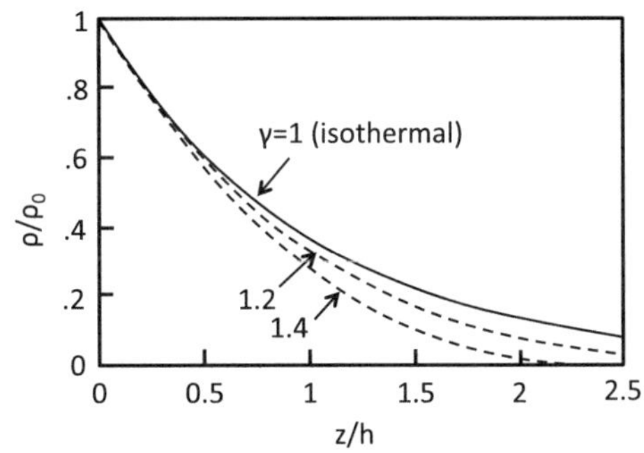

Fig. 9.3 Density profile for the hydrostatic equilibrium under uniform gravity

is employed as the equation of state, the hydrostatic equilibrium is given as

$$\rho(z) = \rho_0 \left(1 - \frac{(\gamma - 1)z}{h}\right)^{\frac{1}{\gamma-1}} . \tag{9.7}$$

When $\gamma > 1$, the gas temperature decreases as z is increased, and thus the spread of the gas is reduced compared with that of the isothermal gas.

9.3.2 *Hydrostatic Equilibrium for the Gravity Obeying the Inverse-Square Law*

Let us consider the case where the gravity depends on the altitude by the inverse-square law,

$$f = -\frac{\rho g r_0^2}{r^2}\boldsymbol{r} , \tag{9.8}$$

where the spherical coordinates are used. The above may be interpreted as the gravitational force for the solar atmosphere, with r_0 being the distance from the center to the photosphere, and $g = GM_s/r_0^2$ the gravitational acceleration evaluated at $r = r_0$, where M_s is the solar mass and G is the gravitational constant. The difference between (9.5) and (9.8) is simply in the spatial scales they describe: in (9.5), representing, for example, the earth's atmosphere, the gravitation force can be approximated as a constant because the scale height turns out to be much less compared with the earth radius. On the other hand, for the sun, the coronal temperature is so high that the constant gravity approximation cannot be used.

Assuming the isothermal gas as before, the density distribution is given as

$$\rho(z) = \rho_0 \exp\left(-\frac{r_0}{h}\left(1 - \frac{r_0}{r}\right)\right) . \tag{9.9}$$

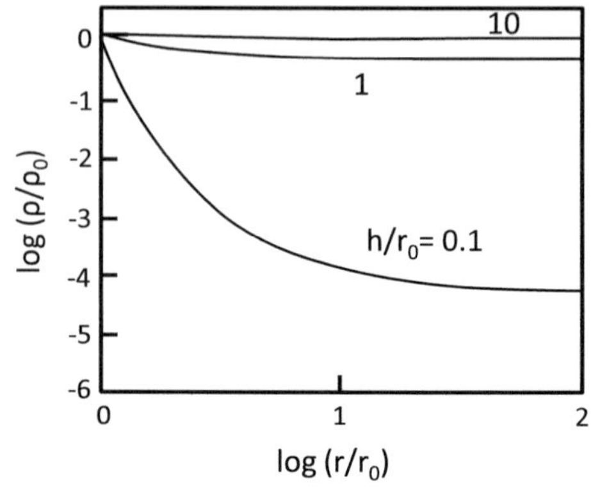

Fig. 9.4 Density profile for the hydrostatic equilibrium with the gravity obeying the inverse-square law

Some numerical solutions are plotted in Fig. 9.4 for various values of h/r_0, a parameter representing how strongly the gravitational force depends on the altitude. In the limit of $h/r_0 \to 0$, (9.9) reduces to (9.6) by writing $r - r_0 = z$. The static solution (9.9) was once proposed as the model of the solar wind, but it is clear now that this is different from the real solar wind.

9.3.3 *Streaming Solution in a Cartesian Geometry*

Next we allow the solution to be streaming, $u \neq 0$. Let $f = -\rho g z$, and $\rho = \rho_0$ and $u = u_0$ (initial velocity) at $z = 0$, and employ again the isothermal gas, the solution for the density is given as

$$\log \frac{\rho}{\rho_0} + \frac{M^2}{2}\left[\left(\frac{\rho}{\rho_0}\right)^2 - 1\right] = -\frac{z}{h}\,, \tag{9.10}$$

where $M = u_0/a$ is the **Mach number** defined for the initial flow speed. Once the density is given, the flow speed can be obtained by

$$\rho u = \rho_0 u_0\,. \tag{9.11}$$

Figure 9.5 shows these solutions for various values of M. Both ρ and u are multi-valued as functions of z, which is unphysical since both of these quantities have to be uniquely specified at any altitude. The only exception is the case with $M = 0$, the hydrostatic solution, in which $u = 0$ everywhere. To summarize, the attempt to introduce the steady streaming of the flow fails for the case with constant gravity in a purely one-dimensional geometry (i.e., no radial expansion).

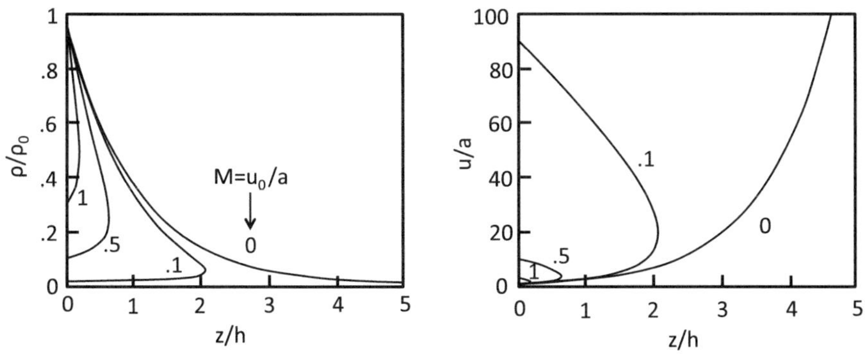

Fig. 9.5 Density and velocity profiles of streaming solutions in a cartesian geometry (unphysical)

9.3.4 Streaming Solution in a Spherical Geometry

Now we show that a steady streaming solution exists in a spherical geometry. Take a spherical coordinate with its origin at the center of the sun, and re-write (9.2) and (9.3) after dropping the time derivative terms $\partial/\partial t = 0$,

$$\frac{d}{dr}(r^2 \rho u) = 0\,, \tag{9.12}$$

$$\rho u \frac{du}{dr} + \frac{dp}{dr} = -\frac{\rho G M_s}{r^2}\,. \tag{9.13}$$

Assuming once again the isothermal gas, a nonlinear ordinary differential equation for u is obtained,

$$\frac{1}{u}\frac{du}{dr} = \frac{2a^2}{r^2}\frac{r - r_c}{u^2 - a^2}\,, \tag{9.14}$$

where $r_c = GM_s/2a^2$ is sometimes referred to as the **critical distance**. Although this equation can be solved analytically by separation of variables (shown below in (9.15)), it is more instructive to construct the solution set graphically. The phase space of r and u is divided into four regions with boundaries at $r = r_c$ and $u = a$. Through (9.14), each region uniquely specifies the sign of du/dr, i.e., whether the stream is accelerated or decelerated as the distance from the sun increases. Solutions to (9.14) can be obtained by smoothly connecting the solution curves in each region at the boundaries (Fig. 9.6).

Mathematical solutions are not necessarily physical solutions. The group of solutions in region I and III are multi-valued for a single r and are unphysical. The solutions in region II represent flows that are accelerated within $r < r_c$ but are decelerated afterwards, and remain subsonic $M = u/a < 1$ everywhere. This does not match the observation and must be discarded. The solutions in region IV must

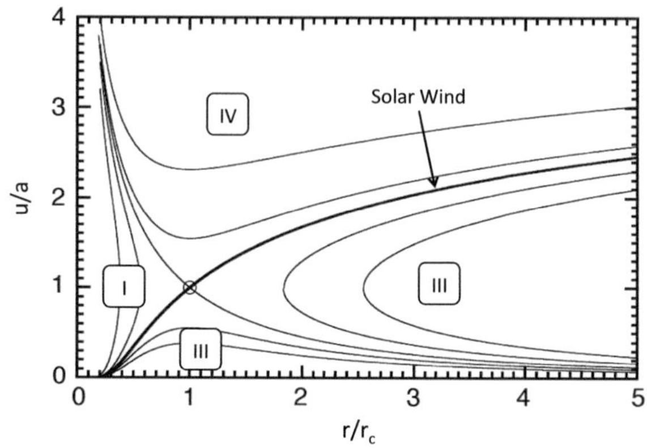

Fig. 9.6 Velocity (Mach number) profile of streaming solutions in a spherical geometry. Isothermal gas is assumed. The *thick line* corresponds to the solar wind solution

have very large initial speeds and again unrealistic (there is no physical process to accelerate the flow at the photosphere). The only solution that explains the observation is the one that starts from an initial subsonic speed, cross the double singular point at $r = r_c$ and $u = a$, and becomes supersonic afterwards. This is the solar wind solution, first discovered by Eugene Parker in 1958. In this solution, the flow is accelerated everywhere.

Let us now evaluate various important parameters using realistic values of the sun. As shown in Table 9.1, the solar radius is $R_s = 6.96 \times 10^8$ m, solar mass $M_s = 1.99 \times 10^{30}$ kg, and we assume the corona constitutes of protons and electrons and we only need to consider acceleration of protons, since much lighter electrons will follow the motion of protons as they are coupled by electrostatic force. Taking the coronal temperature as a million kelvin ($=1.38 \times 10^{-17}$ J), the proton mass $m_p = 1.67 \times 10^{-27}$ kg, and the gravitational constant $G = 6.67 \times 10^{-11}$ m^3/kg s^2, we find the isothermal sound speed $a \sim 91$ km/s, and the critical distance $r_c \sim 8.0 \times 10^9$ m. If $r_c < R_s$, i.e., if the critical distance is within the solar radius, there will be no solar wind solution, but in reality $r_c \sim 10R_s$. Using the parameters above, we can estimate that the wind speed at the earth orbit (1 A.U. = astronomical unit, $r = 1.5 \times 10^{11}$ m) is about $u = 500$ km/s. This roughly matches the observation.

The asymptotic behavior of the isothermal solar wind can be evaluated. Integration of (9.14) yields the following for the solar wind solution

$$\frac{u^2}{a^2} - \log\frac{u^2}{a^2} = 4\log\left(\frac{r}{r_c}\right) + \frac{4r_c}{r} - 3\,. \tag{9.15}$$

By letting $r \rightarrow \infty$ and assuming u is sufficiently larger than a, one finds

$$u(r) \sim 2a\left(\log\left(\frac{r}{r_c}\right)\right)^{1/2}\,. \tag{9.16}$$

Namely, the flow speed monotonically increases as the wind flows further away. On the other hand, since $r^2\rho u$ must be conserved from (9.12), the wind density is approximated as

$$\rho(r) \sim \rho(r_c)\frac{r_c^2 u_c}{r^2 u}\,. \tag{9.17}$$

Imagine a spherical shell with radius r with its center corresponding to the sun. The energy flux by the wind passing this shell per unit time is

$$F_e = 4\pi r^2 u\left(\frac{\rho u^2}{2}\right) \sim \frac{8\pi r_c^2 \rho_c a^2}{u_c^2}\log\left(\frac{r}{r_c}\right)\,. \tag{9.18}$$

Since F_e increases as r is increased, there must exist an external energy source that keeps injecting the energy into the wind according to this model. So, what is this energy source? The answer to this question is in our assumption of the isothermal equation of state. Suppose we release a box in the solar wind that can freely deform its shape according to the motion of the background fluid. Since the solar wind is accelerated everywhere, the size of the box keeps expanding, and due to our isothermal assumption, the internal energy within the box must keep increasing as well. In this model, the energy is assumed to be provided by an external heat sink with infinite capacity. Although this is unrealistic, we learn here that an addition of energy can accelerate the solar wind.

9.3.5 Polytropic Solar Wind

Instead of the isothermal gas, let us consider a polytropic gas, which obeys the equation of state

$$p = p_0\left(\frac{\rho}{\rho_0}\right)^{\Gamma}\,, \tag{9.19}$$

where p_0 and ρ_0 are constants, and Γ is the external parameter. By assigning Γ a value of the ratio of specific heat, one can model the adiabatic change of the ideal gas. For example, for a monoatomic gas, $\Gamma = 5/3$, and for gas with molecules with the number of degrees of freedom f, $\Gamma = (f+2)/f$. Since the temperature of the solar wind is way above the ionization level, the gas can be well assumed to be monoatomic, but by allowing Γ to be different from 5/3, one can model a solar wind that is not completely adiabatic but has a channel of external energy flow into the wind. The range of Γ is $1 \leq \Gamma \leq 5/3$, corresponding to a gas that is between the isothermal and the adiabatic states.

Combining (9.12), (9.13) and (9.19), an equation for the polytropic solar wind is derived

$$\frac{1}{u}\frac{du}{dr} = \frac{2a(u,r)^2}{r^2}\frac{r - r_c(u,r)}{u^2 - \Gamma a(u,r)^2}\,, \tag{9.20}$$

where $a(u,r)$ is the isothermal sound speed (i.e., the real sound speed is $\Gamma^{1/2}a(u,r)$), $r_c(u,r)$ is the critical distance, and they both depend on u and r as below,

$$a(u,r)^2 = \frac{p_0}{\rho_0}\left(\frac{u_0}{u}\right)^{\Gamma-1}\left(\frac{r_0}{r}\right)^{\Gamma-1}, \tag{9.21}$$

$$r_c(u,r) = \frac{GM}{2a^2(u,r)}. \tag{9.22}$$

In the above, ρ_0, p_0, u_0 are the values of ρ, p, u at a certain distance $r = r_0$, which is not necessarily the solar radius. The polytropic solar wind equation (9.20) is similar to the isothermal solar wind equation (9.14), but the right-hand side of the former is extremely nonlinear in u and so its analytic solution cannot be obtained. On the other hand, for the polytropic gas, one can make use of the **energy equation**,[8]

$$\frac{dF_e}{dr} = 0\ , \quad f_e = \frac{F_e}{4\pi r^2 \rho u} = \frac{u^2}{2} + \frac{\Gamma}{\Gamma - 1}\frac{p}{\rho} - \frac{GM}{r}, \tag{9.23}$$

where F_e is the total energy flux per unit time as defined before, and f_e is its density per unit mass flux. Since $4\pi r^2 \rho u$ is constant, so is f_e as well. Using this fact, let us evaluate the asymptotic solar wind speed, u_∞, in the polytropic model. First we choose r_0 to be the critical distance, $r_0 = r_c$, so that $\rho_0 = \rho_c$, $p_0 = p_c$, and $u_0 = u_c$. Then, after simple algebra, the value of f_e at $r = r_c$ is found to be

$$f_e = \frac{5 - 3\Gamma}{4(\Gamma - 1)}\frac{GM}{r_c}. \tag{9.24}$$

On the other hand, at $r \to \infty$, $u \to u_\infty$ and $p/\rho \sim \rho^{\Gamma-1}$ with $\rho \to 0$, so that $p/\rho \to 0$. Since $GM/r \to 0$ as well, we find

$$f_e = \frac{u_\infty^2}{2}. \tag{9.25}$$

Due to the conservation of the energy flux, by equating (9.24) and (9.25), the asymptotic flow speed is found to be

$$u_\infty^2 = \frac{5 - 3\Gamma}{2(\Gamma - 1)}\frac{GM}{r_c}. \tag{9.26}$$

From the above, it is apparent that u_∞ keeps increasing and blows up in the isothermal limit ($\Gamma = 1$), while the acceleration is terminated for purely adiabatic solar wind ($\Gamma = 5/3$). Realistic solar wind speed can be well represented phenomenologically

---

[8] In case of the isothermal gas, $\Gamma = 1$, the energy equation cannot be defined. One can see this in (9.23) where the factor $\Gamma - 1$ is included in the denominator of the enthalpy per unit mass (second term in the definition of F_e).

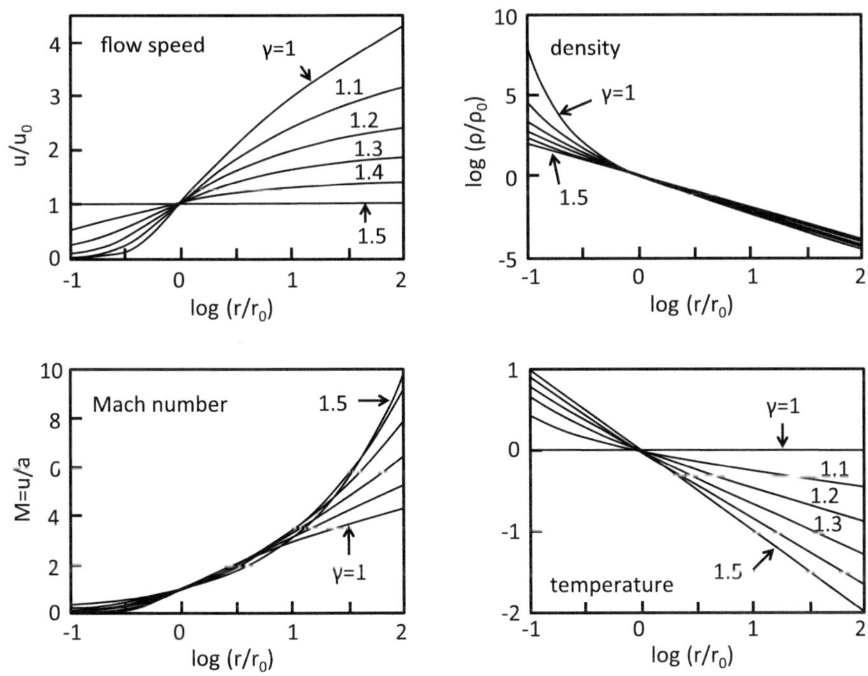

Fig. 9.7 Polytropic solar wind solutions

when $\Gamma = 1.1 \sim 1.2$. In Fig. 9.7, several solar wind solutions are plotted for various values of Γ.

9.3.6 Physical Mechanism of the Solar Wind Acceleration

To understand the acceleration of the solar wind from a physical point of view, let us consider a **nozzle flow** as depicted in Fig. 9.8. For simplicity, we treat the one-dimensional problem: the flow is assumed to be always parallel to the z-axis. Then, the continuity equation and the equation of motion are,

$$\frac{\partial \rho}{\partial t} + \frac{\partial}{\partial z}(A(z)\rho u) = 0\,, \tag{9.27}$$

$$\rho\left(\frac{\partial}{\partial t} + u\frac{\partial}{\partial z}\right)u + \nabla p = 0\,, \tag{9.28}$$

where $A(z)$ is the cross section of the nozzle. Assuming time stationarity ($\partial/\partial t = 0$) and isothermal gas, the evolution equation of u is given as

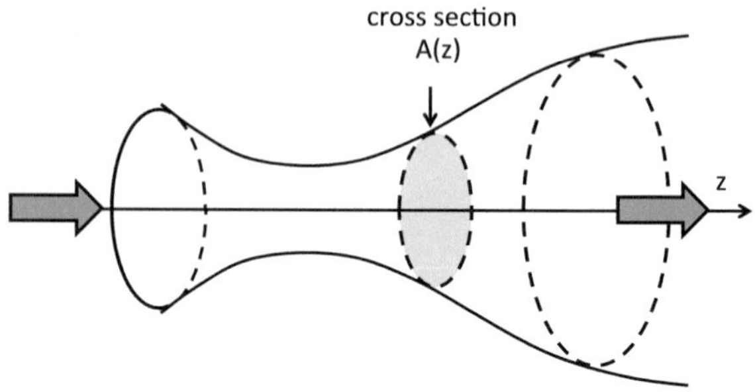

Fig. 9.8 Nozzle flow. This can be made equivalent to the solar wind flow by appropriately choosing the cross section of the nozzle, $A(z)$

$$\frac{1}{u}\frac{du}{dz} = \frac{1}{A}\frac{dA}{dz}\frac{a^2}{u^2 - a^2}. \tag{9.29}$$

Comparison of the above with (9.14) can specify the shape of the nozzle that models the isothermal solar wind.

The critical distance corresponds to the point with $dA/dz = 0$, i.e., the location where the nozzle is at its narrowest. According to the solution corresponding to the solar wind, the flow is subsonic and supersonic, before and after this critical point, respectively. Now let us consider the case where the nozzle becomes narrower as the flow proceeds ($dA/dz < 0$). There are two examples corresponding to this situation with opposite outcomes. One of them is a toothpaste tube. As everyone experiences, toothpaste leaks out quickly from the tube, when the bottom part of the tube is gently pressed. Namely, the fluid of toothpaste is accelerated in this example. The second example is a flow of people rushing towards a single exit. This again must be experienced by everyone, that near the exit people are jammed up, and the average speed of human flow is slowed down as they approach the exit. In this example, what is observed is the deceleration of the fluid. The two examples above share the same types of geometry, $dA/dz < 0$, but their outcome is opposite.

This is due to the difference in relative speed of the flow compared with propagation speed of information in the flow, which in the ordinary gas corresponds to the sound speed. Toothpaste is dense and may be regarded as almost a solid for the internal sound wave to propagate, so that the flow of the toothpaste, triggered by a little push of the tube, is sub-sonic. On the other hand, the flow of people must be regarded as supersonic. This is because, when there are many people, it takes a certain while for the usual person to notice that there is another person in front and react to avoid hitting this person. Typically, the information propagation speed in a dense crowd of people is less than the walking speed of people.[9] As illustrated by the two examples above, the narrowing nozzle produces flow acceleration for the subsonic flow ($M < 1$), and deceleration for the supersonic flow ($M > 1$). Here the Mach number is defined as the ratio of the flow speed to the speed of information

[9] If there are not many people, they all may be able to exit smoothly, because nobody needs to avoid hitting other people, which significantly reduces the reaction time scale. When the human flow is very smooth, the flow speed can increase near the exit.

propagation within the flow. The exact opposite takes place for the divergent geometry of the nozzle, $dA/dz > 1$. Namely, the subsonic and the supersonic flows are decelerated and accelerated, respectively.

The presence of gravity in the case of the solar wind corresponds to narrowing of the nozzle. This is effective only near the sun, since at larger distances from the sun, the gravity fades away due to the inverse square law. Instead, at greater distances, expansion of the coordinate due to spherical symmetry dominates, which corresponds to the widening of the nozzle. The solar wind solution utilizes both regions in such a way that the wind is always accelerated, by being subsonic before the critical point ($dA/dz < 0$ and $M < 1$), and by being supersonic after the critical point ($dA/dz > 0$ and $M > 1$).

9.4 What is a Plasma?

We have just learned in the previous section that the space environment is filled with a plasma. The plasma is sometimes called the fourth state of matter, following the solid, liquid, and the (usual) gas as the first to the third states, respectively. The transition to a different phase can be made by giving or subtracting heat. For example, under normal pressure, ice melts to become water at 0 °C, water to vapor at 100 °C, and vapor to the plasma around ten thousand degrees. In the state of the plasma, the temperature is so high that the electrons and nucleus that comprise the molecules are separated so that they can move freely. Since these particles are electrically charged, their motion is influenced by the electromagnetic field, while the electromagnetic field is generated by their motion. Namely, the plasma and the electromagnetic field interact with each other.[10] Since we cannot see neither the plasma nor the electromagnetic field with our naked eyes, we usually do not notice their presence, but they are everywhere in space.

9.4.1 *Debye Shielding*

The plasma consists of many positively and negatively charged particles. In case of the space plasma, usually the negatively charged particles are the electrons, and the positive ones are mostly the protons (nucleus of the hydrogen atom), plus some minor populations such as ionized helium, oxygen, etc. While most of the space plasma are fully ionized due to the high temperature, plasma in the laboratory, or plasma in the ionosphere, for example, are not fully ionized, and contains a substantial number of neutral particles.

[10]To be precise, here we are talking about the interaction between collective motion of particles and the electromagnetic field at macroscopic scales (spatial scale greater than the Debye scale length, to be defined in a moment).

Hereafter, for simplicity we consider a plasma consisting only of electrons (species $j = e$) and protons ($j = i$). Let their number density (number of particles per unit volume) be n_j, mass m_j, and charge e_j, where $e_e = -e$ and $e_i = e$ with e being the electronic charge, and T_j be the temperature measured in the unit of energy.[11] If the protons (hereafter, ions) and electrons Coulomb collide with each other frequently, their temperatures would be equal. However, due to the low plasma density, the collision is almost negligible, and their temperatures can be independent. Since the plasma is produced via ionization of neutral particles, the number of positive and negative charges must be equal as a whole. On the other hand, there may be a local deviation from neutrality. Let us discuss this quantitatively.

Supposing that both ions and electrons obey the Boltzmann statistics,[12] we should have

$$n_e(\phi) = n_0 \exp\left(\frac{e\phi}{T_e}\right), \quad n_i(\phi) = n_0 \exp\left(-\frac{e\phi}{T_i}\right) \tag{9.30}$$

where $\phi(\boldsymbol{r})$ is the electrostatic potential, which is chosen to vanish (so that $n_i = n_e = n_0$) at infinity. If a point test charge, q, is placed at $\boldsymbol{r} = 0$, how do the ambient particles react? The Poisson equation for this setting is written as

$$\nabla^2\phi = -\frac{1}{\epsilon_0}\left[e(n_i - n_e) + q\delta(\boldsymbol{r})\right], \tag{9.31}$$

where ϵ_0 is the permittivity of vacuum, and inside the parenthesis of the r.h.s. is the charge density. Expanding (9.30) assuming $|e\phi|/T_j \ll 1$ (this can be confirmed afterwards), and also by noting the spherical symmetry of the problem, (9.31) is written as,

$$\frac{1}{r^2}\frac{\partial}{\partial r}\left(r^2\frac{\partial\phi}{\partial r}\right) - \frac{\phi}{\lambda_D^2} = -\frac{q}{\epsilon_0}\delta(\boldsymbol{r}), \tag{9.32}$$

where the **Debye length** is defined as

$$\lambda_D = \left(\frac{n_0 e^2}{\epsilon_0 T_e} + \frac{n_0 e^2}{\epsilon_0 T_i}\right)^{-1/2} = \left(\frac{1}{\lambda_{De}^2} + \frac{1}{\lambda_{Di}^2}\right)^{-1/2}. \tag{9.33}$$

Solution to (9.32) is given by the Yukawa-type potential,

$$\phi(\boldsymbol{r}) = \frac{q}{4\pi\epsilon_0 r}\, e^{-r/\lambda_D}. \tag{9.34}$$

Compared with the electrostatic potential created by a point charge in a vacuum, $\phi(\boldsymbol{r}) = q/4\pi\epsilon_0 r$, the influence of the test charge immersed in a plasma is limited

---

[11] Temperature can be defined in the unit of energy by multiplying the usual temperature in K by the Boltzman constant, $k_B = 1.38 \times 10^{-23}$ J K^{-1}. Temperature of 1 eV is equivalent to 11,600 K.

[12] We have encountered an example of a gas obeying the Boltzmann statistics in (9.6), which can be re-written as $\rho(z) = \rho_0 \exp(-mgh/T)$.

within the scale λ_D. This is due to the shielding effect by the plasma, achieved by displacement of the electrons and the ions. If there are no thermal fluctuations ($T_e = T_i = 0$), the shielding is perfect and $\lambda_D = 0$.

After placing the test charge in the plasma, there is a finite time necessary for each species to move around and complete the shield. To see this, we write the Debye length due to each species as

$$\lambda_{De} = \left(\frac{\epsilon_0 T_e}{n_0 e^2}\right)^{1/2} = \frac{v_{te}}{\omega_e}, \quad \lambda_{Di} = \left(\frac{\epsilon_0 T_i}{n_0 e^2}\right)^{1/2} = \frac{v_{ti}}{\omega_i}, \tag{9.35}$$

where $v_{tj} = (T_j/m_j)^{1/2}$ is the thermal speed, and $\omega_j = (n_0 e^2/\epsilon_0 m_j)^{1/2}$ is the **plasma frequency** for species j. Inverse plasma frequency gives a measure of reaction time scale for each species.[13] Since ω_e is much larger than ω_i, electrons are much quicker to respond to the appearance of the test charge. If the ions cannot participate in the shielding, the Debye shielding distance will be given by λ_{De}.

Now let us evaluate how much the charge neutrality is violated at local scales. If we assume $T_i = 0$ for simplicity,

$$|\frac{e(n_i - n_e)}{en_0}| = |\frac{\epsilon_0 \nabla^2 \phi}{en_0}| = |\frac{\epsilon_0 T_e}{en_0 L^2}\frac{e\phi}{T_e}| \sim \frac{\lambda_D^2}{L^2}, \tag{9.36}$$

where L is the spatial scale of interest, and we have used $e\phi/T_e \sim 1$, i.e., equipartition of the potential and the electron kinetic energy. The above relation suggests that the plasma can be considered neutral for $L > \lambda_D$.

Although the Coulomb force is a long range force (the inverse square law), the presence of the plasma reduces its effective distance down to the Debye shielding scale. If the plasma is non-neutral, the Coulomb force remains long range, and the property of such a system becomes similar to a multi-body gravitational system where there is no negative mass to shield the gravitational force.

9.4.2 *Plasma Parameter*

The number of particles contained in a sphere with radius λ_D,

$$N_D \equiv \frac{4\pi}{3} n \lambda_D^3, \tag{9.37}$$

is called the "plasma parameter". In the discussion of the Debye shielding, we assumed a priori that there exists a large number of electrons and ions near the test charge, i.e., $N_D \gg 1$, so that they can be treated as fluids. When $N_D < 1$,

[13] For example, the skin depth (or the electron inertial length), c/ω_e, where c is the speed of light, gives the scale that a light wave can penetrate into a plasma.

only a small number of electrons and ions can participate in the shielding, so strictly speaking the Debye shielding model we discussed above cannot be applied.

For a plasma with the number density n_0, the average distance between the electrons is of the order of $a \equiv n_0^{-1/3}$. The Coulomb potential energy evaluated at this distance divided by the average kinetic energy of electrons is called the **Coulomb coupling parameter**, and is related to N_D as,

$$\Gamma \equiv \frac{e^2/4\pi\epsilon_0 a}{T_e} \sim N_D^{-2/3}. \tag{9.38}$$

Namely, when $N_D \gg 1$, Coulomb energy is much less than the average particle kinetic energy. Such a plasma is referred to as the "weakly-coupled plasma". The majority of space plasma is in this category. The plasma with $N_D < 1$, on the other hand, is called the "strongly-coupled plasma" since the Coulomb interaction dominates the thermal motion. The condition $N_D < 1$ occurs for a large density and/or a small temperature. Such a plasma may be found in the interior of stars, in inertial fusion experiments, and in the electron gas in metals. Depending on the value of N_D, this type of plasma can behave like a liquid, or even a solid (e.g., Coulomb crystal).

9.4.3 Collisional and Collisionless Plasma

The air in the atmosphere is dense, and the molecules there collide with each other very frequently: typical **collision time scale** between air molecules is $\sim 10^{-10}$ s, and their typical **mean free path** is $\sim 10^{-7}$ m. As a result, air molecules are in thermal equilibrium, and distribution functions of all species (such as nitrogen, oxygen, etc.) are written by the Maxwellian distribution with an equal temperature for all the species. On the other hand, since space plasma in many situations have extremely low density, the Coulomb collisions between particles seldom occur, leading to the emergence of non-equilibrium distributions. In fact, it is safe to say that distribution functions of any species in space are almost always non-Maxwellian.

Let us make a quantitative argument. Suppose that in a vacuum an ion is at rest, and an electron approaches it from a long distance away with velocity v. Since the ion is much heavier than the electron, we can assume that it continues to be at rest during the entire course of the interaction. Let b be the closest distance between the two particles. We will call it a collision, if the Coulomb energy evaluated at this distance equals or is larger than the initial electron kinetic energy,

$$\frac{m_e v^2}{2} \leq \frac{e^2}{4\pi\epsilon_0 b}. \tag{9.39}$$

From the above, the collision cross section is obtained,

$$\sigma = \pi b^2 = \frac{e^4}{4\pi\epsilon_0^2 m_e^2 v^4} . \tag{9.40}$$

A fast particle will pass quickly through the region under influence of the ion charge, and so it will not be much accelerated by the Coulomb force, while a slower particle will stay a longer time near the ion and will receive a strong acceleration.

Now let us replace v by the thermal velocity, $v_e = (2T_e/m_e)^{1/2}$. Then the mean free path of electrons is $L_{mfp} = 1/n_0\sigma$, and the electron - ion collision frequency is estimated as

$$\nu_{ei} = \frac{v_e}{L_{mfp}} = \frac{2^{1/2} n_0 e^4}{4\pi\epsilon_0^2 m_e^{1/2} T_e^{3/2}} . \tag{9.41}$$

A more rigorous calculation taking into account the exact trajectory of the incident electron gives

$$\nu_{ei} = \frac{n_0 e^4}{32\pi^{1/2}\epsilon_0^2 m_e^{1/2}(2T_e)^{3/2}} \log(12\pi n_0 \lambda_D^3) . \tag{9.42}$$

The above two estimates are essentially the same, except for the logarithmic term (Coulomb logarithm) in (9.42). In most cases this term is of the order of 10.

Let the spatial and the temporal scales of the subject of interest be L and τ, respectively. If $L_{mfp} \ll L$ and $\nu_{ei}^{-1} \ll \tau$, then the electron-ion collision occurs frequently enough both in spatial and in temporal scales, and we may consider such a plasma to be in **thermal equilibrium**. On the other hand, if either of the above conditions is not satisfied, the plasma has to be treated as a **non-equilibrium** medium.

Typical parameters in various plasmas in space and laboratories are summarized in Table 9.2. Most of the plasmas satisfy $n_0\lambda_D^3 \gg 1$. In the solar wind, $L_{mfp} \sim 10^{11}$ m, which is almost equal to the distance between the sun and the Earth (1AU $= 1.49 \times 10^{11}$ m). The plasma in the magnetosphere also has a very long mean free path. Typically, the space plasma is a **collisionless plasma**. collisionless plasma can exhibit quite peculiar characteristics, for example, the temperature measured along and perpendicular to the magnetic field can be different (temperature anisotropy), the distribution function may include a beam component, and different species can have different temperatures, to mention a few.

So far, as a means of the collision between the particles, we have only considered the Coulomb interaction. However, there is a channel of effective collision via the electromagnetic field. In this way of collision, first a coherent motion of particles generate electric current and/or charge displacement, which then produce the electromagnetic field, and finally the electromagnetic field just produced accelerates the other group of particles. This type of process, sometimes referred to as **anomalous collision** or **anomalous dissipation**, plays central roles in various space plasma phenomena. Some of the non-equilibrium distributions mentioned above, for example, the plasma with temperature anisotropy or with the beam, can relax to a state closer to the Maxwellian due to the anomalous dissipation. During this course, plasma waves are excited. Another example exhibiting the importance of the anomalous dissipation

Table 9.2 Basic parameters in various plasmas [3]

	n (m^{-3})	T (eV)	λ_D (m)	$n_0\lambda_D^3$	ν_{ee} (s^{-1})	L_{mfp} (m)	L (m)
Solar corona	10^{15}	100	10^{-3}	10^7	10^2	10^5	10^8
Solar wind	10^7	10	10	10^9	10^{-5}	10^{11}	10^{11}
Earth's magnetosphere	10^4	10	100	10^{11}	10^{-8}	10^{14}	10^8
Earth's ionosphere	10^{11}	0.1	10^{-2}	10^4	10^2	10^3	10^5
Magnetic fusion	10^{20}	10^4	10^{-4}	10^7	10^4	10^4	10^1
Laser fusion	10^{31}	10^4	10^{-10}	10^2	10^{14}	10^{-7}	10^{-5}
Laboratory plasma	10^{20}	5	10^{-6}	10^3	10^8	10^{-2}	10^{-1}

is the collisionless shock waves, a ubiquitous entity in space and astrophysical environments. Discovery of the collisionless bow shock in front of the Earth's magnetosphere, with its shock transition scale much less than 1AU, led many plasma scientists to believe that the anomalous dissipation is the key issue in space plasma physics.

9.5 Waves in a Cold Plasma

To understand the nature of the collisionless plasma in space, let us see what kind of waves can exist there. Understanding the linear response (=linear waves) is the first step for any physical systems in dynamical equilibrium.

First we write down the **Maxwell's equations**,

$$\nabla \times \boldsymbol{E} = -\frac{\partial \boldsymbol{B}}{\partial t} \tag{9.43}$$

$$\nabla \cdot \boldsymbol{E} = \frac{\rho_e}{\epsilon_0} \tag{9.44}$$

$$\nabla \times \boldsymbol{B} = \frac{1}{c^2}\frac{\partial \boldsymbol{E}}{\partial t} + \mu_0 \boldsymbol{j} \tag{9.45}$$

$$\nabla \cdot \boldsymbol{B} = 0 \tag{9.46}$$

where $\rho_e = (n_i - n_e)e$ is the charge density, $\boldsymbol{j} = en_i\boldsymbol{u}_i - en_e\boldsymbol{u}_e$ is the current density, ϵ_0 and μ_0 are the permittivity and the magnetic permeability of vacuum, and $c = (\epsilon_0\mu_0)^{-1/2}$ is the speed of light. By eliminating $\boldsymbol{B}$ from (9.43) and (9.45), we have

$$\nabla \times (\nabla \times \boldsymbol{E}) + \frac{1}{c^2}\frac{\partial^2 \boldsymbol{E}}{\partial t^2} = -\mu_0 \frac{\partial \boldsymbol{j}}{\partial t}\,. \tag{9.47}$$

If we can specify $\boldsymbol{j}$ as a function of $\boldsymbol{E}$, the equation is closed. This is done by inspecting how the ions and electrons respond to the electromagnetic field, namely, by looking at the equations of motion of these particles.

$$n_i m_i \frac{d\boldsymbol{u}_i}{dt} = n_i e(\boldsymbol{E} + \boldsymbol{u}_i \times \boldsymbol{B}) \tag{9.48}$$

$$n_e m_e \frac{d\boldsymbol{u}_e}{dt} = -n_e e(\boldsymbol{E} + \boldsymbol{u}_e \times \boldsymbol{B}) \tag{9.49}$$

Here we are assuming both the electrons and ions are cold ($T_e = T_i = 0$) and there is no thermal (random) motion associated with them, so for each species, they all respond to the electromagnetic field in exactly the same way. This assumption of **cold plasma** enables us to treat both electrons and ions as fluids without pressure, and their equations of motion are given simply by writing the equations of motion of individual particles multiplied by the density, n_j, by replacing the time derivative by the Lagrangian derivative ($d/dt = \partial/\partial t + \boldsymbol{u}_j \cdot \nabla$), and by interpreting the velocity $\boldsymbol{u}_j$ as the fluid bulk velocity.

Just as in any linear wave analysis, we take the following steps to proceed. (i) Ensure that the zeroth state is in balance. (ii) Write all the variables as a superposition of the zeroth order quantities and the perturbation, i.e., $\boldsymbol{A} = \boldsymbol{A}_0 + \delta \boldsymbol{A}$. (iii) Keep only the first order perturbations and drop all the terms in higher orders. (iv) Fourier transform the set of equations for the first order perturbations both in time and in space. Namely, replace $\partial/\partial t \to -i\omega$ and $\nabla \to i\boldsymbol{k}$, where ω is the wave angular frequency and $\boldsymbol{k}$ is the wave vector.

9.5.1 When $B_0 = 0$

We start from the case without the background magnetic field, $\boldsymbol{B}_0$. Step (i) above leads to $n_{i0} = n_{e0} \equiv n_0$. By taking an appropriate coordinate, we choose $\boldsymbol{u}_{j0} = 0$, $\boldsymbol{E}_0 = 0$, and $\boldsymbol{j}_0 = 0$.

In Step (ii), we have from (9.48) and (9.49),

$$\delta \boldsymbol{j} = en_0(\delta \boldsymbol{u}_i - \delta \boldsymbol{u}_e) = \frac{in_0 e^2}{\omega}\left(\frac{1}{m_i} + \frac{1}{m_e}\right)\delta \boldsymbol{E} = \sigma_\parallel \delta \boldsymbol{E}\,, \tag{9.50}$$

where $\sigma_\parallel = i\epsilon_0(\omega_e^2 + \omega_i^2)$ is the complex conductivity, or reciprocal of the complex electric resistivity. Note that $\delta \boldsymbol{j}$ and $\delta \boldsymbol{E}$ here are the Fourier transformed variables, which are the functions of ω and $\boldsymbol{k}$, instead of time t and spatial coordinates $\boldsymbol{r}$. We see from (9.50) that the current is induced in a direction parallel to the electric field, with its magnitude inversely proportional to ω. Due to the phase shift between $\delta \boldsymbol{j}$ and $\delta \boldsymbol{E}$, there is no net power (the real part of $\delta \boldsymbol{j} \cdot \delta \boldsymbol{E}$) associated with this current. On substitution of (9.50) to the Fourier transformed version of (9.47),

$$\boldsymbol{k} \times (\boldsymbol{k} \times \delta \boldsymbol{E}) + \frac{\omega^2}{c^2} \delta \boldsymbol{E} = -i\omega\mu_0 \, \delta \boldsymbol{j} \,, \tag{9.51}$$

we have a vector equation for $\delta \boldsymbol{E}$,

$$\left(k^2 - \frac{\omega_p^2 - \omega^2}{c^2} \right) \delta \boldsymbol{E} = \boldsymbol{k}\boldsymbol{k} \cdot \delta \boldsymbol{E} \,, \tag{9.52}$$

where $\omega_p^2 = \omega_i^2 + \omega_e^2 \sim \omega_e^2$ is the squared plasma angular frequency. There are two solutions to (9.52). The eigenvalues and the associated eigenvectors are,

$$\omega^2 = \omega_p^2 \quad \text{with } \delta \boldsymbol{E} \parallel \boldsymbol{k} \tag{9.53}$$

and

$$\omega^2 = \omega_p^2 + c^2 k^2 \quad \text{with } \delta \boldsymbol{E} \perp \boldsymbol{k} \ . \tag{9.54}$$

The first solution (9.53), the **plasma oscillation**, is a longitudinal oscillation caused by the recovering force due to a local imbalance of the electric charge and the electron inertia. In order to understand the nature of this oscillation, let us suppose that the space is filled with a cold plasma with density n_0, and that all the electrons within a region, $x_1 < x < x_2$, are given an equal displacement δx along the x-axis. Near the edges at $x = x_1$ and x_2, there will appear non-neutral regions with thickness δx, one with electron surplus and the other with electron deficiency. Between these regions, due to the planar distribution of charges, the electric field, $\delta E = n_0 e \delta x / \epsilon_0$, is produced which is independent of the position, x. Electrons are accelerated by this electric field, and their motion is given by $m_e \delta \ddot{x} = -e \delta E$ (dots represent the time derivative), which yields the oscillation at $\omega = \omega_e$. Strictly speaking, the ions also move slightly, and with this correction the dispersion relation becomes (9.53). As the argument above shows, the electric field produced by the charge separation is independent of x, and thus the dispersion relation is independent of k. Note that the electric field is not Debye shielded due to the large frequency of the oscillation.

The second solution, (9.54), is the transverse wave corresponding to the **light wave** in a plasma. In fact, in a vacuum ($\omega_p = 0$), (9.54) is reduced to the light wave dispersion relation, $\omega^2 = c^2 k^2$. The wave can propagate only when $\omega^2 > \omega_p^2$. In the opposite regime, $\omega^2 < \omega_p^2$, the electrons can follow and shield the wave electric field, and the wave becomes evanescent.

9.5.2 When $\boldsymbol{B}_0 \neq 0$

When there is a background magnetic field, $\boldsymbol{B}_0 \neq 0$, we can proceed basically the same way to derive the wave dispersion relation. The only essential difference is that the non-zero background field introduces a natural direction in the plasma, so we have to specify the wave vector as a function of not only its magnitude but also of its relative angle to $\boldsymbol{B}_0$. Let us choose, without losing generality, $\boldsymbol{B}_0 = (0, 0, B_0)$, and $\boldsymbol{k} = (k \sin\theta, 0, k\cos\theta)$. A relation equivalent to (9.50) is then,

$$\delta \boldsymbol{j} = \sigma \cdot \delta \boldsymbol{E}\,, \tag{9.55}$$

where σ is the complex conductivity tensor,

$$\sigma = \frac{i\epsilon_0}{\omega} \sum_j \omega_j^2 \begin{pmatrix} \dfrac{\omega^2}{\omega^2 - \Omega_j^2} & \dfrac{i\omega\Omega_j}{\omega^2 - \Omega_j^2} & 0 \\ \dfrac{-i\omega\Omega_j}{\omega^2 - \Omega_j^2} & \dfrac{\omega^2}{\omega^2 - \Omega_j^2} & 0 \\ 0 & 0 & 1 \end{pmatrix}. \tag{9.56}$$

In the above, $\Omega_j = e_j B_0/m_j$ is the angular **gyro-frequency** (angular **cyclotron frequency**) for species j, and e_j is e for protons and $-e$ for electrons. Substitution of (9.55) and (9.56) to (9.51) yields,

$$\boldsymbol{n} \times (\boldsymbol{n} \times \delta\boldsymbol{E}) + \left(1 - \frac{\sigma}{i\omega\epsilon_0}\right) \cdot \delta\boldsymbol{E} = 0\,, \tag{9.57}$$

where $\boldsymbol{n} = c\boldsymbol{k}/\omega$ is the refractive index vector. After some algebra, (9.57) can be written in a matrix form,

$$\mathbf{K} \cdot \delta\boldsymbol{E} = \begin{bmatrix} S - n^2\cos^2\theta & iD & n^2\sin\theta\cos\theta \\ -iD & S - n^2 & 0 \\ n^2\sin\theta\cos\theta & 0 & P - n^2\sin^2\theta \end{bmatrix} \cdot \delta\boldsymbol{E} = 0\,, \tag{9.58}$$

where $S = (R + L)/2$, $D = (R - L)/2$, and

$$R = 1 - \sum_{j=i,e} \frac{\omega_j^2}{\omega(\omega - \Omega_j)}\,, \quad L = 1 - \sum_{j=i,e} \frac{\omega_j^2}{\omega(\omega - \Omega_j)}\,, \quad P = 1 - \sum_{j=i,e} \frac{\omega_j^2}{\omega^2}\,. \tag{9.59}$$

The **dispersion relation** is given by det $\mathbf{K} = 0$, which reduces to

$$D(\boldsymbol{n}, \omega) = An^4 - Bn^2 + C = 0\,, \tag{9.60}$$

where $A = S\sin^2\theta + P\cos^2\theta$, $B = RL\sin^2\theta + PS(1 + \cos^2\theta)$, and $C = PRL$. Since (9.60) is a quadratic equation for n^2, the dispersion relation for a given plasma can be determined for finding n for given ω and θ. If n^2 is negative, ω is imaginary, and the wave is evanescent.[14]

---

[14]One can show that n^2 is either real or pure imaginary.

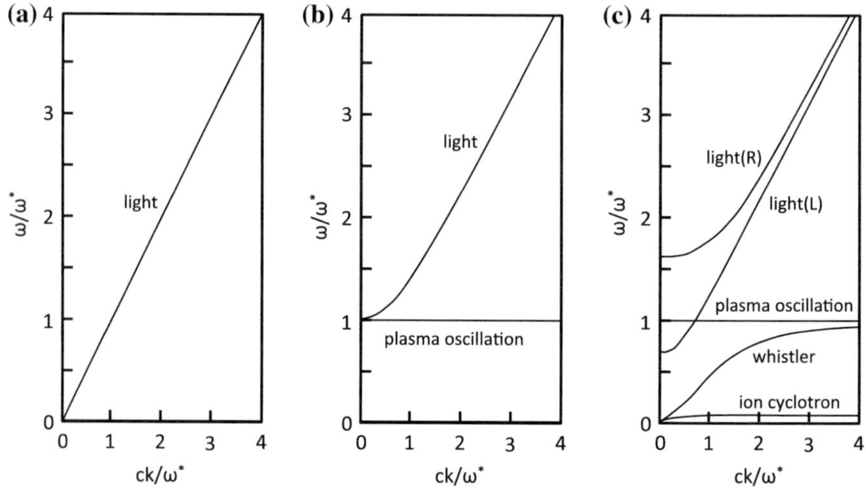

Fig. 9.9 Comparison of dispersion relations in **a** a vacuum, **b** a plasma and **c** a plasma with background magnetic field (only the parallel propagating waves are shown)

Figure 9.9 compares the dispersion relation in (a) a vacuum, (b) a plasma without the background magnetic field, and (c) a plasma with the background magnetic field. Only the part with $\omega > 0$ and $k > 0$ is shown, since the dispersion relation we treat here are symmetric both in the sign change of ω and k (or $\theta \rightarrow \pi - \theta$). In each plot, the frequency and the wave number are normalized using c and ω^*, where ω^* is arbitrary in (a), $\omega^* = \omega_p$ in (b), and $\omega^* = \Omega_e$ in (c). Parameters used in (c) are $\omega_p/\Omega_e = 1$, $\Omega_e/\Omega_i = 10$ and $\theta = 0$.

In Fig. 9.9(a), there exists only the light wave. Although we only see a single line in this plot, there are indeed four light wave modes; two waves are degenerate for $\omega = ck$, and the other two are degenerate also for $\omega = -ck$. The degenerate waves can be distinguished by the choice of $\delta\boldsymbol{E}$ (eigenvector). One can choose $\delta\boldsymbol{E}$ as two independent linearly polarized waves, or two different combinations of the linearly polarized waves to represent the right-hand polarized and the left-hand polarized waves. In (b) we see that the light wave dispersion is bent due to the shielding effect of the plasma. In addition, the plasma oscillation appears at $\omega = \omega_p$. The figure becomes quite complicated in (c), when the background magnetic field is introduced. We see that, in addition to the light waves and the plasma oscillation, a group of waves appears in the lower frequency range. They are the families of the Alfven and the whistler waves, which propagate making use of the magnetic tension as a recovery force and the ion mass (for $\omega < \Omega_i$) or the electron mass (for $\omega < |\Omega_e|$) for the inertia. We also see that the introduction of the zeroth order magnetic field resolves the degeneracy of waves, due to the Zeeman effect. Now it is convenient to specify the two light waves according to their sense of polarization.

Figure 9.10 shows an example of plasma waves measured by a Japanese spacecraft, GEOTAIL, when it was in the magnetotail region (to be introduced in Sect. 9.7)

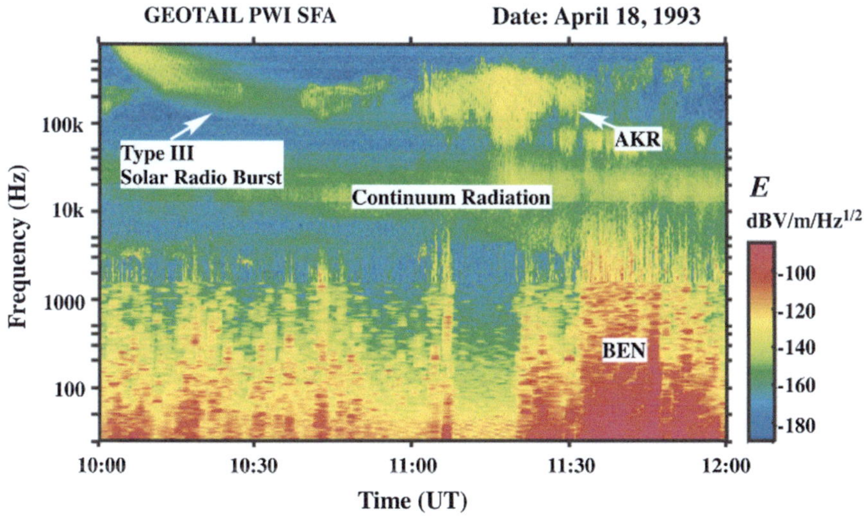

Fig. 9.10 Frequency-time spectrogram of plasma waves observed in the geomagnetic tail region by GEOTAIL [9]

of the Earth's magnetosphere. The vertical axis is the wave frequency (f) in the logarithmic scale, the horizontal axis is the universal time, and the color indicates the spectral intensity of the measured wave electric field. We do not discuss here detailed properties of all the waves observed. Instead, we just mention that, roughly speaking, there are three groups of waves at (i) $f > f_p$, (ii) $f \sim f_p$, and (iii) $f < f_{ce}$, where $f_p = \omega_p/2\pi \sim 1.2 \times 10^4\ \text{s}^{-1}$ is the plasma frequency and $f_{ce} = \Omega_e/2\pi \sim 200\ \text{s}^{-1}$ is the electron cyclotron frequency within the time interval covered in the plot. The high frequency waves (i) are the light waves known to be generated by electron beams associated with solar flares (labeled "Solar Radio Burst") or the light waves again excited by energetic electrons precipitating into the Earth's ionosphere ("AKR", or "Auroral Kilometric Radiation"). Since the light waves can almost freely propagate when $f > f_p$, even the waves originated near the sun can propagate in the solar wind plasma and arrive at GEOTAIL in the Earth's magnetosphere. The waves in the second group ("Continuum Radiation") are again the light waves, and are believed to be produced by mode conversion (linear coupling) of "upper hybrid" waves, which are essentially the plasma oscillation modulated by electron gyro-motion. Finally, the lower frequency waves in the third group are the whistler waves called ("BEN", or "Broadband Electrostatic Noise"). It is known that their broad bandwidth (f extends from f_{ce} down to almost 0) is a consequence of their soliton-like waveform.[15] Since this discovery, these waves are more often referred to as "ESW", or "Electrostatic Solitary Wave".

[15] According to a Fourier analysis, a solitary wave is expressed as a superposition of many monochromatic waves with different wavelengths, i.e., waves with different frequencies if the waves satisfy the whistler mode dispersion relation.

Space plasma waves mentioned here are only a fraction of all the varieties known to exist in space. The wave data obtained by GEOTAIL experiments, including the example spectrogram we treated above, are open to the public,[16] so interested readers are encouraged to visit their site and take a look at the original data.

9.6 Magnetohydrodynamics (MHD)

While the discussions in the previous section covered all the waves in a cold plasma for a wide range of frequencies and the wave numbers, when we are only dealing with large-scale waves and structures, which evolve with a time scale much longer than $1/\Omega_i$, we can simplify the equations and arrive at a model called **magnetohydrodynamics** (MHD). Derivation of the equations here is largely heuristic: for a more rigorous discussion, one should start by writing the kinetic equation, which treats time evolution of particle distribution functions, and by evaluating various moments to derive evolution equations of a series of macroscopic variables such as the density, bulk velocity, and the pressure tensor.[17]

9.6.1 Formulation

Let us start from the "averaged" equations of motion for ions and electrons, where the term "average" refers to taking the average of many particles in a small but finite volume. They are,

$$nm_i \frac{d\boldsymbol{u}_i}{dt} = ne(\boldsymbol{E} + \boldsymbol{u}_i \times \boldsymbol{B}) - \nabla p_i - \frac{nm_e}{\tau}(\boldsymbol{u}_i - \boldsymbol{u}_e) \tag{9.61}$$

$$nm_e \frac{d\boldsymbol{u}_e}{dt} = -ne(\boldsymbol{E} + \boldsymbol{u}_e \times \boldsymbol{B}) - \nabla p_e + \frac{nm_e}{\tau}(\boldsymbol{u}_i - \boldsymbol{u}_e) \tag{9.62}$$

where p_i and p_e are the ion and electron pressures. The last terms in the r.h.s. are introduced to model a collisional momentum exchange between ions and electrons, which is assumed to take place with a typical time scale, τ. Let us now define the bulk velocity of the plasma as $\boldsymbol{u} = (m_i\boldsymbol{u}_i + m_e\boldsymbol{u}_e)/(m_i + m_e)$. Together with the definition of the current density, $\boldsymbol{j} = ne(\boldsymbol{u}_i - \boldsymbol{u}_e)$, we can write, within the accuracy of the order $\epsilon = m_e/m_i$,

---

[16]The frequency-time spectrogram of plasma waves can be obtained from: http://space.rish.kyoto-u.ac.jp/gtlpwi/gtldata.html. General information on GEOTAIL wave experiments can be found at: http://space.rish.kyoto-u.ac.jp/gtlpwi/.

[17]This type of procedure, however, always confront the difficulty of closure: in order to specify the time evolution of the pressure tensor, one has to compute the higher order moment of the distribution function, which, in this case, is the heat flux tensor.

$$\boldsymbol{u}_i = \boldsymbol{u} \quad , \quad \boldsymbol{u}_e = \boldsymbol{u} - \frac{\boldsymbol{j}}{ne} . \tag{9.63}$$

Adding (9.61) and (9.62), we have the equation of motion for the bulk plasma,

$$\rho \frac{d\boldsymbol{u}}{dt} = \boldsymbol{j} \times \boldsymbol{B} - \nabla p , \tag{9.64}$$

where $\rho = nm_i$ is the mass density of the plasma, and $p = p_i + p_e$ is the total pressure. This equation states that the plasma can be accelerated by a combination of the Lorentz force and the pressure gradient. The last terms in the r.h.s. of (9.61) and (9.62) cancel each other, since momentum transfer between the species does not influence the total momentum.

An equation independent to (9.64) is simply given by (9.62), after dropping terms proportional to or higher order than ϵ,

$$\boldsymbol{E} + \boldsymbol{u} \times \boldsymbol{B} = \eta \boldsymbol{j} + \frac{\boldsymbol{j}}{ne} \times \boldsymbol{B} - \frac{1}{ne} \nabla p_e , \tag{9.65}$$

where $\eta = m_e/ne^2\tau$ is the **electric resistivity**. This equation may be viewed as a generalization of the **Ohm's law**, that relates the electric field in a moving frame[18] to the current density $\boldsymbol{j}$.

In many situations the second and the third terms in the r.h.s of (9.65) can be neglected. In fact, we can show that

$$\left|\frac{\boldsymbol{j}/ne \times \boldsymbol{B}}{\boldsymbol{u} \times \boldsymbol{B}}\right| \sim \left|\frac{\nabla \times \boldsymbol{B}}{\mu_0 neu}\right| \sim \left(\frac{V_A}{u}\right)^2 \left|\frac{u/L}{\Omega_i}\right| \tag{9.66}$$

and

$$\left|\frac{\nabla p_e/ne}{\boldsymbol{u} \times \boldsymbol{B}}\right| \sim \left|\frac{p_e}{neuB}\right| \sim \left(\frac{V_s^*}{u}\right)^2 \left|\frac{u/L}{\Omega_i}\right| , \tag{9.67}$$

where $V_A = B_0/\sqrt{\mu_0 \rho}$ is the **Alfven velocity** (described in more detail later), $V_s^* = \sqrt{p_e/m}$ is similar to the **sound velocity** ($V_s = \sqrt{\gamma p/m}$), and L is the spatial scale of the phenomena of interest. When we are dealing with phenomena with large spatial and temporal scales, for example, convection of the solar wind, its interaction with the Earth's magnetic field and the formation of the magnetosphere, the factor $u/L\Omega_i$ is much less than unity, while the characteristic velocity of the phenomena u is about the same order of V_A and V_s. Then the second and the third terms in (9.65) can be dropped, and (9.65) is reduced to

$$\boldsymbol{E} + \boldsymbol{u} \times \boldsymbol{B} = \eta \boldsymbol{j} . \tag{9.68}$$

[18] According to the special theory of relativity, the electric field in a frame moving with velocity $\boldsymbol{u}$ is given approximately by $\boldsymbol{E} + \boldsymbol{u} \times \boldsymbol{B}$, i.e., the motional electric field is superposed to the electric field in the original frame, when $|\boldsymbol{u}|$ is not too close to the speed of light.

Substitution of (9.68) into (9.43) yields

$$\frac{\partial \boldsymbol{B}}{\partial t} = \nabla \times (\boldsymbol{u} \times \boldsymbol{B}) + \eta \nabla^2 \boldsymbol{B}\,, \tag{9.69}$$

where η was assumed constant. This equation is identical to the vorticity equation (1.124) for barotropic fluids, if the magnetic field $\boldsymbol{B}$ and the electric resistivity η are replaced with the vorticity $\boldsymbol{\omega}$ and the viscosity ν, respectively. Thus all the arguments made to the vorticity in Chap. 1 can be applied to the magnetic field in a plasma as well. In particular, when $\eta = 0$ (the ideal plasma), the magnetic field is tied to the plasma, i.e., the **frozen-in** of the magnetic field (see Note 2.1). The presence of finite η makes the magnetic field diffuse as is evident from the form of (9.69).

When the magnetic field is "frozen-in" to the plasma, it is either that the magnetic field follows the motion of the plasma, or that the plasma moves in response to the change of the magnetic field. A handy non-dimensional parameter to describe this is the so-called **beta ratio**,

$$\beta = \frac{P}{\boldsymbol{B}^2/2\mu_0}\,, \tag{9.70}$$

which is the ratio between the plasma pressure to the magnetic pressure. The dynamics of the system is dominated by hydrodynamical effects if $\beta \gg 1$, while magnetic effects are dominant if $\beta \ll 1$. In nuclear fusion devices, β is usually considerably below unity, so that the dominant magnetic field can confine hot plasma. In space, β can take a much wider range of values. For example, for the solar plasma, β can be over 10 in the photosphere, much less at $\sim$0.1 in the chromosphere and corona, and typically, $\beta \sim 1$ in the solar wind.[19]

The continuity equation is given as

$$\frac{\partial \rho}{\partial t} + \nabla \cdot (\rho \boldsymbol{u}) = 0\,. \tag{9.71}$$

This follows directly from the ion fluid continuity equation, using (9.63) and $m_e \ll m_i$.

To derive the energy equation, let us borrow the idea from equilibrium thermodynamics, and write down the change of entropy contained in a small volume (ΔV) in the plasma,

$$dS = \frac{dQ}{T}\,, \tag{9.72}$$

where

$$S = S_0 + C_V \log\left(\frac{P}{\rho^\gamma}\right) \tag{9.73}$$

[19]One should be careful that, in the communities of nuclear fusion and laboratory experiments, the beta ratio is often defined as $\beta = P/(\boldsymbol{B}^2/2\mu_0 + P)$, i.e., the total pressure is used for the denominator.

is the entropy, $C_V = Nk_B/(\gamma - 1)$ is the specific heat capacity at constant volume for the small amount of plasma we consider (number of particles $= N$), k_B is the Boltzmann constant, γ is the ratio of specific heats, S_0 is constant, R is the gas constant, and T is the plasma temperature.[20] Now if we assume that the heat is provided by Joule heating, i.e., $dQ = \eta \boldsymbol{j}^2 dV$, then time evolution of the entropy is

$$\frac{dS}{dt} = C_V\left(\frac{dP/dt}{P} - \gamma\frac{d\rho/dt}{\rho}\right) = C_V\left(\frac{1}{P}\left(\frac{\partial P}{\partial t} + \boldsymbol{u}\cdot\nabla P\right) + \gamma\nabla\cdot\boldsymbol{u}\right) = \frac{\eta \boldsymbol{j}^2}{T}\Delta V\,. \quad (9.74)$$

In deriving the above, (9.71) has been used. Using the equation of state, $P\Delta V = Nk_B T$, (9.74) can be re-written as

$$\frac{\partial}{\partial t}\left(\frac{P}{\gamma - 1}\right) + \nabla\cdot\left(\frac{\gamma P\boldsymbol{u}}{\gamma - 1}\right) = \boldsymbol{u}\cdot\nabla P + \eta\boldsymbol{j}^2\,. \quad (9.75)$$

On the other hand, combining (9.64) and (9.71) yields

$$\frac{\partial}{\partial t}\left(\frac{\rho\boldsymbol{u}^2}{2}\right) + \nabla\cdot\left(\frac{\rho\boldsymbol{u}^2}{2}\boldsymbol{u}\right) = \boldsymbol{u}\cdot\boldsymbol{j}\times\boldsymbol{B} - \boldsymbol{u}\cdot\nabla P \quad (9.76)$$

and from (9.43)–(9.46),

$$\frac{\partial}{\partial t}\left(\frac{\boldsymbol{B}^2}{2\mu_0}\right) + \nabla\cdot\left(\frac{\boldsymbol{E}\times\boldsymbol{B}}{\mu_0}\right) = -\boldsymbol{u}\cdot\boldsymbol{j}\times\boldsymbol{B} - \eta\boldsymbol{j}^2\,. \quad (9.77)$$

Note that the displacement current in (9.45) has been neglected, since

$$\left|\frac{1}{c^2}\frac{\partial \boldsymbol{E}}{\partial t}/\nabla\times\boldsymbol{B}\right| \sim \left|\frac{E}{B}\frac{L}{c^2 T}\right| \sim \left|\frac{L^2}{c^2T^2}\right| \ll 1\,, \quad (9.78)$$

where the scaling $E/B \sim L/T$ due to (9.43) is used.

Adding (9.75)–(9.77), we arrive at the energy equation in a conservation form,

$$\frac{\partial}{\partial t}\left(\frac{\rho\boldsymbol{u}^2}{2} + \frac{P}{\gamma - 1} + \frac{\boldsymbol{B}^2}{2\mu_0}\right) + \nabla\cdot\left(\frac{\rho\boldsymbol{u}^2}{2}\boldsymbol{u} + \frac{\gamma P\boldsymbol{u}}{\gamma - 1} + \frac{\boldsymbol{E}\times\boldsymbol{B}}{\mu_0}\right) = 0\,. \quad (9.79)$$

The three terms within the partial time derivative are bulk energy density, thermal energy density, and the magnetic energy density, respectively, and the three terms within the divergence operator are associated energy flux. Figure 9.11 illustrates the interchange of various energy, via acceleration due to Lorentz force ($\boldsymbol{u}\cdot\boldsymbol{j}\times\boldsymbol{B}$), acceleration due to the pressure gradient force ($-\boldsymbol{u}\cdot\nabla P$), and the Joule heating ($\eta\boldsymbol{j}^2$). Note that the energy flow due to the Joule heating is uni-directional.

[20] The temperature here does not contain the Boltzmann constant.

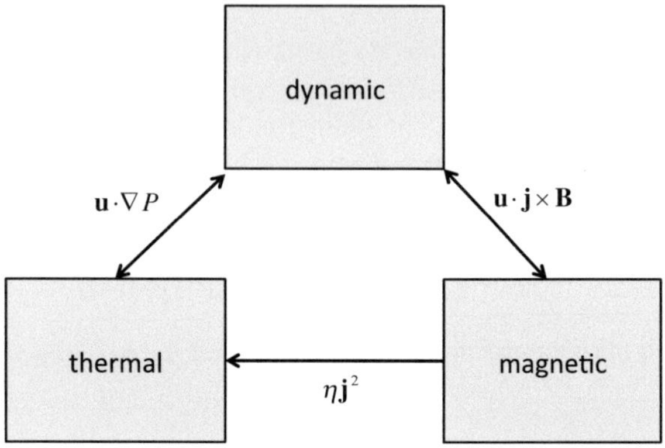

Fig. 9.11 Energy conversion between the dynamic, thermal, and magnetic energies in the MHD system. *Note* that the conversion between the magnetic and the thermal energies is uni-directional

Let us summarize the MHD set of equations here. They are, the continuity equation (9.71), the equation of motion (9.64), Faraday's law (9.69), and the energy equation (9.79). The electric field can be obtained by (9.68), and the current density by (9.45) after omitting the displacement current term. The magnetic field must be divergence free (9.46). In case of the ideal MHD ($\eta = 0$), it is often convenient to use the **adiabatic condition**,

$$\frac{d}{dt}\left(\frac{P}{\rho^\gamma}\right) = 0\,, \tag{9.80}$$

instead of the energy equation (9.79).

The ideal MHD is invariant under a scaling transformation,

$$t \to \alpha t \quad , \quad \boldsymbol{r} \to \alpha \boldsymbol{r} \tag{9.81}$$

for arbitrary (non-zero) α. This means that there is no characteristic time scales or spatial scales in the ideal MHD, and that phenomena with very large temporal and spatial scales may be viewed as the same phenomena with reduced scales. The introduction of non-ideal terms such as the dissipation effect or the Hall-effect (the second term in the r.h.s of (9.65)) violates this scaling similarity.

9.6.2 MHD Waves

As we did for the cold plasma, let us find out what kind of waves can exist in the MHD set of equations. Since we dropped the displacement current in Faraday's law, there is no light wave in this system. Instead, finite pressure will give rise to sound waves with finite sound speed.[21]

[21] In our analysis of the cold plasma, the wave mode corresponding to the sound wave did exist also, but the sound speed was zero and the wave was not propagating.

As an equilibrium state we assume a homogenous plasma with mass density ρ_0, pressure P_0, and the magnetic field $\boldsymbol{B}_0$. We further assume $\eta = 0$ for simplicity. Following the same steps described before, we find an equation for the plasma bulk velocity at the perturbation order,

$$\omega^2 \delta \boldsymbol{u} - V_s^2 \boldsymbol{k}\boldsymbol{k} \cdot \delta \boldsymbol{u} - V_A^2 (\boldsymbol{k} \times (\boldsymbol{k} \times (\delta \boldsymbol{u} \times \boldsymbol{b}_0)) \times \boldsymbol{b}_0) = 0 \,, \tag{9.82}$$

where $\boldsymbol{b}_0 = \boldsymbol{B}_0/B_0$ is the unit vector along $\boldsymbol{B}_0$, $V_s = \sqrt{\gamma P_0/\rho_0}$ is the sound speed and $V_A = \sqrt{B_0^2/\mu_0\rho_0}$ is the Alfven speed. Without losing generality, we take $\boldsymbol{B}_0 = (0, 0, B_0)$ and $\boldsymbol{k} = (k \sin\theta, 0, k\cos\theta)$, and then (9.82) can be written in a matrix form,

$$\begin{bmatrix} k^2V_A^2 + k^2V_s^2 \sin^2\theta - \omega^2 & 0 & k^2V_s^2 \sin\theta\cos\theta \\ 0 & k^2V_A^2\cos^2\theta - \omega^2 & 0 \\ k^2V_s^2 \sin\theta\cos\theta & 0 & k^2V_s^2\sin^2\theta - \omega^2 \end{bmatrix} \cdot \delta\boldsymbol{u} = 0 \,. \tag{9.83}$$

It is clear from (9.83) that the wave with the bulk velocity perturbation $(\delta u_x, 0, \delta u_z)$ and that with $(0, \delta u_y, 0)$ are decoupled. Since

$$\frac{\partial \rho}{\partial t} = -\nabla \cdot (\rho_0 \boldsymbol{u}) \sim -i\rho_0 \boldsymbol{k} \cdot \boldsymbol{u} \,, \tag{9.84}$$

the former wave is associated with density compression (compressional mode) while the latter is non-compressional.

The dispersion relation for the non-compressional mode (**intermediate mode**) is

$$\omega^2 = V_A^2 k^2 \cos^2\theta \tag{9.85}$$

and for the compressional modes (**fast mode** and **slow mode**),

$$\omega^4 - k^2(V_s^2 + V_A^2)\omega^2 + k^4 V_s^2 V_A^2 \cos^2\theta = 0 \,, \tag{9.86}$$

which gives the phase velocities

$$V_{ph}^2 = \frac{\omega^2}{k^2} = \frac{V_s^2 + V_A^2}{2} \pm \left(\left(\frac{V_s^2 + V_A^2}{2} \right)^2 - V_s^2 V_A^2 \cos^2\theta \right)^{1/2} . \tag{9.87}$$

Both (9.85) and (9.87) allow solutions with positive and negative ω, and so there are six propagating wave modes in the MHD system.[22]

[22]To be precise, there is another mode called the entropy wave, in which all the perturbations are zero except for $\delta\rho$ and δT, which are given in such a way that the plasma pressure is constant, i.e., $T_0\delta\rho + \rho_0\delta T = 0$. For the entropy wave, $\omega = 0$ for any $\boldsymbol{k}$ and is non-propagating. Nevertheless, it can carry information by being convected by a plasma flow. Including the entropy wave, there exit

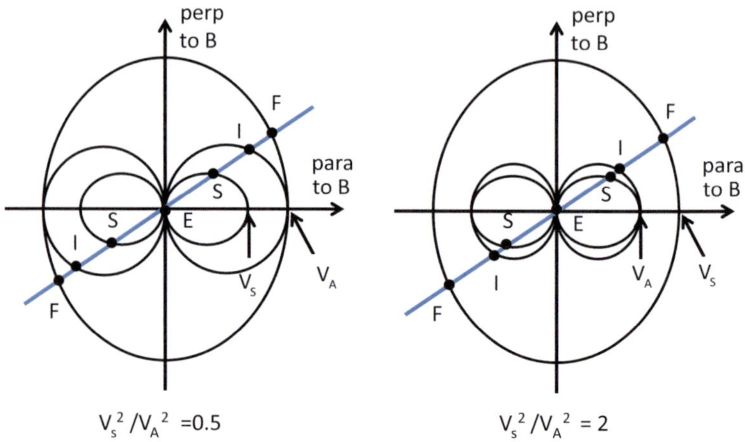

Fig. 9.12 The Friedricks diagrams of the MHD wave phase velocities for two different values of V_s^2/V_A^2. Labels F, I, S, and E stand for the fast, intermediate, slow, and entropy waves, respectively

The dispersion relation of the MHD waves can be concisely shown in a polar plot called the Friedricks diagram (Fig. 9.12). Shown in this figure are the phase velocities plotted as functions of the propagation direction, θ, for two different values of $V_s^2/V_A^2 = \gamma\beta/2$. For parallel propagation ($\theta = 0$), longitudinal sound waves and transverse Alfven waves are completely decoupled. Let us recall that, in general, waves propagate due to a balance between the recovery force and inertia. In case of the sound wave, the recovery force is the pressure gradient force raised by density perturbations. In case of the Alfven wave, it is provided by the tension of the magnetic field. If we re-write (9.64) using (9.45), we have

$$\rho\frac{d\boldsymbol{u}}{dt} = \frac{1}{\mu_0}(\nabla \times \boldsymbol{B}) \times \boldsymbol{B} - \nabla p = -\nabla\left(p + \frac{\boldsymbol{B}^2}{2\mu_0}\right) + \frac{1}{\mu_0}\boldsymbol{B}\cdot\nabla\boldsymbol{B}\,, \tag{9.88}$$

where the first term in the r.h.s. represents the gradient of plasma and magnetic pressures. Since $\boldsymbol{k}\cdot\delta\boldsymbol{B} = 0$, magnetic contribution to this term is small for propagation nearly parallel to the background magnetic field. The second term in the r.h.s. represents the change of $\boldsymbol{B}$ along the magnetic field. Bending of $\boldsymbol{B}$ produces the force that acts to straighten the magnetic field, i.e., the tension of the magnetic field.

For oblique propagations ($\theta \neq 0$), recovery force due to the Lorentz force and the plasma pressure are superposed. The superposition can be either in phase or out of phase: the former leads to the propagation of the "fast wave", and the latter to the "slow wave". "Intermediate wave" (9.85) is non-compressional, and is driven only

(Footnote 22 continued)
seven independent MHD waves for fixed $\boldsymbol{k}$, corresponding to the number of independent variables in the MHD system.

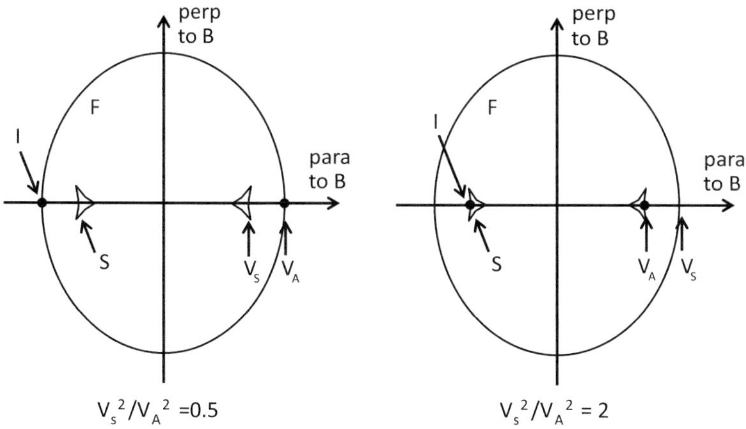

Fig. 9.13 The Friedricks diagrams of the MHD wave group velocities

by the Lorentz force. For perpendicular propagation ($\theta = \pi/2$), only the fast wave can propagate, with the phase velocity $V_{ph} = (V_s^2 + V_A^2)^{1/2}$.

In Fig. 9.13, the group velocity of the MHD waves is plotted in the format of the Friedricks diagram. In the polar coordinates, the group velocity is given by

$$\boldsymbol{V}_{gr} = \frac{\partial \omega}{\partial \boldsymbol{k}} = \frac{\partial \omega}{\partial k}\boldsymbol{e}_k + \frac{1}{k}\frac{\partial \omega}{\partial \theta}\boldsymbol{e}_\theta\,, \tag{9.89}$$

where $\boldsymbol{e}_k$ and $\boldsymbol{e}_\theta$ are unit vectors along $\boldsymbol{k}$ and θ directions, respectively. Group velocity of the intermediate wave is always exactly parallel to $\boldsymbol{B}_0$, suggesting that this wave can be viewed as an oscillation of a single magnetic field line. Neighboring field lines are not influenced by this oscillation. For the slow wave, the propagation direction of the group velocity is also confined to a small angle parallel to $\boldsymbol{B}_0$. In practical terms, it is only the fast wave that can convey energy and information across the magnetic field.

9.7 Magnetosphere

As we discussed in detail before, the solar wind, a fully ionized collisionless plasma, is streaming continuously outward from the sun. As it travels through the solar system, it carries the solar magnetic field that is "frozen-in" to the solar wind plasma. On the other hand, the Earth has its own magnetic field, well approximated by a dipolar magnet placed at the center of the Earth, nearly aligned with the Earth's spin axis. When the solar wind approaches the Earth, it is stopped at a distance where the solar wind dynamic pressure balances the magnetic pressure of the Earth's dipolar field. A layer separating the two regions is called the **magnetopause**. Although the two

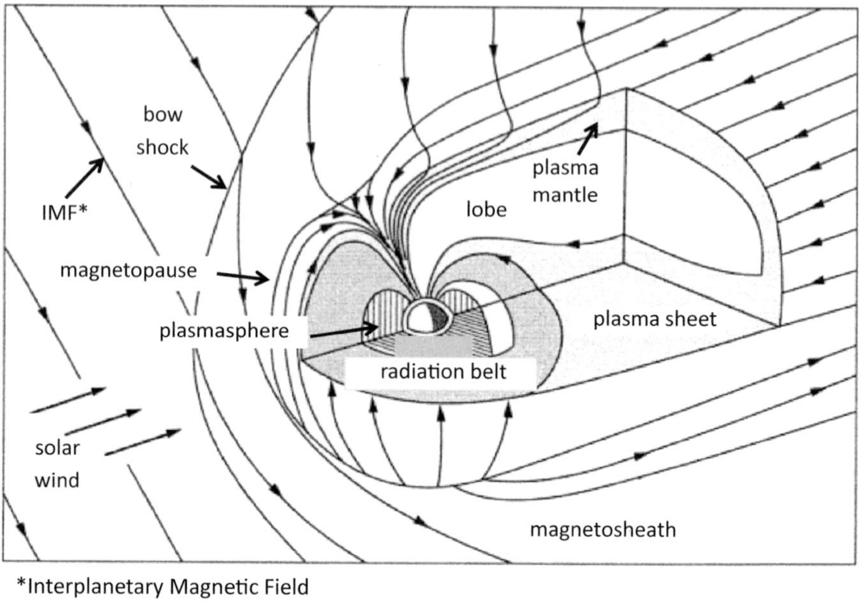

Fig. 9.14 Sketch of the Earth's magnetosphere and its major plasma regimes [17]

regions are in close contact, due to the frozen-in condition, the solar wind plasma cannot penetrate into the region of the Earth field. The cavity created in this way around the Earth is called the Earth's **magnetosphere**.

Inside the magnetosphere is not a vacuum: it is filled with plasma, either of ionospheric origin, or of solar wind origin. The fact that the solar wind plasma can be found inside the magnetosphere[23] suggests that the frozen-in argument is not working perfectly, and there is some leakage of the solar wind plasma coming into the magnetosphere. Inside the magnetosphere, a variety of different plasma regions exists (Fig. 9.14). Since the solar wind is supersonic to both Alfven speed and the sound speed, a bow shock is generated in front of the magnetosphere. Below we take a brief look at these regions one by one.

9.7.1 *Magnetopause*

The boundary between the regions of the solar wind and the Earth's magnetic field, the magnetopause, is defined as an equilibrium between the solar wind dynamic pressure and the pressure of the Earth's magnetic field. The contributions of the magnetic and

[23]One can tell the origin of the plasma by examining the "heavy ion elements" included in the plasma. Plasma from the ionosphere includes such ions as O^+ and He^+, while solar wind plasma contains He^{++}.

thermal pressure of the solar wind, as well as that of the dynamic pressure inside the magnetosphere, are small and can be neglected. The pressure balance right in front of the Earth can be roughly estimated by

$$\rho_{sw} V_{sw}^2 = \frac{B^2}{2\mu_0}, \tag{9.90}$$

where ρ_{sw} and V_{sw} are the mass density and the bulk velocity of the solar wind plasma, and B is the magnitude of the Earth's dipolar field at a distance of R measured from the center of the Earth. Since B is related to the magnetic field at the Earth's surface at the equator, $B_0 \sim 3 \times 10^{-4}$ T, as

$$B = B_0 \left(\frac{R_e}{R}\right)^3, \tag{9.91}$$

where $R_e = 6{,}370$ km is the Earth's radius, the rough estimate of the magnetopause location is

$$R = R_e \left(\frac{B_0^2}{\mu_0 \rho_{sw} V_{sw}^2}\right)^{1/6}. \tag{9.92}$$

which gives the stand-off distance, $R/R_e \sim 10$, or $R \sim 64{,}000$ km. The magnetopause has a three dimensional structure surrounding and enclosing the Earth's magnetic field, and is much elongated toward the direction of the solar wind. Responding to fluctuations in the velocity, density, and the magnetic field of the solar wind, the size and the shape of the magnetosphere change continuously.

If the idea of the frozen-in magnetic field is working perfectly, the solar wind plasma will not penetrate across the magnetopause. Then, the solar wind would confine the Earth's field to a cavity surrounding the planet, forming a "closed"

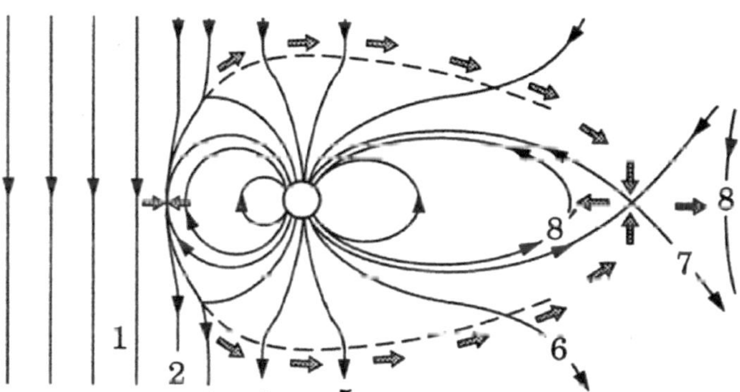

Fig. 9.15 Schematic view of Dungey's open magnetosphere for a purely southward interplanetary magnetic field

magnetosphere. On the other hand, in the early 1960s, a British physicist E. Dungey suggested that the break-down of the frozen-in picture leads to the "open magnetosphere" (Fig. 9.15). When the frozen-in condition is relaxed, the magnetic field will diffuse in the plasma at the magnetopause, allowing the interplanetary and the Earth's magnetic field to merge through the boundary. Dungy called this process **magnetic reconnection**. The reconnected magnetic lines are carried downstream by the magnetosheath flow and stretched into a long cylindrical tail, and eventually they close again by reconnection in the tail. The open magnetosphere model is accepted nowadays since plasma of solar wind origin is detected in the magnetosphere. The overall cycle time of the magnetic field lines is estimated to be about 12 h.

9.7.2 Bow Shock, Magnetosheath, and Foreshock

Since the solar wind is supersonic to the MHD fast wave (typically, the flow speed is about 5 to 10 times faster than the fast wave speed), a **bow shock** is formed about a few Re (Earth radius) ahead of the magnetopause, in a way similar to the shock wave made in front of a moving ship. At the bow shock, the upstream solar wind flow is suddenly decelerated, and its bulk kinetic energy is converted mainly to the thermal energy of the downstream plasma. Unlike a shock wave on the water surface or gas dynamic shock in the air where dissipation is provided by Coulomb collision between molecules, shock waves in space are usually collisionless shocks in which the plasma waves and turbulence generated at the shock play key roles in dissipating the upstream plasma flow energy. The formation of collisionless shocks involves a number of complex plasma processes, including how plasma waves and turbulence are generated at the shock, and how they provide an effective collision between ions and electrons.

Magnetosheath is the region between the bow shock and the magnetopause, where the solar wind plasma is slowed down and deflected. This is a region of shocked solar wind, whose plasma β is very large at $\sim$10 or even larger. Furthermore, the plasma dissipation time scale[24] is very long, and is longer than tens of ion gyro-periods. Thus, the plasma in the magnetosheath is usually in a state far from thermal equilibrium, which can trigger plasma instabilities and can generate a variety of plasma waves. Large amplitude MHD waves and turbulence are actually observed by spacecraft experiments in this region.

Another interesting property of the collisionless shock wave is that some waves and plasma populations can travel upstream. This is in strong contrast to a usual collisional shock wave, where nothing can travel backward into the upstream region. In case of the Earth's bow-shock, it is evidenced by spacecraft observations that some fraction of ions and electrons do travel backward from the shock. They are believed to be a mixture of upstream plasma reflected by the shock and the leakage of downstream plasma. These back-streaming plasmas act as a "beam" in the solar wind plasma, and

[24] Time scale for the plasma to reach thermal equilibrium.

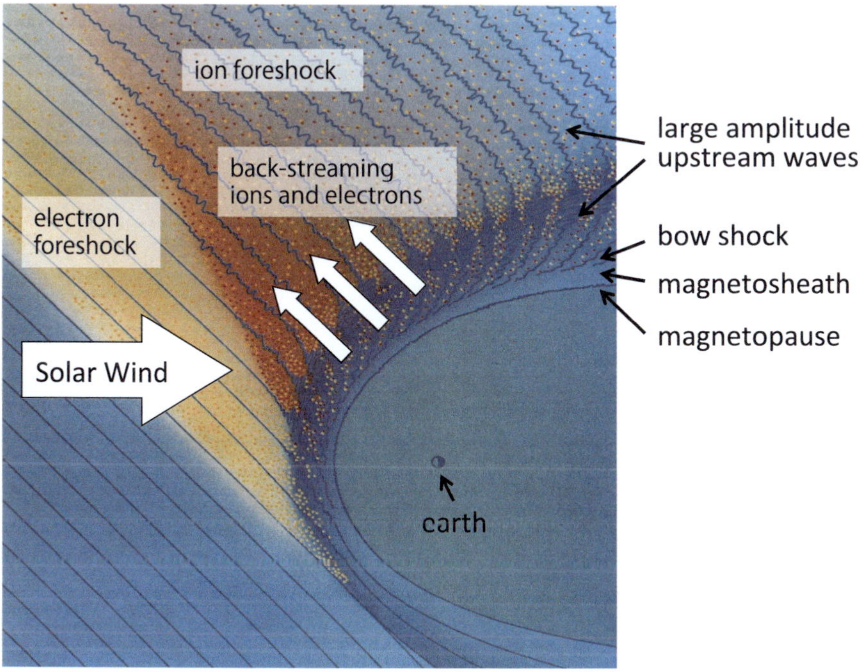

Fig. 9.16 Sketch of the Earth's bowshock and the foreshock region [16]

the interaction between this beam and the solar wind plasma leads to strong plasma instabilities. As a result, large amplitude waves are generated, since the free energy for amplifying the waves is continuously brought in by the back-streaming plasma. This is the region called Earth's **foreshock** (Fig. 9.16). It is sometimes said that the Earth's foreshock is an ideal natural laboratory for studying nonlinear waves, since almost always one can find large amplitude waves in this region, and the spatial scale of the foreshock is so large that the waves can propagate and nonlinearly evolve without being interrupted by boundaries.

Such an example of upstream waves observed by GEOTAIL is shown in Fig. 9.17, where one of the three components of the magnetic field is plotted versus time. The average magnitude of the magnetic field in this interval is about $B_0 = 4.3$ nT, while the peak-to-peak magnetic field wave amplitude is of the order of $2\delta B \sim 20$ nT as seen from the figure. So the maximum wave amplitude normalized to the background, $\delta B/B_0$, well exceeds unity in this particular example, suggesting that these waves have extremely large amplitude and are thus nonlinear. The observed complex waveform is an outcome of nonlinear wave evolution.

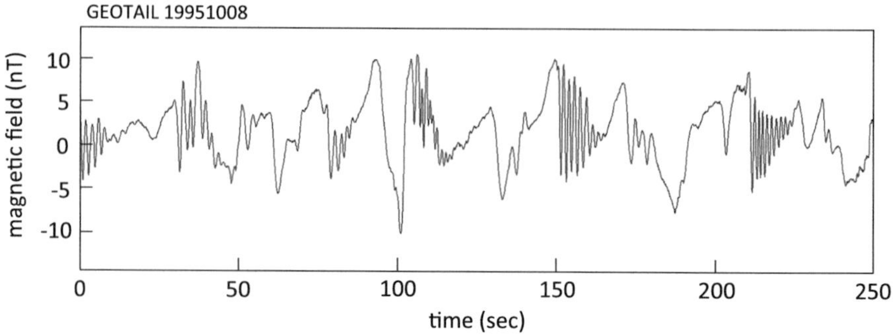

Fig. 9.17 Large amplitude waves observed in the foreshock region

9.7.3 Plasmasphere

The **plasmasphere** is a domain of dense and low energy plasma whose energies are less than 20 eV. It is located at the central core of the magnetosphere. The origin of the plasma in the plasmasphere is the ionosphere, as it is evident from the high O^+/H^+ ratio and the existence of other ion species such as He^+, O^{2+}, N^+ and N^{2+} which cannot be found in the solar wind. The plasmasphere has been regarded as a region of cold plasma with its dynamics dominated by the Earth's magnetic field, and hence, co-rotating with the Earth, although recent spacecraft observations show the presence of temporal variations. The bounding surface of the plasmasphere is called the "plasmapause". The distance of the plasmapause from the center of the Earth is typically 5 Re. At the plasmapause the density of the low-energy plasma drops sharply from the order of 10^2 cm^{-3} to the order of 1 cm^{-3} or less.

9.7.4 Radiation Belt

The **radiation belt** is a region of energetic particles trapped by the magnetic field. Typical energy of the particles is greater than 10 keV and often reaches MeV or more. In the earthward part it overlaps the plasmasphere but the radiation belt occupies much greater space than the plasmasphere. The boundary surface of the radiation belt, sometimes referred to as the trapping boundary, is also covered by magnetic field lines and it makes contact with the magnetopause in the sub-solar meridian. The Earth in fact has two radiation belts, and sometimes others may be created temporarily. The discovery of the belts is credited to James Van Allen, and thus they are sometimes called the Van Allen belts. The majority of the particles in the belts are thought to come from the solar wind and other particles by cosmic rays.

9.7.5 *Magnetotail, Lobe, and Plasma Mantle*

On the nightside, the Earth's magnetosphere has an extended tail called a **magnetotail**. It extends at least to the lunar distance (∼60 Re), and further extension to about 3,000Re has been indicated by spacecraft observations. Although the magnetic field lines in the magnetotail originate from the Earth's geomagnetic field lines, their extended configuration has little resemblance to the dipole field lines. The magnetotail contains two lobes, referred to as the northern and southern tail lobes. The tail lobes are domains with a very tenuous plasma ($\sim 10^{-2}$ cm^{-3}), but with a relatively strong magnetic field (∼10 nT or larger). The magnetic field energy stored in the lobes is believed to be the source of "magnetospheric substorms", a brief release of energy (duration ∼ a few hours) from the magnetotail causing disturbances of the near-Earth magnetic field, acceleration of energetic particles, and a brightening of the auroral arcs.

The plasma sheet is an area separated by the two **lobes**, with a weak magnetic field (a few nT or less), and with a relatively high density plasma (~ 10 cm^{-3}). Arguably, the magnetospheric substorms are triggered by the magnetic reconnection process that takes place in the plasma sheet (see Fig. 9.15). The origin of the magnetotail is believed to be the transfer of momentum from the solar wind to the magnetosphere. Namely, the solar wind exerts a drag on the magnetospheric plasma as it flows by and thus pulls field lines, which are frozen in the plasma, away from the planet. The nature of this "drag" is an important and open issue of magnetospheric physics.

The **plasma mantle** is a thin layer of plasma with energies of the solar wind, suggesting that the plasma in this region is the solar wind plasma penetrated into the magnetopause. The mantle plasma flows along magnetotail field lines.

9.7.6 *Ionosphere*

The **ionosphere** is the layer of the Earth's atmosphere that is partially ionized by the solar radiation in the UV range. It is divided into D, E, and F regions, at altitudes of 60–90, 90–140, and 140–1000 km, respectively, based on the electron density profile (Fig. 9.18). The electron density is maximized at around 300 Km altitude, and it takes a value ranging from $\sim 10^{11}$ to 3×10^{12} m^{-3}, depending on the local time, the season, and the intensity of the solar activity. Since the density of atmospheric neutral particles at the same altitude is $\sim 10^{15}$ m^{-3}, the degree of ionization is as low as ∼0.1% even where the electron density is peaked. Nevertheless, the gas in the ionosphere exhibits the unique properties of a plasma. In particular, the ionosphere can carry an electric current, so it serves as electricity conducting boundaries for magnetohydrodynamic phenomena occurring in the magnetosphere. Furthermore, the ionosphere has a practical importance as it can reflect electromagnetic waves and let radio waves to propagate a long distance. As we learned before, the plasma frequency is

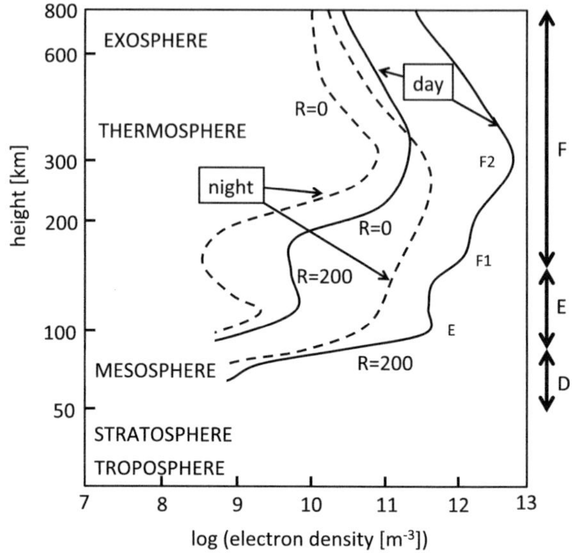

Fig. 9.18 Typical day and night profiles of electron density in the ionosphere. R is the monthly median solar index, which roughly measures the activity of the sun

$$f_p = \frac{\omega_p}{2\pi} = \frac{1}{2\pi}\left(\frac{n_0 e^2}{\epsilon_0 m_j}\right)^{1/2} \sim 9 \times {n_0}^{1/2}\ \mathrm{s}^{-1}, \tag{9.93}$$

where n_0 is the electron density measured in the unit of m^{-3}. Substituting the typical values of n_0 in the ionosphere given above, we find that the radio waves with the frequency below ~10 MHz, i.e., the waves in the HF (high frequency) range or below, are reflected by the ionosphere.

9.8 Space Weather

The Earth's environment is linked to the space environment, which is strongly influenced by the solar activities. A relatively new concept, **space weather**, is given a definition by the US National Space Weather Programme as follows: "Conditions on the Sun and in the solar wind, magnetosphere, ionosphere and thermosphere that can influence the performance and reliability of space-borne and ground-based technological systems, and can endanger human life or health. Adverse conditions in the space environment can cause disruption of satellite operations, communications, navigation, and electric power distribution grids, leading to a variety of socioeconomic losses" [14].

Figure 9.19 summarizes the space weather related phenomena and the links between them [12]. Our sun is the major cause for the space weather effects. The left-most columns are the sources of the mass and the energy from the sun, which

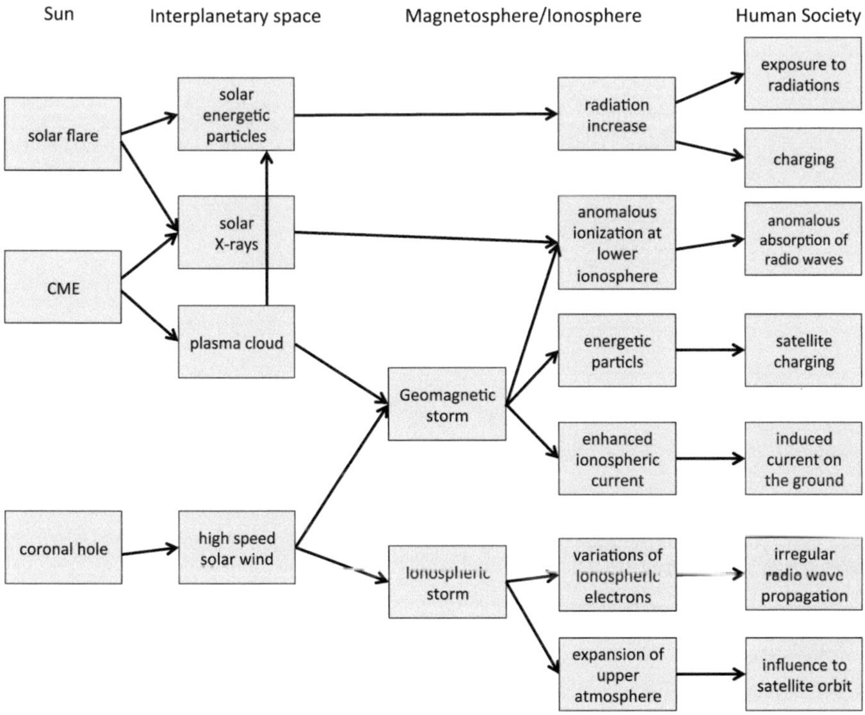

Fig. 9.19 Space weather related phenomena and the links between them [12]

drive subsequent events toward the right. The right-most columns are phenomena on the Earth that can place a significant impact on our society.

9.8.1 Solar Flares, CMEs, and Solar Energetic Particles (SEP)

The sun is at the center of triggering most of the space weather outcomes. In particular, **solar flares** and **coronal mass ejections** (CMEs) are important due to the huge amount of energy and mass they release. A solar flare occurs when magnetic energy stored in the solar atmosphere is suddenly released, and it is recognized as a sudden flash of brightness on the surface of the sun. The energy released during a flare is on the order of 10^{20} J per second, and a total energy released is up to 6×10^{25} J. The first record of a solar flare dates back to Carrington in 1859, who observed a sudden brightening of a sunspot in a white light. After this "explosion", the sunspot structure had changed, and about a day later a violent "geomagnetic storm" with a strong auroral activity was observed. This was the first occasion that a direct influence of the solar activity on Earth was recognized.

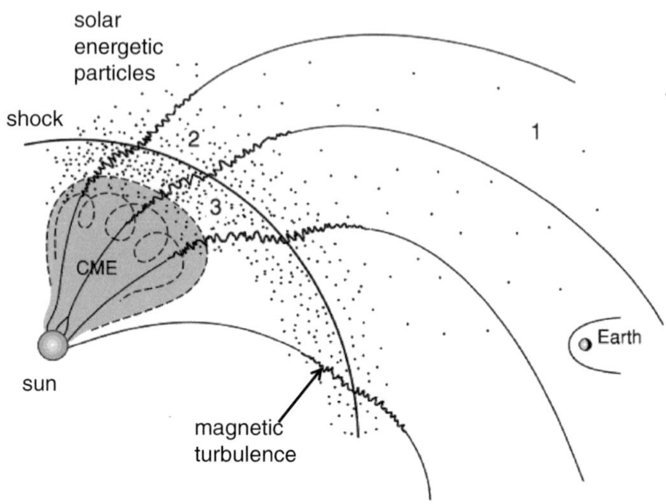

Fig. 9.20 The relationship between the solar flare, the coronal mass ejection (CME), and the CME-driven shock wave [11]

The solar flares are often, but not always, followed by a coronal mass ejection (CME), a massive burst of solar plasma rising above the solar corona and emitted into the interplanetary space. The CME includes an intense magnetic field with a helical structure. These clouds typically reach Earth in one or several days after the event. A causal relationship between the flares and the CMEs has not been established and it remains as one of the hottest subjects of scientific research on solar physics.

Solar energetic particles (SEP) are high-energy (a few tens of keV to GeV) particles coming from the sun, consisting mainly of protons, electrons, or heavier nuclei. They are of particular interest and importance because they can endanger life in outer space. It is believed that the SEPs are generated either by energization at a solar flare site or by shock waves associated with CMEs. However, only about 1 percent of the CMEs produce strong SEP events. A possible relationship between the solar flare, CME, CME-driven shock wave, and SEP is depicted in Fig. 9.20.

9.8.2 *Coronal Hole*

From X-ray observations it was realized that the temperature of the corona is not uniform. The **coronal holes** are the regions with a lower temperature, thus darker looking, and with a lower-density plasma than average. They are particularly prominent near sunspot minimum and near the solar poles. Coronal holes tend to form near the centers of large unipolar magnetic regions. A comparison of the X-ray images with those of magnetic field lines calculated on the assumption that the observed photospheric field line structures extend into the corona as potential fields indicates that they are regions of open (diverging) magnetic fields. Coronal holes are thought to be the origin of a "fast solar wind" (typical flow speed ~800 km/s), which is about twice the speed of the regular solar wind.

9.8.3 Geomagnetic Storm and Ionospheric Storm

A **geomagnetic storm** is a major disturbance of the Earth's magnetosphere caused by a CME or by a fast solar wind originating from a coronal hole. The increase in the solar wind pressure initially compresses the magnetosphere, and the solar wind's magnetic field interacts with the Earth's magnetic field and transfers an increased energy into the magnetosphere. As a consequence, the geomagnetic storms also result in a generation of intense currents in the magnetosphere, modifications in the radiation belts, and changes in the ionosphere, including heating of the ionosphere and the thermosphere.

An **ionospheric storm** is a disturbance in the Earth's ionosphere, triggered mainly by an enhanced flux of energetic electrons produced by solar flares or CMEs

9.8.4 Space Weather Effect on Space Systems

As discussed before, the Earth's geomagnetic field and the ionosphere are acting as shields to stop an invasion of the solar wind and energetic particles to human society, which is developed mostly on the surface of the Earth. In space, however, technological systems, or humans in a manned mission, are extremely vulnerable to these severe conditions.

Spacecraft Anomalies

Spacecrafts malfunction for a variety of reasons in the severe environment of space, and some of the failures can be attributed to space weather. The two most common negative space weather effects on spacecraft are radiation (energetic particles) damage and spacecraft charging. Energetic particles (SEP or galactic cosmic rays) pass through the skin of the spacecraft and strike the electronic components. When the particle hits a sensitive node in a microelectronic device such as in a microprocessor or semiconductor memory, in most cases, it causes an erroneous signal or a one-bit change in memory ("single-event upsets"). Less frequently, the energetic particles destroy a section of the electronics ("single-event latch-up"). Spacecraft charging is the build up of an electronic charge on a non-conducting material on the surface of the spacecraft by low energy particles. A discharge, or a spark, occurs when the charge accumulation reaches a certain threshold. A recent study indicates that spacecraft charging is the primary space weather effect on spacecraft in a geosynchronous orbit.

Spacecraft Orbital Changes

Solar ultraviolet radiation enhanced by geomagnetic storms causes heating and an expansion outward of the atmosphere (thermosphere). Geomagnetic storms also enhance electric currents in the ionosphere, which heat the atmosphere via Joule heating and let it expand. Consequently, an atmospheric drag on spacecraft in low Earth orbits is increased, leading to their orbit decay to lower altitudes. A geomagnetic storm can make an orbit change over a few days that otherwise would occur over a year or more. In extreme cases, the orbit of a spacecraft can decay so much that it will fall out of the orbit and crash to the Earth's surface. The loss of space station Skylab of NASA in 1979 was attributed to this effect.

Effect of Radiation on Humans in Space

The exposure of a human body to energetic particles (radiation) in space has the same harmful effects as a medical X-ray or a nuclear power plant. The degree of the adverse effect depends on the length of exposure and the energy density of the radiation. A short-term exposure to relatively high levels of radiation, such as the SEP, may lead to "a cell depletion of sensitive tissues such as the bone marrow, intestinal epithelium, skin, etc., and may lead to symptoms affecting the health and performance of crews during a mission" (NASA, Strategic Program Plan for Space Radiation Health Research, 1998). A long-term exposure to relatively low levels of radiation may result in "an enhanced probability of cancer and, possibly, changes in the cells of the brain, reproductive organs, or in other tissues" (NASA, ibid, 1998).

9.8.5 Space Weather Effects on Ground Systems

Space weather effects can at times make significant influences on our society. We illustrate a few examples below.

Ground Induced Electric Fields

Geomagnetic storm activity can induce electric fields and currents in the Earth's conducting ground, which can then adversely affect the normal operation of high voltage power transmission grids, the active cathodic protection of pipelines, telecommunication cables, and possibly, railway signaling [4]. Uncontrolled electric currents driven into electric power grids may damage transformers or malfunction protective relays. The great geomagnetic storm of March 1989 caused the complete collapse of the Hydro-Quebec electric-power grid in Canada.

Geophysical Exploration

Geophysical explorations may use a great variety of sensing instruments, and data may be collected from above or below the Earth's surface or from aerial, orbital, or marine platforms. The surveys can be influenced by rapid magnetic field variations during geomagnetic storms. In particular, geomagnetic storms cause data interpretation problems because the magnetic field variations originated by space weather effects are similar in magnitude to those of the sub-surface crustal magnetic field.

Effect of Radiation on Humans at or Near Ground Level

Cosmic rays and SEPs enter the mesosphere and the stratosphere, and collide with atmospheric atoms and molecules. The interaction produces a cascade of lower energy, but still harmful, particles called air shower secondary radiation. These particles can penetrate deep into the atmosphere. All aircraft flying above a 10 km altitude are exposed to a noticeable amount of radiation.

9.8.6 Space Weather Effects on Terrestrial Weather

Quantitatively speaking, the total amount of energy brought into the troposphere and the stratosphere due to all the space weather phenomena is negligible compared to the energy input via sunlight (visible and infra-red radiation). However, some scientists argue that there is a linkage between the 11-year sunspot cycle and the Earth's climate. For example, the Maunder Minimum, a 70-year period almost devoid of sunspots, appears to correlate to a cooling of the Earth's climate. One possible explanation for this is that changes in the cosmic ray flux cause changes in cloud formation. On the other hand, it is also argued that the cosmic ray effect on the concentration of cloud condensation nuclei is too weak to have any detectable climatic influence during a solar cycle or over the last century [7]. The linkage between space weather and the climate has not been demonstrated conclusively.

Note 9.1 (*Literature for further study*) For further studies, you may refer textbooks [2, 3, 5, 6, 8, 10, 13, 15].

References

1. Aschwanden, M.J.: Physics of the Solar Corona. Springer, Berlin (2005)
2. Baumjohann, W., Treumann, R.A.: Basic Space Plasma Physics. Imperial College Press, London (1997)
3. Bellan, P.M.: Fundamentals of Plasma Physics. Cambridge University Press, Cambridge (2006)
4. Clark, T.D.G., Clarke, E.: Space weather services for the offshore drilling industry, in "Space Weather Workshop: Looking Towards a Future European Space Weather Programme," ESTEC, ESA WPP-194, 2001
5. Gurnett, D.A., Bhattacharjee, A.: Introduction to Plasma Physics with Space and Laboratory Applications. Cambridge University Press, Cambridge (2005)
6. Ichimaru, S.: Basic Principles of Plasma Physics. W. A. Benjamin, Reading (1973)
7. IPCC, Climate Change 2013: The Physical Science Basis. Contribution of Working Group I to the Fifth Assessment Report of the Intergovernmental Panel on Climate Change, Stocker, T.F., D. Qin, G.-K. Plattner, M. Tignor, S.K. Allen, J. Boschung, A. Nauels, Y. Xia, V. Bex and P.M. Midgley (eds.), Cambridge University Press, Cambridge, United Kingdom and New York, NY, USA, pp. 1535 (2013)
8. Kivelson, M.G., Russell, C.T.: Introduction to Space Physics. Cambridge University Press, Cambridge (1995)

9. Kojima, H.: In: Oyama, K., Cheng, C.Z. (eds.), An Introduction to Space Instrumentation, Terrapub, Tokyo, pp. 227–237 (2013)
10. Lamers, H., Cassinelli, J.P.: Introduction to Stellar Winds. Cambridge University Press, Cambridge (1999)
11. Mikic, Z., Lee, M.A.: An Introduction to Theory Models of CMEs, Shocks, and Solar Energetic Particles, Space Science Reviews, vol. 123(1) pp. 57–80 (2006)
12. Ondoh, T., Marubashi, K., Obara, T., Tomita, F., Maruyama, T.: Science of Space Environment, Ohmusha Ltd., Tokyo (2001)
13. Stix, T.H.: Waves in Plasma. American Institute of Physics, New York (1992)
14. The National Space Weather Program: The Strategic Plan, FCM-P30-1995, Washington, DC (1995)
15. Treumann, R.A., Baumjohann, W.: Advanced Space Plasma Physics. Imperial College Press, London (2001)
16. Tsurutani, B., Rodriguez, P.: Upstream Waves and Particles: An Overview of ISEE Results, J. Geophys. Res. vol. 86(A6), pp. 4319–4324 (1981)
17. National Institute of Polar Research (ed.): Science in Antarctica 2, Aurora and Polar Upper Atmosphere. Kokon Shoin (1983)

Erratum to: Fluid Dynamics for Global Environmental Studies

Erratum to:
D.E.S.S.T. Kyushu Univ., *Fluid Dynamics for Global Environmental Studies*, DOI 10.1007/978-4-431-56499-7

In the original version of the book, the belated corrections have to be incorporated in Frontmatter and Chapter 9.

In Copyright page of Frontmatter, "Based on a translation from the Japanese language edition: Chikyu Kankyo wo Manabu tameno Ryutai Rikigaku by Department of Earth System Science and Technology, Kyushu University Copyright © SEIZANDO-SHOTEN PUBLISHING CO., LTD. All Rights Reserved" has to be included.

In Chapter 9, new reference [17] has to be added to reference list and a citation for [17] has to be added at the end of Fig. 9.14 caption.

The updated online version of this book can be found at https://doi.org/10.1007/978-4-431-56499-7, https://doi.org/10.1007/978-4-431-56499-7_9

D.E.S.S.T. Kyushu Univ., *Fluid Dynamics for Global Environmental Studies*,
DOI 10.1007/978-4-431-56499-7_10

Appendix

A.1 Tensors

A.1.1 Vector Transformation Rule

Denoting the basic vectors in the Cartesian coordinate system x_1, x_2, x_3 by $\boldsymbol{e}_1, \boldsymbol{e}_2, \boldsymbol{e}_3$ and the basic vector in the rotated Cartesian coordinate system x'_1, x'_2, x'_3 by $\boldsymbol{e}'_1, \boldsymbol{e}'_2, \boldsymbol{e}'_3$, the following relationships hold:

$$\boldsymbol{e}_i \cdot \boldsymbol{e}_j = \delta_{ij} , \quad \boldsymbol{e}'_l \cdot \boldsymbol{e}'_m = \delta_{lm} . \tag{A.1}$$

Here δ_{ij} is the Kronecker delta. Let $\alpha_{il} = \boldsymbol{e}_i \cdot \boldsymbol{e}'_l$ be the direction cosine between the new and original coordinate axes; then,

$$\boldsymbol{e}_i = \alpha_{il}\boldsymbol{e}'_l , \quad \boldsymbol{e}'_l = \alpha_{il}\boldsymbol{e}_i . \tag{A.2}$$

Substituting (A.2) into (A.1) gives

$$\boldsymbol{e}_i \cdot \boldsymbol{e}_j = \alpha_{il}\alpha_{jm}\boldsymbol{e}'_l \cdot \boldsymbol{e}'_m = \alpha_{il}\alpha_{jm}\delta_{lm} = \alpha_{il}\alpha_{jl} .$$

Similarly, as we obtain $\boldsymbol{e}'_l \cdot \boldsymbol{e}'_m = \alpha_{il}\alpha_{im}$, (A.1) results in

$$\alpha_{il}\alpha_{jl} = \delta_{ij} , \quad \alpha_{il}\alpha_{im} = \delta_{lm} . \tag{A.3}$$

This shows that the matrix (α_{il}) is an orthogonal matrix.

Let the component of vector $\boldsymbol{v}$ be v_i in the original coordinate system and that be v'_l in the new coordinate system, then

$$\boldsymbol{v} = v_i\boldsymbol{e}_i = v'_l\boldsymbol{e}'_l . \tag{A.4}$$

Substituting (A.2) into (A.4) gives

D.E.S.S.T. Kyushu Univ., *Fluid Dynamics for Global Environmental Studies*,
DOI 10.1007/978-4-431-56499-7

$$v_i = \alpha_{il} v'_l , \quad v'_l = \alpha_{il} v_i . \tag{A.5}$$

These give the transformation rule of the vector components in the coordinate transformation. Conversely, a vector (or 1st-order tensor) can be defined as a set of three numbers satisfying this transformation rule. Moreover, an object whose magnitude is unchanged by the coordinate transformation is called a scalar (0-order tensor).

According to this definition, it can be proved that if u_i, v_i are vectors, then $u_i + v_i$ is also a vector. Furthermore, it can be proved that if ϕ is a scalar and v_i is a vector, then ϕv_i and $\nabla\phi$ are vectors, and $u_i v_i$ is a scalar.

For example, on the basis of $u'_l = \alpha_{il} u_i$, $v'_l = \alpha_{il} v_i$, the following formulae are immediately obtained.

$$u'_l + v'_l = \alpha_{il} u_i + \alpha_{il} v_i = \alpha_{il}(u_i + v_i) ,$$

$$u'_l v'_l = \alpha_{il}\alpha_{jl} u_i v_j = \delta_{ij} u_i v_j = u_i v_i .$$

The above equations show that $u_i + v_i$ is the vector component and $u_i v_i$ is a scalar. Moreover, from

$$\frac{\partial\phi}{\partial x'_l} = \frac{\partial x_i}{\partial x'_l}\frac{\partial\phi}{\partial x_i} = \alpha_{il}\frac{\partial\phi}{\partial x_i} ,$$

$$\frac{\partial v'_l}{\partial x'_l} = \frac{\partial x_i}{\partial x'_l}\frac{\partial v'_l}{\partial x_i} = \alpha_{il}\alpha_{jl}\frac{\partial v_j}{\partial x_i} = \delta_{ij}\frac{\partial v_j}{\partial x_i} = \frac{\partial v_i}{\partial x_i} ,$$

it can be seen that $\nabla\phi$ is a vector and $\nabla \cdot \boldsymbol{v}$ is a scalar.

A.1.2 Second-Order Tensors

The vectors u_i and v_i can produce nine products of their components:

$$\begin{matrix} u_1 v_1 & u_1 v_2 & u_1 v_3 \\ u_2 v_1 & u_2 v_2 & u_2 v_3 \\ u_3 v_1 & u_3 v_2 & u_3 v_3 \end{matrix}$$

This set of nine products is called a **tensor product** of vectors u_i and v_i, expressed as $u_i v_j$ $(i, j = 1, 2, 3)$ or simply as $u_i v_j$. Each $u_i v_j$ is called a component of the tensor product.

As u_i and v_i obey the transformation rule (A.5) under the coordinate transformation,

$$u_i v_j = \alpha_{il}\alpha_{jm} u'_l v'_m , \quad u'_l v'_m = \alpha_{il}\alpha_{jm} u_i v_j \tag{A.6}$$

are given. This is the transformation equations for the components of the tensor product of vectors. Generally, when a set of nine numbers T_{ij} is transformed as

$$T_{ij} = \alpha_{il}\alpha_{jm}T'_{lm}\,, \quad T'_{lm} = \alpha_{il}\alpha_{jm}T_{ij} \tag{A.7}$$

under the coordinate transformation (A.2), this set of nine numbers T_{ij} is called a **2nd-order tensor**, and each T_{ij} is called the components of the tensor. A tensor with components T_{ij} can be simply referred to as tensor T_{ij}. Now, as two sets of equations of (A.7) can be derived one from the other, only one set of equations needs to be satisfied.

According to this definition, the tensor product of two vectors is a tensor. Moreover, if T_{ij} and S_{ij} are tensors, then

$$T_{ij} + S_{ij}\,, \quad T_{ij} - S_{ij}$$

are also tensors. The former is referred to sum and the latter difference. When ϕ is a scalar,

$$\phi\, T_{ij}$$

is also a tensor. A tensor whose components are all zeros is called the **zero tensor**. The tensor T_{ji} is called a **transposed tensor**. The tensor T_{ij} satisfying $T_{ij} = T_{ji}$ is called a **symmetric tensor**, and the tensor T_{ij} satisfying $T_{ij} = -T_{ji}$ is called an **anti-symmetric tensor**. It is shown that the symmetry and anti-symmetry properties are independent of coordinate transformation. Any tensor can be decomposed into the sum of symmetric and anti-symmetric tensors:

$$T_{ij} = \frac{1}{2}(T_{ij} + T_{ji}) + \frac{1}{2}(T_{ij} - T_{ji})\,.$$

Substituting δ_{ij} into T_{ij} in the second formula of (A.7), we get

$$T'_{lm} = \alpha_{il}\alpha_{jm}\delta_{ij} = \alpha_{il}\alpha_{im} = \delta_{lm}\,.$$

Therefore, the Kronecker delta is components of a tensor. In other words, the tensor δ_{ij} is the tensor of form of the unit matrix in any Cartesian coordinate system. This is called the **unit tensor**.

Moreover, if u_i is a vector, we have

$$\frac{\partial u'_k}{\partial x'_l} = \frac{\partial x_j}{\partial x'_l}\frac{\partial}{\partial x_j}(\alpha_{ik}u_i) = \alpha_{ik}\alpha_{jl}\frac{\partial u_i}{\partial x_j}\,.$$

Therefore, $\partial u_i/\partial x_j$ is a tensor. If u_i is the velocity vector of a fluid, this is a velocity gradient tensor.

Exercise A.1 Show that the symmetry and anti-symmetry properties of a 2nd-order tensor are independent of a coordinate transformation.

A.1.3 Tensors as Linear Operators

If v_i is a vector and T_{ij} is a tensor, then

$$T_{ij}v_j\,,\quad T_{ji}v_j \tag{A.8}$$

are both vectors. Therefore, T_{ij} and T_{ji} are linear operators. Let us prove the first case. Put

$$u_i = T_{ij}v_j\,,$$

and let us show that u_i is a component of a vector. Suppose that coordinate transformation (A.2) transforms v_i and T_{ij} to v'_i and T'_{ij}, respectively, then we put u'_i as

$$u'_i = T'_{ij}v'_j\,.$$

Substitution of $T'_{ij} = \alpha_{ki}\alpha_{lj}T_{kl}$ and $v'_j = \alpha_{mj}v_m$ into the above equation gives

$$u'_i = \alpha_{ki}\alpha_{lj}T_{kl}\alpha_{mj}v_m = \alpha_{ki}\delta_{lm}T_{kl}v_m = \alpha_{ki}(T_{kl}v_l) = \alpha_{ki}u_k\,.$$

Since u_i satisfies the transformation rule, it is a vector.

The second formula in (A.8) can be proven similarly.

A.1.4 Higher-Order Tensors

Consider three vectors u_i, v_i, and w_i and let

$$T_{ijk} = u_i v_j w_k\,.$$

Then, the transformation of the Cartesian coordinates gives

$$u'_l v'_m w'_n = \alpha_{il}u_i\alpha_{jm}v_j\alpha_{kn}w_k = \alpha_{il}\alpha_{jm}\alpha_{kn}u_i v_j w_k\,;$$

that is,

$$T'_{lmn} = \alpha_{il}\alpha_{jm}\alpha_{kn}T_{ijk}\,.$$

A 3rd-order tensor can be defined as a set of 27 numbers satisfying this transformation rule. Higher-order tensors can be similarly defined. For example, the product of a 3rd-order tensor T_{ijk} and a 2nd-order tensor S_{ij} defines a 5th-order tensor

$$T_{ijk}\,S_{lm}\ \ (i, j, k, l, m = 1, 2, 3)\,.$$

The operation that sets two subscripts of a tensor identical and takes a sum from 1 to 3 for the subscript is called a contraction. Generally, one contraction reduces the order of the tensor by 2. For example, the divergence of a vector u_i, $\partial u_i/\partial x_i$, is the zero-order tensor obtained by contracting the 2nd-order tensor $\partial u_i/\partial x_j$; in other words, it is a scalar.

Exercise A.2 Prove that the contraction of the 4th-order tensor T_{ijkl} to T_{ijkk} yields a 2nd-order tensor.

A.1.5 Alternating Tensors ϵ_{ijk}

Along with the Kronecker delta, the 3rd-order alternating tensor ϵ_{ijk} is commonly used. This tensor is defined as follows: $\epsilon_{i,j,k} = 1$ when (i, j, k) is an even permutation of $(1, 2, 3)$; $\epsilon_{i,j,k} = -1$ when (i, j, k) is an odd permutation of $(1, 2, 3)$; and $\epsilon_{ijk} = 0$ otherwise. Even permutations of $(1, 2, 3)$ are $(1, 2, 3)$, $(2, 3, 1)$, $(3, 1, 2)$, and odd permutations are $(1, 3, 2)$, $(3, 2, 1)$, $(2, 1, 3)$. According to this definition, we observe that exchange of two subscripts (for example, i and j) alters the sign of ϵ_{ijk}. Among the various names assigned to ϵ_{ijk} is Eddington's epsilon, the Levi-Civita symbol, or the completely anti-symmetric unit tensor. In terms of ϵ_{ijk}, the vector product of two vectors u_i and v_i becomes

$$(\boldsymbol{u} \times \boldsymbol{v})_i = \epsilon_{ijk} u_j v_k \,. \tag{A.9}$$

Moreover, if u_i is a vector field, and $\partial/\partial x_i$ is expressed as ∂_i, then

$$(\nabla \times \boldsymbol{u})_i = (\mathrm{rot}\boldsymbol{u})_i = \epsilon_{ijk} \partial_j u_k \,. \tag{A.10}$$

Let T_{jk} be any 2nd-order anti-symmetric tensor. If the tensor product of T_{jk} and ϵ_{ijk} is contracted with respect to j and k, a 1st-order tensor is obtained, i.e., a vector. Denoting this vector by w_i, we have

$$w_i = \epsilon_{ijk} T_{jk}, \quad T_{kj} = -T_{jk} \,. \tag{A.11}$$

Then,

$$(w_1, w_2, w_3) = 2\,(T_{23}, T_{31}, T_{12}) \,. \tag{A.12}$$

This relationship can also be formulated as

$$T_{ij} = \frac{1}{2} \epsilon_{ijk} w_k \,. \tag{A.13}$$

These relationships show that the 2nd-order anti-symmetric tensor is equivalent to a single vector.

Exercise A.3 Show that

$$\epsilon_{ijk}\epsilon_{pqr} = \begin{vmatrix} \delta_{ip} & \delta_{iq} & \delta_{ir} \\ \delta_{jp} & \delta_{jq} & \delta_{jr} \\ \delta_{kp} & \delta_{kq} & \delta_{kr} \end{vmatrix} .$$

What are the contractions of this expression $\epsilon_{ijk}\epsilon_{iqr}$, $\epsilon_{ijk}\epsilon_{ijr}$, $\epsilon_{ijk}\epsilon_{ijk}$?

A.1.6 Isotropic Tensors

A tensor that is invariant to coordinate rotation is called an **isotropic tensor**. As already seen, δ_{ij} is a 2nd-order isotropic tensor. Conversely, a 2nd-order isotropic tensor is restricted to scalar multiples of δ_{ij}. A 3rd-order isotropic tensor is a scalar multiple of ϵ_{ijk}. A 4th-order isotropic tensor is expressed as $A\delta_{ij}\delta_{kl} + B\delta_{ik}\delta_{jl} + C\delta_{il}\delta_{jk}$, where A, B, and C are scalars.

Let us prove only that a 2nd-order isotropic tensor is a scalar multiple of δ_{ij}. It is sufficient to consider an infinitesimal rotation of the coordinate system in place of an arbitrary rotation of it. The form of α_{ij} corresponding to an infinitesimal rotation is expressed as

$$\alpha_{ij} = \delta_{ij} + c_{ij}, \quad c_{ji} = -c_{ij} , \tag{A.14}$$

where c_{ij} is negligibly small. To see this, α_{ij} may be substituted into either of the orthogonality conditions (A.3) to ignore the negligible 2nd-order quantity.

Let a 2nd-order isotropic tensor be T_{ij}, and let (A.7) be written in terms of α_{ij} in (A.14). Ignoring the negligible 2nd-order quantity, we obtain

$$T'_{lm} = (\delta_{il} + c_{il})(\delta_{jm} + c_{jm})T_{ij} = T_{lm} + c_{il}T_{im} + c_{jm}T_{lj} .$$

Since T_{ij} is an isotropic tensor, for any l and m, the following must hold.

$$c_{il}T_{im} + c_{jm}T_{lj} = 0 . \tag{A.15}$$

First, suppose that l and m are different, for example, $l = 1$, $m = 2$. Then, as $c_{11} = c_{22} = 0$, $c_{21} = -c_{12}$, we obtain

$$c_{12}(T_{11} - T_{22}) + c_{31}T_{32} + c_{32}T_{13} = 0 .$$

As c_{12}, c_{31}, c_{32} are mutually independent negligible quantities,

$$T_{11} = T_{22}, \quad T_{13} = T_{32} = 0 .$$

If we consider also another different pairs of l and m, then $T_{11} = T_{22} = T_{33}$, $T_{ij} = 0$ $(i \neq j)$ are obtained. When $l = m$, for example, $l = m = 1$, (A.15) becomes

$c_{21}(T_{21} + T_{12}) + c_{31}(T_{31} + T_{13}) = 0$, which is automatically satisfied. Therefore, we obtain $T_{ij} = A\delta_{ij}$.

The 3rd- and 4th-order cases are more complicated; however, they can be proved by the same principle.

A.2 Expressions in Orthogonal Curvilinear Coordinate Systems

Generally, equations are expressed in the Cartesian coordinate system. However, in many real problems, it is often convenient to adopt an appropriate orthogonal curvilinear coordinate system. This section presents the Navier–Stokes equations and their components in various orthogonal curvilinear coordinate systems.

Let ξ_1, ξ_2, ξ_3 be an orthogonal curvilinear coordinate system, and let us define unit vectors parallel to the coordinate lines in the direction of increasing ξ_1, ξ_2, ξ_3 as $\boldsymbol{a}$, $\boldsymbol{b}$, and $\boldsymbol{c}$, respectively. Of course, $\boldsymbol{a}$, $\boldsymbol{b}$, and $\boldsymbol{c}$ are mutually orthogonal. The position vector $\boldsymbol{r}$ can be expressed as

$$\boldsymbol{r} = \boldsymbol{r}(\xi_1, \xi_2, \xi_3)\,. \tag{A.16}$$

(A.16) describes a coordinate transformation. The change in the position vector, as ξ_1, ξ_2, ξ_3 increases, is given by

$$\delta\boldsymbol{r} = h_1\delta\xi_1\boldsymbol{a} + h_2\delta\xi_2\boldsymbol{b} + h_3\delta\xi_3\boldsymbol{c}\,. \tag{A.17}$$

The factors h_1, h_2, and h_3 transform a coordinate increment (not necessarily a length increment) into a length increment, and are known as scale factors. Squaring (A.17), we obtain

$$\delta\boldsymbol{r}\cdot\delta\boldsymbol{r} = (\delta x)^2 + (\delta y)^2 + (\delta z)^2 = (h_1\delta\xi_1)^2 + (h_2\delta\xi_2)^2 + (h_3\delta\xi_3)^2\,, \tag{A.18}$$

from which we can find h_1, h_2, h_3 (>0). Because of space limitations, we skip a general discussion and state only the results of plane polar coordinates (Fig. A.1), cylindrical coordinates (Fig. A.2), spherical coordinates (Fig. A.3), and terrestrial coordinates (Fig. A.4). Detailed explanations are provided in Imai[1] and Bachelor.[2] In curvilinear coordinate systems, it is important to note that the direction of the basis vector $\boldsymbol{a}$, $\boldsymbol{b}$, $\boldsymbol{c}$ depends on its position. This is very different from the basis vector in a Cartesian coordinate system, which is invariant with position. Therefore, when differentiating vectors in spatial curvilinear coordinates, the basis vector itself must be differentiated, not merely its components.

[1] Isao Imai: Fluid Dynamics (First Part), Shokabo (1973, in Japanese).

[2] G.K. Batchelor: An Introduction to Fluid Dynamics, Cambridge University Press (1967).

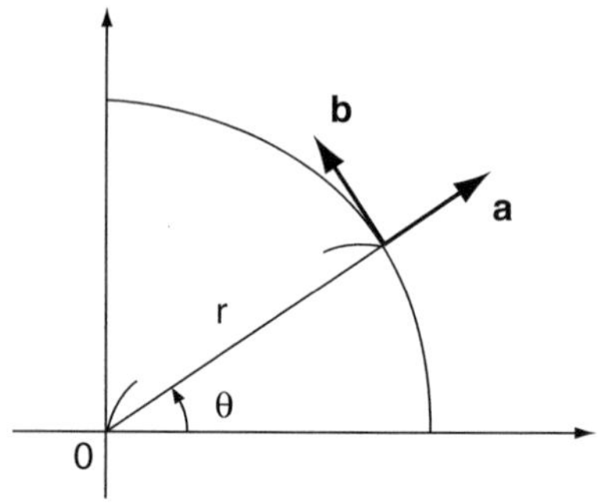

Fig. A.1 Plane polar coordinates (r, θ)

A.2.1 Plane Polar Coordinates (r, θ)

$$\xi_1 = r\,,\ \xi_2 = \theta\,;\ \ h_1 = 1\,,\ h_2 = r\,,$$

$$\frac{\partial \boldsymbol{a}}{\partial r} = 0\,,\ \frac{\partial \boldsymbol{a}}{\partial \theta} = \boldsymbol{b}\,,\ \frac{\partial \boldsymbol{b}}{\partial r} = 0\,,\ \frac{\partial \boldsymbol{b}}{\partial \theta} = -\boldsymbol{a}\,.$$

$$\boldsymbol{A} = A_r \boldsymbol{a} + A_\theta \boldsymbol{b}$$

$$\nabla \Phi = \boldsymbol{a}\frac{\partial \Phi}{\partial r} + \frac{\boldsymbol{b}}{r}\frac{\partial \Phi}{\partial \theta}\,,$$

$$\boldsymbol{n}\cdot\nabla\boldsymbol{A} = \boldsymbol{a}\Big(\boldsymbol{n}\cdot\nabla A_r - \frac{n_\theta A_\theta}{r}\Big) + \boldsymbol{b}\Big(\boldsymbol{n}\cdot\nabla A_\theta + \frac{n_\theta A_r}{r}\Big)\,,$$

where

$$\boldsymbol{n}\cdot\nabla = n_r\frac{\partial}{\partial r} + \frac{n_\theta}{r}\frac{\partial}{\partial \theta}\,.$$

$$\nabla\cdot\boldsymbol{A} = \frac{1}{r}\frac{\partial(rA_r)}{\partial r} + \frac{1}{r}\frac{\partial A_\theta}{\partial \theta}\,,$$

$$\nabla\times\boldsymbol{A} = \Big(\frac{1}{r}\frac{\partial(rA_\theta)}{\partial r} - \frac{1}{r}\frac{\partial A_r}{\partial \theta}\Big)\boldsymbol{a}\times\boldsymbol{b}\,,$$

$$\nabla^2\Phi = \frac{1}{r}\frac{\partial}{\partial r}\left(r\frac{\partial \Phi}{\partial r}\right) + \frac{1}{r^2}\frac{\partial^2\Phi}{\partial\theta^2}\,,$$

$$\nabla^2 A = \boldsymbol{a}\Big(\nabla^2 A_r - \frac{A_r}{r^2} - \frac{2}{r^2}\frac{\partial A_\theta}{\partial \theta}\Big) + \boldsymbol{b}\Big(\nabla^2 A_\theta + \frac{2}{r^2}\frac{\partial A_r}{\partial \theta} - \frac{A_\theta}{r^2}\Big)\,.$$

Rate-of-strain tensor:

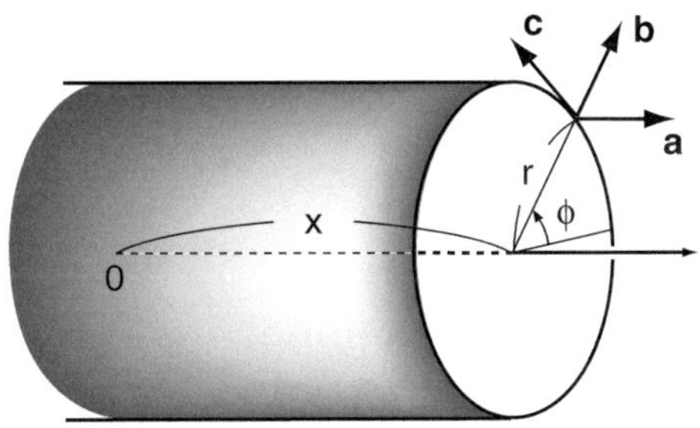

Fig. A.2 Cylindrical coordinates (x, r, ϕ)

$$e_{rr} = \frac{\partial u_r}{\partial r}\,, \; e_{\theta\theta} = \frac{1}{r}\frac{\partial u_\theta}{\partial \theta} + \frac{u_r}{r}\,, \; e_{r\theta} = \frac{r}{2}\frac{\partial}{\partial r}\Big(\frac{u_\theta}{r}\Big) + \frac{1}{2r}\frac{\partial u_r}{\partial \theta}\,.$$

Navier–Stokes equations:

$$\frac{\partial u_r}{\partial t} + \Big(u_r\frac{\partial}{\partial r} + \frac{u_\theta}{r}\frac{\partial}{\partial \theta}\Big)u_r - \frac{u_\theta^2}{r} = -\frac{1}{\rho}\frac{\partial p}{\partial r} + \nu\Big(\nabla^2 u_r - \frac{u_r}{r^2} - \frac{2}{r^2}\frac{\partial u_\theta}{\partial \theta}\Big)\,,$$

$$\frac{\partial u_\theta}{\partial t} + \Big(u_r\frac{\partial}{\partial r} + \frac{u_\theta}{r}\frac{\partial}{\partial \theta}\Big)u_\theta + \frac{u_r u_\theta}{r} = -\frac{1}{\rho r}\frac{\partial p}{\partial \theta} + \nu\Big(\nabla^2 u_\theta + \frac{2}{r^2}\frac{\partial u_r}{\partial \theta} - \frac{u_\theta}{r^2}\Big)\,.$$

A.2.2 Cylindrical Coordinates (x, r, ϕ)

$$\xi_1 = x\,,\; \xi_2 = r\,,\; \xi_3 = \phi\,;\;\; h_1 = 1\,,\; h_2 = 1\,,\; h_3 = r\,.$$

$$\frac{\partial \boldsymbol{a}}{\partial \phi} = 0\,,\; \frac{\partial \boldsymbol{b}}{\partial \phi} = \boldsymbol{c}\,,\; \frac{\partial \boldsymbol{c}}{\partial \phi} = -\boldsymbol{b}\,.$$

Moreover $\boldsymbol{a}$, $\boldsymbol{b}$, $\boldsymbol{c}$ are independent of x, r.

$$\boldsymbol{A} = A_x\boldsymbol{a} + A_r\boldsymbol{b} + A_\phi\boldsymbol{c}$$

$$\nabla\Phi = \boldsymbol{a}\frac{\partial\Phi}{\partial x} + \boldsymbol{b}\frac{\partial\Phi}{\partial r} + \frac{\boldsymbol{c}}{r}\frac{\partial\Phi}{\partial\phi}\,,$$

$$\boldsymbol{n}\cdot\nabla\boldsymbol{A} = \boldsymbol{a}(\boldsymbol{n}\cdot\nabla A_x) + \boldsymbol{b}\left(\boldsymbol{n}\cdot\nabla A_r - \frac{n_\phi A_\phi}{r}\right) + \boldsymbol{c}\Big(\boldsymbol{n}\cdot\nabla A_\phi + \frac{n_\phi A_r}{r}\Big)\,,$$

where

$$\boldsymbol{n}\cdot\nabla = n_x\frac{\partial}{\partial x} + n_r\frac{\partial}{\partial r} + \frac{n_\phi}{r}\frac{\partial}{\partial\phi}\,.$$

$$\nabla \cdot \boldsymbol{A} = \frac{\partial A_x}{\partial x} + \frac{1}{r}\frac{\partial (r A_r)}{\partial r} + \frac{1}{r}\frac{\partial A_\phi}{\partial \phi},$$

$$\nabla \times \boldsymbol{A} = \boldsymbol{a}\Big(\frac{1}{r}\frac{\partial (r A_\phi)}{\partial r} - \frac{1}{r}\frac{\partial A_r}{\partial \phi}\Big) + \boldsymbol{b}\Big(\frac{1}{r}\frac{\partial A_x}{\partial \phi} - \frac{\partial A_\phi}{\partial x}\Big) + \boldsymbol{c}\Big(\frac{\partial A_r}{\partial x} - \frac{\partial A_x}{\partial r}\Big),$$

$$\nabla^2 \Phi = \frac{\partial^2 \Phi}{\partial x^2} + \frac{1}{r}\frac{\partial}{\partial r}\Big(r\frac{\partial \Phi}{\partial r}\Big) + \frac{1}{r^2}\frac{\partial^2 \Phi}{\partial \phi^2},$$

$$\nabla^2 bmA = \boldsymbol{a}(\nabla^2 A_x) + \boldsymbol{b}\Big(\nabla^2 A_r - \frac{A_r}{r^2} - \frac{2}{r^2}\frac{\partial A_\phi}{\partial \phi}\Big) + \boldsymbol{c}\Big(\nabla^2 A_\phi + \frac{2}{r^2}\frac{\partial A_r}{\partial \phi} - \frac{A_\phi}{r^2}\Big).$$

Rate-of-strain tensor:

$$e_{xx} = \frac{\partial u_x}{\partial x}, \quad e_{rr} = \frac{\partial u_r}{\partial r}, \quad e_{\phi\phi} = \frac{1}{r}\frac{\partial u_\phi}{\partial \phi} + \frac{u_r}{r},$$

$$e_{r\phi} = \frac{r}{2}\frac{\partial}{\partial r}\Big(\frac{u_\phi}{r}\Big) + \frac{1}{2r}\frac{\partial u_r}{\partial \phi}, \quad e_{\phi x} = \frac{1}{2r}\frac{\partial u_x}{\partial \phi} + \frac{1}{2}\frac{\partial u_\phi}{\partial x}, \quad e_{xr} = \frac{1}{2}\frac{\partial u_r}{\partial x} + \frac{1}{2}\frac{\partial u_x}{\partial r}.$$

Navier–Stokes equations:

$$\frac{\partial u_x}{\partial t} + \boldsymbol{u} \cdot \nabla u_x = -\frac{1}{\rho}\frac{\partial p}{\partial x} + \nu \nabla^2 u_x,$$

$$\frac{\partial u_r}{\partial t} + \boldsymbol{u} \cdot \nabla u_r - \frac{u_\phi^2}{r} = -\frac{1}{\rho}\frac{\partial p}{\partial r} + \nu\Big(\nabla^2 u_r - \frac{u_r}{r^2} - \frac{2}{r^2}\frac{\partial u_\phi}{\partial \phi}\Big),$$

$$\frac{\partial u_\phi}{\partial t} + \boldsymbol{u} \cdot \nabla u_\phi + \frac{u_r u_\phi}{r} = -\frac{1}{\rho r}\frac{\partial p}{\partial \phi} + \nu\Big(\nabla^2 u_\phi + \frac{2}{r^2}\frac{\partial u_r}{\partial \phi} - \frac{u_\phi}{r^2}\Big).$$

A.2.3 Spherical Coordinates (r, θ, ϕ)

$$\xi_1 = r, \ \xi_2 = \theta, \ \xi_3 = \phi; \quad h_1 = 1, \ h_2 = r, \ h_3 = r\sin\theta.$$

$$\frac{\partial \boldsymbol{a}}{\partial r} = 0, \quad \frac{\partial \boldsymbol{a}}{\partial \theta} = \boldsymbol{b}, \quad \frac{\partial \boldsymbol{a}}{\partial \phi} = \boldsymbol{c}\sin\theta,$$

$$\frac{\partial \boldsymbol{b}}{\partial r} = 0, \quad \frac{\partial \boldsymbol{b}}{\partial \theta} = -\boldsymbol{a}, \quad \frac{\partial \boldsymbol{b}}{\partial \phi} = \boldsymbol{c}\cos\theta,$$

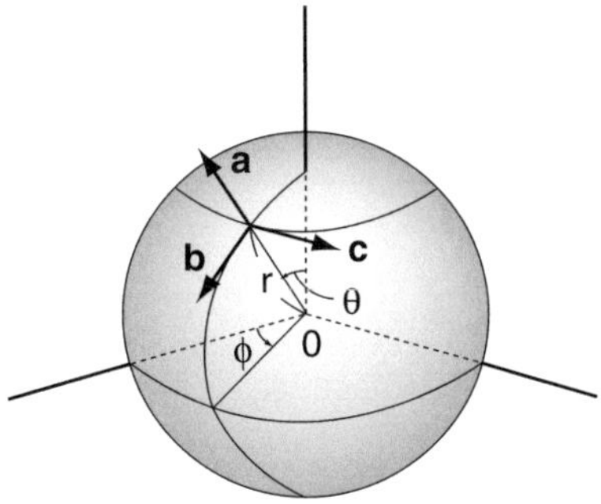

Fig. A.3 Spherical coordinates (r, θ, ϕ)

$$\frac{\partial \boldsymbol{c}}{\partial r} = 0\,, \quad \frac{\partial \boldsymbol{c}}{\partial \theta} = 0\,, \quad \frac{\partial \boldsymbol{c}}{\partial \phi} = -\boldsymbol{a}\sin\theta - \boldsymbol{b}\cos\theta\,.$$

$$\boldsymbol{A} = A_r\boldsymbol{a} + A_\theta\boldsymbol{b} + A_\phi\boldsymbol{c}\,,$$

$$\nabla\Phi = \boldsymbol{a}\frac{\partial\Phi}{\partial r} + \frac{\boldsymbol{b}}{r}\frac{\partial\Phi}{\partial\theta} + \frac{\boldsymbol{c}}{r\sin\theta}\frac{\partial\Phi}{\partial\phi}\,,$$

$$\begin{aligned}\boldsymbol{n}\cdot\nabla\boldsymbol{A} = \boldsymbol{a}\Big(&\boldsymbol{n}\cdot\nabla A_r - \frac{n_\theta A_\theta}{r} - \frac{n_\phi A_\phi}{r}\Big)\\ &+ \boldsymbol{b}\Big(\boldsymbol{n}\cdot\nabla A_\theta - \frac{n_\phi A_\phi}{r}\cot\theta + \frac{n_\theta A_r}{r}\Big)\\ &+ \boldsymbol{c}\Big(\boldsymbol{n}\cdot\nabla A_\phi + \frac{n_\phi A_r}{r} + \frac{n_\phi A_\theta}{r}\cot\theta\Big)\,,\end{aligned}$$

where

$$\boldsymbol{n}\cdot\nabla = n_r\frac{\partial}{\partial r} + \frac{n_\theta}{r}\frac{\partial}{\partial\theta} + \frac{n_\phi}{r\sin\theta}\frac{\partial}{\partial\phi}\,.$$

$$\nabla\cdot\boldsymbol{A} = \frac{1}{r^2}\frac{\partial(r^2A_r)}{\partial r} + \frac{1}{r\sin\theta}\frac{\partial(\sin\theta A_\theta)}{\partial\theta} + \frac{1}{r\sin\theta}\frac{\partial A_\phi}{\partial\phi}\,,$$

$$\begin{aligned}\nabla\times\boldsymbol{A} = &\frac{\boldsymbol{a}}{r\sin\theta}\Big(\frac{\partial(A_\phi\sin\theta)}{\partial\theta} - \frac{\partial A_\theta}{\partial\phi}\Big) + \frac{\boldsymbol{b}}{r}\Big(\frac{1}{\sin\theta}\frac{\partial A_r}{\partial\psi} - \frac{\partial(rA_\phi)}{\partial r}\Big)\\ &+ \frac{\boldsymbol{c}}{r}\Big(\frac{\partial(rA_\theta)}{\partial r} - \frac{\partial A_r}{\partial\theta}\Big),\end{aligned}$$

$$\nabla^2\Phi = \frac{1}{r^2}\frac{\partial}{\partial r}\Big(r^2\frac{\partial\Phi}{\partial r}\Big) + \frac{1}{r^2\sin\theta}\frac{\partial}{\partial\theta}\Big(\sin\theta\frac{\partial\Phi}{\partial\theta}\Big) + \frac{1}{r^2\sin^2\theta}\frac{\partial^2\Phi}{\partial\phi^2}\,,$$

$$\nabla^2 \boldsymbol{A} = \boldsymbol{a}\left(\nabla^2 A_r - \frac{2A_r}{r^2} - \frac{2}{r^2 \sin\theta}\frac{\partial(A_\theta \sin\theta)}{\partial\theta} - \frac{2}{r^2 \sin\theta}\frac{\partial A_\phi}{\partial\phi}\right)$$
$$+ \boldsymbol{b}\left(\nabla^2 A_\theta + \frac{2}{r^2}\frac{\partial A_r}{\partial\theta} - \frac{A_\theta}{r^2 \sin^2\theta} - \frac{2\cos\theta}{r^2 \sin^2\theta}\frac{\partial A_\phi}{\partial\phi}\right)$$
$$+ \boldsymbol{c}\left(\nabla^2 A_\phi + \frac{2}{r^2 \sin\theta}\frac{\partial A_r}{\partial\phi} + \frac{2\cos\theta}{r^2 \sin^2\theta}\frac{\partial A_\theta}{\partial\phi} - \frac{A_\phi}{r^2 \sin^2\theta}\right).$$

Rate-of-strain tensor:

$$e_{rr} = \frac{\partial u_r}{\partial r}, \; e_{\theta\theta} = \frac{1}{r}\frac{\partial u_\theta}{\partial\theta} + \frac{u_r}{r}, \; e_{\phi\phi} = \frac{1}{r\sin\theta}\frac{\partial u_\phi}{\partial\phi} + \frac{u_r}{r} + \frac{u_\theta \cot\theta}{r},$$

$$e_{\theta\phi} = \frac{\sin\theta}{2r}\frac{\partial}{\partial\theta}\left(\frac{u_\phi}{\sin\theta}\right) + \frac{1}{2r\sin\theta}\frac{\partial u_\theta}{\partial\phi}, \; e_{\phi r} = \frac{1}{2r\sin\theta}\frac{\partial u_r}{\partial\phi} + \frac{r}{2}\frac{\partial}{\partial r}\left(\frac{u_\phi}{r}\right),$$

$$e_{r\theta} = \frac{r}{2}\frac{\partial}{\partial r}\left(\frac{u_\theta}{r}\right) + \frac{1}{2r}\frac{\partial u_r}{\partial\theta}.$$

Navier–Stokes equations:

$$\frac{\partial u_r}{\partial t} + \boldsymbol{u}\cdot\nabla u_r - \frac{u_\theta^2}{r} - \frac{u_\phi^2}{r} = -\frac{1}{\rho}\frac{\partial p}{\partial r}$$
$$+ \nu\left(\nabla^2 u_r - \frac{2u_r}{r^2} - \frac{2}{r^2\sin\theta}\frac{\partial(u_\theta\sin\theta)}{\partial\theta} - \frac{2}{r^2\sin\theta}\frac{\partial u_\phi}{\partial\phi}\right),$$

$$\frac{\partial u_\theta}{\partial t} + \boldsymbol{u}\cdot\nabla u_\theta + \frac{u_r u_\theta}{r} - \frac{u_\phi^2\cot\theta}{r} = -\frac{1}{\rho r}\frac{\partial p}{\partial\theta}$$
$$+ \nu\left(\nabla^2 u_\theta + \frac{2}{r^2}\frac{\partial u_r}{\partial\theta} - \frac{u_\theta}{r^2\sin^2\theta} - \frac{2\cos\theta}{r^2\sin^2\theta}\frac{\partial u_\phi}{\partial\phi}\right),$$

$$\frac{\partial u_\phi}{\partial t} + \boldsymbol{u}\cdot\nabla u_\phi + \frac{u_\phi u_r}{r} + \frac{u_\theta u_\phi\cot\theta}{r} = -\frac{1}{\rho r\sin\theta}\frac{\partial p}{\partial\phi}$$
$$+ \nu\left(\nabla^2 u_\phi + \frac{2}{r^2\sin\theta}\frac{\partial u_r}{\partial\phi} + \frac{2\cos\theta}{r^2\sin^2\theta}\frac{\partial u_\theta}{\partial\phi} - \frac{u_\phi}{r^2\sin^2\theta}\right).$$

A.2.4 Terrestrial Coordinates $(\phi, \bar{\theta}, r)$

ϕ and $\bar{\theta}$ denote longitude and latitude, respectively.

Fig. A.4 Terrestrial coordinates $(r, \bar{\theta}, \phi)$

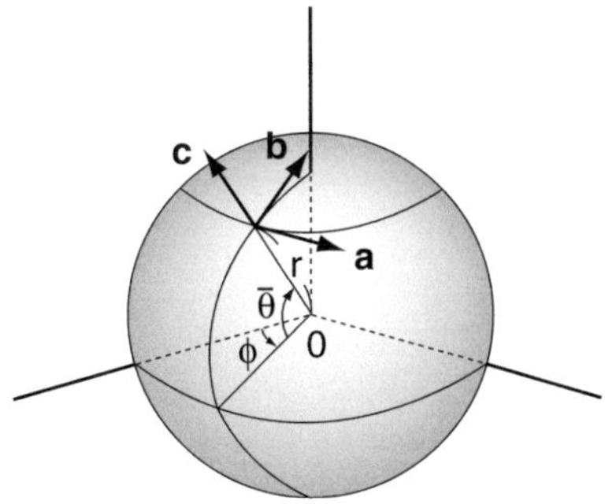

$$\xi_1 = \phi,\ \xi_2 = \bar{\theta},\ \xi_3 = r;\ \ h_1 = r\cos\bar{\theta},\ h_2 = r,\ h_3 = 1\,.$$

Substituting the following expressions into spherical coordinates

$$\boldsymbol{a} \to \boldsymbol{c},\ \boldsymbol{b} \to -\boldsymbol{b},\ \boldsymbol{c} \to \boldsymbol{a},\ \theta \to \frac{\pi}{2} - \bar{\theta}\,,$$

$$A_\theta \to -A_{\bar{\theta}},\ n_\theta \to -n_{\bar{\theta}},\ u_\theta \to -u_{\bar{\theta}},\ \frac{\partial}{\partial\theta} \to -\frac{\partial}{\partial\bar{\theta}}$$

gives the expressions in terrestrial coordinates as follows.

$$\boldsymbol{A} = A_\phi \boldsymbol{a} + A_{\bar{\theta}} \boldsymbol{b} + A_r \boldsymbol{c}\,,$$

$$\nabla\Phi = \frac{\boldsymbol{a}}{r\cos\bar{\theta}}\frac{\partial\Phi}{\partial\phi} + \frac{\boldsymbol{b}}{r}\frac{\partial\Phi}{\partial\bar{\theta}} + \boldsymbol{c}\frac{\partial\Phi}{\partial r}\,,$$

$$\nabla\cdot\boldsymbol{A} = \frac{1}{r\cos\bar{\theta}}\frac{\partial A_\phi}{\partial\phi} + \frac{1}{r\cos\bar{\theta}}\frac{\partial(\cos\bar{\theta}A_{\bar{\theta}})}{\partial\bar{\theta}} + \frac{1}{r^2}\frac{\partial(r^2 A_r)}{\partial r}\,,$$

$$\nabla\times\boldsymbol{A} = \frac{\boldsymbol{a}}{r}\left(\frac{\partial A_r}{\partial\bar{\theta}} - \frac{\partial(rA_{\bar{\theta}})}{\partial r}\right) + \frac{\boldsymbol{b}}{r}\left(\frac{\partial(rA_\phi)}{\partial r} - \frac{1}{\cos\theta}\frac{\partial A_r}{\partial\phi}\right)$$
$$+ \frac{\boldsymbol{c}}{r\cos\bar{\theta}}\left(\frac{\partial A_{\bar{\theta}}}{\partial\phi} - \frac{\partial(A_\phi\cos\bar{\theta})}{\partial\bar{\theta}}\right),$$

$$\nabla^2\Phi = \frac{1}{r^2\cos^2\bar{\theta}}\frac{\partial^2\Phi}{\partial\phi^2} + \frac{1}{r^2\cos\bar{\theta}}\frac{\partial}{\partial\bar{\theta}}\left(\cos\bar{\theta}\frac{\partial\Phi}{\partial\bar{\theta}}\right) + \frac{1}{r^2}\frac{\partial}{\partial r}\left(r^2\frac{\partial\Phi}{\partial r}\right).$$

Rate-of-strain tensor:

$$e_{\phi\phi} = \frac{1}{r\cos\bar{\theta}}\frac{\partial u_\phi}{\partial \phi} + \frac{u_r}{r} - \frac{u_{\bar{\theta}}\tan\bar{\theta}}{r}, \quad e_{\bar{\theta}\bar{\theta}} = \frac{1}{r}\frac{\partial u_{\bar{\theta}}}{\partial \bar{\theta}} + \frac{u_r}{r}, \quad e_{rr} = \frac{\partial u_r}{\partial r},$$

$$e_{\bar{\theta}r} = \frac{r}{2}\frac{\partial}{\partial r}\left(\frac{u_{\bar{\theta}}}{r}\right) + \frac{1}{2r}\frac{\partial u_r}{\partial \bar{\theta}}, \quad e_{\phi\bar{\theta}} = \frac{\cos\bar{\theta}}{2r}\frac{\partial}{\partial \bar{\theta}}\left(\frac{u_\phi}{\cos\bar{\theta}}\right) + \frac{1}{2r\cos\bar{\theta}}\frac{\partial u_{\bar{\theta}}}{\partial \phi},$$

$$e_{r\phi} = \frac{r}{2}\frac{\partial}{\partial r}\left(\frac{u_\phi}{r}\right) + \frac{1}{2r\cos\bar{\theta}}\frac{\partial u_r}{\partial \phi}.$$

Navier–Stokes equations:

$$\frac{\partial u_\phi}{\partial t} + \boldsymbol{u}\cdot\nabla u_\phi + \frac{u_\phi u_r}{r} - \frac{u_\phi u_{\bar{\theta}}\tan\bar{\theta}}{r} = -\frac{1}{\rho r\cos\bar{\theta}}\frac{\partial p}{\partial \phi}$$
$$+ \nu\left(\nabla^2 u_\phi - \frac{u_\phi}{r^2\cos^2\bar{\theta}} - \frac{2\sin\bar{\theta}}{r^2\cos^2\bar{\theta}}\frac{\partial u_{\bar{\theta}}}{\partial \phi} + \frac{2}{r^2\cos\bar{\theta}}\frac{\partial u_r}{\partial \phi}\right).$$

$$\frac{\partial u_{\bar{\theta}}}{\partial t} + \boldsymbol{u}\cdot\nabla u_{\bar{\theta}} + \frac{u_{\bar{\theta}} u_r}{r} + \frac{u_\phi^2\tan\bar{\theta}}{r} = -\frac{1}{\rho r}\frac{\partial p}{\partial \bar{\theta}}$$
$$+ \nu\left(\nabla^2 u_{\bar{\theta}} + \frac{2\sin\bar{\theta}}{r^2\cos^2\bar{\theta}}\frac{\partial u_\phi}{\partial \phi} - \frac{u_{\bar{\theta}}}{r^2\cos^2\bar{\theta}} + \frac{2}{r^2}\frac{\partial u_r}{\partial \bar{\theta}}\right),$$

$$\frac{\partial u_r}{\partial t} + \boldsymbol{u}\cdot\nabla u_r - \frac{u_\phi^2}{r} - \frac{u_{\bar{\theta}}^2}{r} = -\frac{1}{\rho}\frac{\partial p}{\partial r}$$
$$+ \nu\left(\nabla^2 u_r - \frac{2}{r^2\cos\bar{\theta}}\frac{\partial u_\phi}{\partial \phi} - \frac{1}{r^2\cos\bar{\theta}}\frac{\partial(u_{\bar{\theta}}\cos\bar{\theta})}{\partial \bar{\theta}} - \frac{2u_r}{r^2}\right),$$

where,

$$\boldsymbol{u}\cdot\nabla = \frac{u_\phi}{r\cos\bar{\theta}}\frac{\partial}{\partial \phi} + \frac{u_{\bar{\theta}}}{r}\frac{\partial}{\partial \bar{\theta}} + u_r\frac{\partial}{\partial r}.$$

Commentary on Exercises

Chapter 1 Fundamental Equations in Fluid Dynamics

Exercise 1.1 Using the symbols from Fig. 1.2,

$$2\delta S\boldsymbol{n} = \overrightarrow{\mathrm{P_1P_2}} \times \overrightarrow{\mathrm{P_1P_3}} = (\overrightarrow{\mathrm{OP_2}} - \overrightarrow{\mathrm{OP_1}}) \times (\overrightarrow{\mathrm{OP_3}} - \overrightarrow{\mathrm{OP_1}}) = 2\delta S_1\boldsymbol{e}_1 + 2\delta S_2\boldsymbol{e}_2 + 2\delta S_3\boldsymbol{e}_3$$

is obtained. Take the inner product between this and $\boldsymbol{e}_i$.

Exercise 1.2 Consider a vertical column of water with an orthogonal cross section δS. Based on balancing the forces working on the cylindrical portion between the water surface and the horizontal surface at depth h in the column, $p = p_0 + \rho g h$ holds.

Exercise 1.3 If an arbitrary curve C on the surface passing through point P is expressed by an equation $\boldsymbol{r} = \boldsymbol{r}(t)$, then $F(\boldsymbol{r}(t)) =$ constant holds. Differentiating this with respect to the parameter t, we find that $\nabla F \cdot d\boldsymbol{r}/dt = 0$. As $d\boldsymbol{r}/dt = 0$ is a tangent vector of C at point P, ∇F is a normal vector.

Exercise 1.4 Omitted.

Exercise 1.5 In a steady flow, the equation of continuity is $\nabla \cdot (\rho \boldsymbol{u}) = 0$. Integrate this in the region surrounded by the cross sections S and S' and the sides of the stream tube in Fig. 1.9. Convert this result to a surface integral using Gauss' theorem. Since $\boldsymbol{u} \cdot \boldsymbol{n} = 0$, the integral at the sides makes no contribution to the total integral; hence, we have $\iint_S \rho \boldsymbol{u} \cdot \boldsymbol{n}\, dS = \iint_{S'} \rho \boldsymbol{u} \cdot \boldsymbol{n}\, dS$.

Exercise 1.6 Assuming that the surface of an object is S and a region surrounded by the surface is V, the force exerted by the fluid on the object is expressed by $\boldsymbol{F} = -\iint_S p\boldsymbol{n}dS = -\iiint_V \nabla p dV$. With $\boldsymbol{K} = \boldsymbol{g}$, $\boldsymbol{F} = -\iiint_V \rho \boldsymbol{g} dV = -\boldsymbol{g}\rho V$ holds.

D.E.S.S.T. Kyushu Univ., *Fluid Dynamics for Global Environmental Studies*,
DOI 10.1007/978-4-431-56499-7

Exercise 1.7 Prove equations (1.122), (1.126), (1.123), (1.120) and (1.125).

(1.122): $\nabla \times \nabla\phi = 0$

$$(\nabla \times \nabla\phi)_i = \epsilon_{ijk}\partial_j\partial_k\phi = 0$$

(1.126): $\nabla \cdot (\nabla \times \boldsymbol{u}) = 0$

$$\begin{aligned}\nabla \cdot (\nabla \times \boldsymbol{u}) &= \partial_i\epsilon_{ijk}\partial_j u_k \\ &= \epsilon_{ijk}\partial_i\partial_j u_k = 0\end{aligned}$$

(1.123): $\nabla \times (\phi\boldsymbol{u}) = \nabla\phi \times \boldsymbol{u} + \phi\nabla \times \boldsymbol{u}$

$$\begin{aligned}(\nabla \times (\phi\boldsymbol{u}))_i &= \epsilon_{ijk}\partial_j(\phi\, u_k) \\ &= \epsilon_{ijk}(\partial_j\phi)u_k + \phi\,\epsilon_{ijk}\partial_j u_k \\ &= (\nabla\phi \times \boldsymbol{u})_i + (\phi\nabla \times \boldsymbol{u})_i\end{aligned}$$

(1.120): $(\boldsymbol{u} \cdot \nabla)\boldsymbol{u} = \frac{1}{2}\nabla(\boldsymbol{u} \cdot \boldsymbol{u}) - \boldsymbol{u} \times (\nabla \times \boldsymbol{u})$

$$\begin{aligned}(\text{rhs})_i &= \frac{1}{2}\partial_i u_m u_m - \epsilon_{ijk}u_j\epsilon_{klm}\partial_l u_m \\ &= u_m\partial_i u_m - (\delta_{il}\delta_{jm} - \delta_{im}\delta_{lj})u_j(\partial_l u_m) \\ &= u_m\partial_i u_m - u_m\partial_i u_m + u_l\partial_l u_i \\ &= u_l\partial_l u_i = (\text{lhs})_i\end{aligned}$$

(1.125): $\nabla \times (\boldsymbol{u} \times \boldsymbol{v}) = (\boldsymbol{v} \cdot \nabla)\boldsymbol{u} - (\boldsymbol{u} \cdot \nabla)\boldsymbol{v} + (\nabla \cdot \boldsymbol{v})\boldsymbol{u} - (\nabla \cdot \boldsymbol{u})\boldsymbol{v}$

$$\begin{aligned}(\text{lhs})_i &= \epsilon_{ijk}\partial_j\epsilon_{klm}u_l v_m = \epsilon_{ijk}\epsilon_{klm}\partial_j(u_l v_m) \\ &= (\delta_{il}\delta_{jm} - \delta_{im}\delta_{lj})\partial_j(u_l v_m) \\ &= \partial_m(u_i v_m) - \partial_l(u_l v_i) \\ &= (\partial_m v_m)u_i + v_m\partial_m u_i - (\partial_m u_m)v_i - u_m\partial_m v_i \\ &= [(\nabla \cdot \boldsymbol{v})\boldsymbol{u} + (\boldsymbol{v} \cdot \nabla)\boldsymbol{u} - (\nabla \cdot \boldsymbol{u})\boldsymbol{v} - (\boldsymbol{u} \cdot \nabla)\boldsymbol{v}]_i = (\text{rhs})_i\end{aligned}$$

Chapter 2 Dynamics of Perfect Fluids

Exercise 2.1 (2.1) can be obtained from $(\boldsymbol{u} \cdot \nabla)\boldsymbol{u} = -\boldsymbol{u} \times \boldsymbol{\omega} + \nabla(q^2/2)$. Next, rearranging the right-hand side of (2.3), we get $\nabla \times (\boldsymbol{u} \times \boldsymbol{\omega}) = u_i\partial_j\omega_j + \omega_j\partial_j u_i - \omega_i\partial_j u_j - u_j\partial_j\omega_i$. Here $\partial_j\omega_j = \nabla \cdot (\nabla \times \boldsymbol{u}) = 0$. From the equation of continuity, $\partial_j u_j = -(1/\rho)(D\rho/Dt)$. Substituting these results into (2.3) and rearranging, we obtain (2.4).

Exercise 2.2 Assuming that, on the water surface, $p = p_0$ holds and the flow velocity is 0, $p + \rho q^2/2 + \rho g z = p_0 + \rho g h$ is obtained. At the position of the hole, $z = 0$ and $p = p_0$; hence, $q = \sqrt{2gh}$.

Exercise 2.3 Assuming that, at a water surface height of z, the flow velocity is v, we have $v = \sqrt{2gz}$. If the flow volume exiting the small hole with an area a within dts is dQ, the conservation of flow volume gives $dQ = avdt = -Adz$. Therefore,

$$dt = -\frac{A}{a}\frac{dz}{v} = -\frac{A}{a\sqrt{2g}}\frac{dz}{\sqrt{z}}.$$

Integrate this expression under the conditions (1) $z : H \to H/2$ and (2) $z : H \to 0$.

Exercise 2.4 Substituting $r = a$, $x = -a$ into $\partial\Phi/\partial x = 0$ gives $a = \sqrt{m/U}$. Since the flow volume in an infinite downstream equals the volume from the source, we have $4\pi m = U\pi b^2$. Hence, $b = 2\sqrt{m/U} = 2a$.

Exercise 2.5 Velocity potential is obtained by

$$\Phi = Ux - \frac{m}{r_1} + \frac{m}{r_2}, \quad \left.\begin{matrix} r_1 \\ r_2 \end{matrix}\right\} = \sqrt{(x \pm \ell)^2 + y^2 + z^2}.$$

Points A and B are located at $x = \mp a$ and $y = z = 0$, respectively; hence, the solution at the stagnation point, where $\partial\Phi/\partial x = 0$, is $(a^2 - \ell^2)^2 - 4m\ell a/U = 0$. Moreover, the flow amount along the y-axis ($x = 0$) in the Rankine oval equals the amount flowing from the source $x = -\ell$ ($4\pi m$). Therefore, we can find a relational expression with respect to b, $Ub^2\sqrt{b^2 + \ell^2} = 4m\ell$. Maximum velocity also occurs at $x = 0$. Substituting $R = b$ into $\partial\Phi/\partial x$, we obtain the following equation:

$$u_{\max} = U\left\{1 + \frac{b^2}{2(\ell^2 + b^2)}\right\} \xrightarrow{\ell=0} \frac{3}{2}U \quad \text{(in the case of a sphere)}.$$

Exercise 2.6 Substituting $\mathrm{P} = (h, 0, 0)$, $\boldsymbol{r} = (h, 0, -z)$, and $\delta s = dz\boldsymbol{k}$ into the Biot–Savart law, $\delta u_y = (\Gamma/4\pi)hdz/(h^2 + z^2)^{3/2}$. Integrating this expression from $-\infty \to +\infty$ with respect to z gives the required equation.

Exercise 2.7 Substituting $\mathrm{P} = (0, 0, z)$, $\boldsymbol{r} = (-R\cos\theta, -R\sin\theta, z)$, and $\delta s = (-R\sin\theta\, d\theta, R\cos\theta\, d\theta, 0)$, according to the Biot–Savart law, gives $\delta w = (\Gamma/4\pi)R^2\, d\theta/(R^2 + z^2)^{3/2}$. Integrating this expression with respect to θ from $\theta \to 2\pi$ gives $w = (\Gamma/2)R^2/(R^2 + z^2)^{3/2}$. Alternatively, we can find the velocity potential from (2.77):

$$\Phi = \frac{\Gamma}{4\pi}\iint_S \frac{\partial}{\partial z}\left(\frac{1}{r}\right) R\, dR\, d\theta = -\frac{\Gamma}{4\pi}\int_0^{2\pi} d\theta \int_0^R \frac{z}{(R^2 + z^2)^{3/2}} R\, dR,$$

Evaluating this integral and then computing $w = \partial\Phi/\partial z$ from the obtained result, we can verify that the result equals the solution obtained according to the Biot–Savart law.

Exercise 2.8 For semi-infinite bodies, $\Phi = Ux - m/r$ from (2.58); therefore, we can use, for example, the relational expression with respect to the flow velocity for calculation

$$\Psi_a = \frac{1}{2}Uy^2 - m\frac{x}{r} + \text{const.}$$

For a spherical body, assuming that the uniform flow flows in the positive x-axis, $\Phi = Ux - \mu(\partial/\partial x)(1/r), (\mu = Ua^3/2)$. On the basis of the result for a semi-infinite body, we obtain

$$\Psi_a = \frac{1}{2}Uy^2 - \mu\frac{\partial}{\partial x}\left(\frac{x}{r}\right) = \frac{1}{2}Uy^2 - \mu\frac{y^2}{r^3} + \text{const.}$$

For the Rankine oval, $\Phi = Ux - m/r_1 + m/r_2$ from Exercise 2.5; hence,

$$\Psi_a = \frac{1}{2}Uy^2 - m\frac{x+\ell}{r_1} + m\frac{x-\ell}{r_2} + \text{const.}$$

Exercise 2.9 Considering mirror symmetry along $y = 0$, we have $f(z) = -Uz + m\log(z - ia) + m\log(z + ia)$. Evaluate df/dz and substitute $z = x$ to obtain u. The pressure is then given by $p = p_0 - \rho u^2/2$. The position of minimum pressure is that of maximum velocity. Here, as $du/dx = 0$, we find $x = \pm a$. However, as $|u|$ is maximized at $x = -a$, the pressure is minimized at $x = -a$.

Exercise 2.10 Considering mirror symmetry about $y = 0$, the circulation reverses its direction; hence,

$$f(z) = -Uz - \frac{i\Gamma}{2\pi}\log(z - ia) + \frac{i\Gamma}{2\pi}\log(z + ia)\,.$$

The velocity induced at P $= (0, a)$ is contributed by flows other than vortex filaments. Thus, we have

$$\frac{d}{dz}\left\{f(z) + \frac{i\Gamma}{2\pi}\log(z - ia)\right\} = 0 \overset{z=ia}{\longrightarrow} \Gamma = 4\pi aU\,.$$

Exercise 2.11 The complex velocity potential is generated by a vortex of length $d\xi$ at $z = \xi + ia$ is $-(i\gamma\, d\xi/2\pi)\log\{z - (\xi + ia)\}$ and continuously distributed $(-\infty < \xi < \infty)$. From this and its mirror image contribution, we have

$$f(z) = -Uz - \frac{i\gamma}{2\pi}\int_{-\infty}^{\infty} \log\{(x-\xi)+i(y-a)\}\,d\xi$$
$$+ \frac{i\gamma}{2\pi}\int_{-\infty}^{\infty} \log\{(x-\xi)+i(y+a)\}\,d\xi\,.$$

Calculate df/dz and substitute $x = 0$. At $y > a$ and $y < -a$, we obtain $df/dz = -U$. When $|y| < a$, we find that $df/dz = -U+\gamma$. Therefore, the velocity condition at the wall surface $= 0$ is $\gamma = U$.

Exercise 2.12 We have $f(z) = Uz + m\log z$. As $df/dz = 0$ at $z = -a$, we obtain $a = m/U$. Additionally, the flow volume in the infinite downstream equals the volume flowing from the source ($2\pi m$). Thus, we have $U2b = 2\pi m$, from which $b = \pi a$. These values are greater than the corresponding three-dimensional values ($a = \sqrt{m/U}, b = 2a$).

Exercise 2.13 For a complex velocity potential, we replace z in (2.138) by $z\,e^{-i(\pi+\alpha)}$. That is,

$$f(z) = U\left(z\,e^{-i\alpha} + \frac{a^2}{z}\,e^{i\alpha}\right) - i\kappa\log z\,.$$

As $df/dz = 0$ at $z = a$, we obtain $\kappa = -2Ua\sin\alpha$. Assuming that the circulation in the clockwise direction is positive, $\kappa = 2Ua\sin\alpha$.

Exercise 2.14 By differentiating $f(z) = U(z + a^2/z)$ with respect to z and substituting $z = r\,e^{i\theta}$, we obtain

$$u = U\left(1 - \frac{a^2}{r^2}\cos 2\theta\right), \quad v = -U\frac{a^2}{r^2}\sin 2\theta\,.$$

Substituting this expression into $\sqrt{u^2+v^2} = cU$ gives $(c^2-1)r^4 + 2a^2r^2\cos 2\theta = a^4$. The angle is computed from $\tan\alpha = -v/u$, $r^2\sin\alpha = a^2\sin(\alpha+2\theta)$.

Exercise 2.15 On a flat plate, $\sqrt{z^2-a^2} = \pm i\sqrt{a^2-x^2}$ (+ is the top surface and − is the bottom surface); hence, $\Phi_+ = -\sqrt{a^2-x^2}$ and $\Phi_- = \sqrt{a^2-x^2}$. Moreover, since $\Psi = -x$, we have $\partial\Phi/\partial n = \pm\partial\Phi/\partial y = \mp\partial\Psi/\partial x = \pm 1$. From these results, we obtain

$$m = -\rho\int_{-a}^{a}(\Phi_- - \Phi_+)(-1)\,dx = 4\rho\int_0^a \sqrt{a^2-x^2}\,dx = \rho\pi a^2\,.$$

Exercise 2.16 The complex velocity potential around a cylinder of radius R on the ζ-plane is $f(\zeta) = -U(\zeta + R^2/\zeta)$. To reformulate this in the z-plane and remove the disturbed flow component, we must overlap the uniform flow in the positive x direction. Thus, we get $f(\zeta) + Uz = U(R^2-a^2)/\zeta$. The disturbance velocity potential per unit velocity is obtained by substituting $\zeta = R\,e^{i\theta}$ to give $\phi = -\left\{(R^2-a^2)/R\right\}\cos\theta = -B\cos\theta$. From this, the additional mass in the x-axis direction is

$$m_x = -\rho \int_S \phi n_x \, d\ell = -\rho \int_0^{2\pi} (-B\cos\theta)\, B\cos\theta \, d\theta = \rho\pi B^2 .$$

A similar analysis in the y-axis direction yields m_y.

Chapter 3 Theory of Free Surface Waves

Exercise 3.1 A rotating ellipsoid body can be expressed as $F \equiv \left(\frac{x}{a}\right)^2 + \left(\frac{y}{b}\right)^2 + \left(\frac{z}{b}\right)^2 - 1 = 0$. Here

$$\frac{\partial F}{\partial x} = \frac{2}{a}\cos\theta, \quad \frac{\partial F}{\partial y} = \frac{2}{b}\sin\theta\cos\varphi, \quad \frac{\partial F}{\partial z} = \frac{2}{b}\sin\theta\sin\varphi .$$

The required equation is obtained from $\boldsymbol{n} = \nabla F / |\nabla F|$.

Exercise 3.2 The following equation is obtained from (3.1) and (3.4):

$$\left(\frac{\partial}{\partial t} + \nabla\Phi\cdot\nabla\right)(z-\zeta) = \frac{\partial\Phi}{\partial z} - \frac{\partial\zeta}{\partial t} - \nabla\Phi\cdot\nabla\zeta = 0 ,$$

$$\zeta = -\frac{1}{g}\left(\frac{\partial\Phi}{\partial t} + \frac{1}{2}\nabla\Phi\cdot\nabla\Phi\right) .$$

By eliminating ζ from this expression and rearranging, we get

$$\frac{\partial^2\Phi}{\partial t^2} + g\frac{\partial\Phi}{\partial z} + 2\nabla\Phi\cdot\nabla\left(\frac{\partial\Phi}{\partial t}\right) + O(\Phi^3) = 0 \quad \text{on } z=\zeta .$$

This equation should be satisfied at $z=\zeta$. By Taylor expansion of each term on the left-hand side, up to $O(\Phi^2)$, we obtain

$$\frac{\partial^2\Phi}{\partial t^2} + g\frac{\partial\Phi}{\partial z} + 2\nabla\Phi\cdot\nabla\left(\frac{\partial\Phi}{\partial t}\right) - \frac{1}{g}\frac{\partial\Phi}{\partial t}\frac{\partial}{\partial z}\left(\frac{\partial^2\Phi}{\partial t^2} + g\frac{\partial\Phi}{\partial z}\right) + O(\Phi^3) = 0 \text{ on } z=0 ,$$

which is exactly (3.12).

Exercise 3.3 The velocity potential at a limited water depth is (3.29). As $h\to\infty$, $\cosh k(z+h)/\cosh kh \longrightarrow e^{kz}$ and $\Phi = -(ga/\omega)\, e^{kz}\sin(\omega t - kx)$. (The dispersion relation at this point is $k=\omega^2/g$.) By applying this at $z=0$, the following relation is obtained:

$$\frac{\partial\Phi}{\partial t} = -ga\cos(\omega t - kx) , \quad \nabla\Phi\cdot\nabla\Phi = (a\omega)^2, \quad \frac{\partial^2\Phi}{\partial z\partial t} = -a\omega^2\cos(\omega t - kx)$$

so

$$\begin{aligned}\zeta &= a\cos(\omega t - kx) - \frac{1}{2}ka^2 + ka^2\cos^2(\omega t - kx)\\ &= a\cos(\omega t - kx) + \frac{1}{2}ka^2\cos 2(\omega t - kx)\,.\end{aligned}$$

At the crest of the waves, i.e., at $(\omega t - kx) = 0,\ 2\pi,\ \ldots$, ζ is greater than that of linearized waves, whereas, at the troughs of the waves, i.e., at $(\omega t - kx) = \pi, 3\pi, \ldots$, ζ is smaller than that of linearized waves. In other words, the wave crests become peaked, and the wave troughs are flattened.

Exercise 3.4 The construction of the graphs is left to the reader. To maximize (3.61), from $F(x) = \frac{1}{2}\sqrt{\tanh x}\,\{1 + 2x/\sinh 2x\}$, $x = kh$. Assuming $F'(x) = 0$, we obtain $3\sinh 2x - 4x\cosh 2x + 2x = 0$. The value of x satisfying this equation is not necessarily positive, but a numerical solution using a technique such as Newton's method yields $x \simeq 0.852$.

Exercise 3.5

$$\begin{aligned}L &\equiv \mathrm{Re}\big[A\,e^{i\omega t}\big]\,\mathrm{Re}\big[B\,e^{i\omega t}\big]\\ &= \frac{1}{2}\left(A\,e^{i\omega t} + A^* e^{-i\omega t}\right)\frac{1}{2}\left(B\,e^{i\omega t} + B^* e^{-i\omega t}\right)\\ &= \frac{1}{4}\left(AB\,e^{i2\omega t} + A^*B^* e^{-i2\omega t} + A\,B^* + A^*B\right)\,.\end{aligned}$$

The time-averaged value of this expression is $L = \frac{1}{4}\,(AB^* + A^*B) = \frac{1}{2}\mathrm{Re}\,[A\,B^*]$. (3.77) is given as

$$\phi = \frac{iga}{\omega}\frac{\cosh k(z+h)}{\cosh kh}\,e^{-ikx}\,,\quad \frac{\partial\phi^*}{\partial x} = \frac{gak}{\omega}\frac{\cosh k(z+h)}{\cosh kh}\,e^{ikx}\,.$$

Therefore, we have

$$\begin{aligned}\mathcal{L} &= \frac{\rho}{2}\mathrm{Re}\int_{-h}^{0}(i\omega\phi)\frac{\partial\phi^*}{\partial x}\,dz = -\frac{\rho}{2}(ga)^2\frac{k}{\omega}\frac{1}{\cosh^2 kh}\int_{-h}^{0}\cosh^2 k(z+h)\,dz\\ &= -\frac{\rho}{2}(ga)^2\frac{k}{\omega}\frac{1}{\cosh^2 kh}\frac{2kh + \sinh 2kh}{4k} = -\frac{1}{4}\rho g a^2\frac{\omega}{k}\left\{1 + \frac{2kh}{\sinh 2kh}\right\}\,,\end{aligned}$$

where the dispersion relation has been used in the final equation rearrangement.

Exercise 3.6 By expressing both velocity potential and pressure in complex notation, i.e., $\Phi = \mathrm{Re}[\,\phi(x,z)\,e^{i\omega t}\,]$, $P = \mathrm{Re}[\,p(x,z)\,e^{i\omega t}\,]$, $p(x,z) = -\rho i\omega\phi(x,z) + O(\phi^2)$, the condition for pressure $= 0$ is written as $\phi = \phi_I + \phi_R = 0$ at $x = 0$. Here ϕ_I is the velocity potential of the incident wave expressed as

$$\phi_I = \frac{iga}{\omega}\frac{\cosh k(z+h)}{\cosh kh}e^{ikx},$$

and ϕ_R is the velocity potential of the reflected wave. Assuming the unknown amplitude R, the following is obtained.

$$\phi_R = \frac{igR}{\omega}\frac{\cosh k(z+h)}{\cosh kh}e^{-ikx}.$$

By applying the boundary condition $\phi = 0$, we find that $R = -a$.

Exercise 3.7 By expressing the displacement in complex number notation, i.e., $X = \mathrm{Re}[\,\bar{X}\,e^{i\omega t}\,]$, the boundary condition equation at the wall surface is $\partial\phi/\partial n = i\omega\bar{X}$. At the specified point, we have $\bar{X} = \alpha\, p(0, z) = \alpha\{-\rho i\omega\phi(0, z)\}$; hence, $\partial\phi/\partial n = \rho\omega^2\alpha\phi$ at $x = 0$. The amplitude R of the reflected wave is given by $R = -a\{(\rho\omega^2\alpha - ik)/(\rho\omega^2\alpha + ik)\}$. As $\alpha \to 0$, $R = a$, and the result equals (3.80). Moreover, with $\alpha \to \infty$, $R = -a$; hence, the result of Exercise 3.6 is given.

Exercise 3.8 From Exercise 3.7, the reflected wave coefficient is $|R/a| = |\rho\omega^2\alpha - ik|/|\rho\omega^2\alpha + ik| = 1$. The work is calculated as

$$W_D = \frac{\rho}{2}\,\mathrm{Re}\int_{-h}^{0} i\omega\phi^*\frac{\partial\phi}{\partial x}\bigg|_{x=0} dz = \frac{\rho}{2}\,\mathrm{Re}\int_{-h}^{0} i\omega\phi^*\left(\rho\omega^2\alpha\phi\right) dz\,.$$

Since the integrand of the work is purely imaginary, we immediately obtain $W_D = 0$. The fact that the reflective wave coefficient $= 1$ shows that the energy possessed by the reflected wave is the same as that possessed by the incident wave, i.e., energy is neither absorbed nor supplied by the wall. This is also clear from $W_D = 0$.

Chapter 4 Dynamics of Viscous Fluids

Exercise 4.1 The Navier–Stokes equation becomes $\nu\partial^2 u/\partial y^2 = (1/\rho)dp/dx$. Assuming that the boundary condition $u = 0$ at $y = 0$ and $y = h$, we find that $u(y) = (1/2\rho\nu)dp/dx(y - h)y$. From this result, U_{max}, the flow amount $Q \equiv \int_0^h u(y)dy$, and average flow velocity $U_{\mathrm{mean}} \equiv Q/h$ are easily calculated.

Exercise 4.2 Rearrange the Navier–Stokes equation to obtain $\partial u/\partial t = \nu\partial^2 u/\partial y^2$, and assume a solution of the form $u = \mathrm{Re}[f(y)\exp(i\omega t)]$. Then, we obtain $(i\omega/\nu)f = d^2 f/dy^2$. This expression has the general solution $f = A\exp(\lambda y) + B\exp(-\lambda y)$, where $\lambda \equiv (i\omega/\nu)^{1/2} = (1+i)(\omega/2\nu)^{1/2}$. Using boundary conditions that $f \to 0$ when $y \to \infty$ and that $f = U$ at $y = 0$, we can determine coefficients as

$$
\begin{aligned}
u &= \mathrm{Re}[U \exp(-\lambda y) \exp(i\omega t)] \\
&= \mathrm{Re}[U \exp(-(\omega/2\nu)^{1/2}\, y) \exp\{i(\omega t - (\omega/2\nu)^{1/2}\, y)\}] \\
&= U \exp(-(\omega/2\nu)^{1/2}\, y) \cos(\omega t - (\omega/2\nu)^{1/2}\, y)\,.
\end{aligned}
$$

Exercise 4.3 From $\left[\frac{\partial^2}{\partial r^2} + \frac{\sin\theta}{r^2}\frac{\partial}{\partial\theta}\left(\frac{1}{\sin\theta}\frac{\partial}{\partial\theta}\right)\right]\Psi_a^{(n)} = (n-2)(n+1)r^{n-2}\sin^2\theta = (n-2)(n+1)\Psi_a^{(n-2)}$, we have, $\left[\frac{\partial^2}{\partial r^2} + \frac{\sin\theta}{r^2}\frac{\partial}{\partial\theta}\left(\frac{1}{\sin\theta}\frac{\partial}{\partial\theta}\right)\right]^2\Psi_a^{(n)} = (n-2)(n+1)(n-4)(n-1)r^{n-4}\sin^2\theta$, which is 0 only when $n = -1, 1, 2$, or 4.

Assuming that $\Psi_a = (A_{-1}r^{-1} + A_1 r + A_2 r^2 + A_4 r^4)\sin^2\theta$, $u_r = 2(A_{-1}r^{-3} + A_1 r^{-1} + A_2 + A_4 r^2)\cos\theta$, $u_\theta = (A_{-1}r^{-3} - A_1 r^{-1} - 2A_2 - 4A_4 r^2)\sin\theta$, and by determining the coefficients from the boundary condition (4.87), we obtain (4.88).

Exercise 4.4 From $\tau_{rr} = 2\mu e_{rr} = 2\mu\frac{\partial u_r}{\partial r} = 2\mu U\cos\theta\left(-\frac{3a^3}{2r^4} + \frac{3a}{2r^2}\right)$, we obtain $\tau_{rr}|_{r=a} = 0$. From $\tau_{\theta\theta} = 2\mu e_{\theta\theta} = 2\mu\left[\frac{1}{r}\frac{\partial u_\theta}{\partial\theta} + \frac{u_r}{r}\right] = 2\mu U\cos\theta\left(\frac{3a^3}{4r^4} - \frac{3a}{4r^2}\right)$, we also obtain $\tau_{\theta\theta}|_{r=a} = 0$. Furthermore, from $\tau_{\phi\phi} = 2\mu e_{\phi\phi} = 2\mu\left[\frac{1}{r\sin\theta}\frac{\partial u_\phi}{\partial\phi} + \frac{u_r}{r} + \frac{u_\theta\cot\theta}{r}\right] = 2\mu U\cos\theta\left(\frac{3a^3}{4r^4} - \frac{3a}{4r^2}\right)$, we obtain $\tau_{\theta\theta}|_{r=a} = 0$.

Since $\tau_{r\theta} = -\frac{3}{2}\frac{U\mu\sin\theta}{r}\frac{a^3}{r^3}$, is determined from (4.96), $\tau_{r\theta}|_{r=a} = -\frac{3}{2}\frac{U\mu\sin\theta}{a}$ and $\tau_{\theta\phi} = 2\mu e_{\theta\phi} = 2\mu\left[\frac{\sin\theta}{2r}\frac{\partial}{\partial\theta}\left(\frac{u_\phi}{\sin\theta}\right) + \frac{1}{2r\sin\theta}\frac{\partial u_\theta}{\partial\phi}\right] = 0$. Therefore, $\tau_{\theta\phi}|_{r=a} = 0$, $\tau_{\phi r} = 2\mu e_{\phi r} = 2\mu\left[\frac{1}{2r\sin\theta}\frac{\partial u_r}{\partial\phi} + \frac{r}{2}\frac{\partial}{\partial r}\left(\frac{u_\phi}{r}\right)\right] = 0$ and $\tau_{\phi r}|_{r=a} = 0$.

Exercise 4.5 By substituting

$$
\begin{aligned}
\frac{\partial\Psi}{\partial y} &= (\nu x U_0)^{1/2}\frac{\partial f(\eta)}{\partial y} = U_0 f'(\eta)\,, \\
\frac{\partial^2\Psi}{\partial y^2} &= \frac{\partial}{\partial y}\left(U_0 f'(\eta)\right) = U_0\frac{\partial f'(\eta)}{\partial y} = \frac{U_0^{3/2}}{(\nu x)^{1/2}} f''(\eta) \\
\frac{\partial^3\Psi}{\partial y^3} &= \frac{\partial}{\partial y}\left(\frac{U_0^{3/2}}{(\nu x)^{1/2}} f''(\eta)\right) = \frac{U_0^{3/2}}{(\nu x)^{1/2}}\frac{\partial f''(\eta)}{\partial y} = \frac{U_0^2}{\nu x} f'''(\eta) \\
\frac{\partial\Psi}{\partial x} &= \frac{d(\nu x U_0)^{1/2}}{dx} f(\eta) + (\nu x U_0)^{1/2}\frac{\partial f(\eta)}{\partial x} = \frac{(\nu U_0)^{1/2}}{2x^{1/2}}\left(f(\eta) - \eta f'(\eta)\right) \\
\frac{\partial^2\Psi}{\partial x\partial y} &= \frac{\partial}{\partial x}\left(U_0 f'(\eta)\right) = U_0\frac{\partial f'(\eta)}{\partial x} = -U_0 f''(\eta)\frac{\eta}{2x}\,,
\end{aligned}
$$

which may be derived from $\frac{\partial X(\eta)}{\partial y} = \frac{dX(\eta)}{d\eta}\cdot\frac{\partial\eta}{\partial y} = X'(\eta)\cdot\left(\frac{U_0}{\nu x}\right)^{1/2}$, $\frac{\partial X(\eta)}{\partial x} = \frac{dX(\eta)}{d\eta}\cdot\frac{\partial\eta}{\partial x} = -X'(\eta)\frac{\eta}{2x}$, we obtain (4.139).

Exercise 4.6 Assuming that the coefficient of η^m is 0, we have $A_{m+3} = -\frac{m!}{2}\sum_{k=0}^{m}\frac{A_{k+2}}{k!}\frac{A_{m-k}}{(m-k)!}$ $(m \geq 0)$. Consider some $m+3$ $(m \geq 0)$ and assume that it holds for all ks such that $0 \leq k < m+3$. After dividing $m+3$ by 3, if the remainder is 1 or 0, then at least one number from $k+2$ to $m-k$ will be a non-integer with a remainder of 2. Thus, among a combination of A_{k+2} and A_{m-k}, at least one of the values will be 0. Therefore, after dividing $m+3$ by 3, if the remainder is 1 or 0, then A_{m+3} is also 0.

When $m = 2, 5$, or 8, we have

$$A_5 = -\frac{2!}{2}\left(\frac{A_2}{0!}\frac{A_2}{2!}\right) = -\frac{1}{2}A_2^2\,,$$

$$A_8 = -\frac{5!}{2}\left(\frac{A_2}{0!}\frac{A_5}{5!} + \frac{A_5}{3!}\frac{A_2}{2!}\right) = -\frac{1}{2}\,(1+10)\,A_2 A_5 = \left(-\frac{1}{2}\right)^2 \times 11A_2^3\,,$$

$$A_{11} = -\frac{8!}{2}\left(\frac{A_2}{0!}\frac{A_8}{8!} + \frac{A_5}{3!}\frac{A_5}{5!} + \frac{A_8}{6!}\frac{A_2}{2!}\right) = \left(-\frac{1}{2}\right)^3 (11+56+11\times 28)A_2^4$$
$$= \left(-\frac{1}{2}\right)^3 \times 375A_2^4\,.$$

Exercise 4.7 (1) Approximate the integral by

$$I_\delta \equiv \sum_{n=0}^{\infty} f'(n\delta\eta)\left(1 - f'(n\delta\eta)\right)\delta\eta$$

(where $\delta\eta = 0.2$).

$$I_\delta = \left[f'(0.0)\times(1-f'(0.0)) + f'(0.2)\times(1-f'(0.2)) + \cdots\cdots\cdots\right]\times 0.2$$
$$\cong [0.0\times(1.0-0.0) + 0.06641\times(1.0-0.06641) + \cdots + 1.0\times(1.0-1.0)]\times 0.2$$
$$\cong 0.663\,.$$

(2) $\int_0^\infty f'(1-f')d\eta = \left[f(1-f')\right]_0^\infty + \int_0^\infty ff''d\eta = -\int_0^\infty 2f'''d\eta = 2f''(0) = 2\alpha$.

Chapter 5 Turbulence

Exercise 5.1 Using the terms $\partial \boldsymbol{u}'/\partial x = ik_x\hat{\boldsymbol{u}}(y)\exp\left[i(k_x x + k_z z - \omega t)\right]$, $\partial \boldsymbol{u}'/\partial z = ik_z\hat{\boldsymbol{u}}(y)\exp\left[i(k_x x + k_z z - \omega t)\right]$, $\partial \boldsymbol{u}'/\partial t = -i\omega\hat{\boldsymbol{u}}(y)\exp\left[i(k_x x + k_z z - \omega t)\right]$, write out the separate components of the equation:

$$-i\omega\hat{u} + iUk_x\hat{u} + \hat{v}dU/dy = -(1/\rho)ik_x\hat{p} + \nu\left[d^2/dy^2 - (k_x^2+k_z^2)\right]\hat{u}\,, \quad (1)$$
$$-i\omega\hat{v} + iUk_x\hat{v} = -(1/\rho)d\hat{p}/dy + \nu\left[d^2/dy^2 - (k_x^2+k_z^2)\right]\hat{v}\,, \quad (2)$$
$$-i\omega\hat{w} + iUk_x\hat{w} = -(1/\rho)ik_z\hat{p} + \nu\left[d^2/dy^2 - (k_x^2+k_z^2)\right]\hat{w}\,, \quad (3)$$
$$ik_x\hat{u} + d\hat{v}/dy + ik_z\hat{w} = 0\,. \quad (4)$$

Eliminating $(ik_x\hat{u} + ik_z\hat{w})$ in (1) $\times ik_x$ + (3) $\times ik_z$ using Eq. (4), we obtain

$$-i\omega d\hat{v}/dy + iUk_x d\hat{v}/dy - ik_x\hat{v}dU/dy = -(k_x^2+k_z^2)\hat{p}/\rho + \nu\left[d^2/dy^2 - (k_x^2+k_z^2)\right]d\hat{v}/dy\,. \quad (5)$$

Now eliminating $\hat{p}$ from $\partial(5)/\partial y - (2)(k_x^2+k_z^2)$, we can derive the Orr–Sommerfeld equation.

Exercise 5.2 Multiply (5.34) by $2\theta'$, take the average value and expand. The result is

$$\underbrace{\overline{2\theta'\frac{\partial\overline{\theta}}{\partial t}}}_{0} + \underbrace{\overline{2\theta'\frac{\partial\theta'}{\partial t}}}_{\frac{\partial\overline{\theta'^2}}{\partial t}} + \underbrace{\overline{2\theta'\overline{u}_j\frac{\partial\overline{\theta}}{\partial x_j}}}_{0} + \underbrace{\overline{2\theta'\overline{u}_j\frac{\partial\theta'}{\partial x_j}}}_{\overline{u}_j\frac{\partial\overline{\theta'^2}}{\partial x_j}} + \underbrace{\overline{2\theta'u'_j\frac{\partial\overline{\theta}}{\partial x_j}}}_{\overline{2\theta'u'_j}\frac{\partial\overline{\theta}}{\partial x_j}} + \underbrace{\overline{2\theta'u'_j\frac{\partial\theta'}{\partial x_j}}}_{\frac{\partial}{\partial x_j}\left(\overline{\theta'^2u'_j}\right)}$$

$$= \underbrace{2\nu_\theta\overline{\theta'\frac{\partial^2\overline{\theta}}{\partial x_j^2}}}_{0} + \underbrace{2\nu_\theta\overline{\theta'\frac{\partial^2\theta'}{\partial x_j^2}}}_{2\nu_\theta\frac{\partial}{\partial x_j}\left(\overline{\theta'\frac{\partial\theta'}{\partial x_j}}\right) - 2\nu_\theta\overline{\frac{\partial\theta'}{\partial x_j}\frac{\partial\theta'}{\partial x_j}}}\,.$$

The rearrangement of this expression yields (5.43).

Taking the average of (5.33) $\times u'_k$,

$$\underbrace{\overline{u'_k\frac{\partial\overline{u_i}}{\partial t}}}_{0} + \overline{u'_k\frac{\partial u'_i}{\partial t}} + \underbrace{\overline{u'_k\overline{u}_j\frac{\partial\overline{u_i}}{\partial x_j}}}_{0} + \overline{u'_k\overline{u}_j\frac{\partial u'_i}{\partial x_j}} + \underbrace{\overline{u'_ku'_j\frac{\partial\overline{u_i}}{\partial x_j}}}_{\overline{u'_ku'_j}\frac{\partial\overline{u_i}}{\partial x_j}} + \overline{u'_ku'_j\frac{\partial u'_i}{\partial x_j}}$$

$$= -\frac{1}{\rho}\underbrace{\overline{u'_k\frac{\partial\overline{p}}{\partial x_i}}}_{0} - \frac{1}{\rho}\overline{u'_k\frac{\partial p'}{\partial x_i}} + \underbrace{\nu\overline{u'_k\frac{\partial^2\overline{u_i}}{\partial x_j^2}}}_{0} + \nu\overline{u'_k\frac{\partial^2u'_i}{\partial x_j^2}}\,.$$

Swapping i and k and take the sum of them yields

$$\underbrace{\overline{u'_k\frac{\partial u'_i}{\partial t}} + \overline{u'_i\frac{\partial u'_k}{\partial t}}}_{\frac{\partial}{\partial t}\left(\overline{u'_iu'_k}\right)} \underbrace{+\overline{u'_k\overline{u}_j\frac{\partial u'_i}{\partial x_j}} + \overline{u'_i\overline{u}_j\frac{\partial u'_k}{\partial x_j}}}_{\overline{u}_j\frac{\partial}{\partial x_j}\left(\overline{u'_iu'_k}\right)} + \overline{u'_ku'_j}\frac{\partial\overline{u_i}}{\partial x_j}$$

$$+\overline{u'_iu'_j}\frac{\partial\overline{u_k}}{\partial x_j} \underbrace{+\overline{u'_ku'_j\frac{\partial u'_i}{\partial x_j}} + \overline{u'_iu'_j\frac{\partial u'_k}{\partial x_j}}}_{\frac{\partial}{\partial x_j}\left(\overline{u'_iu'_ju'_k}\right)}$$

$$= \underbrace{-\frac{1}{\rho}\overline{u'_k\frac{\partial p'}{\partial x_i}}}_{-\frac{1}{\rho}\frac{\partial}{\partial x_i}\left(\overline{u'_kp'}\right)+\frac{1}{\rho}\overline{p'\frac{\partial u'_k}{\partial x_i}}} \quad \underbrace{-\frac{1}{\rho}\overline{u'_i\frac{\partial p'}{\partial x_k}}}_{-\frac{1}{\rho}\frac{\partial}{\partial x_k}\left(\overline{u'_ip'}\right)+\frac{1}{\rho}\overline{p'\frac{\partial u'_i}{\partial x_k}}} \quad + \underbrace{\nu\overline{u'_k\frac{\partial^2u'_i}{\partial x_j^2}} + \nu\overline{u'_i\frac{\partial^2u'_k}{\partial x_j^2}}}_{\nu\frac{\partial^2}{\partial x_j^2}\left(\overline{u'_ku'_i}\right)-2\nu\overline{\frac{\partial u'_i}{\partial x_j}\frac{\partial u'_k}{\partial x_j}}}$$

from which (5.44) is obtained.

Taking the average of (5.33) $\times\theta'$, we obtain

$$\underbrace{\overline{\theta'\frac{\partial \overline{u_i}}{\partial t}}}_{0} + \overline{\theta'\frac{\partial u_i'}{\partial t}} + \underbrace{\overline{\theta'\overline{u_j}\frac{\partial \overline{u_i}}{\partial x_j}}}_{0} + \overline{\theta'\overline{u_j}\frac{\partial u_i'}{\partial x_j}} + \underbrace{\overline{\theta' u_j'\frac{\partial \overline{u_i}}{\partial x_j}}}_{\overline{\theta' u_j'}\frac{\partial \overline{u_i}}{\partial x_j}} + \overline{\theta' u_j'\frac{\partial u_i'}{\partial x_j}}$$

$$= -\frac{1}{\rho}\overbrace{\overline{\theta'\frac{\partial \overline{p}}{\partial x_i}}}^{0} - \frac{1}{\rho}\overline{\theta'\frac{\partial p'}{\partial x_i}} + \overbrace{\nu\overline{\theta'\frac{\partial^2 \overline{u_i}}{\partial x_j^2}}}^{0} + \nu\overline{\theta'\frac{\partial^2 u_i'}{\partial x_j^2}}\,.$$

Meanwhile, the average of (5.34) $\times u_i'$ gives

$$\underbrace{\overline{u_i'\frac{\partial \overline{\theta}}{\partial t}}}_{0} + \overline{u_i'\frac{\partial \theta'}{\partial t}} + \underbrace{\overline{u_i'\overline{u_j}\frac{\partial \overline{\theta}}{\partial x_j}}}_{0} + \overline{u_i'\overline{u_j}\frac{\partial \theta'}{\partial x_j}} + \underbrace{\overline{u_i u_j'\frac{\partial \overline{\theta}}{\partial x_j}}}_{\overline{u_i' u_j'}\frac{\partial \overline{\theta}}{\partial x_j}} + \overline{u_i' u_j'\frac{\partial \theta'}{\partial x_j}}$$

$$= \underbrace{\nu_\theta\overline{u_i'\frac{\partial^2 \overline{\theta}}{\partial x_j^2}}}_{0} + \nu_\theta\overline{u_i'\frac{\partial^2 \theta'}{\partial x_j^2}}\,.$$

By rearranging the sum of both equations,

$$\underbrace{\overline{\theta'\frac{\partial u_i'}{\partial t}} + \overline{u_i'\frac{\partial \theta'}{\partial t}}}_{\frac{\partial}{\partial t}\left(\overline{\theta' u_i'}\right)} \underbrace{+\overline{\theta'\overline{u_j}\frac{\partial u_i'}{\partial x_j}} + \overline{u_i'\overline{u_j}\frac{\partial \theta'}{\partial x_j}}}_{\overline{u_j}\frac{\partial}{\partial x_j}\left(\overline{\theta' u_i'}\right)} + \overline{\theta' u_j'}\frac{\partial \overline{u_i}}{\partial x_j} + \overline{u_i' u_j'}\frac{\partial \overline{\theta}}{\partial x_j}$$

$$\underbrace{+\overline{\theta' u_j'\frac{\partial u_i'}{\partial x_j}} + \overline{u_i' u_j'\frac{\partial \theta'}{\partial x_j}}}_{\frac{\partial}{\partial x_j}\left(\overline{\theta' u_i' u_j'}\right)} = \underbrace{-\frac{1}{\rho}\overline{\theta'\frac{\partial p'}{\partial x_i}}}_{-\frac{1}{\rho}\left[\frac{\partial}{\partial x_i}\left(\overline{\theta' p'}\right) - \overline{p'\frac{\partial \theta'}{\partial x_i}}\right]} + \nu\overline{\theta'\frac{\partial^2 u_i'}{\partial x_j^2}} + \nu_\theta\overline{u_i'\frac{\partial^2 \theta'}{\partial x_j^2}}$$

we obtain (5.45).

Exercise 5.3 (1) Given $E(k_i, t) = \exp(-2\nu k_i^2 t)$ and by substituting $k_2 = 2k_1$, the total energy is

$$E(t) = \exp(-2\nu k_1^2 t) + \exp(-8\nu k_1^2 t)\,.$$

(2) Substitute $k_2 = 2k_1$ and $\alpha = 4\nu k_1^2$, $E_+(t) \equiv 2E(k_1, t) + E(k_2, t)$, $E_-(t) \equiv -E(k_1, t) + 2E(k_2, t)$ as a basis,

$$\frac{d}{dt}\begin{pmatrix} E_+(t) \\ E_-(t) \end{pmatrix} = \begin{pmatrix} -4\nu k_1^2 & 0 \\ 0 & -14\nu k_1^2 \end{pmatrix}\begin{pmatrix} E_+(t) \\ E_-(t) \end{pmatrix}.$$

Based on the initial conditions;

$$E_+(0) = 2E(k_1, 0) + E(k_2, 0) = 3,$$
$$E_-(0) = -E(k_1, 0) + 2E(k_2, 0) = 1,$$
$$E_+(t) = 3\exp(-4\nu k_1^2 t)\,,\; E_-(t) = \exp(-14\nu k_1^2 t)$$

then

$$E(k_1, t) = (6/5)\exp(-4\nu k_1^2 t) - (1/5)\exp(-14\nu k_1^2 t)\,,$$
$$E(k_2, t) = (3/5)\exp(-4\nu k_1^2 t) + (2/5)\exp(-14\nu k_1^2 t)\,.$$

Therefore, the total energy is $E(t) = (9/5)\exp(-4\nu k_1^2 t) + (1/5)\exp(-14\nu k_1^2 t)$.
(3) With $t \to \infty$, we find that (1): $E(t) \sim \exp(-2\nu k_1^2 t)$, (2): $E(t) \sim (9/5)$ $\exp(-4\nu k_1^2 t)$; hence, the latter solution quickly attenuates. Solutions (1) and (2) differ only by the existence of $T(k)$ in (2), whose function is to redistribute energy. This term does not directly cause dissipation, but its existence in (2) ensures that the total energy is effectively dissipated.

Exercise 5.4 The dimension of $E(k)$ is $[\mathrm{L}^3\mathrm{T}^{-2}]$, the dimension of ε is $[\mathrm{L}^2\mathrm{T}^{-3}]$, and the dimension of k is $[\mathrm{L}^{-1}]$. Thus, for the dimension $[\mathrm{L}^{2\alpha-\beta}\mathrm{T}^{-3\alpha}]$ of $\varepsilon^\alpha k^\beta$ to equal that of $E(k)$, we require the relations $3 = 2\alpha - \beta$ and $-2 = -3\alpha$. Solving these expressions gives $\alpha = 2/3$ and $\beta = -5/3$.

Exercise 5.5 Multiplying both sides of (5.97) by ω and regionally integrating, we get $\frac{d}{dt}\int \frac{1}{2}\omega^2 d\boldsymbol{r} + \int \frac{\partial}{\partial x_j}\left(\frac{1}{2}\omega^2 u_j\right) d\boldsymbol{r} = \nu \int \frac{\partial}{\partial x_i}\left(\omega \frac{\partial \omega}{\partial x_i}\right) d\boldsymbol{r} - \nu \int \left(\frac{\partial \omega}{\partial x_i}\right)^2 d\boldsymbol{r}$. The second term on the left-hand side and the first term on the right-hand side can be eliminated from the above expression under appropriate boundary conditions using Gauss' theorem.

Exercise 5.6 The dimension of $E(k)$ is $[\mathrm{L}^3\mathrm{T}^{-2}]$, the dimension of η is $[\mathrm{T}^{-3}]$, and the dimension of k is $[\mathrm{L}^{-1}]$. Thus, for the dimension $[\mathrm{L}^{-\beta}\mathrm{T}^{-3\alpha}]$ of $\eta^\alpha k^\beta$ to equal that of $E(k)$, we require the relations $3 = -\beta$ and $-2 = -3\alpha$. Solving these expressions gives $\alpha = 2/3$ and $\beta = -3$.

Chapter 6 Dynamics of Stratified Fluids

Exercise 6.1 (1) We have $p = p(0) - \rho g z$. If $p - p(0) = 100$ atmospheric pressure, at $z = -h$, then $h = (p - p(0))/\rho g \cong 100 \cdot 10^3 \cdot 10^2\ \mathrm{Pa}/(10^3\ \mathrm{kg\ m^{-3}} \cdot 10\ \mathrm{m\ s^{-2}}) \simeq$ 1,000 m.
(2) By solving $dp/dz = -(g/RT)p$, we obtain $p = p(0)exp(-(g/RT)z)$. Therefore, $H = RT/g$. For the atmosphere, $H = RT/g \cong 2.9 \times 10^2\ \mathrm{J\ kg^{-1}K^{-1}} \cdot 270\ \mathrm{K}/10\ \mathrm{m\ s^{-2}} \cong 7.8 \times 10^3$ m.

Exercise 6.2 From (6.19), we find that $N^2 < 0$ when $dT/dz < -g/c_p \cong -10\ \mathrm{m\ s^{-2}}/1.0 \times 10^3\ \mathrm{J\ kg^{-1}K^{-1}} \simeq -1.0 \times 10^{-2}\ \mathrm{Km^{-1}}$. In other word, if the temperature reduces by more than 1.0 K for every 100 m rise in height, the fluid is unstable.

Exercise 6.3 Given

$$-c_p\theta\nabla\Pi + \boldsymbol{g} = -\underbrace{c_p\theta_0\nabla\Pi_h}_{\boldsymbol{g}} - (\theta'/\theta_0)\underbrace{c_p\theta_0\nabla\Pi_h}_{\boldsymbol{g}} - c_p\theta_0\nabla\Pi_d - (\theta'/\theta_0)c_p\theta_0\nabla\Pi_d + \boldsymbol{g}$$
$$= -(\theta'/\theta_0)\boldsymbol{g} - c_p\theta_0\nabla\Pi_d - (\theta'/\theta_0)c_p\theta_0\nabla\Pi_d$$

Since the last term can be neglected with respect to the second term, the left hand side is similarly equal to

$$-(\theta'/\theta_0)\boldsymbol{g} - c_p\theta_0\nabla\Pi_d\,.$$

Exercise 6.4 Given that $\omega = 0$, we must resolve the denominator of $U = \mp N/(k_x^2 + k_z^2)^{1/2}$. Therefore,

$$c_{gx} - U = -Nk_z^2/(k_x^2 + k_z^2)^{3/2} < 0\,,$$
$$c_{gx} = N/(k_x^2 + k_z^2)^{1/2} - Nk_z^2/(k_x^2 + k_z^2)^{3/2} = Nk_x^2/(k_x^2 + k_z^2)^{3/2} > 0\,.$$

Moreover, given $c_{gz} = Nk_zk_x/(k_x^2 + k_z^2)^{3/2} > 0$, k_x and k_z have the same sign.

Exercise 6.5 The derivative $dRa/d\tilde{k}_h = 2(\pi^2 + \tilde{k}_h^2)^2(2\tilde{k}_h^2 - \pi^2)/\tilde{k}_h^3$ is minimized when $\tilde{k}_h = \pi/2^{1/2}$. Here $Ra\,(\tilde{k}_h{=}\pi/2^{1/2}) = (\pi^2 + \pi^2/2)^3/(\pi^2/2) = 27\pi^4/4$.

Exercise 6.6 $w' = 0$ is self-evident. From (6.101), it is also easily seen that $T' = 0$ implies that $(\partial/\partial\tilde{t} - \tilde{\nabla}^2)\tilde{\nabla}^2 w' = 0$. Based on the adhesiveness condition, at any point on the boundary in the $\tilde{x}$ and $\tilde{y}$ directions, $u' = v' = 0$. Hence, $\partial u'/\partial\tilde{x} = \partial v'/\partial\tilde{y} = 0$ and $\partial w'/\partial\tilde{z} = 0$ from the equation of continuity.

Chapter 7 Dynamics of Rotating Fluids

Exercise 7.1 As the Coriolis force is orthogonal to the flow velocity, the inner product of the Coriolis force and the flow velocity is 0. Hence, no work is done.

Exercise 7.2 Provided that the scale of the flow velocity is U and the Earth's radius is R, the ratio of the metric acceleration and the Coriolis' acceleration is 1 when $\frac{U^2}{R}/2\Omega U = U/(2\Omega R) \simeq 1$, i.e., when U is approximately $2\Omega R$. Given that $\Omega = 2\pi/24/3600$ rad s^{-1} and $R = 6.4 \times 10^6$ m, we evaluate $U \gg 9.3 \times 10^2$ m s^{-1}.

Exercise 7.3 If the flow velocity scale U is sought to satisfy $|fU| = g$, the minimum required velocity is $g/2\Omega$. Given g= 9.8 m s^{-2} and $\Omega = 2\pi/24/3600$ rad s^{-1}, the required velocity is as large as 6.7×10^4 m s^{-1}.

Exercise 7.4 As the absolute vorticity is preserved, we have $\frac{D_{\bar{h}}}{Dt}\left(\frac{f+\bar{\zeta}}{h}\right) = 0$. If the horizontal scale of the flow is large, the vorticity $|\bar{\zeta}|$ can be negligible at medium to high latitudes where $|f|$ is not small; namely, the vorticity is approximately $f + \zeta \simeq f$. Meanwhile, in deep oceanic regions, $h = H + \eta \simeq H$, and the absolute vorticity is approximately f/H. As f and H are not functions of time, we have $\frac{D_h}{Dt}\left(\frac{f}{H}\right) = \boldsymbol{u} \cdot \nabla \left(\frac{f}{H}\right) = 0$, and therefore the isoline of f/H and flow u are parallel to each other.

Exercise 7.5 1. Using $f = 2\Omega \sin\varphi$ and $\beta = (2\Omega/R)\cos\varphi$, and given $\Omega = 2\pi/24/3600$ rad s^{-1} and $R = 6.4 \times 10^6$ m, we have $f_0 = 0$ s^{-1} and $\beta_0 = 2.3 \times 10^{-11}$ m^{-1}s^{-1} at the Equator, and $f_0 = 1.5 \times 10^{-4}$ s^{-1} and $\beta_0 = 0$ m^{-1}s^{-1} at the North Pole.

2. By substituting f and β in the given inequality and dividing both sides by $\cos\varphi$ (>0), we obtain $\tan\varphi > L/R$. Therefore, for $L \simeq 1000$ km, the condition is satisfied at all latitudes φ except in low latitude regions, 8.8° south and north of the Equator.

Exercise 7.6 Provided that the flow velocity and width are U and L, respectively, the amplitude of η is f_0UL/g. Substituting the specified quantities into this expression, the amplitude is obtained as 0.7 m.

Exercise 7.7 Set $f_0 = -|f_0|$. As $\sqrt{-i} = (1-i)/\sqrt{2}$, the sign of the coefficient of $\boldsymbol{B}$'s imaginary number is reversed; consequently, the sign of $\boldsymbol{A}$'s exponent is reversed. Thus, "clockwise" becomes "counterclockwise."

Exercise 7.8 Entering the solution of the wave motion into the basic equation and algebraically rearranging, we obtain

$$\begin{pmatrix} -i\omega & -f_0 & igk \\ f_0 & -i\omega & igl \\ ik & il & -\frac{i\omega}{H} \end{pmatrix} \begin{pmatrix} \boldsymbol{U} \\ \boldsymbol{V} \\ \boldsymbol{Z} \end{pmatrix} = \begin{pmatrix} 0 \\ 0 \\ 0 \end{pmatrix}.$$

To obtain a non-trivial result in terms of $\boldsymbol{U}$, $\boldsymbol{V}$, and $\boldsymbol{Z}$, the determinant of this 3×3 matrix must be 0, so that

$$-\frac{i\omega}{H}\left[f^2 - \omega^2 + gH(k^2 + l^2)\right] = 0.$$

For a wave solution ($\omega \neq 0$), we can derive the dispersion relation.

Exercise 7.9 Clearly, the general solution is $u = A\sin f_0t + B\cos f_0t$, where A and B are constants. Meanwhile, from $f_0v = \partial u/\partial t$, we have $v = -B\sin f_0t + A\cos f_0t$. Substituting the initial values at $t = 0$ gives $B = U_0$ and $A = 0$. Thus, we have $u = U_0\cos f_0t$, $v = -U_0\sin f_0t$, and $\eta = 0$. Moreover, since $u = \partial x/\partial t$, by integrating u and v with respect to t, the object at the origin satisfies $x = U_0/f_0 \sin f_0t$ and $y = U_0/f_0(\cos f_0t - 1)$, which describes a circular orbit with radius $|U_0/f_0|$.

Exercise 7.10 As the direction of the Ekman transport against the wind in the Southern Hemisphere is opposite to that in the Northern Hemisphere, the Ekman transport diverges at the central part. Therefore, the water column in the interior region of the central part stretches, and its pressure tends to decline. From the law of conservation of potential vorticity, changes occur in a sense to increase the absolute value of absolute vorticity; namely, $f_0 + \zeta \simeq f_0 < 0$ becomes more negative.

In the f-plane, this implies that the relative vorticity ζ is decreased and the clockwise circulation is enhanced. This is consistent with the reduced pressure at the central part, because the geostrophic current flows in the direction of lower pressure to its right.

In contrast, in the β-plane, southward currents are expected that reduce the planetary vorticity. Therefore, as in the Northern Hemisphere, the circulation is strengthened on the west coast.

Appendix

Exercise A.1 To prove symmetry, swap the indices l and m, in the second equation in (A.7) and next swap the indices i and j. Using $T_{ji} = T_{ij}$, we find that $T'_{ml} = \alpha_{im}\alpha_{jl}T_{ij} = \alpha_{jm}\alpha_{il}T_{ji} = \alpha_{il}\alpha_{jm}T_{ij} = T'_{lm}$. Anti-symmetry is proved similarly.

Exercise A.2 T_{ijkl} satisfies the transformation rule $T'_{pqrs} = \alpha_{ip}\alpha_{jq}\alpha_{kr}\alpha_{ls}T_{ijkl}$. Here let $s = r$ and contract T'_{pqrr}. Using the first equation of (A.3), we obtain $T'_{pqrr} = \alpha_{ip}\alpha_{jq}\alpha_{kr}\alpha_{lr}T_{ijkl} = \alpha_{ip}\alpha_{jq}\delta_{kl}T_{ijkl} = \alpha_{ip}\alpha_{jq}T_{ijkk}$, which is the desired equation.

Exercise A.3 Arranging rows (i, j, k) in the order of $(1, 2, 3)$ and next arranging columns (p, q, r) in the order of $(1, 2, 3)$, we obtain

$$\text{Right-hand side} = \epsilon_{ijk}\begin{vmatrix} \delta_{1p} & \delta_{1q} & \delta_{1r} \\ \delta_{2p} & \delta_{2q} & \delta_{2r} \\ \delta_{3p} & \delta_{3q} & \delta_{3r} \end{vmatrix} = \epsilon_{ijk}\epsilon_{pqr}\begin{vmatrix} \delta_{11} & \delta_{12} & \delta_{13} \\ \delta_{21} & \delta_{22} & \delta_{23} \\ \delta_{31} & \delta_{32} & \delta_{33} \end{vmatrix} = \text{Left-hand side.}$$

Therefore,

$$\epsilon_{ijk}\epsilon_{iqr} = \begin{vmatrix} \delta_{ii} & \delta_{iq} & \delta_{ir} \\ \delta_{ji} & \delta_{jq} & \delta_{jr} \\ \delta_{ki} & \delta_{kq} & \delta_{kr} \end{vmatrix} = \begin{vmatrix} 1 & 0 & 0 \\ 0 & \delta_{jq} & \delta_{jr} \\ 0 & \delta_{kq} & \delta_{jr} \end{vmatrix} = \delta_{jq}\delta_{kr} - \delta_{jr}\delta_{kq},$$

$$\epsilon_{ijk}\epsilon_{ijr} = \delta_{jj}\delta_{kr} - \delta_{jr}\delta_{kj} = 3\delta_{kr} - \delta_{kr} = 2\delta_{kr},$$

$$\epsilon_{ijk}\epsilon_{ijk} = 2\delta_{kk} = 6.$$

Index

D.E.S.S.T. Kyushu Univ., *Fluid Dynamics for Global Environmental Studies*,
DOI 10.1007/978-4-431-56499-7

FSC
www.fsc.org
MIX
Papier aus verantwortungsvollen Quellen
Paper from responsible sources
FSC® C105338

If you have any concerns about our products,
you can contact us on
ProductSafety@springernature.com

In case Publisher is established outside the EU,
the EU authorized representative is:
Springer Nature Customer Service Center GmbH
Europaplatz 3, 69115 Heidelberg, Germany

Printed by Libri Plureos GmbH
in Hamburg, Germany